Y0-AFY-111

PROGRESS IN BRAIN RESEARCH

VOLUME 155

VISUAL PERCEPTION, PART 2

FUNDAMENTALS OF AWARENESS: MULTI-SENSORY INTEGRATION AND HIGH-ORDER PERCEPTION

Other volumes in PROGRESS IN BRAIN RESEARCH

Volume 122: The Biological Basis for Mind Body Interactions, by E.A. Mayer and C.B. Saper (Eds.) – 1999, ISBN 0-444-50049-9.
Volume 123: Peripheral and Spinal Mechanisms in the Neural Control of Movement, by M.D. Binder (Ed.) – 1999, ISBN 0-444-50288-2.
Volume 124: Cerebellar Modules: Molecules, Morphology and Function, by N.M. Gerrits, T.J.H. Ruigrok and C.E. De Zeeuw (Eds.) – 2000, ISBN 0-444-50108-8.
Volume 125: Transmission Revisited, by L.F. Agnati, K. Fuxe, C. Nicholson and E. Syková (Eds.) – 2000, ISBN 0-444-50314-5.
Volume 126: Cognition, Emotion and Autonomic Responses: the Integrative Role of the Prefrontal Cortex and Limbic Structures, by H.B.M. Uylings, C.G. Van Eden, J.P.C. De Bruin, M.G.P. Feenstra and C.M.A. Pennartz (Eds.) – 2000, ISBN 0-444-50332-3.
Volume 127: Neural Transplantation II. Novel Cell Therapies for CNS Disorders, by S.B. Dunnett and A. Björklund (Eds.) – 2000, ISBN 0-444-50109-6.
Volume 128: Neural Plasticity and Regeneration, by F.J. Seil (Ed.) – 2000, ISBN 0-444-50209-2.
Volume 129: Nervous System Plasticity and Chronic Pain, by J. Sandkühler, B. Bromm and G.F. Gebhart (Eds.) – 2000, ISBN 0-444-50509-1.
Volume 130: Advances in Neural Population Coding, by M.A.L. Nicolelis (Ed.) – 2001, ISBN 0-444-50110-X.
Volume 131: Concepts and Challenges in Retinal Biology, by H. Kolb, H. Ripps, and S. Wu (Eds.) – 2001, ISBN 0-444-506772.
Volume 132: Glial Cell Function, by B. Castellano López and M. Nieto-Sampedro (Eds.) – 2001. ISBN 0-444-50508-3.
Volume 133: The Maternal Brain. Neurobiological and neuroendocrine adaptation and disorders in pregnancy and post partum, by J.A. Russell, A.J. Douglas, R.J. Windle and C.D. Ingram (Eds.) – 2001, ISBN 0-444-50548-2.
Volume 134: Vision: From Neurons to Cognition, by C. Casanova and M. Ptito (Eds.) – 2001, ISBN 0-444-50586-5.
Volume 135: Do Seizures Damage the Brain, by A. Pitkänen and T. Sutula (Eds.) – 2002, ISBN 0-444-50814-7.
Volume 136: Changing Views of Cajal's Neuron, by E.C. Azmitia, J. DeFelipe, E.G. Jones, P. Rakic and C.E. Ribak (Eds.) – 2002, ISBN 0-444-50815-5.
Volume 137: Spinal Cord Trauma: Regeneration, Neural Repair and Functional Recovery, by L. McKerracher, G. Doucet and S. Rossignol (Eds.) – 2002, ISBN 0-444-50817-1.
Volume 138: Plasticity in the Adult Brain: From Genes to Neurotherapy, by M.A. Hofman, G.J. Boer, A.J.G.D. Holtmaat, E.J.W. Van Someren, J. Verhaagen and D.F. Swaab (Eds.) – 2002, ISBN 0-444-50981-X.
Volume 139: Vasopressin and Oxytocin: From Genes to Clinical Applications, by D. Poulain, S. Oliet and D. Theodosis (Eds.) – 2002, ISBN 0-444-50982-8.
Volume 140: The Brain's Eye, by J. Hyönä, D.P. Munoz, W. Heide and R. Radach (Eds.) – 2002, ISBN 0-444-51097-4.
Volume 141: Gonadotropin-Releasing Hormone: Molecules and Receptors, by I.S. Parhar (Ed.) – 2002, ISBN 0-444-50979-8.
Volume 142: Neural Control of Space Coding, and Action Production, by C. Prablanc, D. Pélisson and Y. Rossetti (Eds.) – 2003, ISBN 0-444-509771.
Volume 143: Brain Mechanisms for the Integration of Posture and Movement, by S. Mori, D.G. Stuart and M. Wiesendanger (Eds.) – 2004, ISBN 0-444-513892.
Volume 144: The Roots of Visual Awareness, by C.A. Heywood, A,D. Milner and C. Blakemore (Eds.) – 2004, ISBN 0-444-50978-X.
Volume 145: Acetylcholine in the Cerebral Cortex, by L. Descarries, K. Krnjević and M. Steriade (Eds.) – 2004, ISBN 0-444-511253.
Volume 146: NGF and Related Molecules in Health and Disease, by L. Aloe and L. Calzà (Eds.) – 2004, ISBN 0-444-51472-4.
Volume 147: Development, Dynamics and Pathology of Neuronal Networks: From Molecules to Functional Circuits, by J. Van Pelt, M. Kamermans, C.N. Levelt, A. Van Ooyen, G.J.A. Ramakers and P.R. Roelfsema (Eds.) – 2005, ISBN 0-444-51663-8.
Volume 148: Creating Coordination in the Cerebellum, by C.I. De Zeeuw and F. Cicirata (Eds.) – 2005, ISBN 0-444-51754-5.
Volume 149: Cortical Function: A View From The Thalamus, by V.A. Casagrande, R.W. Guillery and S.M. Sherman (Eds.) – 2005, ISBN 0-444-51679-4.
Volume 150: The Boundaries of Consciousness: Neurobiology and Neuropathology, by Steven Laureys (Ed.) – 2005, ISBN 0-444-51851-7
Volume 151: Neuroanatomy of the Oculomotor System, by J.A. Büttner-Ennever (Ed.) – 2006, ISBN 0-444-516964.
Volume 152: Autonomic Dysfunction After Spinal Cord Injury, by L.C. Weaver and C. Polosa (Eds.) – 2006, ISBN 0-444-51925-4.
Volume 153: Hypothalamic Integration of Energy Metabolism, by A. Kalsbeek, E. Fliers, M.A. Hofman , D.F. Swaab, E.J.W. Van Someren and R. M. Buijs (Eds.) – 2006, ISBN 978-0-444-52261-0.
Volume 154: Visual Perception, Part 1, Fundamentals of Vision: Low and Mid-Level Processes in Perception by S. Martinez-Conde, S.L. Macknik, L.M. Martinez, J.M. Alonso and P.U. Tse (Eds.) – 2006, ISBN 0-444-52966-7.

PROGRESS IN BRAIN RESEARCH

VOLUME 155

VISUAL PERCEPTION, PART 2

FUNDAMENTALS OF AWARENESS: MULTI-SENSORY INTEGRATION AND HIGH-ORDER PERCEPTION

EDITED BY

S. MARTINEZ-CONDE

Department of Neurobiology, Barrow Neurological Institute, Phoenix, AZ 85013, USA

S.L. MACKNIK

Departments of Neurosurgery and Neurobiology, Barrow Neurological Institute, Phoenix, AZ 85013, USA

L.M. MARTINEZ

Departamento de Medicina, Facultade de Ciencias da Saúde, Campus de Oza, Universidade da Coruña, 15006, A Coruña, Spain

J.-M. ALONSO

Department of Biological Sciences, State University of New York – Optometry, New York, NY 10036, USA

P.U. TSE

Department of Psychological and Brain Sciences, Dartmouth College, Hanover, NH 03755, USA

ELSEVIER

AMSTERDAM – BOSTON – HEIDELBERG – LONDON – NEW YORK – OXFORD
PARIS – SAN DIEGO – SAN FRANCISCO – SINGAPORE – SYDNEY – TOKYO

Elsevier
Radarweg 29, PO Box 211, 1000 AE Amsterdam, The Netherlands
The Boulevard, Langford Lane, Kidlington, Oxford OX5 1GB, UK

First edition 2006

Library of Congress Cataloging-in-Publication Data
A catalog record for this book is available from the Library of Congress

British Library Cataloguing in Publication Data

European Conference on Visual Perception (28th : 2005 :
A Coruna, Spain)
Visual Perception
Part 2: Fundamentals of awareness: multi-sensory integration and high-order perception. - (Progress in brain research; v. 155)
1. Vision – Congresses 2. Physiological optics – Congresses
3. Visual perception – Congresses
I. Title II. Martinez-Conde, S.
612.8'4

ISBN-13: 9780444519276
ISBN-10: 0444519270

ISBN-13: 978-0-444-51927-6 (this volume)
ISBN-10: 0-444-51927-0 (this volume)
ISBN-13: 978-0-444-52966-4 (vol. 154; Part 1)
ISBN-10: 0-444-52966-7 (vol. 154; Part 1)
ISBN-13: 978-0-444-80104-3 (series)
ISBN-10: 0-444-80104-9 (series)
ISSN: 0079-6123

Printed and bound in The Netherlands

06 07 08 09 10 10 9 8 7 6 5 4 3 2 1

Contents

List of Contributors

D. Alais, Department of Physiology and Institute for Biomedical Research, School of Medical Science, Auditory Research Laboratory, University of Sydney, Sydney, NSW 2006, Australia

E. Aminoff, MGH Martinos Center for Biomedical Imaging, Harvard Medical School, 149 Thirteenth Street, Charlestown, MA 02129, USA

S. Anstis, Department of Psychology, UCSD, 9500 Gilman Drive, La Jolla, CA 92093-0109, USA

M. Bar, MGH Martinos Center for Biomedical Imaging, Harvard Medical School, 149 Thirteenth Street, Charlestown, MA 02129, USA

P. Berbel, Instituto de Neurociencias de Alicante UMH-CSIC, Campus de San Juan, Apartado 18, 03550 San Juan de Alicante, Spain

N.P. Bichot, McGovern Institute for Brain Research, Massachusetts Institute of Technology, Bldg. 46-6121, Cambridge, MA 02139, USA

J.W. Bisley, Center for Neurobiology and Behavior, Columbia University, 1051 Riverside Drive, Unit 87, New York, NY 10032, USA

S. Blau, Department of Neuroscience, Brown University, Providence, RI 02912, USA

D. Burr, Dipartimento di Psicologia, Università degli Studi di Firenze, Via S. Nicolò, Florence, Italy and Istituto di Neuroscience del CNR, Via Moruzzi 1, Pisa 56100, Italy

I.C. Cuthill, School of Biological Sciences, University of Bristol, Woodland Road, Bristol BS8 1UG, UK

B. de Gelder, Cognitive and Affective Neurosciences Laboratory, Department of Psychology, Tilburg University, PO Box 90153, Tilburg, 5000 LE, The Netherlands

V.F. Descalzo, Instituto de Neurosciencias de Alicante, Universidad Miguel Hernandez-CSIC, Apartado 18, 03550 San Juan de Alicante, Spain

R. Desimone, McGovern Institute for Brain Research, Massachusetts Institute of Technology, 77 Massachusetts Avenue, Bldg. 46-3160, Cambridge, MA 02139, USA

M.J. Fenske, MGH Martinos Center for Biomedical Imaging, Harvard Medical School, 149 Thirteenth Street, Charlestown, MA 02129, USA

E. Gallego, Department of Psychology, University of La Coruña, Campus de Elviña, La Coruña, 15071, Spain

R. Gallego, Instituto de Neurosciencias de Alicante UMH-CSIC, Campus de San Juan, Apartado 18, 03550 San Juan de Alicante, Spain

J.V. Garcia-Velasco, Instituto de Neurociencias de Alicante UMH-CSIC, Campus de San Juan, Apartado 18, 03550 San Juan de Alicante, Spain

M.E. Goldberg, Center for Neurobiology and Behavior, Columbia University, 1051 Riverside Drive, Unit 87, New York, NY 10032, USA

J. Gottlieb, Center for Neurobiology and Behavior, Columbia University, 1051 Riverside Drive, Unit 87, New York, NY 10032, USA

J.M. Groh, Center for Cognitive Neuroscience, Department of Pschycology and Neuroscience, and Department of Neurobiology, Duke University, LSRC Rm B203, Durham, NC 27708, USA

N. Gronau, MGH Martinos Center for Biomedical Imaging, Harvard Medical School, 149 Thirteenth Street, Charlestown, MA 02129, USA

X. Huang, Keck Center, Department of Physiology, University of California, San Francisco, CA 94143-0444, USA
S. Kastner, Department of Psychology, Center for the Study of Brain, Mind and Behavior, Princeton University, Green Hall, Princeton, NJ 08544, USA
A. Kingstone, Department of Psychology, University of British Columbia, 2136 West Mall, Vancouver, BC V6T 1Z4, Canada
S.P. MacEvoy, Department of Neurobiology, Duke University Medical Center, Durham, NC 27710, USA
S.L. Macknik, Departments of Neurosurgery and Neurobiology, Barrow Neurological Institute, 350 W. Thomas Road, Phoenix, AZ 85013, USA
H.K.M. Meeren, Cognitive and Affective Neurosciences Laboratory, Department of Psychology, Tilburg University, PO Box 90153, Tilburg, 5000 LE, The Netherlands
L.G. Nowak, Centre de Recherche "Cerveau et Cognition", CNRS-Université Paul Sabatier, 133 Route de Narbonne, 31062 Toulouse Cedex, France
A. Oliva, Department of Brain and Cognitive Sciences, Massachusetts Institute of Technology, 77 Massachusetts Avenue, Building 46-4065, Cambridge, MA 02139, USA
M.A. Paradiso, Department of Neuroscience, Brown University, Providence, RI 02912, USA
C.A. Párraga, Department of Experimental Psychology, University of Bristol, 8 Woodland Road, Bristol BS8 1TN, UK
K.K. Porter, University of Alabama, School of Medicine, Birmingham, AL 35294, USA
K.D. Powell, Laboratory of Sensorimotor Research, National Eye Institute, National Institutes of Health, Bethesda, MD 20892, USA
R. Righart, Cognitive and Affective Neurosciences Laboratory, Department of Psychology, Tilburg University, PO Box 90153, Tilburg, 5000 LE, The Netherlands
A.F. Rossi, Department of Psychology, Vanderbilt University, Nashville, TN 37203, USA
N. Sagiv, Centre for Cognition and Neuroimaging, Brunel University, Uxbridge, Middlesex UB8 3PH, UK
M.V. Sanchez-Vives, Instituto de Neurociencias de Alicante UMH-CSIC, Campus de San Juan, Apartado 18, 03550 San Juan de Alicante, Spain
K.A. Schneider, Department of Psychology, Center for the Study of Brain, Mind and Behavior, Princeton University, Green Hall, Princeton, NJ 08544, USA
G. Shalev, Department of Neuroscience, Brown University, Providence, RI 02912, USA
S. Soto-Faraco, Hospital Saint Joan de Déu (Edifici Docent), C/Santa Rosa 39-57, Planta 4a, 08950 Esplugues de Llobregat, Barcelona, Spain
C. Spence, Department of Experimental Psychology, University of Oxford, South Parks Road, Oxford OX1 3UD, UK
L. Spillmann, Neurozentrum, University Hospital, Breisacher Street 64, 79106 Freiburg, Germany
M. Stevens, School of Biological Sciences, University of Bristol, Woodland Road, Bristol BS8 1UG, UK
P. Stoerig, Institute of Experimental Psychology II, Heinrich-Heine-University, Building 23.03, Universitätsstr 1, D-40225 Dusseldorf, Germany
M. Tamietto, Department of Psychology, University of Turin, Turin, Italy
A. Torralba, Computer Science and Artificial Intelligence Laboratory, Massachusetts Institute of Technology, 77 Massachusetts Avenue, 32-D462, Cambridge, MA 02139, USA
T. Troscianko, Department of Experimental Psychology, University of Bristol, 8 Woodland Road, Bristol BS8 1TN, UK
F. Valle-Inclán, Department of Psychology, University of La Coruña, Campus de Elviña, La Coruña 15071, Spain
W.A.C. Van De Riet, Cognitive and Affective Neurosciences Laboratory, Department of Psychology, Tilburg University, P.O. Box 90153, Tilburg, 5000 LE, The Netherlands

J. van Den Stock, Cognitive and Affective Neurosciences Laboratory, Department of Psychology, Tilburg University, PO Box 90153, Tilburg, 5000 LE, The Netherlands

J. Ward, Department of Psychology, University College London, 26 Bedford Way, London WC1H 0AP, UK

K. Wunderlich, Department of Psychology, Center for the Study of Brain, Mind and Behavior, Princeton University, Green Hall, Princeton, NJ 08544, USA

General Introduction

"Visual Perception" is a two-volume series of Progress in Brain Research, based on the symposia presented during the 28th Annual Meeting of the European Conference on Visual Perception (ECVP), the premier transnational conference on visual perception. The conference took place in A Coruña, Spain, in August 2005. The Executive Committee members of ECVP 2005 edited this volume, and the symposia speakers provided the chapters herein.

The general goal of these two volumes is to present the reader with the state-of-the-art in visual perception research, with a special emphasis in the neural substrates of perception. "Visual Perception (Part 1)" generally addresses the initial stages of the visual pathway, and the perceptual aspects than can be explained at early and intermediate levels of visual processing. "Visual Perception (Part 2)" is generally concerned with higher levels of processing along the visual hierarchy, and the resulting percepts. However, this separation is not very strict, and several of the chapters encompass both early and high-level processes.

The current volume "Visual Perception (Part 2) — Fundamentals of Awareness, Multi-Sensory Integration and High-Order Perception" contains 18 chapters, organized into 4 general sections, each addressing one of the main topics in vision research today: "The role of context in recognition"; "From perceptive fields to Gestalt. A tribute to Lothar Spillmann"; "The neural bases of visual awareness and attention, and "Crossmodal interactions in visual perception". Each section includes a short introduction and four to five related chapters. The topics are tackled from a variety of methodological approaches, such as single-neuron recordings, fMRI and optical imaging, psychophysics, eye movement characterization and computational modeling. We hope that the contributions enclosed will provide the reader with a valuable perspective on the current status of vision research, and more importantly, with some insight into future research directions and the discoveries yet to come.

Many people helped to compile this volume. First of all, we thank all the authors for their contributions and enthusiasm. We also thank Shannon Bentz, Xoana Troncoso and Jaime Hoffman, at the Barrow Neurological Institute, for their assistance in obtaining copyright permissions for several of the figures reprinted here. Moreover, Shannon Bentz transcribed Lothar Spillmann's lecture, and provided general administrative help. Xoana Troncoso was heroic in her effort to help us to meet the submission deadline by collating and packing all the chapters, and preparing the table of contents. We are indebted to Johannes Menzel and Maureen Twaig, at Elsevier, for all their encouragement and assistance; it has been wonderful working with them.

Finally, we thank all the supporting organizations that made the ECVP 2005 conference possible: Ministerio de Educación y Ciencia, International Brain Research Organization, European Office of Aerospace Reseach and Development of the USAF, Consellería de Educación, Industria e Comercio-Xunta de Galicia, Elsevier, Pion Ltd., Universidade da Coruña, Sociedad Española de Neurociencia, SR Research Ltd., Consellería de Sanidade-Xunta de Galicia, Mind Science Foundation, Museos Científicos Coruñeses, Barrow Neurological Institute, Images from Science Exhibition, Concello de A Coruña, Museo Arqueolóxico e Histórico-Castillo de San Antón, Caixanova, Vision Science, Fundación Pedro Barrié de la Maza, and Neurobehavioral Systems.

Susana Martinez-Conde
Executive Chair, European Conference on Visual Perception 2005

On behalf of ECVP 2005's Executive Committee: Stephen Macknik, Luis Martinez, Jose-Manuel Alonso and Peter Tse

SECTION I

The Role of Context in Recognition

Introduction

Predator or prey? Big or little? City skyline or the latest line of kitchen cabinetry? These questions may seem random, but it all depends on the context in which they are posed. In fact, they are all questions that become coherent and can be answered only if the context is known. Well, the skyline/cabinetry question may still seem out of the blue, but it will all snap into focus as you read the following four chapters that explore how visual object recognition critically depends on the context that the objects lie in, and their relevance to the observer.

Fenske, Aminoff, Gronau, and Bar start the section with a chapter that discusses how top-down facilitation modifies the differential contributions of object-based and context-based object recognition. Oliva and Torralba discuss the importance of context to our ability to recognize the gist of a scene at a single glance: without considering context, we might not be able to tell apart the city skyline from the kitchen cabinets. De Gelder, Meeren, Righart, Van den Stock, van de Riet, and Tamietto show how context also plays a critical role in face recognition. Stevens, Cuthill, Parraga, and Troscianko discuss how disruptive coloration in camouflage serves to conceal objects within the context of their surroundings.

Stephen L. Macknik

Martinez-Conde, Macknik, Martinez, Alonso & Tse (Eds.)
Progress in Brain Research, Vol. 155
ISSN 0079-6123

CHAPTER 1

Top-down facilitation of visual object recognition: object-based and context-based contributions

Mark J. Fenske, Elissa Aminoff, Nurit Gronau and Moshe Bar*

MGH Martinos Center for Biomedical Imaging, Harvard Medical School, 149 Thirteenth Street, Charlestown, MA 02129, USA

Abstract: The neural mechanisms subserving visual recognition are traditionally described in terms of bottom-up analysis, whereby increasingly complex aspects of the visual input are processed along a hierarchical progression of cortical regions. However, the importance of top-down facilitation in successful recognition has been emphasized in recent models and research findings. Here we consider evidence for top-down facilitation of recognition that is triggered by early information about an object, as well as by contextual associations between an object and other objects with which it typically appears. The object-based mechanism is proposed to trigger top-down facilitation of visual recognition rapidly, using a partially analyzed version of the input image (i.e., a blurred image) that is projected from early visual areas directly to the prefrontal cortex (PFC). This coarse representation activates in the PFC information that is back-projected as "initial guesses" to the temporal cortex where it presensitizes the most likely interpretations of the input object. In addition to this object-based facilitation, a context-based mechanism is proposed to trigger top-down facilitation through contextual associations between objects in scenes. These contextual associations activate predictive information about which objects are likely to appear together, and can influence the "initial guesses" about an object's identity. We have shown that contextual associations are analyzed by a network that includes the parahippocampal cortex and the retrosplenial complex. The integrated proposal described here is that object- and context-based top-down influences operate together, promoting efficient recognition by framing early information about an object within the constraints provided by a lifetime of experience with contextual associations.

Keywords: object recognition; top-down; feedback; orbitofrontal cortex; low spatial frequencies; visual context; parahippocampal cortex; retrosplenial cortex; visual associations; priming

Successful interaction with the visual world depends on the ability of our brains to recognize visual objects quickly and accurately, despite infinite variations in the appearance of objects and the settings in which they are encountered. How does the visual system deal with all of this information in such a fluent manner? Here we consider the cortical mechanisms and the type of information that they rely on to promote highly efficient visual recognition through top-down processes. The evidence we review, from studies by our lab and others, suggests that top-down facilitation of recognition can be achieved through an object-based mechanism that generates predictions about an object's identity through rapidly analyzed, coarse information. We also review evidence that top-down facilitation of recognition can be achieved through the predictive information provided by contextual associations between an object or scene

*Corresponding author. Tel.: +1-617-726-7467; Fax: +1 617-726-7422; E-mail: bar@nmr.mgh.harvard.edu

DOI: 10.1016/S0079-6123(06)55001-0

and the other objects that are likely to appear together in a particular setting. In the following sections, we first consider each of these forms of top-down enhancement of object recognition separately, and then consider how these object- and context-based mechanisms might operate together to promote highly efficient visual recognition.

An object-based cortical mechanism for triggering top-down facilitation

The traditional view regarding visual processing is that an input image is processed in a bottom-up cascade of cortical regions that analyze increasingly complex information. This view stems from the well-defined functional architecture of the visual cortex, which has a clear hierarchical structure. However, several models propose that both bottom-up and top-down analyses can occur in the cortex simultaneously (Grossberg, 1980; Kosslyn, 1994; Ullman, 1995; Desimone, 1998; Engel et al., 2001; Friston, 2003; Lee and Mumford, 2003), and recent evidence suggests that top-down mechanisms play a significant role in visual processing (e.g., Kosslyn et al., 1993; Humphreys et al., 1997; Barceló et al., 2000; Hopfinger et al., 2000; Miyashita and Hayashi, 2000; Gilbert et al., 2001; Pascual-Leone and Walsh, 2001; Mechelli et al., 2004; Ranganath et al., 2004a). Nevertheless, the way in which such top-down processing is initiated remains an important outstanding issue. The crux of the issue concerns the fact that top-down facilitation of perception requires high-level information to be activated before some low-level information.

Recently, Bar (2003) proposed a model that specifically addresses the question of how top-down facilitation of visual object recognition might be triggered. The gist of this proposal concerns a cortical (or subcortical) "short-cut" of early information through which a partially analyzed version of the input image, comprising the low spatial frequency (LSF) components (i.e., a blurred image), is projected rapidly from early visual areas directly to the prefrontal cortex (PFC). This coarse representation is subsequently used to activate predictions about the most likely interpretations of the input

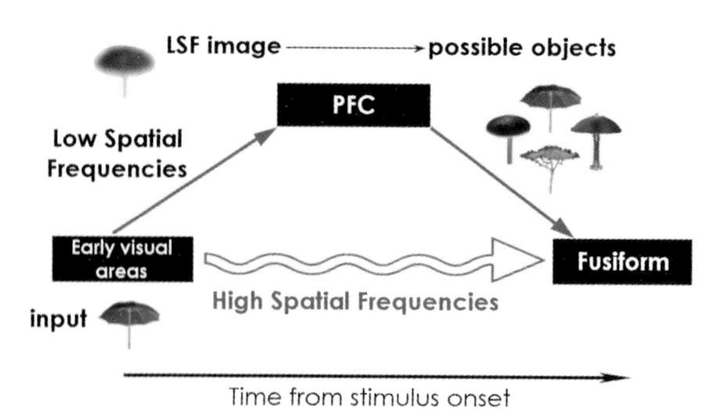

Fig. 1. Schematic illustration of a cortical mechanism for triggering top-down facilitation in object recognition (Bar, 2003). A low spatial frequency (LSF) representation of the input image is projected rapidly, possibly via the magnocellular dorsal pathway, from early visual cortex to the prefrontal cortex (PFC), in addition to the systematic and relatively slower propagation of information along the ventral visual pathway. This coarse representation is sufficient for activating a minimal set of the most probable interpretations of the input, which are then integrated with the relatively slower and detailed bottom-up stream of analysis in object-processing regions of the occipito-temporal cortex (e.g., fusiform gyrus).

image in the temporal cortex. For example, if an image of an oval blob on top of a narrow long blob were extracted from the image initially (Fig. 1), then object representations sharing this global profile would be activated (e.g., an umbrella, a tree, a mushroom, a lamp). Combining these top-down "initial guesses" with the systematic bottom-up analysis could thereby facilitate visual recognition by substantially limiting the number of object representations that need to be tested.

Orbitofrontal involvement in top-down facilitation of recognition

The proposal that object recognition might benefit from a cortical "short-cut" of early, cursory information about the input image implies that there should be an additional cortical region that shows recognition-related activity before other object-processing regions. The cortical regions most often associated with visual object recognition are situated in the occipito-temporal cortex (Logothetis et al., 1996; Tanaka, 1996). Within this region, the fusiform gyrus and the lateral occipital cortex are especially crucial for visual recognition in humans

(Kosslyn et al., 1995; Martin et al., 1996; Bar et al., 2001; Grill-Spector et al., 2001; Malach et al., 2002). The PFC has also been shown to be involved in visual recognition (e.g., Bachevalier and Mishkin, 1986; Wilson et al., 1993; Parker et al., 1998; Freedman et al., 2001), and recent evidence suggests that the orbitofrontal cortex (OFC) might be specifically related to top-down visual processing (Bar et al., 2001, 2006; Bar, 2003). Bar et al. (2001), for example, used functional magnetic resonance imaging (fMRI) to compare the cortical activity elicited by trials in which objects were successfully recognized with that elicited by the same pictures under identical conditions when they were not recognized. As expected, this contrast showed differential activity in the occipito-temporal regions previously associated with object recognition. However, successful recognition was also associated with increased activity in a site within the OFC (Fig. 2). The connection between activity in the OFC and successful object recognition makes it a prime candidate for being involved in top-down facilitation of visual recognition.

Rapid recognition-related activity in orbitofrontal cortex

If the recognition-related region of the OFC identified in the Bar et al. (2001) study is critical for early top-down facilitation, then recognition-related activity should develop in this site before other object-processing regions of occipito-temporal cortex. To test this prediction, Bar et al. (2006) used the same object recognition task as Bar et al. (2001) while obtaining magnetoencephalography (MEG) recordings, which provide millisecond-resolution measurements of cortical activity. As predicted, the contrast between recognized and not-recognized trials in this study revealed differential activation in the OFC 50 ms before it developed in the object-processing regions of the occipito-temporal cortex (Fig. 3). Moreover, a time-frequency, trial-by-trial covariance analysis of the MEG data demonstrated strong synchrony between occipital visual regions and the OFC at a relatively early stage (beginning at approximately 80 ms after stimulus onset), and a strong synchrony between the OFC and the fusiform gyrus activity at a relatively later stage (130 ms after stimulus onset). Taken together, these results provide strong converging support for the proposal that this frontal region plays a critical role in top-down facilitation of object recognition.

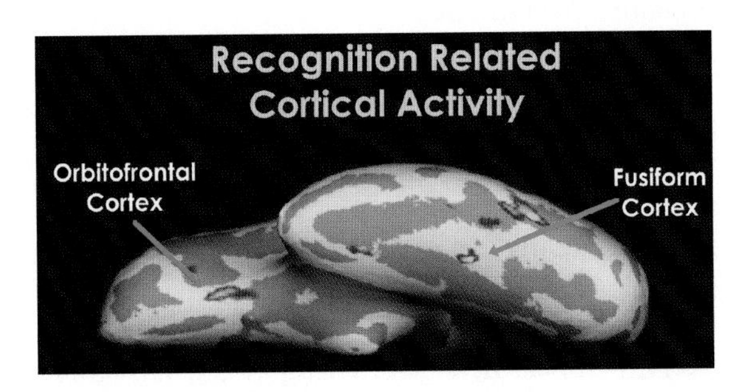

Fig. 2. A group averaged statistical activation map showing recognition-related cortical activity (adapted from Bar et al., 2001) on the left hemisphere (ventral view) of a brain that has been "inflated" to expose both sulci (dark gray) and gyri (light gray). Successful object recognition produced greater activity than unsuccessful recognition attempts in inferior regions of both the occipito-temporal (e.g., fusiform gyrus) and the frontal (e.g., orbitofrontal cortex) lobes.

Early orbitofrontal activity is triggered by low spatial frequencies

The finding that early activity in the OFC is related to successful object recognition is consistent with the model of top-down facilitation proposed by Bar (2003). This model posits that early activity in the OFC is related to the direct projection from early visual areas of an LSF representation of the input image (Bar, 2003). The projection of this early and rudimentary information to the OFC would thereby allow it subsequently to sensitize the representation of the most likely candidate objects in the occipito-temporal cortex as "initial guesses." The possibility that early OFC activity is driven by LSF information is supported by physiological findings that the magnocellular pathway conveys LSF information (i.e., a blurred image) early and rapidly (Shapley, 1990; Merigan and Maunsell, 1993; Bullier and Nowak, 1995; Grabowska and Nowicka, 1996; Leonova et al., 2003), and by evidence from anatomical studies showing bidirectional connections between early visual

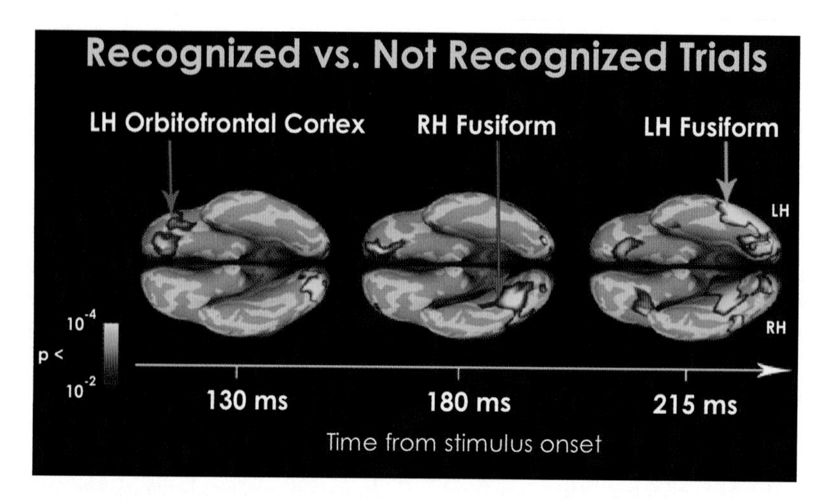

Fig. 3. Cortical dynamics of object recognition. Bar et al. (2006) found that recognition-related activity in the orbitofrontal cortex (OFC) precedes that in the temporal cortex. Anatomically (MRI) constrained group averaged statistical activation maps were calculated from MEG at three different latencies from stimulus onset. Recognition-related activation (recognized vs. not-recognized) peaked in the left OFC 130 ms from stimulus onset, 50 ms before it peaked in recognition-related regions in the temporal cortex.

areas and the prefrontal cortex (e.g., humans: Oenguer and Price, 2000; macaque: Rempel-Clower and Barbas, 2000) (although information about direct connections between occipital visual areas and the OFC is still lacking). These findings suggest that the neural infrastructure is indeed in place for LSF information to trigger prefrontal activity subserving top-down facilitation.

To test the prediction that early recognition-related activity in the OFC should depend on the availability of LSF information in the image, Bar et al. (in press) used both fMRI and MEG to compare activity in the OFC elicited by filtered images of objects (Fig. 4) containing predominantly LSFs or predominantly high spatial frequencies (HSFs). They found that the same region of the OFC that shows early recognition-related activity (Fig. 3) also shows early differential activity for LSF vs. HSF images. This differential activity associated with spatial frequency content was found to peak around 115 ms from stimulus onset. Moreover, a trial-by-trial covariance analysis of the MEG data indicated that there was a clear interaction between the occipital visual regions and the OFC, and between the OFC and the fusiform gyrus for LSF images, as suggested by phase synchrony, but no significant synchrony existed between these regions for the HSF stimuli.

Taken together, the results reviewed in this section provide strong support for the proposal that visual object recognition benefits from the rapid analysis and projection of coarse LSF information of an input image to orbitofrontal regions of prefrontal cortex. Importantly, this object-based mechanism allows early information about an object to act as the catalyst for top-down enhancement of its own bottom-up cortical processing.

A context-based cortical mechanism for triggering top-down facilitation

In addition to the top-down benefit provided by prefrontal analysis of cursory object information, recognition efficiency can be increased through processes that take advantage of naturally occurring regularities in the environment (Gibson, 1969). A lifetime of visual experience can thereby guide expectations about which objects are likely to appear in a given setting to aid subsequent recognition of those objects through their contextual associations (Biederman, 1972, 1981; Palmer, 1975; Biederman et al., 1982; Bar and Ullman, 1996; Davenport and Potter, 2004). A beach setting, for example, typically contains objects such as beach chairs and beach umbrellas. Recognizing a scene with ocean surf breaking on the

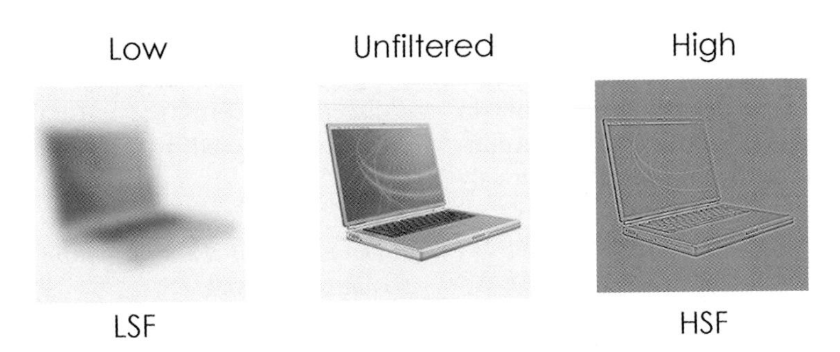

Fig. 4. Examples of the visual stimuli used in the Bar et al. (2006) MEG and fMRI studies. Unfiltered stimuli (middle) were used as intact controls for comparison with activity for low filtered (left) and high filtered (right) stimuli.

sand, combined with this prior knowledge about the scene's typical content, may act to rapidly sensitize the representations of these contextually related objects and facilitate their subsequent recognition. Indeed, several studies have shown that scenes can be recognized in a "glance" and that their content is extracted very rapidly (Biederman et al., 1974; Schyns and Oliva, 1994; Oliva and Torralba, 2001), lending support to the notion that early analysis of the surrounding contextual information may facilitate object recognition. Other studies have directly examined the interaction of scene and object processing, showing that objects are recognized faster when located in a typical environment than in an unusual setting. Palmer (1975), for instance, showed that recognition of briefly presented objects following the presentation of a contextual scene was significantly more accurate when the target object was contextually consistent with the preceding scene (e.g., kitchen — bread) than when it was inconsistent (e.g., kitchen — mailbox). In another series of experiments, Biederman et al. (1982) demonstrated that violating various types of contextual relations (e.g., relative position or size) between an object and its environment hinders subjects' ability to detect the object. Similar effects were reported by Davenport and Potter (2004). Taken together, these findings suggest that objects and their settings are processed interactively, supporting the view that contextual information directly facilitates the perceptual processes involved in visual object recognition (but see Hollingworth and Henderson, 1998, for a different perspective).

Cortical analysis of contextual information

To understand how cortical analysis of contextual associations might generally facilitate perception, and object recognition in particular, it is useful to first consider how the brain analyzes these associations. Bar and Aminoff (2003) recently addressed this issue in a series of fMRI studies. They identified those regions of the brain that are sensitive to contextual information by comparing cortical activity during recognition of individual objects with strong contextual associations with that for objects with weak contextual associations. For example, the association of a toaster with a kitchen setting is relatively stronger than the association of a camera with the various contexts in which it tends to occur. Bar and Aminoff found that such objects with strong contextual associations elicited greater activity than those with weak contextual associations, and did so primarily in the parahippocampal cortex (PHC) and the retrosplenial complex (RSC).[1] They found similar context-related activation in these regions for each subject and across several related experiments, showing clearly that the PHC and the RSC comprise a cortical "context

[1]The cortical region we refer to as the retrosplenial complex extends beyond the callosal sulcus, which is where recent cytoarchitectonic investigations have shown the retrosplenial cortex to be almost entirely located (Vogt et al., 2001). The retrosplenial region that is involved in contextual analysis (Bar and Aminoff, 2003) has a broader extent that is typically centered near the junction of the callosal sulcus, the precuneus, the caudal-ventral portion of the posterior cingulate gyrus, and the ventral-medial portion of the subparietal sulcus.

network" that processes contextual associations during object recognition.

In addition to this newly established role in processing contextual associations (Bar and Aminoff, 2003), the PHC and RSC have both previously been shown to mediate the processing of spatial, place-related information (e.g., Aguirre et al., 1996, 1998; Epstein and Kanwisher, 1998; Levy et al., 2001). Indeed, a region within the PHC typically shows a preferential response to pictures of scenes and topographical landmarks, and has therefore been termed the parahippocampal place area (PPA, Epstein and Kanwisher, 1998). Interestingly, in Bar and Aminoff's (2003) studies, the same region of PHC showed robust context-related activity for individual objects presented in isolation (see Fig. 5). How could individual objects activate a region of the cortex that is typically sensitive to scenes and landmarks? One possibility is that the perception of a strongly contextual object (e.g., a bowling pin) indirectly activates the representation of the place with which it is associated (e.g., a bowling alley). However, it is also possible that the PHC and RSC process contextual associations more generally. Thus, while places and landmarks may often correspond with specific sets of associations, context-related activity in the PHC and RSC might not be restricted to these highly spatial, place-specific contexts. This issue directly concerns the nature of the information that is processed by the context network.

The nature of the information processed by the context network

Does context-related activity in the PHC and RSC reflect the processing of contextual associations in general, or the processing of particular associations with spatial, place-specific contexts? To distinguish between these alternatives, Bar and Aminoff (2003) used objects with weak contextual associations as a baseline and then compared the relative increase in fMRI activation for objects with strong contextual associations with spatial, place-related contexts (e.g., farm) or more abstract nonspatial contexts (e.g., romance). As before, the PHC and RSC produced significantly greater activity for strongly contextual objects than for weakly contextual objects. Importantly, robust context-related activity was found for both types of contexts (Fig. 6). These results provide strong support for the hypothesis that the context network mediates the processing of contextual

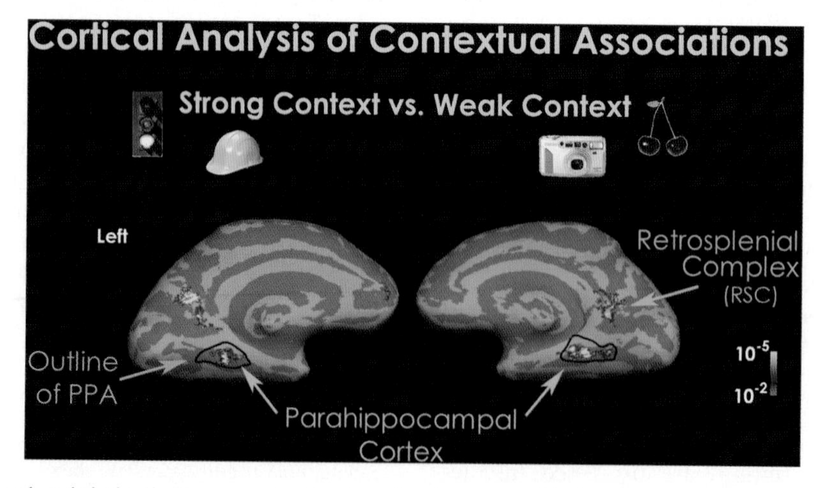

Fig. 5. Group averaged statistical activation maps (medial view) showing greater fMRI response in the parahippocampal cortex (PHC) and the retrosplenial complex (RSC) to individual objects with strong contextual associations (strong context) than to objects with weak contextual associations (weak context). The black outlines in the PHC show the boundaries of the parahippocampal place area (PPA), which was defined in a separate set of subjects by comparing cortical activation for indoor and outdoor scenes with that for objects, faces, and scrambled items (i.e., as in Epstein et al., 2003). Robust context-related activity occurs within the PPA for strong context vs. weak context individual objects.

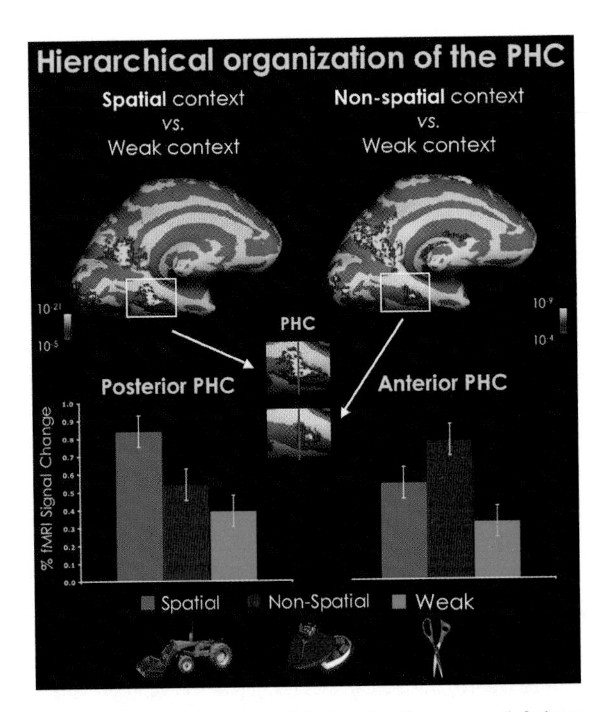

Fig. 6. Group averaged statistical activation maps (left hemisphere, medial view) showing greater fMRI response for objects strongly associated with spatial and nonspatial contexts than for objects with weak contextual associations. Bar graphs show the results of a region of interest analysis for the PHC. The magnitude of fMRI signal change was largest for spatial context objects in posterior PHC and for nonspatial context objects in anterior PHC.

associations in general, and that the representation of contextual associations in these regions is not confined to spatial, place-related contexts. Finding robust activation for objects from nonspatial contexts clearly shows that context-related activity in the PHC in response to individual objects is not merely a reflection of the indirect activation of the specific places with which they are associated.

Bar and Aminoff's (2003) comparison of cortical activity elicited by objects associated with spatial vs. nonspatial contexts revealed an interesting characteristic of the functional organization of the PHC. That is, the activity elicited by objects associated with nonspatial contexts was found to be confined to the anterior half of the PHC, while activity elicited by the objects associated with spatial contexts extended into the posterior half of the PHC (Fig. 6). This pattern of activation suggests that the representation of contextual associations in the PHC is organized along a hierarchy of increasing spatial specificity for more posterior representations. Additional evidence for this type of hierarchical organization has since been found in subsequent studies showing similar differences in the sensitivity of posterior and anterior regions of the medial temporal lobe to spatial information (Düzel et al., 2003; Jackson and Schacter, 2004; Pihlajamaki et al., 2004). This characteristic pattern of cortical activity during contextual analysis indicates that the nature of the information processed by the context network involves both spatial and nonspatial associations.

The involvement of spatial information in contextual representations might go beyond mere associations, and extend to representing the typical spatial relations between objects sharing the same context. A traffic light, for example, is usually located relatively high above a street and is typically fixed to a standard that rises from the corner of a sidewalk, with crosswalk markings and pedestrians below it, and cars driving underneath or stopped in front of it. Are these spatial relations represented in the associations for spatial contexts? The results of several behavioral studies indicate that this type of spatial information can indeed be associated with an object's representation and thereby impact its recognition (Biederman, 1972, 1981; Palmer, 1975; Biederman et al., 1982; Bar and Ullman, 1996). Bar and Ullman (1996), for example, demonstrated that spatial contextual information is important when identifying an otherwise ambiguous object. Thus, a squiggly line that cannot be identified in isolation can readily be identified as a pair of eyeglasses when presented in the appropriate spatial configuration with a hat (Fig. 7). These results support the hypothesis that spatial relations between objects are part of contextual representations that bind information about typical members of a context as well as the typical spatial relations between them, termed *context frames*.

Disentangling the processing and representation of spatial and nonspatial contextual associations is complicated by the fact that real-world objects are encountered in specific places and spatial locations, even those that are otherwise most strongly associated with nonspatial contexts. For example, while a heart-shaped box of chocolates might be

Fig. 7. Example of the importance of spatial context in object recognition. An ambiguous line (A) cannot be interpreted unequivocally in isolation, but (B) is readily identified as a pair of eyeglasses when shown in the appropriate location below a drawing of a hat.

most strongly related to the context of romance, people may also associate it with the store where it was sold or the room where the chocolates were eaten. Nevertheless, if the PHC is sensitive to both spatial and nonspatial associations through visual experience, such sensitivity should also develop in highly controlled situations using novel, unfamiliar contextual associations.

This type of experimental approach can maximize control over spatial and nonspatial contextual associations, and will be important for further examining the nature of the contextual information that is processed and represented in the PHC (Aminoff et al., submitted). For example, contextual associations could be formed through an extensive training phase in which subjects are repeatedly exposed to stimuli from each of three different conditions: spatial, nonspatial, and nonassociative (Fig. 8A). Under the spatial condition, individual groups of three shapes would be shown together in a consistent spatial configuration and with each shape always appearing in the same display location. Individual groups of three shapes would also be shown together under the nonspatial condition, but in a random spatial configuration on each trial. Finally, under a nonassociative condition, single shapes would be presented individually in a randomly determined location on each trial. These individual shapes would therefore not be associated with any other shape, nor any particular location. Consequently, these individual shapes could serve as control stimuli, analogous to the weak contextual objects used in the previous experiments.

After such a training session, fMRI could be used to detect differences in cortical activity for the contextual shapes relative to that for the individual, nonassociative shapes. To do this, a single

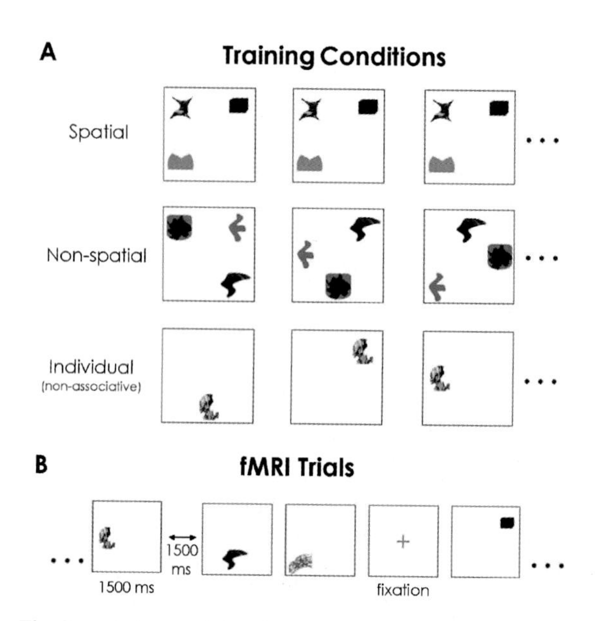

Fig. 8. Isolating spatial and nonspatial components in the formation of contextual associations. (A) During training, groups of three colored abstract shapes (examples shown in grayscale) are repeatedly presented together in the same specific spatial locations (spatial) or in random locations (nonspatial). A nonassociative control condition (individual) is formed by presenting single objects in random locations. (B) fMRI trials involve individual presentations of a single shape from each training-session display, regardless of condition.

shape from the previous training session could be presented during each trial in the scanner. Thus, the only difference for these individual shapes from the different conditions during scanning would be the prior experience the subject had with each shape (see Fig. 8B). In line with our previous results, we would expect the results of such a study to reveal a robust fMRI response in both the PHC and the RSC when subjects viewed shapes with newly formed contextual associations (spatial and nonspatial) relative to that for the nonassociative control stimuli. Importantly, the same anterior-to-posterior progression in spatial specificity in context-related activity of the PHC would be evident if the magnitude of the fMRI response were greatest in the posterior portion of the PHC for shapes from the spatial condition, and greatest in the anterior portion of the PHC for shapes from the nonspatial condition. Finding a similar anterior-to-posterior progression of activity in the PHC for nonspatial and spatial associations using otherwise novel abstract stimuli would provide converging evidence that the representation of contextual information in the PHC is organized along a hierarchy of spatial specificity.

The general conclusion drawn from the results of the experiments reviewed in this section is that the PHC and RSC together mediate the processing of contextual associations, and that the nature of these associations involve both spatial and nonspatial information. Characterizing the role of these regions in associative processing provides a critical foundation for exploring the cognitive and neural mechanisms underlying contextual facilitation of object recognition. The evidence that the PHC and RSC mediate associative processing also provides a framework for bridging previous findings about the function of the PHC and RSC. Specifically, in addition to contextual analysis, the PHC and RSC have both been implicated in the processing of spatial or navigational information (Aguirre et al., 1998; Epstein and Kanwisher, 1998; Maguire, 2001), as well as in episodic memory (Valenstein et al., 1987; Brewer et al., 1998; Wagner et al., 1998; Davachi et al., 2003; Morcom et al., 2003; Wheeler and Buckner, 2003; Kirwan and Stark, 2004; Ranganath et al., 2004b). Thus, at first glance it would seem that the same cortical regions have been assigned several different and seemingly unrelated functions. This potential conflict is resolved, however, when one considers that all of these processes rely on associations as their building blocks. Place-related and navigational processing requires analysis of the associations between the objects and landmarks appearing in a specific place, as well as the spatial relations among them. Episodic memories also rely on familiar associations. For example, an episodic memory of last night's dinner is a conjunction of associated constituents (e.g., the location, the company, the food, the dishes, etc.). Thus, describing the key role of the PHC and RSC in terms of associative processing, rather than limiting this role to spatial or episodic information, provides a common foundation from which various functions can be derived. As we discuss below, the ability of these regions to rapidly process such associations is critical for triggering top-down contextual facilitation of visual object recognition.

Contextual facilitation of object recognition

In the preceding sections we have considered evidence that object recognition can be facilitated in the presence of contextually related information. We have also described the cortical network that is specifically involved in analyzing contextual associations. But how does the processing of contextual associations lead to facilitated visual recognition? We propose that the contextual associations analyzed during the recognition of an object or scene are represented and activated in a corresponding "context frame" in the PHC. Context frames can be considered to be contextual representations (or schema) that include information about the objects that typically appear within that context (see also Bar and Ullman, 1996; Bar, 2004). The precise nature of the information represented in context frames is still being explored, but the results reviewed above suggest that it includes both spatial and nonspatial aspects of contextual relations. In this way, the PHC might serve as a switchboard of associations between items that are represented in detail elsewhere (Bar, 2004). The specific associations mediated by an activated context frame might act to sensitize, or prime, representations of other contextually related objects (Fig. 9). Spreading activation of contextually related object representations might thereby help to facilitate subsequent recognition of such objects as a function of their contextual associations. Interestingly, while the functional role of such efficient representation of contextual associations is hypothesized to benefit perception, evidence from visual false memory studies suggests that certain memory distortions can arise as a byproduct of activating these associations. For example, when tested for their memory of a previously shown picture, subjects in these studies often "remember" having seen objects that are contextually related to the picture but that were not actually in the picture (Miller and Gazzaniga, 1998).

Bar and Aminoff's (2003) studies showed that robust context-related activity can be elicited by a single key object from a particular context. This finding suggests that recognizing a single object might be sufficient to activate a corresponding context frame, prime the associated representations

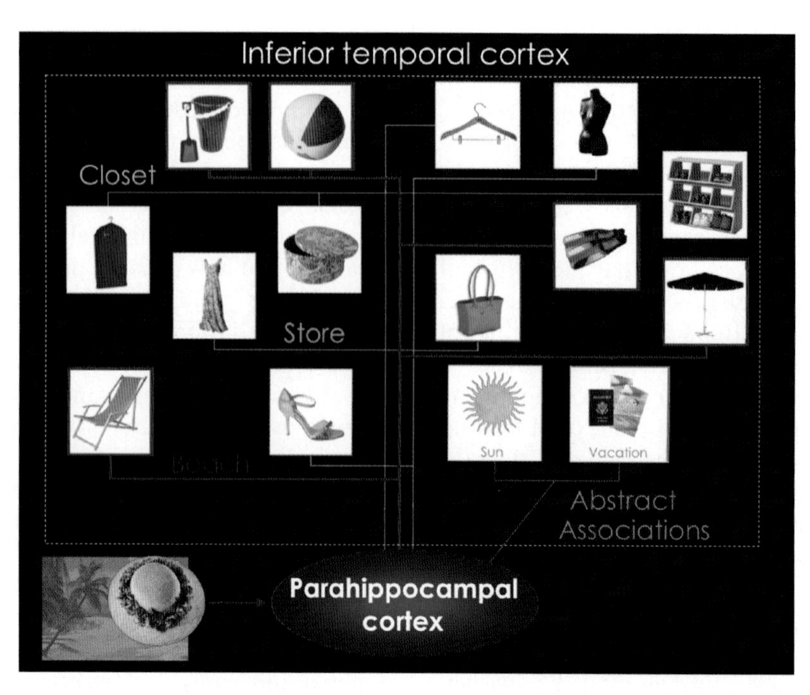

Fig. 9. A model of how contextual associations in the PHC might activate visual representations of contextually related objects in inferior temporal cortex to facilitate object recognition (adapted from Bar, 2004). Individual objects (e.g., a straw hat) can be associated with multiple context frames (e.g., a beach, a store, a closet). The experience-based set of associations represented in a specific context frame is activated as soon as the context has been established through recognition of a strongly contextual object (e.g., a palm tree) or through other contextual cues from the scene.

and thereby facilitate recognition of the other contextually related objects. The hypothesis that initial recognition of even a single strongly contextual object can facilitate subsequent recognition of a different, but contextually related, object was tested directly by Fenske et al. (submitted). While their study focused primarily on the neural substrates of contextual facilitation, it is interesting to note that no prior study had addressed even the behavioral characteristics of this type of object-to-object contextual priming using foveal presentations (cf., Henderson et al., 1987; Henderson, 1992).

Fenske et al. (submitted) presented individual objects with strong contextual associations within a visual priming paradigm, and asked subjects to make a simple size judgment about each object. Recognition improvement due to contextual priming effects was assessed by comparing response times (RTs) and the corresponding change in fMRI signal for objects that were immediately preceded by a contextually related object relative to that for objects from a new unprimed context (Fig. 10). The important result of this study was robust contextual facilitation of recognition-related RTs and corresponding fMRI response reductions in both context- and object-processing regions. Cortical response reductions were found in bilateral PHC and in the left anterior fusiform, lateral occipito-temporal, and inferior frontal cortices. At the behavioral level, these findings replicate previous studies showing more efficient and rapid recognition of contextually related, than unrelated, items. At a neural level, the novel and important finding of the Fenske et al. study was the robust cortical response reduction obtained in the context network and object-processing regions. Such cortical response reductions are important, as traditional priming studies have shown them to be a hallmark of enhanced recognition following prior exposure to the same, or a related, stimulus (Schacter and Buckner, 1998; Koutstaal et al., 2001; Vuilleumier et al., 2002; Henson, 2003; Simons et al., 2003; Lustig and Buckner, 2004; Maccotta and Buckner, 2004; Zago et al., 2005).

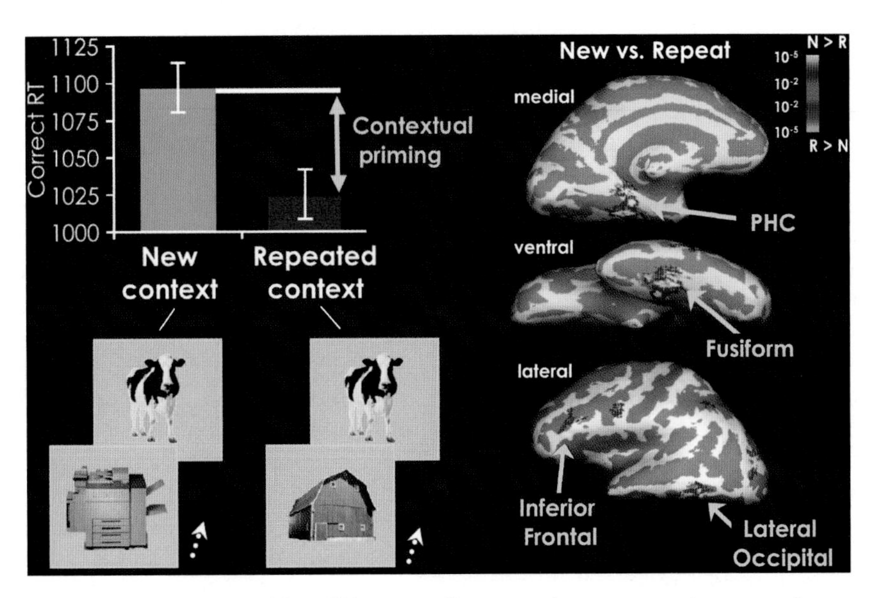

Fig. 10. Contextual facilitation of object recognition. Objects preceding repeated-context target items were always contextually related; objects preceding new-context items were never contextually related. After each picture presentation, participants were required to respond "bigger than a shoebox," or "smaller than a shoebox," by a key-press. Contextual priming was reflected by faster response times (RT) and reduced cortical fMRI response for repeated-context objects than for new-context objects. Contextual priming related fMRI response reductions occurred in the parahippocampal component of the context network, and in object-related regions of occipito-temporal and inferior frontal cortices.

Finding a response reduction in a primary component of the context network (i.e., PHC) supports the hypothesis that context-related facilitation of object recognition results from an enhanced processing of an object's contextual associations. Likewise, the response reduction seen in the left anterior fusiform gyrus implies that encountering a strongly contextual object activates representations of other contextually related objects (such as the contextually related targets), while the response reduction in the lateral occipital cortex, an area implicated in the processing of perceptual features, reflects activation of specific perceptual representations associated with these target objects. Finally, the response reduction in left inferior frontal cortex presumably reflects sensitization of semantic associations between context-related objects. The selective contextual priming effects found in these specific regions therefore provide important insight into the types of representations involved in the contextual facilitation of visual recognition. Importantly, these results suggest that there are many different types of information that are connected by the proposed context frames stored in the PHC, but that the various types of representations (perceptual, conceptual, etc.) are stored elsewhere.

The results of the Fenske et al. (submitted) study provide support for our proposal that contextual analysis during initial recognition of a highly contextual object serves to activate a corresponding context frame and, through the inherent associative connections, sensitizes the representations of other contextually related objects. Recognition is therefore enhanced when these objects whose representations have been primed in this manner are subsequently encountered. However, objects rarely appear in isolation, so it makes sense that context frames might also be activated by the information contained in a scene image, per se. Indeed, just as a coarse, LSF representation of an input image is often sufficient for rapid object recognition, the LSF content in a scene image might also be sufficient for deriving a reliable guess about the context frame that needs to be activated. The work of Schyns, Oliva, and Torralba (Schyns and Oliva, 1994; Oliva and Torralba, 2001) supports the proposal that LSF information from a typical scene is sufficient for a successful categorization of its context (e.g., a street, a beach), and the statistics of these images can help produce expectations about what and where is likely to appear in the scene. This work converges with psychophysical and computational evidence that LSFs are extracted from scene images earlier than HSFs (Palmer, 1975; Metzger and Antes, 1983; Bar, 2003). The proposal that emerges from this is that contextual facilitation of object recognition, just like object-based top-down facilitation, can be triggered rapidly on the basis of early cursory analysis of an input scene or object image.

While the results of the Fenske et al. (submitted) study clearly show that recognition of a strongly contextual object can facilitate the subsequent recognition of other contextually related objects, and that this contextual priming effect is associated with changes in activity in the PHC component of the context network, this study does not address the relative top-down influence of different types of contextual associations (e.g., spatial vs. nonspatial) on subsequent recognition. Nevertheless, the finding that contextual facilitation of object recognition is also associated with cortical response reductions in cortical regions associated with different aspects of object processing (e.g., perceptual, semantic, etc.) suggests that contextual facilitation may indeed be mediated through various types of object-related representations. To maximize contextual facilitation, the context frames that maintain the corresponding associations must therefore accommodate various types of representations. How are context frames organized for this to be achieved? We consider this question in the following section.

Contextual facilitation and the representation of spatial and nonspatial associations

The context network is proposed to mediate the processing and representation of various types of contextual associations through context frames. The nature of these context frames and how they represent different types of associations remains an open and exciting area for future research. Bar and Ullman (1996) suggested that contextual frames contain information about not only the identity of objects that tend to co-occur in scenes but also their typical spatial relations. Indeed, our experiments have repeatedly shown that the PHC component of the context network is involved in the analysis of both spatial and nonspatial associations during recognition of individual objects (e.g., Bar and Aminoff, 2003). In addition, there is clear behavioral evidence that spatial contextual information can facilitate object recognition. Several studies have shown that objects are recognized more efficiently and accurately when located in contextually consistent (expected) locations than in contextually inconsistent (unexpected) locations (Biederman, 1972; Mandler and Johnson, 1976; Hock et al., 1978; Biederman, 1981; Biederman et al., 1982; Cave and Kosslyn, 1993; Bar and Ullman, 1996; Chun and Jiang, 2003). For instance, when viewing a scene of an office, a table lamp would typically be recognized faster when appearing on top of a desk than when appearing beneath it, suggesting that prior knowledge about the spatial constraints of objects in scenes contributes to object recognition.

However, it remains unclear whether nonspatial (e.g., semantic) information about the specific identities of objects in scenes is indeed integrated, as Bar and Ullman (1996) have suggested, with knowledge about the spatial relations between these objects. In other words, does context-based top-down facilitation of object recognition rely on spatial and nonspatial associations that are linked within unified context frames or that are represented independently? This question is critical for understanding the nature of contextual representation and the mechanisms through which associative information influences visual perception. Consider that while spatial analysis is an inherent part of visual processing and the interpretation of real-life scenes, it may nevertheless rely on different forms of cortical representation than those involved in the analysis of the nonspatial content of these scenes. That is, nonspatial and spatial context effects may in fact be mediated by separate, independent mechanisms. Thus, when viewing a desk in an office, a nonspatial context frame may generate predictions regarding the semantic information in the scene (e.g., the likelihood of certain objects to appear with the desk), sensitizing the representation of a "lamp" regardless of its spatial position relative to the desk. At the same time, an independent spatial context frame might constrain the spatial relations between these objects, enhancing the representation of the lamp's *location*, regardless of any particular object identity (i.e., based on the knowledge that objects are typically positioned on top of desks and not beneath them).

An alternate view is that context frames contain unified representations for both nonspatial and spatial contextual information. Thus, when viewing a desk in an office, the representation of a desk lamp would *necessarily* include information about its expected location on top of the desk because of the strong connections between this type of spatial and nonspatial knowledge within the contextual schema of an "office." According to this view, the contextual representation of a particular type of scene is maintained as a single unit (i.e., a single context frame). Any violation of the spatial relations between objects in the scene should also impact analysis of the nonspatial associations between these objects (e.g., their expected identities) and the interpretation of the scene as a whole. Thus, seeing a desk lamp *under* a desk would not only violate expectations regarding the most probable *location* of objects in general (as objects typically appear on top of desks), but also violate expectations regarding the specific *identity* or role of the lamp (as desk lamps appear, by definition, on desks). In other words, in a unified context frame, both spatial and nonspatial knowledge jointly contribute to visual comprehension, and thus any spatial inconsistencies between objects may affect the interpretation of these objects' identity, or meaning, altogether.

The hypothesis that nonspatial and spatial contextual associations are represented independently is supported by evidence that an object's identity is typically analyzed through a ventral anatomical route including inferior occipito-temporal cortex, while its spatial location is typically analyzed through more dorsal fronto-parietal brain regions (Goodale and Milner, 1992). The existence of distinct neural systems for analysis of object's identity and spatial location raises the possibility that different cognitive mechanisms underlie facilitation of object recognition through nonspatial and spatial contextual associations. Indeed, one interpretation of the results of Fenske et al. (submitted) study is that the recognition of contextually related objects is facilitated by a priming mechanism that sensitizes the nonspatial representations of objects that are strongly associated with a contextual cue. In contrast, the results of studies investigating spatial contextual facilitation suggest that recognition of "properly" positioned objects is facilitated by a spatial attention mechanism that enhances the processing of information at a specific location (Hollingworth and Henderson, 1998; Chun and Jiang, 2003; Gordon, 2004). Prior experience with a specific scene or fixed spatial configuration, for example, can guide spatial attention to the location that is most likely to contain a target of interest, thereby facilitating visual search and target recognition within that scene or configuration (Chun and Jiang, 2003).

The two views presented here concerning the underlying structure of context frames and the representation of spatial and nonspatial contextual associations (i.e., unified vs. independent representations) should not necessarily be considered to be mutually exclusive. Indeed, contextual facilitation of object recognition might ultimately be found to involve the top-down influence of both independent and unified representations of spatial and nonspatial contextual associations. Investigating this possibility will require a study that examines the effects of both nonspatial and spatial contextual factors on subsequent object recognition. To determine whether the two types of associative knowledge operate within separate contextual representations, one needs to test whether they can influence perception simultaneously without interacting. Thus, most importantly, an orthogonal design is required in which the two factors are manipulated independently, such that the unique effect of each factor, as well as their joint (interactive) effects on object recognition, can be directly assessed (cf., Sternberg, 2001). To the extent that the two types of associative knowledge are linked within a combined context frame, we would anticipate an interaction in reaction times to target objects. Specifically, one might expect the benefit of identifying an object in a contextually consistent location relative to a contextually inconsistent location to be significantly greater for objects whose identities have been contextually primed than for contextually unprimed objects, suggesting a unified representation for both nonspatial and spatial factors. In this situation, a similar interaction might also be expected for corresponding cortical activity in brain regions associated with contextual processing (i.e., the PHC and/or RSC) and in object-related processing areas (e.g., fusiform gyrus), with the largest differential cortical activation when the target is consistent with both spatial and nonspatial contextual information. If, however, nonspatial and spatial contextual representations operate independently, their contribution to response latencies, as well as to brain activation, will be additive. In addition, a possible anatomical distinction within the contextual network may be found between the two types of representations (as found previously with spatial and nonspatial associations, Bar and Aminoff, 2003). With such an outcome, one could conclude that the two types of information reside in separate representational "stores," suggesting dissociable influences in context-based top-down facilitation of object recognition. Finally, because spatial and nonspatial contextual associations may involve the top-down influence of both independent and unified representations of spatial and nonspatial contextual associations, it is also possible that the effects of these representations will be additive in context- and object-related regions, but will interact at higher-level brain regions associated with postrecognition processes. Future research is needed to address these issues, and further increase our understanding of the nature of contextual representations and their effect on visual object recognition.

Integrated object- and context-based top-down facilitation of recognition

In this overview, we have described how top-down facilitation of recognition can be achieved either (1) through an object-based mechanism that generates "initial guesses" about an object's identity using rapidly analyzed coarse information about the input image or (2) through the predictive information provided by contextual associations between an object or a scene and the other objects that are likely to appear together in a particular setting. However, it is clear that objects do not appear in isolation, and that both object-based and context-related information is typically available to the visual system during recognition. It therefore makes sense that an optimized system might take advantage of both forms of top-down facilitation to maximize the efficiency of recognition processes. How is this achieved? In this section, we consider how both object-based and context-based influences might operate together to facilitate object recognition.

Central to this discussion is the observation that recognition of an object can be enhanced through neural mechanisms that sensitize the cortical representations of the most likely candidate interpretations of that particular object before information processed through the bottom-up stream has sufficiently accumulated. We propose two key mechanisms through which this "sensitization" is achieved. As reviewed in the first part of this paper, an object-based mechanism is proposed to capitalize on a rapidly derived LSF representation of the object itself to generate "initial guesses" about its identity. The back-projection of these candidate interpretations to object-processing regions thereby sensitizes the corresponding cortical representations in advance of the bottom-up information that continues to activate the correct representation. The context-based mechanism, reviewed in the second part of this chapter, is proposed to sensitize the representations of contextually related objects whose associations are activated through context frames stored in the PHC (see Fig. 9). Importantly, we propose that context frames can be activated following prior recognition of strongly contextual object (as in Fenske et al., submitted), or through early, coarse information about a scene or an object. This includes the possibility that contextual analysis begin promptly, even before a scene or an object is fully recognized. Considered together, the mechanisms for top-down facilitation of object recognition that we describe include one that is based on information from the "to-be-recognized" object itself and the other based on information about the context in which the object appears. Given the intricate links between objects and the settings in which they appear, these mechanisms typically should not be expected to operate in isolation. Indeed, provided that sufficient information is available for activating the appropriate context frame, and that an LSF image of a single target object is sufficient for limiting its possible interpretations, the intersection of these two sources of information would result in a unique, accurate identification (Bar, 2004). An exciting direction for future research will be to assess how information about additional objects or the scene in which a target object appears may be processed in parallel to increase the opportunities for such valuable intersections of object- and context-based information.

The interactive nature of the two sources of top-down facilitation of object recognition that we have described emerges when the input to either the object- or context-based mechanism is ambiguous. For example, when the coarse object-related input to the prefrontal cortex is ambiguous (e.g., a blurred image of an umbrella also resembles a mushroom, lamp, and parasol), the benefit of having activated the appropriate context frame will be relatively greater than if the LSF profile of an object is completely diagnostic of the object's identity (e.g., a blurred image of a giraffe only resembles a giraffe). In addition, if ambiguous information is projected to the PHC, then this can result in the activation of multiple context frames. From this possibility emerges a rather counterintuitive prediction of our model. Consider, for instance, that a picture of a gun, when projected rapidly in a blurred (i.e., LSF) form to the PHC, may be interpreted also as a drill and a hairdryer (Fig. 11).

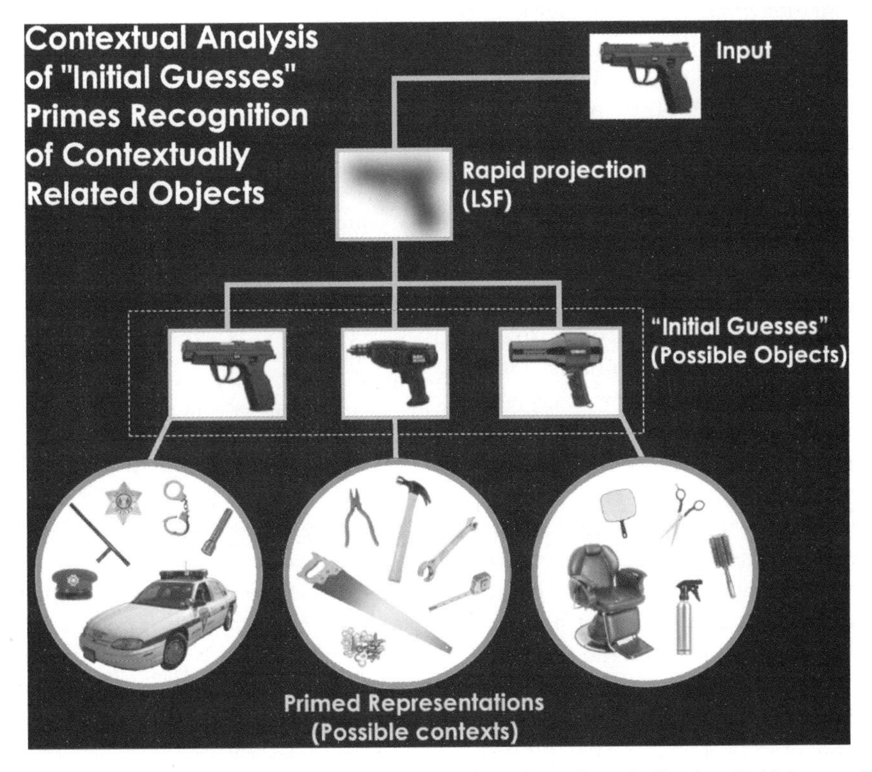

Fig. 11. In the proposed interactive model of object- and context-based top-down down facilitation, "initial guesses" about an object's identity are rapidly generated from cursory analysis of its LSF information. These "initial guesses" are processed by the context network to help determine the possible context, and thereby facilitate the recognition of other objects that may also be present. Contextual information helps to further constrain the "initial guesses" about an object's identity, just as early object-based information helps to determine which context frames should be activated. This interactive model accounts for the otherwise counterintuitive finding that a brief presentation of a gun can facilitate the subsequent recognition of a hairbrush (Fenske and Bar, submitted), despite the lack of any other perceptual, semantic, or contextual relation between these items.

These three objects are associated with three different context frames, and will subsequently trigger the activation of three sets of objects. Consequently, a gun will not only prime the recognition of a police car (i.e., contextual priming), but also the recognition of a hairbrush (i.e., a member of the context frame activated by the hairdryer). Importantly, because there is a complete lack of any other perceptual, semantic, or contextual relation between a gun and a hairbrush, finding significant priming for this condition is best explained through the interactive model of top-down facilitation that we propose. Indeed, we found significant priming for this condition (Fenske and Bar, submitted), and found further that this form of indirect priming existed for relatively short prime exposures (120 ms) but not for longer prime exposures (2400 ms). This finding underscores the interactive nature of the object- and context-based mechanisms we have described, and supports our notion that the arrival of additional information leaves active only the most relevant "initial guess" about an object's identity and only the most relevant context frame.

In conclusion, the models and research findings reviewed here emphasize the importance of top-down facilitation in visual object recognition. Building on the previous work (Bar, 2003), we have examined evidence for an object-based mechanism that rapidly triggers top-down facilitation of visual recognition using a partially analyzed version of the input image (i.e., a blurred image) that generates reliable "initial guesses" about the object's identity in the OFC. Importantly, we have also shown that this form of top-down facilitation does not operate in isolation. Work from our lab indicates that contextual associations between objects and scenes are analyzed by a network including the PHC and the RSC, and that the predictive information provided by these associations can also constrain the "initial guesses" about an objects' identity. We propose that these mechanisms operate together to promote efficient recognition by framing early information about an object within the constraints provided by a lifetime of experience with contextual associations.

Abbreviations

fMRI	functional magnetic resonance imaging
MEG	magnetoencephalography
OFC	orbitofrontal cortex
PFC	prefrontal cortex
PHC	parahippocampal cortex
PPA	parahippocampal place area
RSC	retrosplenial complex
RT	response time

Acknowledgments

This work was supported by NINDS R01-NS44319 and RO1-NS050615, NCRR P41-RR14075, and the MIND Institute.

References

Aguirre, G.K., Detre, J.A., Alsop, D.C. and D'Esposito, M. (1996) The parahippocampus subserves topographical learning in man. Cereb. Cortex, 6: 823–829.

Aguirre, G.K., Zarahn, E. and D'Esposito, M. (1998) An area within human ventral cortex sensitive to "building" stimuli: evidence and implications. Neuron, 21: 373–383.

Aminoff, E. Gronau, N. and Bar, M. (submitted for publication). The parahippocampal cortex mediates spatial and nonspatial associations.

Bachevalier, J. and Mishkin, M. (1986) Visual recognition impairment follows ventromedial but not dorsolateral prefrontal lesions in monkeys. Behav. Brain Res., 20: 249–261.

Bar, M. (2003) A cortical mechanism for triggering top-down facilitation in visual object recognition. J. Cogn. Neurosci., 15: 600–609.

Bar, M. (2004) Visual objects in context. Nat. Rev. Neurosci., 5: 617–629.

Bar, M. and Aminoff, E. (2003) Cortical analysis of visual context. Neuron, 38: 347–358.

Bar, M., Kassam, K.S., Ghuman, A.S., Boshyan, J., Schmidt, A.M., Dale, A.M., Hamalainen, M.S., Marinkovic, K., Schacter, D.L., Rosen, B.R. and Halgren, E. (2006) Top-down facilitation of visual recognition. Proceedings of the National Academy of Science, 103: 449–454.

Bar, M., Tootell, R., Schacter, D., Greve, D., Fischl, B., Mendola, J., Rosen, B. and Dale, A. (2001) Cortical mechanisms of explicit visual object recognition. Neuron, 29: 529–535.

Bar, M. and Ullman, S. (1996) Spatial context in recognition. Perception, 25: 343–352.

Barceló, F., Suwazono, S. and Knight, R.T. (2000) Prefrontal modulation of visual processing in humans. Nat. Neurosci., 3: 399–403.

Biederman, I. (1972) Perceiving real-world scenes. Science, 177: 77–80.

Biederman, I. (1981) On the semantic of a glance at a scene. In: Kubovy, M. and Pomerantz, J.R. (Eds.), Perceptual Organization. Erlbaum, Hillsdale, NJ, pp. 213–253.

Biederman, I., Mezzanotte, R.J. and Rabinowitz, J.C. (1982) Scene perception: detecting and judging objects undergoing relational violations. Cogn. Psychol., 14: 143–177.

Biederman, I., Rabinowitz, J.C., Glass, A.L. and Stacy, E.W. (1974) On the information extracted from a glance at a scene. J. Exp. Psychol., 103: 597–600.

Brewer, J.B., Zhao, Z., Desmond, J.E., Glover, G.H. and Gabrieli, J.D. (1998) Making memories: brain activity that predicts how well visual experience will be remembered. Science, 281: 1185–1187.

Bullier, J. and Nowak, L.G. (1995) Parallel versus serial processing: new vistas on the distributed organization of the visual system. Curr. Opin. Neurobiol., 5: 497–503.

Cave, C.B. and Kosslyn, S.M. (1993) The role of parts and spatial relations in object identification. Perception, 22: 229–248.

Chun, M.M. and Jiang, Y. (2003) Implicit, long-term spatial contextual memory. J. Exp. Psychol. Learn. Mem. Cogn., 29: 224–234.

Davachi, L., Mitchell, J. and Wagner, A. (2003) Multiple routes to memory: distinct medial temporal lobe processes build item and source memories. Proc. Natl. Acad. Sci. USA., 100: 2157–2162.

Davenport, J.L. and Potter, M.C. (2004) Scene consistency in object and background perception. Psychol. Sci., 15: 559–564.

Desimone, R. (1998) Visual attention mediated by biased competition in extrastriate visual cortex. Philos. Trans. R. Soc. Lond. B Biol. Sci., 353: 1245–1255.

Düzel, E., Habib, R., Rotte, M., Guderian, S., Tulving, E. and Heinze, H.J. (2003) Human hippocampal and parahippocampal activity during visual associative recognition memory for spatial and nonspatial stimulus configurations. J. Neurosci., 23: 9439–9444.

Engel, A.K., Fries, P. and Singer, W. (2001) Dynamic predictions: oscillations and synchrony in top-down processing. Nat. Rev. Neurosci., 2: 704–716.

Epstein, R., Graham, K.S. and Downing, P.E. (2003) Viewpoint-specific scene representations in human parahippocampal cortex. Neuron, 37: 865–876.

Epstein, R. and Kanwisher, N. (1998) A cortical representation of the local visual environment. Nature, 392: 598–601.

Fenske, M.J., Boshyan, J. and Bar, M. (submitted) Can a gun prime a hairbrush? The "initial guesses" that drive top-down contextual facilitation of object recognition. Paper presented at the 5th Annual Meeting of the Vision Sciences Society, Sarasota, FL.

Fenske, M.J. and Bar, M. (submitted for publication). Can A Gun Prime A Haitbrush? "Initial Guesses" that Mediate Contextual Facilitation of Object Recognition.

Freedman, D.J., Riesenhuber, M., Poggio, T. and Miller, E.K. (2001) Categorical representation of visual stimuli in the primate prefrontal cortex. Science, 291: 312–316.

Friston, K. (2003) Learning and inference in the brain. Neural Networks, 16: 1325–1352.

Gibson, E.J. (1969) Principles of Perceptual Learning and Development. Appleton-Century-Crofts, New York.

Gilbert, C.D., Sigman, M. and Crist, R.E. (2001) The neural basis of perceptual learning. Neuron, 31: 681–697.

Goodale, M.A. and Milner, A.D. (1992) Separate visual pathways for perception and action. Trends Neurosci, 15: 20–25.

Gordon, R.D. (2004) Attentional allocation during the perception of scenes. J. Exp. Psychol. Hum. Percept. Perform., 30: 760–777.

Grabowska, A. and Nowicka, A. (1996) Visual-spatial-frequency model of cerebral asymmetry: a critical survey of behavioral and electrophysiological studies. Psychol. Bull., 120: 434–449.

Grill-Spector, K., Kourtzi, Z. and Kanwisher, N. (2001) The lateral occipital complex and its role in object recognition. Vision Res., 41: 1409–1422.

Grossberg, S. (1980) How does a brain build a cognitive code? Psychol. Rev., 87: 1–51.

Henderson, J.M. (1992) Identifying objects across saccades: effects of extrafoveal preview and flanker object context. J. Exp. Psychol. Learn. Mem. Cogn., 18: 521–530.

Henderson, J.M., Pollatsek, A. and Rayner, K. (1987) Effects of foveal priming and extrafoveal preview on object identification. J. Exp. Psychol. Hum. Percept. Perform., 13: 449–463.

Henson, R.N. (2003) Neuroimaging studies of priming. Prog. Neurobiol., 70: 53–81.

Hock, H.S., Romanski, L., Galie, A. and Williams, C.S. (1978) Real-world schemata and scene recognition in adults and children. Mem. Cogn., 6: 423–431.

Hollingworth, A. and Henderson, J.M. (1998) Does consistent scene context facilitate object perception? J. Exp. Psychol. Gen., 127: 398–415.

Hopfinger, J.B., Buonocore, M.H. and Mangun, G.R. (2000) The neural mechanisms of top-down attentional control. Nat. Neurosci., 3: 284–291.

Humphreys, G.W., Riddoch, M.J. and Price, C.J. (1997) Top-down processes in object identification: evidence from experimental psychology, neuropsychology and functional anatomy. Philos. Trans. R. Soc. Lond. B Biol. Sci., 352: 1275–1282.

Jackson, O. and Schacter, D.L. (2004) Encoding activity in anterior medial temporal lobe supports subsequent associative recognition. Neuroimage, 21: 456–462.

Kirwan, C.B. and Stark, C.E. (2004) Medial temporal lobe activation during encoding and retrieval of novel face-name pairs. Hippocampus, 14: 919–930.

Kosslyn, S.M. (1994) Image and Brain. MIT Press, Cambridge, MA.

Kosslyn, S.M., Alpert, N.M. and Thompson, W.L. (1995) Identifying objects at different levels of hierarchy: a positron emission tomography study. Hum. Brain Mapping, 3: 107–132.

Kosslyn, S.M., Alpert, N.M., Thompson, W.L., Chabris, C.F., Rauch, S.L. and Anderson, A.K. (1993) Visual mental imagery activates topographically organized visual cortex: PET investigations. J. Cogn. Neurosci., 5: 263–287.

Koutstaal, W., Wagner, A.D., Rotte, M., Maril, A., Buckner, R.L. and Schacter, D.L. (2001) Perceptual specificity in visual object priming: functional magnetic resonance imaging evidence for a laterality difference in fusiform cortex. Neuropsychologia, 39: 184–199.

Lee, T.S. and Mumford, D. (2003) Hierarchical Bayesian inference in the visual cortex. J. Opt. Soc. Am., 20: 1434–1448.

Leonova, A., Pokorny, J. and Smith, V.C. (2003) Spatial frequency processing in inferred PC- and MC-pathways. Vision Res., 43: 2133–2139.

Levy, I., Hasson, U., Avidan, G., Hendler, T. and Malach, R. (2001) Center-periphery organization of human object areas. Nat. Neurosci., 4: 533–539.

Logothetis, N.K., Leopold, D.A. and Sheinberg, D.L. (1996) What is rivalling during binocular rivalry? Nature, 380: 621–624.

Lustig, C. and Buckner, R.L. (2004) Preserved neural correlates of priming in old age and dementia. Neuron, 42: 865–875.

Maccotta, L. and Buckner, R.L. (2004) Evidence for neural effects of repetition that directly correlate with behavioral priming. J. Cogn. Neurosci., 16: 1625–1632.

Maguire, E.A. (2001) The retrosplenial contribution to human navigation: a review of lesion and neuroimaging findings. Scand. J. Psychol., 42: 225–238.

Malach, R., Levy, I. and Hasson, U. (2002) The topography of high-order human object areas. Trends Cogn. Sci., 6: 176–184.

Mandler, J.M. and Johnson, N.S. (1976) Some of the thousand words a picture is worth. J. Exp. Psychol. [Hum. Learn.], 2: 529–540.

Martin, A., Wiggs, C.L., Ungerleider, L.G. and Haxby, J.V. (1996) Neural correlates of category-specific knowledge. Nature, 379: 649–652.

Mechelli, A., Price, C.J., Friston, K.J. and Ishai, A. (2004) Where bottom-up meets top-down: neuronal interactions during perception and imagery. Cereb. Cortex, 14: 1256–1265.

Merigan, W.H. and Maunsell, J.H. (1993) How parallel are the primate visual pathways? Annu. Rev. Neurosci., 16: 369–402.

Metzger, R.L. and Antes, J.R. (1983) The nature of processing early in picture perception. Psychol. Res., 45: 267–274.

Miller, M.B. and Gazzaniga, M.S. (1998) Creating false memories for visual scenes. Neuropsychologia, 36: 513–520.

Miyashita, Y. and Hayashi, T. (2000) Neural representation of visual objects: encoding and top-down activation. Curr. Opin. Neurobiol., 10: 187–194.

Morcom, A.M., Good, C.D., Frackowiak, R.S. and Rugg, M.D. (2003) Age effects on the neural correlates of successful memory encoding. Brain, 126: 213–229.

Oenguer, D. and Price, J.L. (2000) The organization of networks within the orbital and medial prefrontal cortex of rats, monkeys and humans. Cereb. Cortex, 10: 206–219.

Oliva, A. and Torralba, A. (2001) Modeling the shape of a scene: a holistic representation of the spatial envelope. Int. J. Comp. Vis., 42: 145–175.

Palmer, S.E. (1975) The effects of contextual scenes on the identification of objects. Mem. Cogn., 3: 519–526.

Parker, A., Wilding, E. and Akerman, C. (1998) The Von Restorff effect in visual object recognition memory in humans and monkeys. The role of frontal/perirhinal interaction. J. Cogn. Neurosci., 10: 691–703.

Pascual-Leone, A. and Walsh, V. (2001) Fast backprojections from the motion to the primary visual area necessary for visual awareness. Science, 292: 510–512.

Pihlajamaki, M., Tanila, H., Kononen, M., Hanninen, T., Hamalainen, A., Soininen, H. and Aronen, H.J. (2004) Visual presentation of novel objects and new spatial arrangements of objects differentially activates the medial temporal lobe subareas in humans. Eur. J. Neurosci., 19: 1939–1949.

Ranganath, C., DeGutis, J. and D'Esposito, M. (2004a) Category-specific modulation of inferior temporal activity during working memory encoding and maintenance. Brain Res. Cogn. Brain Res., 20: 37–45.

Ranganath, C., Yonelinas, A.P., Cohen, M.X., Dy, C.J., Tom, S.M. and D'Esposito, M. (2004b) Dissociable correlates of recollection and familiarity within the medial temporal lobes. Neuropsychologia, 42: 2–13.

Rempel-Clower, N.L. and Barbas, H. (2000) The laminar pattern of connections between prefrontal and anterior temporal cortices in the rhesus monkey is related to cortical structure and function. Cereb. Cortex, 10: 851–865.

Schacter, D.L. and Buckner, R.L. (1998) Priming and the brain. Neuron, 20: 185–195.

Schyns, P.G. and Oliva, A. (1994) From blobs to boundary edges: evidence for time- and spatial-dependent scene recognition. Psychol. Sci., 5: 195–200.

Shapley, R. (1990) Visual sensitivity and parallel retinocortical channels. Annu. Rev. Psychol., 41: 635–658.

Simons, J.S., Koutstaal, W., Prince, S., Wagner, A.D. and Schacter, D.L. (2003) Neural mechanisms of visual object priming: evidence for perceptual and semantic distinctions in fusiform cortex. Neuroimage, 19: 613–626.

Sternberg, S. (2001) Separate modifiability, mental modules, and the use of pure and composite measures to reveal them. Acta Psychol. (Amst.), 106: 147–246.

Tanaka, K. (1996) Inferotemporal cortex and object vision. Annu. Rev. Neurosci., 19: 109–139.

Ullman, S. (1995) Sequence seeking and counter streams: A computational model for bidirectional information flow in the visual cortex. Cereb. Cortex, 1: 1–11.

Valenstein, E., Bowers, D., Verfaellie, M., Heilman, K.M., Day, A. and Watson, R.T. (1987) Retrosplenial amnesia. Brain, 110: 1631–1646.

Vogt, B.A., Vogt, L.J., Perl, D.P. and Hof, P.R. (2001) Cytology of human caudomedial cingulate, retrosplenial, and caudal parahippocampal cortices. J. Comp. Neurol., 438: 353–376.

Vuilleumier, P., Henson, R.N., Diver, J. and Dolan, R.J. (2002) Multiple levels of visual object constancy revealed by event-related fMRI of repetition priming. Nat. Neurosci., 5: 491–499.

Wagner, A.D., Schacter, D.L., Rotte, M., Koutstaal, W., Maril, A., Dale, A.M., Rosen, B.R. and Buckner, R.L. (1998) Building memories: remembering and forgetting of verbal experiences as predicted by brain activity. Science, 281: 1188–1191.

Wheeler, M.E. and Buckner, R.L. (2003) Functional dissociation among components of remembering: control, perceived oldness, and content. J. Neurosci., 23: 3869–3880.

Wilson, F.A.W., Scalaidhe, S.P.O. and Goldman-Rakic, P.S. (1993) Dissociation of object and spatial processing domains in primate prefrontal cortex. Science, 260: 1955–1958.

Zago, L., Fenske, M.J., Aminoff, E. and Bar, M. (2005) The rise and fall of priming: how visual exposure shapes cortical representations of objects. Cereb. Cortex, 15: 1655–1665.

Martinez-Conde, Macknik, Martinez, Alonso & Tse (Eds.)
Progress in Brain Research, Vol. 155
ISSN 0079-6123

CHAPTER 2

Building the gist of a scene: the role of global image features in recognition

Aude Oliva[1,*] and Antonio Torralba[2]

[1]*Department of Brain and Cognitive Sciences, Massachusetts Institute of Technology, 77 Massachusetts Avenue, Cambridge, MA 02139, USA*
[2]*Computer Science and Artificial Intelligence Laboratory, Massachusetts Institute of Technology, 77 Massachusetts Avenue, Cambridge, MA 02139, USA*

Abstract: Humans can recognize the *gist* of a novel image in a single glance, independent of its complexity. How is this remarkable feat accomplished? On the basis of behavioral and computational evidence, this paper describes a formal approach to the representation and the mechanism of scene gist understanding, based on scene-centered, rather than object-centered primitives. We show that the structure of a scene image can be estimated by the mean of global image features, providing a statistical summary of the spatial layout properties (Spatial Envelope representation) of the scene. Global features are based on configurations of spatial scales and are estimated without invoking segmentation or grouping operations. The scene-centered approach is not an alternative to local image analysis but would serve as a feed-forward and parallel pathway of visual processing, able to quickly constrain local feature analysis and enhance object recognition in cluttered natural scenes.

Keywords: scene recognition; gist; spatial envelope; global image feature; spatial frequency; natural image

Introduction

One remarkable aspect of human visual perception is that we are able to understand the meaning of a complex novel scene very quickly even when the image is blurred (Schyns and Oliva, 1994), or presented for only 20 ms (Thorpe et al., 1996). Mary Potter (1975, 1976, see also Potter et al., 2004) demonstrated that during a rapid presentation of a stream of images, observers were able to identify the semantic category of each image as well as a few objects and their attributes. This rapid understanding phenomenon can be experienced while looking at modern movie trailers which utilize many fast cuts between scenes: with a mere glimpse of each picture, you can identify each shot's meaning, the actors and the emotion depicted in each scene (Maljkovic and Martini, 2005) even though you will not necessarily remember the details of the trailer. The amount of perceptual and semantic information that observers comprehend within a glance (about 200 ms) refers to the *gist* of the scene (for a review, Oliva, 2005). In this paper, we discuss two main questions related to rapid visual scene understanding: what visual information is perceived during the course of a glance, and which mechanisms could account for the efficiency of scene gist recognition.

Research in scene understanding has traditionally treated objects as the atoms of recognition. However, behavioral experiments on fast scene perception suggest an alternative view: that we do not need to perceive the objects in a scene to

*Corresponding author. Tel.: +1-617-452-2492; Fax: +1-617-258-8654; E-mail: oliva@mit.edu

DOI: 10.1016/S0079-6123(06)55002-2

identify its semantic category. The semantic category of most real-world scenes can be inferred from their spatial layout (e.g., an arrangement of basic geometrical forms such as simple *Geons* clusters, Biederman, 1995; the spatial relationships between regions or blobs of particular size and aspect ratio, Schyns and Oliva, 1994; Sanocki and Epstein, 1997; Oliva and Schyns, 2000). Fig. 1 illustrates the importance of the spatial arrangement of regions for scene and object recognition. When looking at the image on the left, viewers describe the scene as a street with cars, buildings and the sky. Despite the fact that the local information available in the image is insufficient for reliable object recognition, viewers are confident and highly consistent in their descriptions. Indeed, the blurred scene has the spatial layout of a street. When the image is shown in high resolution, new details reveal that the image has been manipulated and that the buildings are in fact pieces of furniture. Almost 30% of the image pixels correspond to an indoor scene. The misinterpretation of the low-resolution image is not a defect of the visual system. Instead, it illustrates the strength of spatial layout information in constraining the identity of the objects in normal conditions, which is especially evident in degraded conditions in which object identities cannot be inferred based only on local information (Schyns and Oliva, 1994).

In this paper, we examine what is the initial representation of a complex, real-world scene image that allows for its rapid recognition. According to the global precedent hypothesis advocated by Navon (1977) and validated in numerous studies since (for a review see Kimchi, 1992), the processing of the global structure and the spatial relationships between components precede the analysis of local details. The global precedence effect is particularly strong for images constituted of many element patterns (Kimchi, 1998), as it is the case of most real-world scene pictures.

To clarify the terminology we will be using in this article, in the same way that "red" and "vertical" are local feature values of an object (Treisman and Gelade, 1980), a specific configuration of local features defines a global feature value of a scene or an object. For instance, an image composed of vertical contours on the right side and horizontal contours on the left side could be estimated by one global feature-receptive field tuned to respond to that specific "Horizontal–Vertical" configuration. Global feature inputs are estimated by summations of local feature values but they encode *holistic* properties of the scene as they convey the spatial relationships between components. On the basis of behavioral and computational experiments, we show the relevance of using a low-dimensional code of the spatial layout of a scene, termed *global image features*, to represent the meaning of a scene. Global features capture the diagnostic structure of the image, giving an impoverished and coarse version of the principal

Fig. 1. Illustration of the effect of a coarse layout (at a resolution of 8 cycles/image) on scene identification and object recognition. Despite the lack of local details in the left blurred scene, viewers are confident in describing the spatial layout of a street. However, the high-resolution image reveals that the buildings are in fact furniture. This misinterpretation is not an error of the visual system. Instead, it illustrates the strength of the global spatial layout in constraining the identities of the local image structures (Navon, 1977).

contours and textures of the image that is still detailed enough to recognize the image's *gist*. One of the principal advantages of the global image coding described here lies in its computational efficiency: there is no need to parse the image or group its components in order to represent the spatial configuration of the scene.

In this paper, we examine (1) the possible content of the global structure of a natural scene image, based on experimental results from the scene recognition literature; (2) how the global scene structure can be modeled and (3) how the global features could participate in real-world scene categorization.

The role of global image features on scene perception: experimental evidence

There is considerable evidence that visual input is processed at different spatial scales (from low to high spatial frequency), and psychophysical and computational studies have shown that different spatial scales offer different qualities of information for recognition purpose. On the one hand, the shape of an object is more precisely defined at high spatial frequencies but the object boundaries are interleaved by considerable noise, which requires extensive processing to be filtered out (among others, Marr and Hildreth, 1980; Shashua and Ullman, 1988). On the other hand, low-scale resolution is more contrasted and might be privileged in terms of temporal processing than finer scale (Navon, 1977; Sugase et al., 1999), but this perceptual advantage might be offset by higher uncertainty about the identity of the blobs.

In a series of behavioral experiments, Oliva and Schyns evaluated the role that different spatial frequencies play in fast scene recognition. They created a novel kind of stimuli, termed hybrid images (see Fig. 2), by superimposing two images at two different spatial scales: the low spatial scale is obtained by filtering one image with a low-pass filter (keeping spatial frequencies up to 8 cycles/image), the high spatial scale is obtained by filtering a second image with a high-pass filter (frequencies above 24 cycles/image). The final hybrid image is composed by adding these two different filtered images (the filters are designed in such a way that there is no overlapping between the two images in the frequency domain). The examples in Fig. 2 show hybrid images combining a beach scene and a street scene.

The experimental results using hybrid stimuli showed that for short presentation time (30 ms,

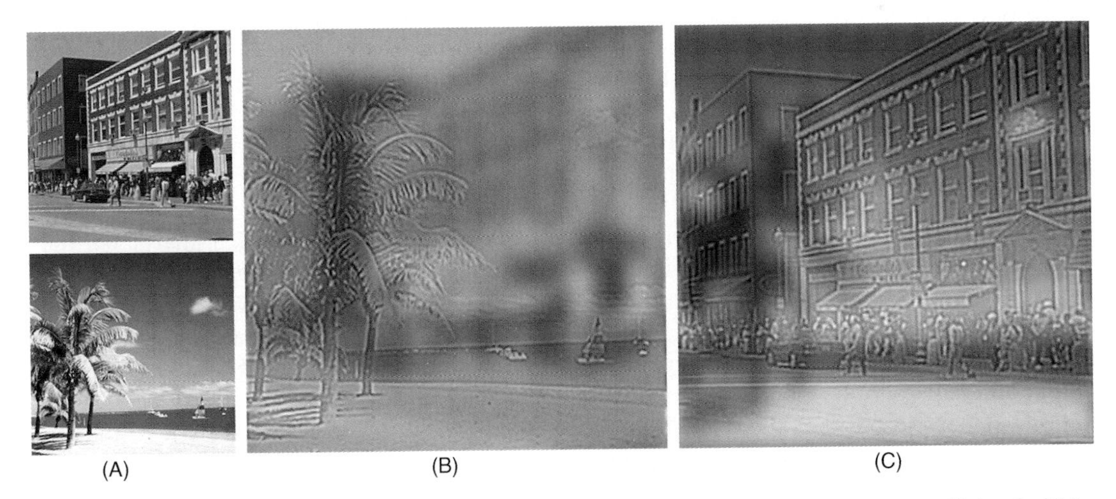

Fig. 2. (A) The two original images used to build the hybrid scenes shown above. (B) A hybrid image combining the high spatial frequency (HSF, 24 cycles/image) of the beach and the low spatial frequency (LSF, 8 cycles/image) of the street scene. If you squint, blink, or defocus, the street scene should replace the beach scene (if this demonstration fails, step back from the image until your perception changes). (C) The complementary hybrid image, with the street scene in HSF and the beach scene in LSF (cf. Schyns and Oliva, 1994; Oliva and Schyns, 1997).

followed by a mask, Schyns and Oliva, 1994), observers used the low spatial frequency part of hybrids (street in Fig. 2B) when solving a scene recognition task, whereas for longer (150 ms) durations of the same image, observers categorized the image on the basis of the high spatial frequencies (e.g., beach in Fig. 2B). In both cases, participants were unaware that the stimuli had two interpretations. It is important to stress that this result is not a evidence for a preference of the low spatial frequencies in the early stages of visual processing: additional experiments (Oliva and Schyns, 1997; Schyns and Oliva, 1999) showed that, in fact, the visual system can select which spatial scale to process depending on task constraints (e.g., if the task is determining the type of emotion of a face, participants will preferentially select the low spatial frequencies, but when the task is determining the gender of the same set of faces, participants used either low or high spatial frequencies). Furthermore, priming studies showed that within a 30-ms exposure, both low and high spatial frequency bands from a hybrid image were registered by the visual system [1] (Parker et al., 1992, 1996; Oliva and Schyns, 1997, Exp.1), but that the requirements of the task determined which scale, coarse or fine, was preferentially selected for covert processing. This suggests that the full range of spatial frequency scales is available with only 30 ms of image exposure, although the resolution at which the local features are analyzed and preattentively combined, when embedded in cluttered natural images, is unknown.

However, hybrid images break one important statistical property of real-world natural images, i.e., the spatial scale contiguity. To the contrary of hybrid images, contours of a natural image are correlated across scale space: a contour existing at low spatial frequency exists also at high spatial frequency. Moreover, statistical analysis of the distributions of orientations in natural images has shown that adjacent contours tend to have similar orientations whereas segments of the same contour that are further apart tend to have more disparate orientations (Geisler et al., 2001). The visual system could take advantage of spatial and spectral contiguities of contours to rapidly construct a sketch of the image structure. Boundary edges that would persist across the scale space are likely to be important structures of the image (Linderberg, 1993), and would define an initial skeleton of the image, fleshed out later by finer structures existing at higher spatial frequency scales (Watt, 1987; Linderberg, 1993; Yu, 2005). Most of the contours in natural scenes need selective attention to be bound together to form a shape of higher complexity (Treisman and Gelade, 1980; Wolfe and Bennet, 1997; Wolfe et al., 2002), but contours persistent through the scale space might need fewer attentional (or computational) resources to be represented early on. Therefore, one cannot dismiss the possibility that the analysis of fine contours and texture characteristics could be performed at the very early stage of scene perception, either because low spatial frequency luminance boundaries bootstrap the perceptual organization of finer contours (Lindeberg, 1993), or because the sparse detection of a few contours is sufficient to predict the orientation of the neighborhood edges (Geisler et al., 2001), or because selective attention was attending to information at a finer scale (Oliva and Schyns, 1997).

Within this framework, the analysis of visual information for fast scene understanding proceeds in a global to local manner (Navon, 1977; Treisman and Gelade, 1980), but not necessarily from low to high spatial frequencies. In other words, when we say "global and local" we do not mean "low and high" spatial frequencies. All spatial frequencies contribute to an early global analysis of the scene layout information, but organized at a rather coarse layout. Fine image edges, like long contours, are available, but their spatial organization is not encoded in a precise way. In the rest of this section we discuss some of the possible mechanisms used for performing the global analysis of the scene.

A simple and reliable global image feature for scene recognition is obtained by encoding the organization of color blobs in the image (under this representation a view of a landscape corresponds to a blue blob on the top, a green blob on the bottom and a brownish blob in the center, e.g., Lipson et al.,

[1] A hybrid scene presented for 30 ms and then masked would prime the recognition of a subsequent related scene, matching either the low- or the high-spatial scale of the hybrid (Oliva and Schyns, 1997, Exp. 1).

1997; Oliva and Schyns, 2000; Carson et al., 2002). Despite the simplicity of such a representation, it is remarkable to note the reliability of scene recognition achieved by human observers when shown a very low-resolution scene picture. Human observers are able to identify most of real-world scene categories based on a resolution as low as 4 cycles/images, but only when the blurred image is in color. If the images are presented in gray levels, performance drop and participants need to see higher-resolution images before achieving the same recognition performance: the performance with a color image at 4 cycles/image is achieved at a resolution of 8 cycles/image for a grayscale image (Oliva and Schyns, 2000, Exp. 3).

However, color blobs are not equally important for all the scenes. The diagnosticity of colored surfaces in an image seems to be a key element of fast scene recognition (Oliva and Schyns, 2000; Goffaux et al., 2005). In order to study the importance of color information, color images were altered by transforming their color modes (e.g., red surfaces became green, yellow surfaces became blue). This provides a way of understanding if color is helping as a grouping cue (and therefore the specific color is not important) or if it is diagnostic for the recognition (the color is specific to the category). For presentation time as short as 30 ms, Oliva and Schyns (2000) observed that altering colors impaired scene recognition when color was a diagnostic feature of the scene category (e.g., forests are greenish, coasts are bluish) but it had no detrimental effect for the recognition of scenes for which color was no diagnostic (e.g., some categories of urban scenes). The naming of a colored scene, relative to a grayscale scene image, was faster if it belonged to a category from which the colors distributions did not vary greatly across exemplars (for natural scenes like forest, coast, canyons), than for scene categories where color distribution varied (for indoors scenes, urban environments, see also Rousselet et al., 2005). Colored surfaces, in addition to providing useful segmentation cues for parsing the image (Carson et al., 2002), also informs about semantic properties of a place, such as its probable temperature (Greene and Oliva, 2005). The neural correlates of the role of color layout has been recently investigated by Goffaux et al. (2005), who have observed an Event-Related Potential (ERP) frontal signal 150 ms after image onset (a well-documented temporal marker of image categorization, Thorpe et al., 1996; Van Rullen and Thorpe, 2001), when observers identified normally colored scene pictures (e.g., a green forest, a red canyon) compared to their grayscale or abnormally colored version (e.g., a purple forest, a bluish canyon). In a similar vein, Steeves et al. (2004) have shown that an individual with a profound visual form agnosia (i.e., incapable of recognizing objects based on their shape) could still identify scene pictures from colors and texture information only. Their fMRI study revealed higher activity in the parahippocampal place area (Epstein and Kanwisher, 1998) when the agnostic patient was viewing normally colored scenes pictures than when she was viewing black and white pictures.

In addition to color, research has shown that the configuration of contours is also a key diagnostic cue of scene categories (Baddeley, 1997; Oliva and Torralba, 2001; Torralba and Oliva, 2003; McCotter et al., 2005) and can help to predict the presence or absence of objects in natural images (Torralba, 2003a; Torralba and Oliva, 2003). Basic-level classes of environmental scenes (forest, street, highway, coast, etc.) as well as global properties of the three-dimensional (3D) space (e.g., in perspective, cluttered) can be determined with a high probability from a diagnostic set of low-level image features (Oliva and Torralba, 2001; Walker Renninger and Malik, 2004; Fei-Fei and Perona, 2005). For instance in urban environments, an estimation of the volume that a scene subtends is well predicted by the layout of oriented contours and texture properties. As the volume of scene space increases, the perceived image on the retina changes from large surfaces to smaller pieces, increasing the high spatial frequency content (Torralba and Oliva, 2002). A different pattern is observed when looking at a natural scene: with increasing distance from the observer, natural surfaces become larger and smoother, so for a given region in the image, the texture becomes coarser.

In the following section, we suggest an operational definition of global image features. The global features proposed encode a coarse representation of the organization of low and high spatial frequencies in the image.

Building a scene representation from global image features

High-level properties of a scene such as the degree of perspective or the mean depth of the space that the scene subtends have been found to be correlated with the configuration of low-level image features (Torralba and Oliva, 2002, 2003). Evidence from the psychophysics literature suggest that our visual system analyzes global statistical summary of the image in a preselective stage of visual processing, or at least with minimal attentional resources (mean orientation, Parkes et al., 2001; mean of set of objects, Ariely, 2001; Chong and Treisman, 2003). By pooling together the activity of local low-level feature detectors across large regions of the visual field, we can build a holistic and low-dimensional representation of the structure of a scene that does not require explicit segmentation of image regions and objects (as in Oliva and Torralba, 2001) and therefore require very low computational (or attentional) resources. This suggests that a reliable scene representation can be built, in a feed-forward manner, from the same low-level features used for local neural representations of an image (receptive fields of early visual areas, Hubel and Wiesel, 1968).

For instance, in a forest-scene picture, the shape of a leaf can be estimated by a set of local receptive fields (encoding oriented edges). The shape of the whole forest picture can be summarized by the configuration of many small-oriented contours, distributed everywhere in the image. In the case of the forest scene, a global features encoding "fine-grained texture everywhere in the image" will provide a good summary of the texture qualities found in the image. In the case of a street scene, we will need a variety of global features encoding the perspective, the level of clutter, etc. Fig. 3 illustrates a global receptive field that would respond maximally to scenes with vertical structures at the top part and horizontal components at the bottom part (as in the case of a street scene).

Given the variability of layout and feature distribution in the visual world, and given the variability of viewpoints that an observer can have on any given scene, most real-world scene structures will need to be estimated not only by one, but by a collection of global features. The number of global features that can be computed is quite high. The most effective global features will be those that reflect the global structures of the visual world. Several methods of image analysis can be used to learn a suitable basis of global features (Vailaya et al., 1998; Oliva and Torralba, 2001; Vogel and Schiele, 2004; Fei-Fei and Perona, 2005) that capture the statistical regularities of natural-scene images. In the modeling presented here, we only consider global features of receptive fields measuring orientations and spatial frequencies of image components that have a spatial resolution between 1 and 8 cycles/image (see Fig. 5). We employed a basis derived by principal component analysis to

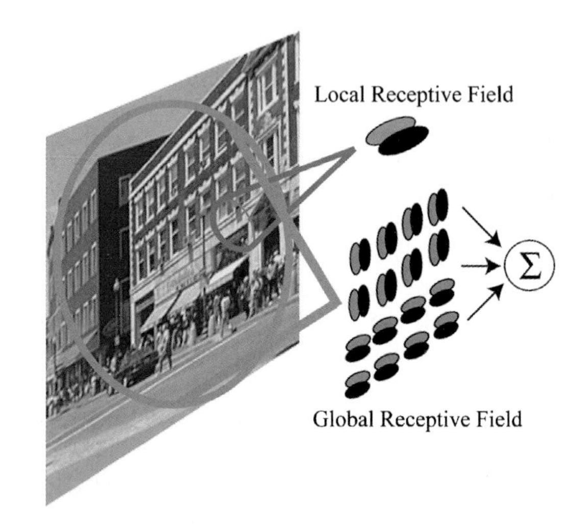

Fig. 3. Illustration of a local receptive field and a global receptive field (RF). A local RF is tuned to a specific orientation and spatial scale, at a particular position in the image. A global RF is tuned to a spatial pattern of orientations and scales across the entire image. A global RF can be generated as a combination of local RFs and can, in theory, be implemented from a population of local RFs like the ones found in the early visual areas. Larger RFs, which can be selective to global scene properties, could be found in higher cortical areas (V4 or IT). The global feature illustrated in this figure is tuned to images with vertical structures at the top part and horizontal component at the bottom part, and will reply strongly to the scene street image.

perform on a database of thousands of real-world images.

We summarize here the steps performed for learning a set of global features corresponding to the statistical configuration of orientation and spatial frequencies existing in the real world. Each global feature value is a weighted combination of the output magnitude of a bank of multiscale-oriented filters. In order to set the weights, we use principal components analysis (PCA). Due to the high dimensionality of images, applying PCA directly to the vector composed by the concatenation of the output magnitudes of all the filters will be very expensive computationally. Several regularization techniques can be used. Here, we decided to reduce the dimensionality of the vector of features by first downsampling each filter output to a size $N \times N$ (with N ranging from 2 to 16 in the computation performed here). All the filter outputs were downsampled to the same image size, independently of the scale of the filter. As a result, each image was represented by a vector of $N \times N \times K$ values (K is the number of different orientation and scales, and $N \times N$ is the number of samples used to encode, in low resolution, the output magnitude of each filter). This gives, for each image, a vector with a relatively small dimensionality (few hundreds of elements). The dimensionality of this vector space is then reduced by applying PCA to a collection of 22,000 images (the image collection includes scenes at all ranges of views, from closeup to panoramic, for both man-made and natural environments, similar to Oliva and Torralba, 2001).

Fig. 4 shows the first principal components of the output magnitude of multiscale-oriented filters for the luminance channel for a spatial resolution of 2 and 4 cycles/image (this resolution refers to the resolution at which the magnitude of each filter output is reduced before applying the PCA. A resolution of 4 cycles/image corresponds to averaging the output of each filter over $N \times N = 8 \times 8$ nonoverlapping windows, and 2 cycles/image corresponds to $N \times N = 4 \times 4$). Each principal component defines the weights used to compute each global feature. At each spatial location on the image, the polar plot shows the weighing of the spatial frequency at each orientation, with the lowest spatial frequencies in the center and the highest spatial frequencies along the maximum radius. In the following, we will refer to this visualization of the principal component weights (shown in Fig. 4) as a global feature *template*. In Fig. 4, the first template responds positively for images with more texture (seen in the mid- and high-frequency range) in the bottom half than in the upper half of the image and responds negatively for images with more texture in the upper half than in the bottom (e.g., a landscape with trees in the background, with

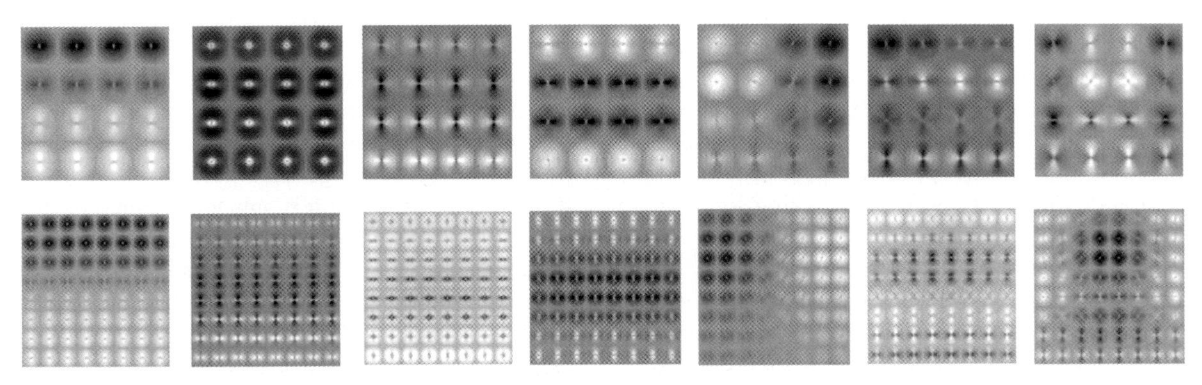

Fig 4. The principal components of natural image statistics define the weights used to compute the global features. The set of weights are obtained by applying principal component analysis (PCA) to the responses of multiscale-oriented filters to a large collection of natural images. The top row shows the 2nd to the 8th principal components for a spatial resolution of 2 cycles/image (4×4 regions). The first component behaves as a global average of the output of all orientations and scales, and therefore it is not shown. The bottom row shows the PCs for a resolution of 4 cycles/image (8×8 regions). For each PC, each subimage shows, in a polar plot (low spatial frequencies are in the center of the plot), how the spatial scale and orientations are weighted at each spatial location. The white corresponds to positive value and the black to negative value. Here we refer to the PCs as global feature templates.

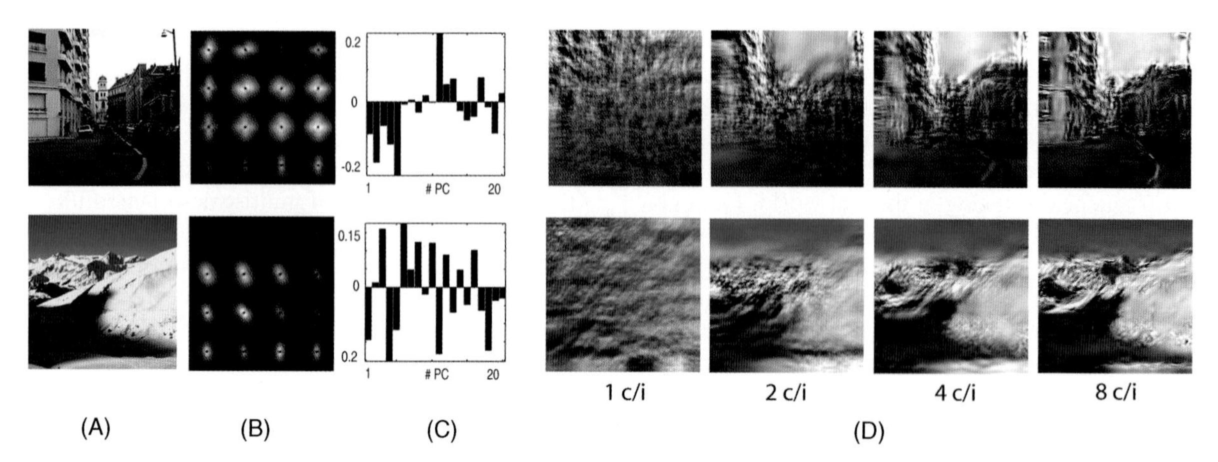

Fig. 5. (A) This figure illustrates the information preserved by the global features for two images. (B) The average of the output magnitude of the multiscale-oriented filters on a polar plot. Each average is computed locally by splitting the image into 4×4 nonoverlapping windows. (C) The coefficients (global features) obtained by projecting the averaged output filters into the first 20 principal components. In order to illustrate the amount of information preserved by this representation, (D) shows noise images that are coerced to have the same color blobs and the same global features ($N = 100$) than the target image. The very low frequency components (colored blobs) of the synthetic images are the same as from the original image. The high spatial frequencies are obtained by adding noise with the constraint that the resulting image should have the same global features as the target image (this only affects the luminance channel). This constraint is imposed by an iterative algorithm. The algorithm starts from white noise. At each iteration, the noise is decomposed using the bank of multiscale-oriented filters and the magnitude output of the filters is modified to match the global features of the target image. From left to right, the spatial resolution (number of windows used to average the filter outputs and the resolution of the color blobs) increases from 2×2, 4×4, 8×8 and 16×16. Note that despite the fact that the 2×2 image provides a poor reconstruction of the detailed structure of the original image, the texture contained in this representation is still relevant for scene categorization (e.g., open, closed, indoor, outdoor, natural or urban scenes).

no view of the sky and snow on the ground). Beyond the first component, the global feature templates increase in complexity and cannot be easily described. Note that principal components are used here as an illustration of an orthogonal basis for generating global features, but they are not the only possibility. For instance, other bases could be obtained by applying independent component analysis (Bell and Sejnowski, 1997) or searching for sparse codes (Olshausen and Field, 1996).

Fig. 5C shows the values of the 20 first global features (according to the ordering of principal components) for coding the structure of the street and the mountain scene. By varying the spatial resolution of the global features, we can manipulate the degree to which local features will be appropriately localized in the image. In order to illustrate the amount of information preserved by a set of global features at various resolution, Fig. 5D shows noise images that are coerced to have the same color blobs (here the color information is added by projecting the image into the principal components of the color channels, and retaining only the first 32 coefficients) and the same global features ($N = 100$) as the street and the mountain scenes. The global feature scene representation looks like a sketch version of the scene in which most of the contours and spatial frequencies from the original image have been conserved, but their spatial organization is only loosely preserved: a sketch at a resolution of 1 cycle/image (pulling local features from a 2×2 grid applied on image) is not informative of the spatial configuration of the image, but keeps the texture characteristics of the original scene so that we could probably decide whether the scene is a natural or man-made environment (Oliva and Torralba, 2001). For higher resolution, we can define the layout of the image and identify regions with different texture qualities, and recognize the probable semantic category of the scene (Oliva and Torralba, 2001, 2002).

Building the gist of the scene from global features: the Spatial Envelope model

How can we infer the semantic *gist* of a scene from the representation generated by the global image features? The *gist* refers to the meaningful information that an observer can identify from a glimpse at a scene (Potter, 1975; Oliva, 2005). The gist description usually includes the semantic label of the scene (e.g., a kitchen), a few objects and their surface characteristics (Rensink, 2000), as well as the spatial layout (e.g., the volume the scene subtends, its level of clutter, perspective) and the semantic properties related to the function of the scene. Therefore, a model of scene gist should go beyond representing the principal contours or objects of the image or classifying an image into a category: it should include a description of semantic information that human observers comprehend and infer about the scene (Oliva, 2005).

In Oliva and Torralba (2001), we introduced a holistic approach to scene recognition not only permitting to categorize the scene in its superordinate (e.g., urban, natural scene) and basic-level categories (e.g., street, mountain), but also describing its spatial layout in a meaningful way. There are many interesting properties of a real-world scene that can be defined independently of the objects. For instance, a forest scene can be described in terms of the degree of roughness and homogeneity of its textural components. These properties are in fact meaningful to a human observer who may use them for comparing similarities between two forest images (cf. Rao and Lohse, 1993; Heaps and Handel, 1999 for a similar account in the domain of textures).

Because a scene is inherently a 3D entity, Oliva and Torralba (2001) proposed that fast scene recognition mechanisms might initially be based on global properties diagnostic of the space that the scene subtends and not necessarily the objects that the scene contains. A variety of spatial properties like "openness" or "perspective" (e.g., a coast is an "open" environment) have indeed a direct transposition into global features of two-dimensional (2D) surfaces (e.g., a coast has a long-horizon line). This permits to evaluate the degree of openness or mean depth of an image by measuring the distribution of local-image features (Torralba and Oliva, 2002, 2003). To determine a vocabulary of spatial layout properties useful for scene recognition, we asked observers to describe real-world scene images according to spatial layout and global appearance characteristics. The vocabulary given by observers (naturalness, openness, expansion, depth, roughness, complexity, ruggedness, symmetry) served to establish an initial *scene-centered description* of the image (based on spatial layout properties, Oliva and Torralba, 2002) offering an alternative to object-centered description (where a scene is identified from labeling the objects or regions, Barnard and Forsyth, 2001; Carson et al., 2002). Similar to the vocabulary used in architecture to portray the spatial properties of a place, we proposed to term the scene-centered description the *Spatial Envelope* of a scene.

Fig. 6 illustrates the framework of the Spatial Envelope model (details can be found in Oliva and Torralba, 2001). For simplicity, the Spatial Envelope model is presented here as a combination of four global scene properties (Fig. 6A). Object identities are not represented in the model. Within this framework, the structure of a scene is characterized by the properties of the boundaries of the space (e.g., the size of the space, its degree of openness and perspective) and the properties of its content (e.g., the style of the surface, natural or man-made, the roughness of these surfaces). Any scene image can be described by the values it takes along each spatial envelope property. For instance, to describe the degree of openness of a given environment, we could refer to a "panoramic", "open", "closed" or "enclosed" scene. A forest would be described as "an enclosed environment, with a dense isotropic texture" and a street scene would be a "man-made outdoor scene, with perspective, and medium level of clutter" (Oliva and Torralba, 2001, 2002). This level of description is meaningful to observers who can infer the probable semantic category of the scene, by providing a conceptual summary of the gist of the scene.

Computational modeling demonstrated that each spatial envelope property (naturalness, openness, expansion, etc.) could be estimated from a collection of global features templates (Fig. 4) measuring how natural, open, expanded, rough the scene image is. The principal structure of a

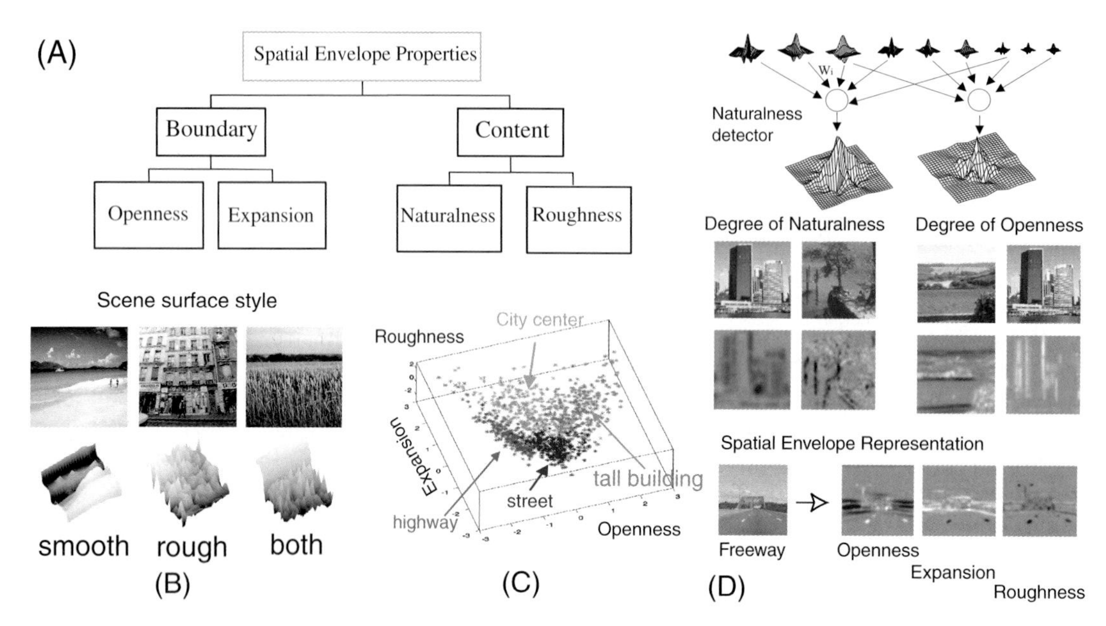

Fig. 6. Schematic representation of the Spatial Envelope model as in Oliva and Torralba (2001). (A) Spatial envelope properties can be classified into properties of boundaries and properties of content. For simplicity, only four properties are represented. (B) Illustration of a scene as a single surface, with different "roughness" qualities. The spatial envelope does not explicitly represent objects; therefore, "roughness" refers to the surface quality of the whole scene. (C) Projection of ~1200 pictures of typical urban scenes onto three spatial envelope axes (openness, roughness, expansion) as in Oliva and Torralba (2001). Semantic categories emerge, showing that the spatial envelope representation carries information about the semantic class of a scene. (D) Illustration of an implementation of the Spatial Envelope model in the form of "scene filters" applied onto the image. A complex "scene filter" can be computed as a linear combination of Gabor-like filters, and as a combination of global feature templates. Features of *openness* are shown in black and features of *closedness* are shown in white.

scene image is initially represented by a combination of global features, on the basis of which the spatial envelope properties can be estimated: each scene is described as a vector of meaningful values indicating the image's degree of naturalness, openness, roughness, expansion, mean depth etc. This description refers to the spatial envelope representation of the scene image. Therefore, if spatial envelope properties capture the diagnostic structure of a scene, two images with similar spatial envelopes should also belong to the same scene semantic categories. Indeed, Oliva and Torralba observed that scenes' images judged by observers to have the same categorical membership (street, highway, forest, coastline, etc.) were projected close together in a multidimensional space whose axes correspond to the Spatial Envelope dimensions (Fig. 6C). Neighborhood images in the spatial envelope space corresponded to images with similar spatial layout and similarconceptual description (cf. Fig. 7 for exemplars of scenes and their nearest neighbors in a spatial envelope space of urban environments). Note that the spatial envelope properties (e.g., openness, naturalness, expansion, symmetry) are implemented here as a weighted combination of global features, but spatial envelope properties could also be derived from other basis of low- or intermediate-level features (Ullman et al., 2002). By providing semantic classification at both superordinate (e.g., open, natural scene) and basic levels (e.g., beach, forest) of description, the Spatial Envelope model provides a theoretical and computational framework for the representation of a meaningful global scene structure, and a step toward understanding the representation and mechanisms of the *gist* of a scene.

Conclusion

Research over the last decade has made substantial progress toward understanding the brain

Fig. 7. Examples of urban scenes sharing the same spatial envelope representation (for a resolution of global features of $2c/i$). Similar scenes were retrieved as the nearest neighbors of the first image of each row, in a five-dimensional spatial envelope representation (naturalness, openness, mean depth, expansion and roughness). On the left, the scenes on each row pertain clearly to the same semantic category. On the right, the spatial envelope similarities are less representative of basic-level categories per se; however, the global structure of the image (coarse layout organization and levels of details) is very similar. There are other important global scene properties that are not shown here (For instance, visual complexity is not represented here (Oliva et al., 2004), and color is also not taken into account.

mechanisms underlying human object recognition (Kanwisher, 2003; Grill-Spector and Malach, 2004) and its modeling (Riesenhuber and Poggio, 1999; Ullman et al., 2002; Torralba et al., 2004; Serre et al., 2005). Converging evidence from behavioral, imaging and computational studies suggest that, at least in early stages of processing, mechanisms involved in natural scene recognition may be independent from those involved in recognizing objects (Schyns and Oliva, 1994; Oliva and Torralba, 2001; Li et al., 2002; Fei Fei and Perona, 2004; Marois et al., 2004; McCotter et al., 2005). On the basis of a review of behavioral and computational work, we argue that fast scene recognition does not need to be built on top of the processing of objects, but can be analyzed in parallel by scene-centered mechanisms. In our framework, a scene image is initially processed as a single entity and local information about objects and parts comes into play at a later stage of visual processing. We propose a formal basis of global features permitting to estimate quickly and in a feed-forward manner a meaningful representation of the scene structure. Global image feature values provide a summary of the layout of real-world images that may precede and constrain the analysis of features of higher complexity. On the basis of a global spatial representation of the image, the Spatial Envelope model (Oliva and Torralba, 2001) provides a conceptual framework for the representation and the mechanisms of fast scene gist interpretation. Global image features and the spatial envelope representation are not meant to be an alternative to local image analysis but serve as a parallel pathway that can, on the one hand, quickly constrain local analysis, narrowing down the search for object in cluttered, real-world scenes (global contextual priming, Torralba 2003a) and, on the other hand, provide a formal instance of a feed-forward mechanism for scene context evaluation, for the guidance of attention and eye movements in the scene (Oliva et al., 2003; Torralba, 2003a,b; Torralba et al., 2006).

Evidence in favor of distinct neural mechanisms supporting scene and object recognition, at least at an earlier stage of visual processing, comes from the pioneer work of Epstein and Kanwisher (1998). They found a region of cortex referred to as the parahippocampal place area (PPA) that responds more strongly to pictures of intact scenes (indoors, outdoors, closeup views), than to objects alone (Epstein et al., 2000). Furthermore, the PPA seems to be sensitive to holistic properties of the scene layout, but not to its complexity in terms of quantity of objects (Epstein and Kanwisher, 1998). The neural independence between scenes- and object-recognition mechanisms was recently strengthened by Goh et al., (2004). They observed activation of different parahippocampal regions when pictures of scenes were processed alone compared to pictures containing a prominent object, consistent within that scene. In a related vein, Bar (Bar and Aminoff, 2003; Bar, 2004) found specific cortical regions (a network relating regions in the parahippocampal region and the retrosplenial cortex) involved in the analysis of the context of objects. The neural underpinnings of the global features, the spatial envelope properties or the gist of a scene remain open issues: the global features are originally built as combinations of local low-level filters of the type found in early visual areas. Lateral and/or feedback connections could combine this information locally to be read out by higher visual areas. Receptive fields in the inferior temporal cortex and parahippocampal region cover most of the useful visual field (20–40°) and thus are also capable, in theory, of encoding scene layout information like the global features and the spatial envelope properties. Clearly, the mechanisms by which scene understanding occurs in the brain remain to be found.

Acknowledgments

The research was funded by an NIMH grant (1R03MH068322-01) and an award from NEC Corporation Fund for research in computers and communications was given to A.O. Thanks to Michelle Greene, Barbara Hidalgo-Sotelo, Naomi Kenner, Talia Konkle and Thomas Serre for comments on the manuscript.

References

Ariely, D. (2001) Seeing sets: representation by statistical properties. Psychol. Sci., 12: 157–162.

Baddeley, R. (1997) The correlational structure of natural images and the calibration of spatial representations. Cogn. Sci., 21: 351–372.

Bar, M. (2004) Visual objects in context. Nat. Neurosci. Rev., 5: 617–629.

Bar, M. and Aminoff, E. (2003) Cortical analysis of visual context. Neuron, 38: 347–358.

Barnard, K. and Forsyth, D.A. (2001). Learning the semantics of words and pictures. Proceedings of the International Conference of Computer Vision, Vancouver, Canada, pp. 408–415.

Bell, A.J. and Sejnowski, T.J. (1997) The 'Independent components' of natural scenes are edge filters. Vision Res., 37: 3327–3338.

Biederman, I. (1995) Visual object recognition. In: Kosslyn, M. and Osherson, D.N. (Eds.), An Invitation to Cognitive Science: Visual Cognition (2nd edn.). MIT Press, Cambridge, MA, pp. 121–165.

Carson, C., Belongie, S., Greenspan, H. and Malik, J. (2002) Blobworld: image segmentation using expectation-maximization and its expectation to image querying. IEEE Trans. Pattern Anal. Mach. Intell., 24: 1026–1038.

Chong, S.C. and Treisman, A. (2003) Representation of statistical properties. Vision Res., 43: 393–404.

Epstein, R. and Kanwisher, N. (1998) A cortical representation of the local visual environment. Nature, 392: 598–601.

Epstein, R., Stanley, D., Harris, A. and Kanwisher, N. (2000) The parahippocampal place area: perception, encoding, or memory retrieval? Neuron, 23: 115–125.

Fei-Fei, L. and Perona, P. (2005) A Bayesian Hierarchical Model for learning natural scene categories. IEEE Proc. Comp. Vis. Pattern Rec., 2: 524–531.

Geisler, W.S., Perry, J.S., Super, B.J. and Gallogly, D.P. (2001) Edge co-occurrence in natural images predicts contour grouping performances. Vision Res., 41: 711–724.

Goffaux, V., Jacques, C., Mouraux, A., Oliva, A., Schyns, P.G. and Rossion, B. (2005) Diagnostic colours contribute to early stages of scene categorization: behavioural and neurophysiological evidence. Vis. Cogn., 12: 878–892.

Goh, J.O.S., Siong, S.C., Park, D., Gutchess, A., Hebrank, A. and Chee, M.W.L. (2004) Cortical areas involved in object, background, and object-background processing revealed with functional magnetic resonance adaptation. J. Neurosci., 24: 10223–10228.

Greene, M.R. and Oliva, A. (2005) Better to run than to hide: the time course of naturalistic scene decisions [Abstract]. J. Vis., 5: 70a.

Grill-Spector, K. and Malach, R. (2004) The human visual cortex. Annu. Rev. Neurosci., 27: 649–677.

Heaps, C. and Handel, C.H. (1999) Similarity and features of natural textures. J. Exp. Psychol. Hum. Percept. Perform., 25: 299–320.

Hubel., D.H. and Wiesel, T.N. (1968) Receptive fields and functional architecture of monkey striate cortex. J. Physiol., 195: 215–243.

Kanwisher, N. (2003) The ventral visual object pathway in humans: evidence from fMRI. In: Chalupa, L. and Werner, J. (Eds.), The Visual Neurosciences. MIT Press, Cambridge, MA, pp. 1179–1189.

Kimchi, R. (1992) Primacy of wholistic processing and global/local paradigm: a critical review. Psychol. Bull., 112: 24–38.

Kimchi, R. (1998) Uniform connectedness and grouping in the perceptual organization of hierarchical patterns. J. Exp. Psychol. Hum. Percept. Perform., 24: 1105–1118.

Li, F.F., VanRullen, R., Koch, C. and Perona, P. (2002) Rapid natural scene categorization in the near absence of attention. Proc. Natl. Acad. Sci. USA, 99: 9596–9601.

Lindeberg, T. (1993) Detecting salient blob-lie images structures and their spatial scales with a scale-space primal sketch: a method for focus of attention. Int. J. Comp. Vis., 11: 283–318.

Lipson, P., Grimson, E. and Sinha, P. (1997) Configuration-based scene classification and image indexing. IEEE Comp. Soc. Conf., 1: 1007–1013.

Maljkovic, V. and Martini, P. (2005) Short-term memory for scenes with affective content. J. Vis., 5: 215–229.

Marois, R., Yi, D.J. and Chun, M. (2004) The neural fate of consciously perceived and missed events in the attentional blink. Neuron, 41: 465–472.

Marr, D. and Hildreth, E.C. (1980) Theory of edge detection. Proc. R. Soc. Lond. B, 207: 187–217.

McCotter, Gosselin, F., Sowden, P. and Schyns, P.G. (2005) The use of visual information in natural scenes. Vis. Cogn., 12: 938–953.

Navon, D. (1977) Forest before trees: the precedence of global features in visual perception. Cogn. Psychol., 9: 353–383.

Oliva, A. (2005) Gist of the scene. In: Itti, L., Rees, G. and Tsotsos, J.K. (Eds.), Neurobiology of Attention. Elsevier, San Diego, CA, pp. 251–256.

Oliva, A., Mack, M.L., Shrestha, M. & Peeper, A. (2004). Identifying the perceptual dimensions of visual complexity in scenes. Proceedings of the 26th Annual Meeting of the Cognitive Science Society, Chicago, August.

Oliva, A. and Schyns, P.G. (1997) Coarse blobs or fine edges? Evidence that information diagnosticity changes the perception of complex visual stimuli. Cogn. Psychol., 34: 72–107.

Oliva, A. and Schyns, P.G. (2000) Diagnostic colors mediate scene recognition. Cogn. Psychol., 41: 176–210.

Oliva, A. and Torralba, A. (2001) Modeling the shape of the scene: a holistic representation of the spatial envelope. Int. J. Comp. Vis., 42: 145–175.

Oliva, A. and Torralba, A. (2002). Scene-centered description from spatial envelope properties. Lecture note in computer science series. Proceedings of the 2nd Workshop on Biologically Motivated Computer Vision, Tubingen, Germany.

Oliva, A., Torralba, A., Castelhano, M.S. and Henderson, J.M. (2003) Top-down control of visual attention in object detection. Proc. IEEE Int. Conf. Image Process., 1: 253–256.

Olshausen, B.A. and Field, D.J. (1996) Emergence of simple-cell receptive field properties by learning a sparse code for natural images. Nature, 381: 607–609.

Parker, D.M., Lishman, J.R. and Hughes, J. (1992) Temporal integration of spatially filtered visual images. Perception, 21: 147–160.

Parker, D.M., Lishman, J.R. and Hughes, J. (1996) Role of coarse and fine information in face and object processing. J. Exp. Psychol. Hum. Percept. Perform., 22: 1448–1466.

Parkes, L., Lund, J., Angelucci, A., Solomon, J.A. and Morgan, M. (2001) Compulsory averaging of crowded orientation signals in human vision. Nat. Neurosci., 4: 739–744.

Potter, M.C. (1975) Meaning in visual scenes. Science, 187: 965–966.

Potter, M.C. (1976) Short-term conceptual memory for pictures. J. Exp. Psychol. [Hum. Learn.], 2: 509–522.

Potter, M.C., Staub, A. and O' Connor, D.H. (2004) Pictorial and conceptual representation of glimpsed pictures. J. Exp. Psychol. Hum. Percept. Perform., 30: 478–489.

Rao, A.R. and Lohse, G.L. (1993) Identifying high-level features of texture perception. GMIP, 55: 218–233.

Rensink, R.A. (2000) The dynamic representation of scenes. Vis. Cogn., 7: 17–42.

Riesenhuber, M. and Poggio, T. (1999) Hierarchical models of object recognition in cortex. Nat. Neurosci., 2: 1019–1025.

Rousselet, G.A., Joubert, O.R. and Fabre-Thorpe, M. (2005) How long to get to the "gist" of real-world natural scenes? Vis. Cogn., 12: 852–877.

Sanocki, T. and Epstein, W. (1997) Priming spatial layout of scenes. Psychol. Sci., 8: 374–378.

Serre, T., Wolf, L. and Poggio, T. (2005) Object recognition with features inspired by visual cortex. Proceedings of IEEE CVPR, IEEE Computer Society Press, San Diego, June.

Shashua, A. and Ullman, S. (1988) Structural saliency: the detection of globally salient structures using a locally connected network. Proceedings of the 2nd International Conference on Computer Vision, Tempa, FL, pp. 321–327.

Schyns, P.G. and Oliva, A. (1994) From blobs to boundary edges: evidence for time- and spatial-scale-dependent scene recognition. Psychol. Sci., 5: 195–200.

Schyns, P.G. and Oliva, A. (1999) Dr. Angry and Mr. Smile: when categorization flexibly modifies the perception of faces in rapid visual presentations. Cognition, 69: 243–265.

Steeves, J.K.E., Humphrey, G.K., Culham, J.C., Menon, R.S., Milner, A.D. and Goodale, M.A. (2004) Behavorial and neuroimaging evidence for a contribution of color and texture information to scene classification in a patient with visual form agnosia. J. Cogn. Neurosci., 16(6): 955–965.

Sugase, Y., Yamane, S., Ueno, S. and Kawano, K. (1999) Global and fine information coded by single neurons in the temporal visual cortex. Nature, 400: 869–873.

Thorpe, S., Fize, D. and Marlot, C. (1996) Speed of processing in the human visual system. Nature, 381: 520–522.

Torralba, A. (2003a) Modeling global scene factors in attention. J. Opt. Soc. Am. A, 20: 1407–1418.

Torralba, A. (2003b) Contextual priming for object detection. Int. J. Comput. Vis., 53: 153–167.

Torralba, A., Murphy, K. P. and Freeman, W. T. (2004). Sharing features: efficient boosting procedures for multiclass object detection. Proceedings of the IEEE CVPR, pp. 762–769.

Torralba, A. and Oliva, A. (2002) Depth estimation from image structure. IEEE Pattern Anal. Mach. Intell., 24: 1226–1238.

Torralba, A. and Oliva, A. (2003) Statistics of natural images categories. Network: Comput. Neural Systems, 14: 391–412.

Torralba, A., Oliva, A., Castelhano, M.S. and Henderson, J.M. (2006). Contextual guidance of attention in natural scenes: the role of global features on object search. Psychol. Rev., in press.

Treisman, A. and Gelade, G.A. (1980) Feature integration theory of attention. Cogn. Psychol., 12: 97–136.

Ullman, S., Vidal-Naquet, M. and Sali, E. (2002) Visual features of intermediate complexity and their use in classification. Nat. Neurosci., 5: 682–687.

Vailaya, A., Jain, A. and Zhang, H.J. (1998) On image classification: city images vs. landscapes. Patt. Recogn., 31: 1921–1935.

Van Rullen, R. and Thorpe, S.J. (2001) The time course of visual processing: from early perception to decision making. J. Cogn. Neurosci., 13: 454–461.

Vogel, J. and Schiele, B. (2004). A semantic typicality measure for natural scene categorization. Proceedings of Pattern Recognition Symposium DAGM, Tubingen, Germany.

Walker Renninger, L. and Malik, J. (2004) When is scene identification just texture recognition? Vision Res., 44: 2301–2311.

Watt, R.J. (1987) Scanning from coarse to fine spatial scales in the human visual system after onset of a stimulus. J. Opt. Soc. Am. A, 4: 2006–2021.

Wolfe, J.M. and Bennett, S.C. (1997) Preattentive object files: shapeless bundles of basic features. Vision Res., 37: 25–44.

Wolfe, J.M., Oliva, A., Butcher, S. and Arsenio, H. (2002) An unbinding problem: the desintegration of visible, previously attended objects does not attract attention. J. Vis., 2: 256–271.

Yu, S. X. (2005). Segmentation induced by scale invariance. Proceedings of the IEEE Conference on Computer Vision Pattern Recognition, San Diego.

Martinez-Conde, Macknik, Martinez, Alonso & Tse (Eds.)
Progress in Brain Research, Vol. 155
ISSN 0079-6123

CHAPTER 3

Beyond the face: exploring rapid influences of context on face processing

Beatrice de Gelder[1,2,3,*], Hanneke K.M. Meeren[1,2,3], Ruthger Righart[1], Jan van den Stock[1], Wim A.C. van de Riet[1,3] and Marco Tamietto[4]

[1]*Cognitive and Affective Neurosciences Laboratory, Tilburg University, Tilburg, The Netherlands*
[2]*Martinos Center for Biomedical Imaging, Massachusetts General Hospital, Harvard Medical School, Charlestown, MA, USA*
[3]*F.C. Donders Centre for Cognitive Neuroimaging, Radboud University in Nijmegen, NL-6500 HB Nijmegen, The Netherlands*
[4]*Department of Psychology, University of Turin, Turin, Italy*

Abstract: Humans optimize behavior by deriving context-based expectations. Contextual data that are important for survival are extracted rapidly, using coarse information, adaptive decision strategies, and dedicated neural infrastructure. In the field of object perception, the influence of a surrounding context has been a major research theme, and it has generated a large literature. That visual context, as typically provided by natural scenes, facilitates object recognition as has been convincingly demonstrated (Bar, M. (2004) Nat. Rev. Neurosci., 5: 617–629). Just like objects, faces are generally encountered as part of a natural scene. Thus far, the facial expression literature has neglected such context and treats facial expressions as if they stand on their own. This constitutes a major gap in our knowledge. Facial expressions tend to appear in a context of head and body orientations, body movements, posture changes, and other object-related actions with a similar or at least a closely related meaning. For instance, one would expect a frightened face when confronted to an external danger to be at least accompanied by withdrawal movements of head and shoulders. Furthermore, some cues provided by the environment or the context in which a facial expression appears may have a direct relation with the emotion displayed by the face. The brain may even fill in the natural scene context typically associated with the facial expression. Recognition of the facial expression may also profit from processing the vocal emotion as well as the emotional body language that normally accompany it. Here we review the emerging evidence on how the immediate visual and auditory contexts influence the recognition of facial expressions.

Keywords: fear; body; scene; voice; P1; N170

Introduction

It is surprising that, except for a few isolated studies, the literature on face recognition has not yet addressed the issue of context. So far, much of the face recognition literature has been dominated by the issue of face modularity, or the notion that our ability to process faces reflects a functional and neurobiological specialization. From the viewpoint of face specificity theorists, face processing may be immune to surrounding context recognition processes because faces are uniquely salient and attention-grabbing signals. If so, context

*Corresponding author. Tel.: +31-13-466-24-95; Fax: +31-13-466-2067; E-mail: B.deGelder@uvt.nl

DOI: 10.1016/S0079-6123(06)55003-4

influence may just be another dimension on which face and object processing differ considerably, as face processing may not be sensitive to context.

Emotion researchers have predominantly used isolated facial expressions, rather than contextual posture, movement, and voice or scenes. This may or may not be justified. On the one hand, facial expressions may be very special indeed. For highly social species like humans, facial expressions may be by far the most salient carriers of emotional information, dwarfing objects or natural scenes as cues of emotional significance. In that case, the meaning of facial expressions will be computed automatically irrespective of other cues present together with the face like emotional body language, arousing context or emotional voice expressions. On the other hand, facial expressions and their behavioral consequences may be influenced by the context in which they appear. This may be because facial expressions are often ambiguous, and additional environment may be required to compute the meaning of a facial expression.

In this chapter we review recent studies of three contexts in which facial expressions are frequently encountered: whole bodies, natural scenes and emotional voices. Recent shifts in the theoretical perspective of the cognitive and affective neurosciences have converged on important notions like embodiment, affect programs and multisensory-based perception integration. This opens a new perspective by which context plays a crucial role, even for highly automated processes such as the recognition of facial expressions. We briefly sketch this background before presenting recent findings on the context of face processing that we deem essential for an ecologically valid theory of facial expressions.

Background

Since its reintroduction in experimental psychology, emotion research has focused mainly on *visual* processes associated with seeing emotional stimuli, de facto facial expressions. Recent findings point to close links between the visual and the sensorimotor system and to the role of the body in perception, such as in research on embodied cognition (Barsalou, 1999). The leading perspectives, that is now approachable due to novel methods, is that individuals embody the emotional gestures of other people, including facial expressions, posture and vocal affect. Imitative behavior produces a corresponding state in the perceiver, leading to the general suggestion that embodied knowledge produces corresponding emotional states.

In the early stages of processing core emotions (Ekman, 1992), bodily resonance is automatic and reflex-like, while in the later, more cognitive and conscious processing stages, it is under strategic control and influenced by higher order knowledge. The notion of embodiment in a more general meaning has also come to the foreground of emotion theories again with the proposals made by Damasio (1994, 1999).

From a more evolutionary-inspired perspective, emotions and facial expressions are closely related to actions, and therefore likely to involve the whole body (Schmidt and Cohn, 2001). Emotion provoking stimuli trigger affect programs (Darwin, 1872; Tomkins, 1963; Frijda, 1986; Panksepp, 1998; Russell and Feldman Barrett, 1999), which produce an ongoing stream of neurophysiological change (or change in a person's homeostatic state) and are associated with evolutionary-tuned behaviors for dealing with stimuli of significant value. Along with the orbitofrontal cortex (OFC) and amygdala, the insula and somatosensory cortex are involved in the modulation of emotional reactions involving the body via connections to brain stem structures (Damasio, 1994, 1999; LeDoux, 1996). This function of the insula and somatosensory cortex may underlie their important role in emotion perception (Adolphs et al., 2000; Winston et al., 2003; Heberlein and Adolphs, 2004). Processes engaging somatosensory cortex and insula may involve simulating the viewed emotional state via the generation of a somatosensory image of the associated body state.

Recognition of faces and facial expressions

A great deal of effort has been devoted in trying to establish that faces constitute a particular category of stimuli processed with dedicated behavioral skills, in specific cortical areas of the brain and possibly with shorter latencies than other stimuli. In the modular model proposed by Kanwisher et al. (1997), a small region in the fusiform gyrus, the so-called fusiform face area (FFA), is specialized in face perception (cf. Gauthier et al., 1998, 1999; Gauthier and Nelson, 2001). This view seems only to concern the neural basis of personal identity cues as provided by the face and neither the facial expression nor the context in which faces typically appear (body, scene, and voice).

These caveats are accounted for in the distributed models for face perception (de Gelder and Rouw, 2000; Haxby et al., 2000, 2002; de Gelder et al., 2003), which also consider other aspects of faces besides person identity (Haxby et al., 1994, 1996, 2000; Puce et al., 1996; Adolphs et al., 2000; de Gelder and Rouw, 2000; Hoffman and Haxby, 2000; Adolphs, 2002; de Gelder et al., 2003). In distributed models, different areas of the brain process different attributes of the face separately, such as identity (FFA and the occipital face area [OFA]), gaze direction (superior temporal sulcus [STS]), and expression and/or emotion (OFC, amygdala, anterior cingulate cortex, premotor cortex, somatosensory cortex). Several of these structures (e.g., OFC, amygdala, and somatosensory cortex) have clearly direct and indirect connections with visceral, autonomic and muscular centers (Adolphs, 2002), thereby influencing the affective homeostasis and making the body part of the perceptual process, i.e., embodiment.

Within this multitude of regions, there is a division of labor. The first route, a subcortical pathway to the amygdala via the superior colliculus (SC) and pulvinar, is concerned with fast and more coarse but subconscious processing (Morris et al., 1998b, 2001; de Gelder et al., 1999b, 2001; Pegna et al., 2005) in case of highly salient, especially threatening stimuli, while the second route, via the lateral geniculate nucleus (LGN) and striate cortex to cortical regions like STS, OFA and FFA, is more concerned with detailed and fine-grained processing in case stimuli are ambiguous and full blown awareness of the perceived face is necessary.

These regions of the parallel routes interact (de Gelder and Rouw, 2000; Adolphs, 2002) and modulate each other with feedforward and feedback projections in order to establish a fine-grained percept composed of identity and emotional aspects of the face, which can be accessible to consciousness. Especially the amygdala has strong functional and structural connections with several cortical regions like FFA, STS and OFC (functional connectivity: Morris et al., 1998a; Iidaka et al., 2001; Vuilleumier et al., 2004; structural connectivity: Carmichael and Price, 1995); or with striate cortex (structural connectivity: Amaral and Price, 1984; Catani et al., 2003).

Electrophysiological studies have shed light on the temporal characteristics of neuronal processing of faces. Two early components that can be readily identified in the waveform of visual event-related potentials (ERP) or magnetic fields (ERF), i.e., the P1 and N170, show sensitivity to faces, hinting at that dedicated systems are attuned to the processing of faces. The first component would point to the involvement in global encoding, i.e., categorizing a face as such (Liu et al., 2002), while the second deflection would reflect configural perceptual processing subserving face identification (Bentin et al., 1996).

The face-sensitive N170 waveform shows a robust face-sensitive "inversion" effect indicative of configural processing, i.e., it is enhanced and delayed to faces that are presented upside down, but not to inverted objects (Watanabe et al., 2003; Stekelenburg and de Gelder, 2004). Controversy exists about the underlying neuronal source of the N170. Some studies point to the STS as generator (Henson et al., 2003; Itier and Taylor, 2004b), while others propose the fusiform gyrus, where the FFA resides, as possible candidate (Halgren et al., 2000; Pizzagalli et al., 2002; Shibata et al., 2002). Whether the N170 is generated in the fusiform gyrus or STS may depend on the exact nature of the task and the stimuli being used. The N170 amplitude is affected by biological motion (Jokisch et al., 2005), eye gaze (Watanabe et al., 2002), facial motion (Puce et al., 2003), facial expressions (Batty and Taylor, 2003; Stekelenburg and de Gelder, 2004), expressional change (Miyoshi et al., 2004) and affective facial features (Pizzagalli et al., 2002).

Recent studies challenge the N170 as earliest marker of selective face processing and draw attention to an earlier component peaking between 100 and 130 ms post-stimulus. The P1 ERP component (or its magnetoencephalography (MEG) equivalent) is mainly generated in "early" extrastriate visual areas (Linkenkaer-Hansen et al., 1998; Di Russo et al., 2005) and it is commonly thought to reflect processing of the low-level features of a stimulus. A few recent studies however suggest that higher order visual processing can already occur at this early stage. Successful categorization of stimuli as faces was found to correlate with an early MEG component at 100–120 ms after onset (Liu et al., 2002). Both the MEG and the ERP components show an inversion effect (Linkenkaer-Hansen et al., 1998; Itier and Taylor, 2002, 2004a), suggesting that some configurational processing already takes place at this early stage. In addition, this component appears to be sensitive to facial likeability (Pizzagalli et al., 2002) and emotional facial expressions in contrast to neutral expression, but not between emotional expressions (Halgren et al., 2000; Batty and Taylor, 2003; Eger et al., 2003, 2004).

Facial expressions in the context of whole bodies

Perception of bodies is a relatively new field as is perception of bodily expressions of emotion. Recent research on neutral and instrumental body postures and movements has set out to raise some of the familiar questions of face researchers. Are the perceptual characteristics of faces and bodies alike? Is one specific brain region dedicated to body perception (modularity hypothesis), or are multiple brain regions involved (distributed model hypothesis)? Or does perception of face and body expression share an underlying common neural basis?

Evidence from single-cell recordings suggests a degree of specialization for either face or neutral body images (Rizzolatti et al., 1996). This view is corroborated by studies reporting that neurons of monkey posterior STS react selectively to body posture and by the fMRI study of Downing and co-workers (Downing et al., 2001) in which a region near the middle occipital gyrus, the so-called extrastriate body area (EBA), reacted selectively to body form and body parts but showed little activation to isolated faces.

However, a recent electrophysiological investigation in humans lends support for common configural perceptual processing mechanisms for faces and bodies. A typical but slightly faster N170 component commonly obtained for faces was also found for the perception of human bodies (Stekelenburg and de Gelder, 2004). Most interestingly, the N170 showed an inversion effect for bodies, comparable to the inversion effect earlier found for faces (Stekelenburg and de Gelder, 2004).

In the studies of Tamietto and co-workers, the simultaneous presentation to both visual hemifields of two emotionally congruent faces (Tamietto et al., 2006) or two emotionally congruent bodies (Tamietto et al., 2005b) leads to shorter latencies for stimulus detection as compared to the unilateral presentation of the same stimuli to either the left or right hemifield. Additionally, patients with hemineglect and visual extinction, who typically fail to report the presence of a contralesional stimulus under conditions of bilateral stimulation, could more easily detect a contralesional happy or angry facial expression than a neutral facial expression (Vuilleumier and Schwartz, 2001). This finding was replicated with emotional bodily expressions in the study of Tamietto and colleagues (Tamietto et al., 2005a), in which fearful bodily expressions were more easily detected than neutral bodily expressions for the contralesional field. These findings indicate similarities in perceptual properties between faces and bodies, and the ability of emotional biological stimuli to attract attention in unattended visual space.

There appear to be also large similarities between emotional bodily and facial expressions at the neural level. A striking finding (Hadjikhani and de Gelder, 2003; de Gelder et al., 2004a) is that observing bodily expressions activates two well-known face areas, such as FFA and amygdala, predominantly associated with processing faces but also linked to biological movement (Bonda et al., 1996). These activations in face-related areas may result from mental imagery (O'Craven and

Kanwisher, 2000) or alternatively – and more probably – from context-driven high-level perceptual mechanisms filling in the face information missing from the input. However, this is unlikely to be the only explanation for similarities between fearful facial expressions and bodily expressions (cf. Cox et al., 2004). The finding of Hadjikhani and de Gelder (2003) was supported by the studies of de Gelder and colleagues (de Gelder et al., 2004a) for bodily expressions, and by Peelen and Downing (2005) for neutral body postures.

Since there is as of yet no literature on how recognition of facial expression is affected by emotional body contexts, we have recently started to explore this critical issue. We used photographs of fearful and angry faces and bodies to create realistically looking face–body compound images, with either matched or mismatched emotional expressions. Fear and anger were selected because they are both emotions with a negative valence and each is associated with evolutionary relevant threat situations. A short stimulus presentation time was used (200 ms), requiring observers to judge the faces on the basis of a "first impression" and to rely on global processing rather than on extensive analysis of separate facial features. Participants attended to the face and made judgments about the facial expression. The recognition of the emotion conveyed by the face was found to be systematically influenced by the emotion expressed by the body (Meeren et al., 2005). Observers made significantly better (81% correct) and faster (774 ms) decisions when faces were accompanied by a matching bodily expression than when the bodily expression did not match the facial expression (67% and 840 ms). The fact that a reliable influence was obtained in an implicit paradigm in which the bodies were not task relevant nor explicitly attended to suggests that the influence they exercise is rapid and automatic. To further test the automatic processing hypothesis we recorded EEG while subjects performed the task. An enlargement of the occipital P1-component as early as 115 ms after presentation onset was found for incongruent face-body combinations (Meeren et al., 2005). This points to the existence of an ultrarapid neural mechanism sensitive to the degree of agreement between simultaneously presented facial and bodily emotional expressions, even when the latter are unattended.

Facial expressions in the context of scenes

Faces routinely appear as part of natural scenes. Hierarchical models of perception tend to assimilate scene effects with semantic effects occurring relatively late at higher cognitive centers (Bar, 2004). However, the processing of objects is influenced by the properties of a scene at an early level. It has been reported that the rapid extraction of the gist of a scene appears to be based on low spatial frequency coding (Oliva and Schyns, 1997) Brief exposure to a known scene activates a representation of its layout that contributes to subsequent processing of spatial relations across the scene (Sanocki, 2003). Segmentation of object from background scenes occurs rapidly, during the first 100 ms of processing (Lamme, 1995), and object detection is faster when presented in an implicitly learned context configuration (Olson et al., 2001). The results support the role of feedback modulations in an early level of processing in animal (Lamme and Roelfsema, 2000) and human studies (Foxe and Simpson, 2002).

The effects of semantic contexts on object processing occur in a much later stage of processing. Objects that are congruent with their context are identified better (Davenport and Potter, 2004) and faster (Ganis and Kutas, 2003), and the interaction occurs at about 390 ms after stimulus-onset (i.e., the N400 component), which is assumed to be a high level of semantic representation of object and scene (Ganis and Kutas, 2003). In an fMRI study, it was found that the parahippocampal cortex (PHC) and retrosplenial cortex (RSC) are involved in a system that associates objects with contexts (Bar and Aminoff, 2003; Bar, 2004).

The effects of emotional contexts may occur on a much earlier level than semantic effects and may involve different neural systems (Hariri et al., 2002). We recently investigated how emotional visual scenes influence face processing. Event-related potentials were recorded for faces (fearful/neutral) embedded in scenes (fearful/neutral) while participants performed an orientation-decision task (face upright/inverted). Thus, the task condition was kept irrelevant to the emotion in context and face. Increased structural encoding, as indicated by the N170 response to faces, was found when faces were perceived in a fearful context as opposed to a neutral context (Righart and de Gelder, 2005). This N170 response was even more increased for fearful faces in a fearful context, possibly as a consequence of congruency. Preliminary behavioral data substantiate these congruency effects, as it was found that facial expressions (e.g., a disgust expression) were recognized faster when they were accompanied by a congruent emotional context (e.g., a rubbish dump). A control condition showed that the increased response on the N170 could not be attributed to the exclusive presence of the context, as the amplitudes did not differ between fearful and neutral contexts without a face.

The N170 of faces, particularly fearful faces, in a threatening context, may be increased in order to enhance structural encoding. In a potentially dangerous situation, it is important to analyze instantly what is happening. The results may be consistent with the model proposed by Haxby et al. (2000). Source analysis studies suggest that an enlarged N170 may be indicative of increased activation in fusiform gyrus or STS (Pizzagalli et al., 2002; Shibata et al., 2002; Henson et al., 2003; Itier and Taylor, 2004b). Fearful faces and contexts may activate the amygdala and modulate activity in the fusiform gyrus (Lang et al., 1998; Morris et al., 1998a; Surguladze et al., 2003), and in this way influence face processing by enhancing the N170 amplitude. Alternatively, activity in the STS has been related to the perception of social cues (Allison et al., 2000). This functional interpretation accords with the findings that the N170 amplitude is profoundly affected by biological motion (Jokisch et al., 2005), eye gaze (Watanabe et al., 2002), facial motion (Puce et al., 2003), facial expressions (Batty and Taylor, 2003; Stekelenburg and de Gelder, 2004), expressional change (Miyoshi et al., 2004) and affective facial features (Pizzagalli et al., 2002).

It is not clear yet whether congruent emotions engage the system that includes the PHC, similar to semantic associations between object and context (Bar and Aminoff, 2003). Future studies should determine whether processing of emotional relations in contexts should be distinguished from semantic relations in context. Our recent data show that the time courses differ (Ganis and Kutas, 2003; Righart and de Gelder, 2005), but no fMRI data are as yet available as to what neural systems are involved in processing faces in emotional contexts.

Further, an interesting question is through which mechanism fearful contexts enhance the perceptual analysis of faces. According to the model of Bar (2004), the gist of the scene is extracted by perception of the low spatial frequencies (Oliva and Schyns, 1997), which provides a rough and quick image of the scene information, on which the high spatial frequencies provide the detailed fill-in of the object. Low spatial frequencies in the face increase amygdala responses for fearful faces (Vuilleumier et al., 2003). If low spatial frequencies are important for fear processing in general then the model could explain why N170 amplitudes are increased for faces in fearful contexts. In such a model, fearful contexts may provide a first coarse template on which the perceptual analysis of faces is interpreted.

Enhanced N170 amplitudes for faces in fearful contexts may be related to enhanced encoding of identity, which may improve recognition memory for faces. Data of prosopagnosia patients indicate that impaired structural encoding, as reflected in the N170, may disrupt facial identification (Eimer, 2000; de Gelder and Stekelenburg, 2005), and that facial expressions may improve their performance on face recognition (de Gelder et al., 2003). It has already been shown that object recognition memory (e.g., tools, furniture, and clothing) is better for objects that were presented in a positive-valenced context background than in a negative- or neutral-valenced background (Smith et al., 2004). Similar increases in accuracy were obtained for words primed by positive background (Erk et al., 2003). An interesting question is whether recognition memory is also improved for faces that are presented in emotional contexts.

Facial expressions in the context of voices

Human cognition and emotion researchers tend to focus on how organisms process information from one sensory system at a time (usually the visual system), but information processing in everyday life is typically multisensory. In many higher species, communication involves multiple sensory systems often in combination. Animal researchers are traditionally more interested in co-occurring behavioral signals, and a number of studies have explored the close link between vocal and visual communication (Parr, 2004) and discovered synergies between the evolutionary history and the functionality of visual and auditory communication signals (Cooper and Goller, 2004). Audiovisual vocalizations are ethologically relevant and thus may tap into specialized neural mechanisms (Ghazanfar and Santos, 2004).

As stated above, the traditional emphasis is on visual processes, foremost facial expressions. In comparison with the processing of facial expressions, there have been only a few attempts to identify the specific neural sites for processing emotions in the voice (George et al., 1996; Ross, 2000; de Gelder et al., 2004b). Available research shows that listeners can readily recognize a speakers' emotion from his tone of voice. Rapid recognition of affect in auditory expressions happens within the first 100–150 ms of stimulus presentation (Bostanov and Kotchoubey, 2004; Goydke et al., 2004) and is based primarily on voice characteristics.

The ability to decode emotional cues in prosody and facial expressions may have a common processing and/or representational substrate in the human brain (Borod et al., 2000; Pourtois et al., 2002; de Gelder and Bertelson, 2003), facilitating processing and integration of these distinct but often calibrated sources of information. Most of the studies on multisensory emotion perception have focused on the integration of facial expression with information in the voice (Massaro and Egan, 1996; de Gelder et al., 1999a; de Gelder and Vroomen, 2000; Pourtois et al., 2000). Judging the emotional state of a speaker is possible via facial or vocal cues (Scherer et al., 1991; Banse and Scherer, 1996) alone but both judgment accuracy and speed seem to benefit from combining the modalities, e.g., response accuracy increases and response speed decreases when a face is paired with a voice expressing the same emotion. This improvement of performance occurs even when participants are instructed to ignore the voice and rate only the face, suggesting that extracting affective information from a voice may be automatic and/or mandatory (de Gelder and Vroomen, 2000). The fact that prosodic and facial expressions of emotion frequently correlate suggests that the underlying cognitive mechanisms are highly sensitive to shared associations activated by cues in each channel (de Gelder et al., 1999a; Massaro et al., 1996).

To assess how emotional judgments of the face are biased by prosody, Massaro and Egan (1996) presented computer-generated faces expressing a *happy, angry* or *neutral* emotion that accompanied a word spoken in one of the three emotional tones. De Gelder and Vroomen (2000) presented photographs taken from the Ekman and Friesen's (1976) series with facial expressions rendered emotionally ambiguous by "morphing" the expressions between *happy* and *sad* as the two endpoints. The emotional prosody tended to facilitate how accurately and quickly subjects rate an emotionally *congruent as compared to an incongruent* face. These findings indicate that the emotional value of prosody-face events is registered and somehow integrated during perceptual tasks, affecting behavioral responses according to the emotion congruity of the combined events. Moreover, these crossmodal influences appear to be resistant to increased attentional demands induced by a dual task, implying that combining the two forms of input may be mandatory (Vroomen et al., 2001). The conclusion of mandatory integration is now considerably strengthened in a study using patients who could recognize facial expressions without being aware of the visual stimuli presented (hemianopic patients suffering from loss of primary visual cortex exhibiting affective blindsight) (de Gelder et al., 2002, 2005).

Our current knowledge of bimodal integration of visual and auditory primate vocal signals in the brain is derived almost exclusively from human neuroimaging studies of audiovisual speech. STS and superior temporal gyrus are consistently activated by bimodal speech signals and often show enhanced activity over unimodal-induced signals (Stein and Meredith, 1993; Calvert et al., 2000; Callan et al., 2003) but audiovisual perception of ecologically valid stimuli may not follow the rules derived from firing patterns of cells with audiovisual receptive fields and superadditivity may not be the correct criterion (de Gelder and Bertelson, 2003; Ghazanfar and Santos, 2004).

A few studies have explored brain areas involved in processing faces in the context of emotional voices. The classical candidate is multisensory convergence in heteromodal cortex (Mesulam, 1998). Cortical areas like STS (Baylis et al., 1987) and ventral premotor cortex (Kohler et al., 2002) appear to play an important role. A recent study in rhesus monkeys has confirmed such integration in the STS at the level of single units for biologically meaningful actions (Barraclough et al., 2005). In a positron emission tomography (PET) study, we found enhanced activity for bimodal stimuli compared to unimodal stimuli situated in the left lateral temporal cortex. Separate analysis for positive and negative emotions showed supplementary convergence area's anteriorly in the left and right hemisphere, respectively (Pourtois et al., 2005). Subcortical audiovisual emotion convergence sites have been found in the amygdala and SC in fMRI studies (Dolan et al., 2001; de Gelder et al., 2005). These subcortical nuclei might play a more important role than hitherto expected in part also because of their role in orienting to novel and highly significant stimuli in the environment.

Information about time course may be more critical than anything else to clarify processing properties. All our EEG studies so far (de Gelder et al., 1999b; Pourtois et al., 2000, 2002; de Gelder, 2005) point in the direction of early interaction between the facial expression and the emotion in the voice.

Conclusions

Recent data show that different types of context influence the recognition of facial expression. When a face is accompanied by a body or voice expressing the same emotion, or when it is presented in a congruent emotional scene, the recognition of facial expression typically improves, i.e., both the judgment accuracy and speed increase. Hence, both the immediate visual and auditory contexts function to disambiguate the signals of facial expression. Our behavioral and electrophysiological data suggest that this perceptual integration of information does not require high-level semantic analysis occurring relatively late at higher cognitive centers. Instead, the integration appears to be an automatic and mandatory process, which takes place very early in the processing stream, before full structural encoding of the stimulus and conscious awareness of the emotional expression is established.

Abbreviations

EBA	extrastriate body area
FFA	fusiform face area
LGN	lateral geniculate nucleus
OFA	occipital face area
OFC	orbitofrontal cortex
PHC	parahippocampal cortex
RSC	retrosplenial cortex
SC	superior colliculus
STS	superior temporal sulcus
P1	an event-related potential component with positive deflection occurring at about 100 ms after stimulus onset
N170	an event-related potential component with negative deflection occurring at about 170 ms after stimulus onset

References

Adolphs, R. (2002) Recognizing emotion from facial expressions: psychological and neurological mechanisms. Behav. Cogn. Neurosci. Rev., 1: 21–61.

Adolphs, R., Damasio, H., Tranel, D., Cooper, G. and Damasio, A.R. (2000) A role for somatosensory cortices in the visual recognition of emotion as revealed by three-dimensional lesion mapping. J. Neurosci., 20: 2683–2690.

Allison, T., Puce, A. and McCarthy, G. (2000) Social perception from visual cues: role of the STS region. Trends Cogn. Sci., 4: 267–278.

Amaral, D.G. and Price, J.L. (1984) Amygdalo-cortical projections in the monkey (*Macaca fascicularis*). J. Comp. Neurol., 230: 465–496.

Banse, R. and Scherer, K.R. (1996) Acoustic profiles in vocal emotion expression. J. Pers. Soc. Psychol., 70: 614–636.

Bar, M. (2004) Visual objects in context. Nat. Rev. Neurosci., 5: 617–629.

Bar, M. and Aminoff, E. (2003) Cortical analysis of visual context. Neuron, 38: 347–358.

Barraclough, N.E., Xiao, D., Baker, C.I., Oram, M.W. and Perrett, D.I. (2005) Integration of visual and auditory information by superior temporal sulcus neurons responsive to the sight of actions. J. Cogn. Neurosci., 17: 377–391.

Barsalou, L.W. (1999) Perceptual symbol systems. Behav. Brain Sci., 22: 577–609; discussion 610–660.

Batty, M. and Taylor, M.J. (2003) Early processing of the six basic facial emotional expressions. Cogn. Brain Res., 17: 613–620.

Baylis, G.C., Rolls, E.T. and Leonard, C.M. (1987) Functional subdivisions of the temporal lobe neocortex. J. Neurosci., 7: 330–342.

Bentin, S., Allison, T., Puce, A., Perez, E. and McCarthy, G. (1996) Electrophysiological studies of face perception in humans. J. Cogn. Neurosci., 8: 551–565.

Bonda, E., Petrides, M., Ostry, D. and Evans, A. (1996) Specific involvement of human parietal systems and the amygdala in the perception of biological motion. J. Neurosci., 16: 3737–3744.

Borod, J.C., Rorie, K.D., Pick, L.H., Bloom, R.L., Andelman, F., Campbell, A.L., Obler, L.K., Tweedy, J.R., Welkowitz, J. and Sliwinski, M. (2000) Verbal pragmatics following unilateral stroke: emotional content and valence. Neuropsychology, 14: 112–124.

Bostanov, V. and Kotchoubey, B. (2004) Recognition of affective prosody: continuous wavelet measures of event-related brain potentials to emotional exclamations. Psychophysiology, 41: 259–268.

Callan, D.E., Jones, J.A., Munhall, K., Callan, A.M., Kroos, C. and Vatikiotis-Bateson, E. (2003) Neural processes underlying perceptual enhancement by visual speech gestures. Neuroreport, 14: 2213–2218.

Calvert, G.A., Campbell, R. and Brammer, M.J. (2000) Evidence from functional magnetic resonance imaging of crossmodal binding in the human heteromodal cortex. Curr. Biol., 10: 649–657.

Carmichael, S.T. and Price, J.L. (1995) Limbic connections of the orbital and medial prefrontal cortex in macaque monkeys. J. Comp. Neurol., 363: 615–641.

Catani, M., Jones, D.K., Donato, R. and Ffytche, D.H. (2003) Occipito-temporal connections in the human brain. Brain, 126: 2093–2107.

Cooper, B.G. and Goller, F. (2004) Multimodal signals: enhancement and constraint of song motor patterns by visual display. Science, 303: 544–546.

Cox, D., Meyers, E. and Sinha, P. (2004) Contextually evoked object-specific responses in human visual cortex. Science, 304: 115–117.

Damasio, A.R. (1994) Descartes' Error: Emotion, Reason, and the Human Brain. G.P. Putnam's Sons, New York.

Damasio, A.R. (1999) The Feeling of What Happens: Body and Emotion in the Making of Consciousness. Hartcourt Brace, New York.

Darwin (1872) The Expression of the Emotion in Man and Animals. J. Murray, London.

Davenport, J.L. and Potter, M.C. (2004) Scene consistency in object and background perception. Psychol. Sci., 15: 559–564.

de Gelder, B. (2005) Nonconscious emotions: New findings and perspectives on nonconscious facial expression recognition and its voice and whole body contexts. In: Feldman Barrett, L., Niedenthal, P.M. and Winkielman, P. (Eds.), Emotion and Consciousness. The Guilford Press, New York-London, pp. 123–149.

de Gelder, B. and Bertelson, P. (2003) Multisensory integration, perception and ecological validity. Trends Cogn. Sci., 7: 460–467.

de Gelder, B., Bocker, K.B., Tuomainen, J., Hensen, M. and Vroomen, J. (1999a) The combined perception of emotion from voice and face: early interaction revealed by human electric brain responses. Neurosci. Lett., 260: 133–136.

de Gelder, B., Frissen, I., Barton, J. and Hadjikhani, N. (2003) A modulatory role for facial expressions in prosopagnosia. Proc. Natl. Acad. Sci. USA, 100: 13105–13110.

de Gelder, B., Morris, J.S. and Dolan, R.J. (2005) Unconscious fear influences emotional awareness of faces and voices. Proceedings of the National Academy of Sciences of the USA, 102: 18682–18687.

de Gelder, B., Pourtois, G., van Raamsdonk, M., Vroomen, J. and Weiskrantz, L. (2001) Unseen stimuli modulate conscious visual experience: evidence from inter-hemispheric summation. Neuroreport, 12: 385–391.

de Gelder, B., Pourtois, G. and Weiskrantz, L. (2002) Fear recognition in the voice is modulated by unconsciously recognized facial expressions but not by unconsciously recognized affective pictures. Proc. Natl. Acad. Sci. USA, 99: 4121–4126.

de Gelder, B. and Rouw, R. (2000) Configural face processes in acquired and developmental prosopagnosia: evidence for two separate face systems? Neuroreport, 11: 3145–3150.

de Gelder, B., Snyder, J., Greve, D., Gerard, G. and Hadjikhani, N. (2004a) Fear fosters flight: a mechanism for fear contagion when perceiving emotion expressed by a whole body. Proc. Natl. Acad. Sci. USA, 101: 16701–16706.

de Gelder, B. and Stekelenburg, J.J. (2005) Naso-temporal asymmetry of the N170 for processing faces in normal viewers but not in developmental prosopagnosia. Neurosci. Letts., 376: 40–45.

de Gelder, B. and Vroomen, J. (2000) The perception of emotions by ear and by eye. Cogn. Emotion, 14: 289–311.

de Gelder, B., Vroomen, J. and Pourtois, G. (2004b) Multisensory perception of affect, its time course and its neural basis. In: Calvert, G., Spence, C. and Stein, B.E. (Eds.), Handbook of Multisensory Processes. MIT, Cambridge, MA, pp. 581–596.

de Gelder, B., Vroomen, J., Pourtois, G. and Weiskrantz, L. (1999b) Non-conscious recognition of affect in the absence of striate cortex. Neuroreport, 10: 3759–3763.

Di Russo, F., Pitzalis, S., Spitoni, G., Aprile, T., Patria, F., Spinelli, D. and Hillyard, S.A. (2005) Identification of the neural sources of the pattern-reversal VEP. Neuroimage, 24: 874–886.

Dolan, R.J., Morris, J.S. and de Gelder, B. (2001) Crossmodal binding of fear in voice and face. Proc. Natl. Acad. Sci. USA, 98: 10006–10010.

Downing, P.E., Jiang, Y., Shuman, M. and Kanwisher, N. (2001) A cortical area selective for visual processing of the human body. Science, 293: 2470–2473.

Eger, E., Jedynak, A., Iwaki, T. and Skrandies, W. (2003) Rapid extraction of emotional expression: evidence from evoked potential fields during brief presentation of face stimuli. Neuropsychologia, 41: 808–817.

Eimer, M. (2000) Event-related brain potentials distinguish processing stages involved in face perception and recognition. Clin. Neurophysiol., 111: 694–705.

Ekman, P. (1992) An argument for basic emotions. Cogn. Emotion, 6: 169–200.

Ekman, P. and Friesen, W.V. (1976) Pictures of Facial Affects. Consulting Psychologists Press, Palo Alto.

Erk, S., Kiefer, M., Grothe, J., Wunderlich, A.P., Spitzer, M. and Walter, H. (2003) Emotional context modulates subsequent memory effect. Neuroimage, 18: 439–447.

Foxe, J.J. and Simpson, G.V. (2002) Flow of activation from V1 to frontal cortex in humans. A framework for defining "early" visual processing. Exp. Brain Res., 142: 139–150.

Frijda, N.H. (1986) The Emotions. Cambridge University Press, Cambridge.

Ganis, G. and Kutas, M. (2003) An electrophysiological study of scene effects on object identification. Brain Res. Cogn. Brain Res., 16: 123–144.

Gauthier, I. and Nelson, C.A. (2001) The development of face expertise. Curr. Opin. Neurobiol., 11: 219–224.

Gauthier, I., Tarr, M.J., Anderson, A.W., Skudlarski, P. and Gore, J.C. (1999) Activation of the middle fusiform 'face area' increases with expertise in recognizing novel objects. Nat. Neurosci., 2: 568–573.

Gauthier, I., Williams, P., Tarr, M.J. and Tanaka, J. (1998) Training 'greeble' experts: a framework for studying expert object recognition processes. Vision Res., 38: 2401–2428.

George, M.S., Parekh, P.I., Rosinsky, N., Ketter, T.A., Kimbrell, T.A., Heilman, K.M., Herscovitch, P. and Post, R.M. (1996) Understanding emotional prosody activates right hemisphere regions. Arch. Neurol., 53: 665–670.

Ghazanfar, A.A. and Santos, L.R. (2004) Primate brains in the wild: the sensory bases for social interactions. Nat. Rev. Neurosci., 5: 603–616.

Goydke, K.N., Altenmuller, E., Moller, J. and Munte, T.F. (2004) Changes in emotional tone and instrumental timbre are reflected by the mismatch negativity. Brain Res. Cogn. Brain Res., 21: 351–359.

Hadjikhani, N. and de Gelder, B. (2003) Seeing fearful body expressions activates the fusiform cortex and amygdala. Curr. Biol., 13: 2201–2205.

Halgren, E., Raij, T., Marinkovic, K., Jousmaki, V. and Hari, R. (2000) Cognitive response profile of the human fusiform face area as determined by MEG. Cereb. Cortex, 10: 69–81.

Hariri, A.R., Tessitore, A., Mattay, V.S., Fera, F. and Weinberger, D.R. (2002) The amygdala response to emotional stimuli: a comparison of faces and scenes. Neuroimage, 17: 317–323.

Haxby, J.V., Hoffman, E.A. and Gobbini, M.I. (2000) The distributed human neural system for face perception. Trends Cogn. Sci., 4: 223–233.

Haxby, J.V., Hoffman, E.A. and Gobbini, M.I. (2002) Human neural systems for face recognition and social communication. Biol. Psychiatry, 51: 59–67.

Haxby, J.V., Horwitz, B., Ungerleider, L.G., Maisog, J.M., Pietrini, P. and Grady, C.L. (1994) The functional organization of human extrastriate cortex: a PET-rCBF study of selective attention to faces and locations. J. Neurosci., 14: 6336–6353.

Haxby, J.V., Ungerleider, L.G., Horwitz, B., Maisog, J.M., Rapoport, S.I. and Grady, C.L. (1996) Face encoding and recognition in the human brain. Proc. Natl. Acad. Sci. USA, 93: 922–927.

Heberlein, A.S. and Adolphs, R. (2004) Impaired spontaneous anthropomorphizing despite intact perception and social knowledge. Proc. Natl. Acad. Sci. USA, 101: 7487–7491.

Henson, R.N., Goshen-Gottstein, Y., Ganel, T., Otten, L.J., Quayle, A. and Rugg, M.D. (2003) Electrophysiological and haemodynamic correlates of face perception, recognition and priming. Cereb. Cortex, 13: 793–805.

Hoffman, E.A. and Haxby, J.V. (2000) Distinct representations of eye gaze and identity in the distributed human neural system for face perception. Nat. Neurosci., 3: 80–84.

Iidaka, T., Omori, M., Murata, T., Kosaka, H., Yonekura, Y., Okada, T. and Sadato, N. (2001) Neural interaction of the amygdala with the prefrontal and temporal cortices in the processing of facial expressions as revealed by fMRI. J. Cogn. Neurosci., 13: 1035–1047.

Itier, R.J. and Taylor, M.J. (2002) Inversion and contrast polarity reversal affect both encoding and recognition processes of unfamiliar faces: a repetition study using ERPs. Neuroimage, 15: 353–372.

Itier, R.J. and Taylor, M.J. (2004a) Effects of repetition learning on upright, inverted and contrast-reversed face processing using ERPs. Neuroimage, 21: 1518–1532.

Itier, R.J. and Taylor, M.J. (2004b) Source analysis of the N170 to faces and objects. Neuroreport, 15: 1261–1265.

Jokisch, D., Daum, I., Suchan, B. and Troje, N.F. (2005) Structural encoding and recognition of biological motion: evidence from event-related potentials and source analysis, Behav. Brain Res., 157: 195–204.

Kanwisher, N., McDermott, J. and Chun, M.M. (1997) The fusiform face area: a module in human extrastriate cortex specialized for face perception. J. Neurosci., 17: 4302–4311.

Kohler, E., Keysers, C., Umilta, M.A., Fogassi, L., Gallese, V. and Rizzolatti, G. (2002) Hearing sounds, understanding actions: action representation in mirror neurons. Science, 297: 846–848.

Lamme, V.A. (1995) The neurophysiology of figure-ground segregation in primary visual cortex. J. Neurosci., 15: 1605–1615.

Lamme, V.A. and Roelfsema, P.R. (2000) The distinct modes of vision offered by feedforward and recurrent processing. Trends Neurosci, 23: 571–579.

Lang, P.J., Bradley, M.M., Fitzsimmons, J.R., Cuthbert, B.N., Scott, J.D., Moulder, B. and Nangia, V. (1998) Emotional arousal and activation of the visual cortex: an fMRI analysis. Psychophysiology, 35: 199–210.

LeDoux, J.E. (1996) The Emotional Brain: The Mysterious Underpinnings of Emotional life. Simon & Schuster, New York.

Linkenkaer-Hansen, K., Palva, J.M., Sams, M., Hietanen, J.K., Aronen, H.J. and Ilmoniemi, R.J. (1998) Face-selective processing in human extrastriate cortex around 120 ms after stimulus onset revealed by magneto- and electroencephalography. Neurosci. Lett., 253: 147–150.

Liu, J., Harris, A. and Kanwisher, N. (2002) Stages of processing in face perception: an MEG study. Nat. Neurosci., 5: 910–916.

Massaro, D.W., Cohen, M.M. and Smeele, P.M. (1996) Perception of asynchronous and conflicting visual and auditory speech. J. Acoust. Soc. Am., 100: 1777–1786.

Massaro, D.W. and Egan, P.B. (1996) Perceiving affect from the voice and the face. Psychon. Bull. Rev., 3: 215–221.

Meeren, H.K.M., van Heijnsbergen, C.C.R.J. and de Gelder, B. (2005) Rapid perceptual integration of facial expression and emotional body language. Proceedings of the National Academy of Sciences of the USA, 102: 16518–16523.

Mesulam, M.M. (1998) From sensation to cognition. Brain, 121(Pt 6): 1013–1052.

Miyoshi, M., Katayama, J. and Morotomi, T. (2004) Face-specific N170 component is modulated by facial expressional change. Neuroreport, 15: 911–914.

Morris, J.S., de Gelder, B., Weiskrantz, L. and Dolan, R.J. (2001) Differential extrageniculostriate and amygdala responses to presentation of emotional faces in a cortically blind field. Brain, 124: 1241–1252.

Morris, J.S., Friston, K.J., Büchel, C., Frith, C.D., Young, A.W., Calder, A.J. and Dolan, R.J. (1998a) A neuromodulatory role for the human amygdala in processing emotional facial expressions. Brain, 121(Pt 1): 47–57.

Morris, J.S., Öhman, A. and Dolan, R.J. (1998b) Conscious and unconscious emotional learning in the human amygdala. Nature, 393: 467–470.

O'Craven, K.M. and Kanwisher, N. (2000) Mental imagery of faces and places activates corresponding stimulus-specific brain regions. J. Cogn. Neurosci., 12: 1013–1023.

Oliva, A. and Schyns, P.G. (1997) Coarse blobs or fine edges? Evidence that information diagnosticity changes the perception of complex visual stimuli. Cogn. Psychol., 34: 72–107.

Olson, I.R., Chun, M.M. and Allison, T. (2001) Contextual guidance of attention: human intracranial event-related potential evidence for feedback modulation in anatomically early temporally late stages of visual processing. Brain, 124: 1417–1425.

Panksepp, J. (1998) Affective Neuroscience: The Foundation of Human and Animal Emotions. Oxford University Press, New York.

Parr, L.A. (2004) Perceptual biases for multimodal cues in chimpanzee (*Pan troglodytes*) affect recognition. Anim. Cogn., 7: 171–178.

Peelen, M.V. and Downing, P.E. (2005) Selectivity for the human body in the fusiform gyrus. J. Neurophysiol., 93: 603–608.

Pegna, A.J., Khateb, A., Lazeyras, F. and Seghier, M.L. (2005) Discriminating emotional faces without primary visual cortices involves the right amygdala. Nat. Neurosci., 8: 24–25.

Pizzagalli, D.A., Lehmann, D., Hendrick, A.M., Regard, M., Pascual-Marqui, R.D. and Davidson, R.J. (2002) Affective judgments of faces modulate early activity (not similar to 160 ms) within the fusiform gyri. Neuroimage, 16: 663–677.

Pourtois, G., de Gelder, B., Bol, A. and Crommelinck, M. (2005) Perception of facial expressions and voices and of their combination in the human brain. Cortex, 41: 49–59.

Pourtois, G., de Gelder, B., Vroomen, J., Rossion, B. and Crommelinck, M. (2000) The time-course of intermodal binding between seeing and hearing affective information. Neuroreport, 11: 1329–1333.

Pourtois, G., Debatisse, D., Despland, P.A. and de Gelder, B. (2002) Facial expressions modulate the time course of long latency auditory brain potentials. Brain Res. Cogn. Brain. Res., 14: 99–105.

Puce, A., Allison, T., Asgari, M., Gore, J.C. and McCarthy, G. (1996) Differential sensitivity of human visual cortex to faces, letterstrings, and textures: a functional magnetic resonance imaging study. J. Neurosci., 16: 5205–5215.

Puce, A., Syngeniotis, A., Thompson, J.C., Abbott, D.F., Wheaton, K.J. and Castiello, U. (2003) The human temporal lobe integrates facial form and motion: evidence from fMRI and ERP studies. Neuroimage, 19: 861–869.

Righart, R. and de Gelder, B. (2005) Context influences early perceptual analysis of faces. An electrophysiological study. Cerebral Cortex. In press.

Rizzolatti, G., Fadiga, L., Gallese, V. and Fogassi, L. (1996) Premotor cortex and the recognition of motor actions. Brain Res. Cogn. Brain. Res., 3: 131–141.

Ross, E.D. (2000) Affective prosody and the aprosodias. In: Mesulam, M. (Ed.), Principles of Behavioral and Cognitive Neurology. Oxford University Press, Oxford, England, pp. 316–331.

Russell, J.A. and Feldman Barrett, L. (1999) Core affect, prototypical emotional episodes, and other things called emotion: dissecting the elephant. J. Pers. Soc. Psychol., 76: 805–819.

Sanocki, T. (2003) Representation and perception of scenic layout. Cogn. Psychol., 47: 43–86.

Scherer, K.R., Banse, R., Wallbott, H.G. and Goldbeck, T. (1991) Vocal cues in emotion encoding and decoding. Motivation Emotion, 15: 123–148.

Schmidt, K.L. and Cohn, J.F. (2001) Human facial expressions as adaptations: evolutionary questions in facial expression research. Am. J. Phys. Anthropol. Suppl., 33: 3–24.

Shibata, T., Nishijo, H., Tamura, R., Miyamoto, K., Eifuku, S., Endo, S. and Ono, T. (2002) Generators of visual evoked potentials for faces and eyes in the human brain as determined by dipole localization. Brain Topogr., 15: 51–63.

Smith, A.P.R., Henson, R.N.A., Dolan, R.J. and Rugg, M.D. (2004) fMRI correlates of the episodic retrieval of emotional contexts. Neuroimage, 22: 868–878.

Stein, B.E. and Meredith, M.A. (1993) The Merging of the Senses. MIT Press, Cambridge, MA.

Stekelenburg, J.J. and de Gelder, B. (2004) The neural correlates of perceiving human bodies: an ERP study on the body-inversion effect. Neuroreport, 15: 777–780.

Surguladze, S.A., Brammer, M.J., Young, A.W., Andrew, C., Travis, M.J., Williams, S.C. and Phillips, M.L. (2003) A preferential increase in the extrastriate response to signals of danger. Neuroimage, 19: 1317–1328.

Tamietto, M., de Gelder, B., Genero, R. and Geminiani, G. (2005a, May 31–June 3) Automatic encoding of emotional body language in spatial neglect and visual extinction. Paper presented at the 4th Dutch Endo-Neuro-Psycho Meeting, Doorwerth, The Netherlands.

Tamietto, M., Geminiani, G. and de Gelder, B. (2005) Interhemispheric interaction for bodily emotional expressions: Is the right hemisphere superiority related to facial rather than emotional processing? Perception 34: 205–206 suppl.

Tamietto, M., Corazzini, L.L., de Gelder, B. and Geminiani, G. (2006) Functional asymmetry and interhemispheric cooperation in the perception of emotions from facial expressions. Exp. Brain Res, 171: 389–404.

Tomkins, S.S. (1963) Affect, Imagery Consciousness, Vol. 2. The Negative Effects, Springer, New York.

Vroomen, J., Driver, J. and de Gelder, B. (2001) Is cross-modal integration of emotional expressions independent of attentional resources? Cogn. Affect. Behav. Neurosci., 1: 382–387.

Vuilleumier, P., Armony, J.L., Driver, J. and Dolan, R.J. (2003) Distinct spatial frequency sensitivities for processing faces and emotional expressions. Nat. Neurosci., 6: 624–631.

Vuilleumier, P., Richardson, M.P., Armony, J.L., Driver, J. and Dolan, R.J. (2004) Distant influences of amygdala lesion on visual cortical activation during emotional face processing. Nat. Neurosci., 7: 1271–1278.

Vuilleumier, P. and Schwartz, S. (2001) Emotional facial expressions capture attention. Neurology, 56: 153–158.

Watanabe, S., Kakigi, R. and Puce, A. (2003) The spatiotemporal dynamics of the face inversion effect: a magneto- and electro-encephalographic study. Neuroscience, 116: 879–895.

Watanabe, S., Miki, K. and Kakigi, R. (2002) Gaze direction affects face perception in humans. Neurosci. Letts., 325: 163–166.

Winston, J.S., O'Doherty, J. and Dolan, R.J. (2003) Common and distinct neural responses during direct and incidental processing of multiple facial emotions. Neuroimage, 20: 84–97.

Martinez-Conde, Macknik, Martinez, Alonso & Tse (Eds.)
Progress in Brain Research, Vol. 155
ISSN 0079-6123

CHAPTER 4

The effectiveness of disruptive coloration as a concealment strategy

Martin Stevens[1], Innes C. Cuthill[1], C. Alejandro Párraga[2] and Tom Troscianko[3,*]

[1]*School of Biological Sciences, University of Bristol, Woodland Rd, Bristol BS8 1UG, UK*
[2]*Department of Experimental Psychology, University of Bristol, 8 Woodland Rd, Bristol BS8 1TN, UK*
[3]*Department of Experimental Psychology, University of Bristol, 8 Woodland Rd, Bristol BS8 1TN, UK*

Abstract: Our understanding of camouflage has been developing for over 100 years. Several underlying principles have emerged. Background pattern matching, or crypsis, is insufficient to conceal objects because of edge information. Other strategies exist to disrupt the continuity of extended edges. These strategies are reviewed. We pay particular attention to the theory of disruptive coloration, which predicts that high-contrast elements located at the object edge will mask the perception of a target as belonging to a certain category of object, in spite of the fact that the edge elements are independently visible. Although this strategy has long been assumed to be effective, there has been a lack of supportive data involving the perception of targets by nonhuman animals. We present evidence, from a field study, in support of the notion that disruptive coloration reduces the chances of bird predation of artificial "moths."

Keywords: camouflage; object perception; bird vision; disruptive coloration; crypsis; contour perception

Introduction

The detection of an object in a natural setting strongly depends on the extent to which a visual system can differentiate it from other objects, which make up the "background." By "background" we mean the scene, at whatever level of visual complexity, in the absence of the object. So, for example, a hawk may not see a mouse on the ground if it is similar in color, texture, and lightness to its general surroundings.

The property of an object that renders it difficult to detect by virtue of its similarity to its environment is known as "camouflage." An awareness of the rules of camouflage arose from attempts to apply knowledge from natural history and art to military applications at the beginning of the 20th century. Abbott Thayer is generally regarded as pivotal in the establishment of modern military camouflage research (Thayer, 1896, 1909). Prior to World War I, military uniforms were brightly colored, at least in part so that they might appear more menacing to the enemy (Scott, 1961). However, with the advent of long-range weapons, and guerrilla fighting tactics as employed, for example, against the British Army in the Boer War (1899–1902), the conspicuousness of these uniforms resulted in heavy losses and various armies began to consider the issue of camouflage seriously. Thayer was a painter, and the "Camouflage Units" set up by the French, British, and American armies brought together an "eclectic mix of artists, biologists, and military strategists" (Sherratt et al., 2005), thus being perhaps the earliest, and arguably one of the most important, collaborations between artists and scientists. World War II produced further examples of such collaboration. Of note here

*Corresponding author. Tel.: +44-117-928-8565;
Fax +44-117-928-8588; E-mail: tom.troscianko@bristol.ac.uk

DOI: 10.1016/S0079-6123(06)55004-6

is the contribution of British zoologist and artist Hugh Cott who served as chief instructor to the British Army's Middle East Camouflage School during that war (Cott, 1940).

Crypsis was realized by Darwin and many of his contemporaries to be an important factor in the apparent design, through evolution by natural selection, of biological systems as well. Many animals avoid predation by concealing themselves, and the benefits confirmed by camouflage survival provided some of the earliest and most convincing examples of Darwinism. In most studies of concealment or crypsis, the level of camouflage is measured by how well an animal represents a random sample of the background, in terms of shape and color, at the place and time of the highest predation risk, as Endler (1978, 1984, 1991) has frequently argued. This is termed "background matching" (Cott, 1940; Edmunds, 1974; Ruxton et al., 2004). Endler's theory assumes that all random samples of the background will be equally cryptic. Given the widespread acceptance of background matching it is surprising that, until recently, there have been few empirical tests of the theory that both quantify background matching as perceived by the predator and measure its efficacy in terms of survival value (see Merilaita and Lind, 2005).

Although crypsis provides a powerful method of reducing the probability of detection of a target, it is clear that there are situations in which it cannot be totally effective. The problem is that the environment changes from location to location; a static object designed to mimic the environment in one place may not be as effectively concealed in another place. For example, the left-hand part of Fig. 1 shows pebbles on a beach. The right-hand part of Fig. 1 shows the same image from which a rectangle has been copied and pasted into the center. In terms of crypsis, the rectangle in the right-hand side of Fig. 1 is optimal because it is a perfect copy of part of the scene. However, even in the absence of relative motion, the object becomes visible because of the discontinuities in boundaries thus produced. Of course, relative motion would make the cryptic object even more visible.

From the above, we see that the property of an object that makes it detectable even in conditions of optimal background matching is the presence of extended contours, which give the object its shape. Arguably, this is where "recognition" rather than simple "detection" becomes the correct procedure, since the object is recognized (in this case, as a rectangle) by virtue of its occluding contour. We know that occluding contours are effective drivers of object recognition processes, e.g., from the study by Marr (1982). It is apparent that something more than just crypsis/background matching is required to prevent the object being recognized *as an object of interest*. Thayer (1909) realized that breaking up the contour of the object to be detected plays an important role in preventing the process of recognition of the target as such. One of the key arguments from Thayer's (1909) theories was the concept of distracting or "dazzling" markings (termed "ruptive" by Thayer, but now

Fig. 1. On the left, an image of pebbles on a beach; on the right, a section of pebbles has been copied from the top-left part of the original image and pasted into another location. The location of the pasted image segment is visible (with inspection), in spite of the fact that this stimulus has near-perfect crypsis.

commonly called disruptive markings; Cott, 1940) (Fig. 2).

These markings are generally too small to perceive, until in near view, when their sharp, isolated, and "noncommittal conspicuousness" tend to draw and hold the attention of the receiver, so that the receiver less readily discerns the body of the animal bearing the markings (Thayer, 1909). Many of these markings are apparently found on lepidopteran wings, and break up or mask the contours of the body.

Interestingly, and detrimentally for the credibility of his theories to many at the time, Thayer went to extremes with some of his ideas. For example, he argued that every aspect of a peacock's coloration was used in concealment (Fig. 3), and that the eyespots on the tail blended into the background vegetation, with the smallest and dimmest spots close to the body, growing bigger and brighter toward the end of the tail, leading attention away from the body. This was, and is, in direct contrast to the idea that many apparently conspicuous markings, such as those of the peacock, are used in sexual selection.

Disruptive coloration is used in military camouflage as well. Fig. 4 shows the "dazzle" coloration of HMS Belfast, which prevents its outline resembling that of a large warship, and supposedly made it difficult to assess its range and speed (Berhens, 2002). This camouflage worked so well that it led to accidental collisions between ships of the Royal Navy (Scott, 1961).

Most of the references to disruptive coloration quote Cott's classic 1940 book. This is, in part, because Cott set out a detailed discussion of the theory, with a series of testable hypotheses. Cott (1940), like Thayer (1909), stated that most animals are active and their movements bring them over a range of changing backgrounds, which themselves are rarely uniform in color or pattern. In addition, the light falling on the background and the animals changes in its intensity and composition. Therefore, even with the processes of countershading (Thayer, 1896; Kiltie, 1988; Edmunds and Dewhirst, 1994), a uniformly colored animal presents to the receiver

Fig. 3. An artistic representation of a peacock in its natural habitat, appearing well concealed. (From Thayer, 1909.)

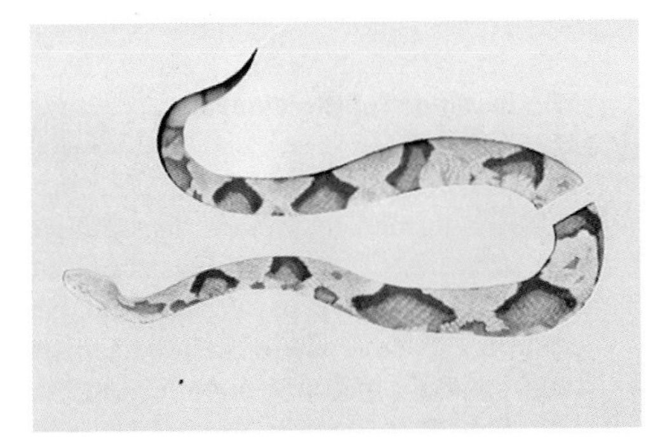

Fig. 2. Artistic drawings of a snake in isolation (left), bearing apparently conspicuous colors, and on the right, placed in the context of the wider environment and looking effectively concealed. (From Thayer, 1909.)

Fig. 4. The "dazzle" coloration of HMS Belfast. Photograph courtesy of the Trustees of the Imperial War Museum.

Fig. 5. A drawing of a snake, showing how the continuous outline of the animal's body reveals its presence and form. (from Cott, 1940.)

a continuous patch or area, which would stand out strongly from the background. It is the outline of the body that enables a receiver to identify the presence of an animal (Fig. 5). Therefore, for concealment to be effective, the perception of the animal's form must be prevented.

The function of disruptive coloration is to prevent or delay the visual recognition of the object. When the surface of an animal is covered with irregular patches of contrasting colors or tones, the patches catch the eye of the receiver and draw their attention away from the outline of the animal that bears them (Cott, 1940). The patches concentrate the attention on themselves, while also passing for part of the general environment. This is coupled to the additional effect of breaking up the continuity of the body outline so that it is difficult to determine the form of the animal possessing the markings.

Differential blending

The first key component of effective disruptive coloration that Cott (1940) identified is the process of differential blending, where the patterns that break up the outline of the body, into a series of unrecognizable units, consist of darker and lighter colors (Fig. 6). Some of these patches blend into the background, while others stand out strongly from it. With this arrangement, some pattern elements will fade into the background while others stand out from it strongly. Provided that the animal is seen against a broken background, it is probably true that any pattern of darker and lighter colors and tones will tend to hinder recognition by destroying the animal's outline, though certain colors, arrangements, and degrees of contrast will be more effective than others. Some of the pattern elements should match the colors found in the general background environment.

Maximum disruptive contrast

The second key component outlined by Cott (1940) is termed maximum disruptive contrast. In this instance, Cott hypothesized that the most effective disruptive patterns would be those that contain elements where the adjacent patterns contrast strongly in tone, since it is important to realize the role played by tone as well as that played by color. The strongly contrasting tones of adjacent patches cause the receiver to distinguish the

Fig. 6. An illustration of the process of differential blending (from Cott, 1940). The second image from the left has an outline that is less obvious than the image on the far left, but is still easy to determine. However, once one of the pattern elements begins to blend into the background, it becomes far harder to determine the animal's outline, as shown by the two moths on the right of the figure.

two patches as separate objects. Generally, light markings on an otherwise dark object, and dark markings on an otherwise light object will be most effective in creating a disruptive effect, provided that the patterns broadly conform to the background environment (Fig. 7). Cott (1940) also argued that distractive markings need to be used in moderation in relation to the whole area of the body, or else they will fail in their effect. Overall, maximally contrasting pattern elements' essential function is to break up the continuity of the surface by means of strongly contrasting tones.

Fig. 7. On the left, an Oleander hawk moth (*Daphnis nerii*) (top) showing apparent differential blending. Image courtesy of David L. Mohn, copyright Light Creations 1993–2005. On the right, a lime hawk moth (*Mimas tiliae*) (bottom) showing maximum disruptive contrast. Image courtesy of Mike J. Rubin, copyright M. J. Rubin.

Coincident disruptive coloration

A third component proposed by Cott (1940) is termed coincident disruptive coloration. This describes the continuous patterns that range over the appendages and the main body, masking the telltale presence of the otherwise conspicuous appendages. This technique could join separate parts of the body, and unite what are actually discontinuous surfaces (Fig. 8). Cott (1940) argued that coincident markings are particularly common in insects, such as moths, where the patterns span the forewings, hind wings, and the body of the individual when at rest.

Differences between background matching and disruptive coloration

Merilaita (1998) used Cott's (1940) theorem to compose a set of testable predictions about the

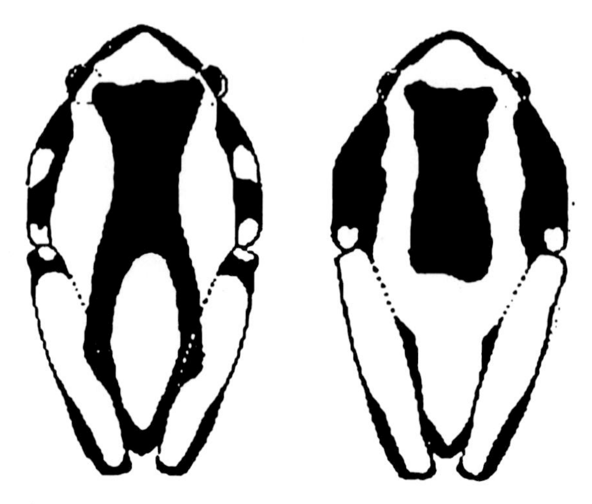

Fig. 8. An illustration of coincident disruptive coloration in the tree frog *Hyla leucophyllata* (from Cott, 1940).

distribution, the geometry, and the colors of patterns in disruptive coloration:

1. The elements of background matching patterns should be distributed in a manner that matches the distribution of the pattern elements as found in the background environment. In comparison, a higher number of markings would be expected to touch the outline of the body if forming part of a disruptive effect.
2. The size and shape of pattern elements should be similar to those found in the background environment when forming part of a background matching pattern. In contrast, disruptive elements should be complex and highly variable in size to give the impression of separate objects.
3. The colors of background matching patterns should be similar to those found in the background environment, whereas in disruptive patterns, the elements should also have patterns that are highly contrasting.

Evidence for and against the disruptive coloration hypothesis

Cott described disruptive coloration as "certainly the most important set of principles relating to concealment", yet in truth, experimental support has been scarce since its proposal.

Invertebrates

The neotropical nymphalid butterfly *Anartia fatima* exhibits strong polymorphism in the wing banding color on the dorsal surface of males (Emmel, 1973). According to Emmel (1973), mate selection experiments have indicated that both males with the white and males with the yellow bands prefer females with a white band in a ratio of about 2:1, and so the elimination of the yellow band in females must be prevented by some other counterbalancing selective force. That force could be disruptive coloration.

Silberglied et al., (1980) conducted a field study and eliminated the apparently disruptive wing stripe in some palatable *Anartia fatima* butterflies found in central America and then compared their survival to unaltered individuals in the field. Since the study was conducted on an island, it was possible to capture and mark the majority of the population of butterflies, with high recapture rate, often several times per individual. It was therefore possible to determine the minimum age of every butterfly at each capture and a minimum longevity for each individual in the population. Butterflies were also measured for general condition and a score of wing damage. They found that butterflies without the wing stripe survived equally well as those with the stripe over a 21-week period of testing, and the condition of individuals in the two groups did not differ. However, the vertical wing stripe was obliterated by applying black felt-tip marking pens — these could have influenced the palatability of modified individuals, so that even though they no longer had a vertical wing stripe, they may have had reduced palatability. It should also be noted that there is no evidence that the wing stripes are disruptive in function, and so it is quite possible that they have another unrelated function. A further problem is that the methods may have unintentionally made the butterflies more similar to a cooccurring unpalatable species (Waldbauer and Sternburg, 1983). Additionally, Endler (1984) suggested that obliterating the white stripe simply converted the pattern from one background matching type to another background matching type of equal crypsis.

In other butterfly species, the presence of wing stripes has also been assumed to be disruptive in function (e.g. *Limenitis arthemus*; Platt and Brower, 1968; Platt, 1975), adding to the concealment of the individuals by breaking up the outline of the butterfly, especially in flight, but again, with no evidence as support for their claims, especially given that the stripe only touches the body outline at two locations.

Merilaita (1998) appears to have undertaken the only study that directly tests for the presence of disruptive coloration in an animal, as opposed to simple background matching. Merilaita (1998) found that spots mimicking the background were more likely to be found at the edge of the body than would be expected by chance in the white-spotted phenotype *albafusca* of the polymorphic isopod *Idotea baltica*. The isopods appear to be cryptic on the brown alga *Fucus vesiculosus* with its white-colored epizooties *Electra crustulenta* and *Balanus imporvisus*. Crypsis via the process of background matching would require that the shapes and sizes of the pattern elements should closely match those of the background, whereas disruptive coloration would involve more marginal elements than expected by the pattern element distribution in the background. Merilaita (1998) recorded for each individual the total number of spots, the number of marginal spots, plus the size and shape of the spots. An index of roundness was also created $I_R = (P^2/4\pi A)-1$, where P is the perimeter and A the area of the spot. When the index equals zero the spot is a perfect circle, and the more complex the shape is and the longer the perimeter in relation to the area, the larger are the values produced. Merilaita (1998) used a random distribution as a null hypothesis for the distribution of the spots on the isopod, which is debatable since the distribution of the spots in the background environment may not have been random, but rather could be clumped. Additionally, the isopod markings may have an additional function, such as providing structural support.

Merilaita (1998) found that spots on the isopods were significantly smaller than *E. crustulenta* colonies and *B. improvisus* individuals, plus the shape of the spots on the males and females was significantly more complex than the shape of the *E. crustulenta* colonies and *B. improvisus* individuals. The spots on the isopods were also on average more variable than *E. crustulenta* colonies and *B. improvisus* individuals in terms of shape and size. Importantly, the number of marginal spots found on the isopods was significantly more than would be expected by chance. Therefore, evidence indicates that *I. baltica* achieves camouflage based on disruptive coloration, though background matching may also be important.

Disruptive coloration has also been proposed in the gyrodactylid monogenean parasite *Macrogyrodactylus polypteri*, of the African freshwater fish *Polypterus senegalus* (Cable et al., 1997). The black-banded gut of the parasite is due to melanin derived from the host epithelial cells in the diet and may be disruptive in making the individual more difficult to detect against the pigmented scales of the host (Cable et al., 1997). Otherwise, the gut of the parasite is colorless and pigmentation of the gut is lost after 5 days without feeding (Cable et al., 1997). The most likely explanation for the banding, suggested by Cable et al., (1997), could be in disruptive concealment from visually hunting predators, though it should be noted that selection pressure on the visual appearance of the parasites in the murky waters may be limited.

Disruptive coloration has been hypothesized in some species of snails, such as the dark stripe down the mid-dorsal part of the body of *Clavator moreleti* (Emberton, 1994). Again, evidence for a role of disruptive coloration was limited, with little correlation between shell disruptive coloration and aestivation site, with exposure rate to visual predators not investigated (Emberton, 1994). Again, there is no evidence that the stripes are used in concealment, and that they are disruptive, rather than a form of background matching.

Cephalopods have on various occasions been described as possessing disruptive patterns at times (since they are often capable of rapid adaptive color changes by neurally controlled chromatophores). For instance, the squid *Loligo pealei* is capable of producing a "banded bottom sitting" pattern that apparently produces disruptive coloration against the substrate (Hanlon et al., 1999).

The pattern is commonly produced and consists of bands with dorsal iridophore "splotches," and serves to break up the longitudinal shape of the squid (Hanlon et al., 1999). Hanlon and Messenger (1988) proposed that disruptive effects were one of the main types of patterns observed in cuttlefish and are characterized by bold transverse or longitudinal components that can be expressed at different strengths according to the level of contrast between the dark and light patches. Hanlon and Messenger (1988) stated that the principles of disruptive coloration are excellently demonstrated in young cuttlefish, in breaking up the outline of the body. These markings, especially prevalent when the substrate particle size is large, obliterate the appearance of a body well, making concealment highly effective to the human eye (Hanlon and Messenger, 1988). According to Hanlon and Messenger (1988), Cott's (1940) principle of differential blending is achieved well in the cuttlefish *Sepia* by some chromatic components blending with the substrate, while others contrast sharply with it, allowing some body parts to stand out strongly, while others fade into the background. Maximum disruptive contrast can also occur when the adjacent components have a high contrast in terms of tone, and there exists a set of patterns of irregular shapes and sizes that are highly contrasting (Hanlon and Messenger, 1988). Even the process of coincident disruptive coloration appears to be present in *Sepia*, with the eyes concealed with a dark anterior head bar, and the mantle and head can be joined optically by other bars and bands, plus a general color resemblance (Hanlon and Messenger, 1988). Hanlon and Messenger's study is important since it is a rare example of a detailed consideration of Cott's (1940) hypotheses, with explanations as to why the authors believe the subjects to be demonstrating disruptive coloration and not just background matching. Chiao and Hanlon (2001) tested *Sepia* for various forms of camouflage on various sizes of black and white checkerboard squares as substrate. They argued that the results they found supported Edmunds' (1974) definition of crypsis, in that the animals resembled specific parts of their background, and also that the disruptive patterns fitted into Endler's (1978) definition that an animal is cryptic if it resembles a random sample of the background. Chiao and Hanlon (2001) stated that achievement of a disruptive effect is brought about by making chromatic components of their body appear as random samples of the background. Unfortunately this illustrates that Chiao and Hanlon have misunderstood the key principles about disruptive coloration: that disruptive patterns *do not* represent random samples of the background, but employ a series of adjacent highly contrasting shapes, which are more likely to be found at the margins of the body than would be expected if the animal were actually matching the background. Disruptive coloration and background matching are two logically distinct methods of concealment.

A more recent study in *Sepia officinalis* by Chiao et al., (2005) extended previous research by investigating camouflage of the cuttlefish on both natural substrate of rocks and laminated images of the substrate. As with the previous studies, the authors failed to show that the markings of the cuttlefish were definitively disruptive as opposed to background matching, but the general findings of the study are interesting for a variety of reasons. Firstly, the animals displayed the disruptive pattern on both the true three-dimensional substrate but also on the laminated images, showing that visual cues are the main (or at least one of the main) cues for producing the appropriate color pattern. Perhaps more interesting, however, was the experiment in which Chiao et al., (2005) manipulated the digital images by removing certain spatial frequencies (such as low-pass and high-pass filterings) to remove certain information about the background, such as edges between objects. Results showed that the cuttlefish did not show disruptive patterns on the high-pass, low-pass, or contrast-enhanced images, leading the authors to conclude that cuttlefish requires both edges and local contrast to recognize the objects. However, it should not be overlooked that the images may have looked strange to the subjects, and that the sample size here was only two.

Fish

Disruptive coloration has also been documented in fish. For instance, the southern mouth brooder *Pseudocrenilabrus philander*, a type of cichlid, is thought to have disruptive coloration (Holden and Bruton, 1994). Also, many unrelated North American benthic freshwater fishes have a color pattern consisting of four dark dorsal saddles that continue down the sides of the body as bars (Armbruster and Page, 1996). These bars are usually black, brown, or dark gray, with the spaces between much lighter in color (Armbruster and Page, 1996). It is unlikely that the bars function as warning devices since the fishes are equipped with minimal mechanisms, and so the most likely explanation for these bars is that of a concealing function (Armbruster and Page, 1996). It is possible that the bars break up the body of the animal into a series of units that appear to be separate objects, provided that the background is not uniform in composition as this would make the bearer more conspicuous (Armbruster and Page, 1996). Samples of substrates on which fishes were found were scored for their apparent complexity (Armbruster and Page, 1996), though this relied on human subjective assessment. Armbruster and Page (1996) found that the fishes that had saddle patterns were always found on gravel (uneven) substrates, and so proposed that the pattern was disruptive in that the dark bars mimicked the shadows found between rocks and the lighter intermediate regions mimicked the rocks themselves. It is also apparent that pelagic fishes may develop saddles or bars when they become benthic (Armbruster and Page, 1996). Therefore, the fishes are apparently mimicking a set of rocks of variable sizes.

Birds

Götmark and Hohlfält (1995) argued that it is unwise to assume that bright plumage in male passerine birds makes them easier to detect, increasing the risk of predation, while in the females, duller plumage is used in camouflage. These are assumptions that have not been tested in the wild. Götmark and Hohlfält (1995) used humans to assess the conspicuousness of pied flycatchers *Ficedula hypoleuca* and chaffinches *Fringilla coelebs* by placing pairs (male and female) of mounted specimens against various forms of vegetation. Götmark and Hohlfält (1995) found that in the chaffinches, males were detected more easily in the trees, though both sexes were detected equally on the ground. They also found that male and female pied flycatchers did not differ in detectability in the trees whereas, on more homogeneous backgrounds on the ground, the male pied flycatchers were more easily detected than the females. Götmark and Hohlfält (1995) argued that this is evidence for the presence of disruptive coloration in male pied flycatchers since the leaves and twigs of trees formed a contrasting mosaic of dark and light backgrounds, which made males difficult to detect. However, the only evidence that the markings on the male flycatchers were disruptive comes from qualitative assessment, and so, it is not clear that the flycatcher males' markings are not simply background matching. Furthermore, Götmark and Hohlfält (1995) acknowledged that there are problems in using humans as "predators" but argued that their study accounted for the fact that many birds can perceive ultraviolet light (reviewed by Cuthill et al., 2000), since they took measurements of the UV reflectance in both sexes, which showed no serious bias. However, it is not clearly known as to what level of bias in the amount of UV reflectance would be discernable by birds, particularly predatory species. Furthermore, birds do not just vary in the ability to perceive UV or violet light, but also have four cone types, as opposed to three in humans, and so are likely to be able to perceive a greater range of hues than humans can (Burkhardt, 1989; Bennett et al., 1994). Additionally, the spectral sensitivities of the long-wave, medium-wave, and short-wave cones in birds and humans are not the same, particularly as birds have oil droplets that selectively absorb wavelengths of light before they reach the photopigment, adding to the overall differences in human and avian visual perception (Vorobyev et al., 1998; Vorobyev, 2003). The statement that the branches and leaves formed a mosaic of light and dark patches, making detection difficult, may also imply a case of background matching instead of disruptive coloration (Ruxton et al., 2004).

Mammals

The adaptive significance of coloration in ungulates has only recently been addressed systematically, yet it seems that background matching, disruptive coloration and countershading may be found in the artiodactyls (Stoner et al., 2003). Stoner et al., (2003) used information on the overall body coloration, seasonal coloration, coloration of face and body parts, and combined the data with ecological variation to determine the possible function of coloration patterns in various species of artiodactyls. They postulated that side bands may function in disruptive coloration by breaking up the shape of the body and may be particularly useful in species that are diurnal and found in open, sunny environments where other forms of concealment are difficult. They found that artiodactyls that had prominent side bands were largely diurnal and were found in open environments. However, these results were not significant after controlling for phylogeny (the problem that closely related species share features through common ancestry as well as, or instead of, common adaptation, and so are statistically nonindependent; statistical methods have been developed to deal with this; e.g., Harvey and Pagel, 1991; Harvey and Nee, 1997). Stoner et al., (2003) did point out that their ecological classifications were very broad, and that variations within each habitat might also be broad, plus detailed data for each species was lacking. Again, the analysis bases the possible occurrence of disruptive coloration on qualitative accounts, rather than on experimental evidence of a disruptive role.

Snakes

Camin and Ehrlich (1958) provided evidence that selection pressures in the small, but not totally isolated, populations of the water snake *Natrix sipedon* on Lake Erie islands had allowed the presence of unbanded individuals to become present in the population. They argued that the unbanded individuals appeared to be better camouflaged than their banded counterparts when they are found on the limestone rocks of the islands peripheries. This is interesting since the banding has been argued as being an important component of a disruptive coloration pattern (Beatson, 1976). This could outline the possibility that under some circumstances, the markings that act disruptively to conceal an animal may become conspicuous. Beatson (1976) found in a different population of *Natrix sipedon* that snakes of all ages generally exhibited disruptive coloration as juveniles *but* not as adults. The disruptive patterns could fade with age, to be replaced by a uniform ground matching color. Generally, it appears as if there is a greater importance in highly contrasting patterns at a young age, and that differential predation in adults was not influenced greatly by the extent of disruptive patterns, but instead by the overall hue imparting resemblance to the background (Beatson, 1976). By contrast, juveniles have a greater range of predators, and different predator groups may exert different degrees of predation pressure at different life stages, which may be accompanied by a change in the visual perception of the main predators at different stages. This is a very interesting theory but, again, one in which a clear distinction between disruptive coloration and background pattern matching needs to be tested.

Other tests of disruptive coloration

One recent study has aimed to determine if there is an advantage of disruptive coloration and different forms of background matching in laboratory trials with artificial prey and great tits *Parus major* (Merilaita and Lind, 2005). Merilaita and Lind (2005) tested Endler's (1978) theory of background matching and the theory of disruptive coloration. They showed that different random samples of the artificial background (a foraging board covered with different shapes) did not give rise to equally good crypsis. They also showed that disruptive coloration would allow for at least equally good crypsis as the best forms of background matching (but not better).

The experiment consisted of six different prey types: two types consisted of background matching samples that were subjectively assumed to be relatively easy to detect, two background matching types assumed to be relatively difficult to detect, and two disruptively marked prey types. In half of the treatments (one type of each treatment pair) the background elements were often cut off by the outline of the body, whereas in the other half of the treatment types, the background shapes were modified so that they were intact and nonoverlapping. One difference in the design of the prey types in Merilaita and Lind's experiment compared to our own (Cuthill et al., 2005) is that Merilaita and Lind designed their disruptive treatments so that the markings on them would act as "distracters" and draw the attention of the predators away from the body toward the shapes, and so markings were placed at angles to the true outline of the "body." Our design was subtly different in that our disruptive treatments had markings, which intersected the body outline disrupting the shape and appearance of the body, potentially masking the presence of any relevant object indicating a group of unimportant unrelated objects. Merilaita and Lind measured the search time to find each item presented in trials and used this as an inverse estimate of the probability of detection.

Effective search time varied significantly between the different prey types representing different random samples of the background, with the prey categorized as easier to find located in shorter times (true for both treatment types). This finding disagrees with Endler's hypothesis. No difference was found between the difficult to detect random sampled prey and the disruptive prey. Merilaita and Lind argued that this provides empirical support for the idea of disruptive coloration. A cautious interpretation, however, is that disruptive coloration in this instance was not more effective in concealment than background matching, but only equally effective as the more difficult background matching forms. A critic could argue that given the subjective method of classifying "difficult" background matching items and the simplicity of the artificial background, the disruptive patterns themselves were also a form of difficult-to-detect background matching patterns and vice versa. Indeed, it has been suggested that the boundary between background matching and disruptive coloration is not always distinct, and that there may be times when the two strategies coexist (e.g. Endler, 1984; Merilaita, 1998). As Merilaita and Lind acknowledged, it is impossible to determine from their experiment to what extent the probability of detection of the disruptive prey came from the principle of disruptive coloration.

Endler (1984) argued that disruptive coloration is merely a special case of crypsis, the definition of crypsis being a random sample of the background; the edges of the random sample of the background are likely to cut across patches of variously contrasting colors. Endler (1984) went on to say that disruptive coloration may work with conspicuous colors, but no studies of disruptive coloration have attempted to control for the background against which the prey is seen by predators. It does seem, however, that Endler (1984) is not fully right here — the main point about disruptive coloration is that it is distinct from background matching, since simply being a random sample of the background would still leave the conspicuous presence of the body outline; therefore, shapes and markings should be found at the body margins at a higher frequency than would be expected by just matching the background.

Disruptive coloration has been suggested in such a wide range of animals without experimental confirmation that experiments along these lines are in great need (Sherratt et al., 2005). Also, as Merilaita and Lind (2005) argued, no explicit mechanistic basis related to visual information processing has been suggested for disruptive coloration — this would help to clarify exactly how disruptive coloration works over and above other methods of concealment. Another interesting idea from Merilaita and Lind is that optimization of disruptive coloration should be related to the shape of the prey, whereas background matching should occur independently of the body shape. Therefore, two species with very different body shapes using the same habitat may be expected to evolve different concealing coloration if they are using disruptive strategies, but may have similar patterns if they rely on background matching. There is also the theoretical question of whether disruptive coloration enables animals to exploit a greater range of habitats (Thayer, 1909; Sherratt et al., 2005), and if disruptive coloration is more common in more active animals (Thayer, 1909). Furthermore, how does the size of the pattern elements affect the effectiveness of disruptive coloration? Small markings offer little advantage, while markings that are too big may highlight the outline of the animal's body.

Many studies label a variety of animals as exhibiting disruptive coloration, with no experimental support to substantiate their claims. Overall, Silberglied et al., (1980) may have made one of the most prudent statements in saying that: "Few concepts in the theory of adaptive coloration are as well accepted, but as poorly documented, as the theory of disruptive coloration." This is in contrast to the all too common statements of others, such as "one powerful and widespread technique used by animals for crypsis is patterning, especially the various forms of disruptive patterns…" (Marshall and Messenger 1996). Until more *experimental* support for this theory is found, statements like this should be avoided.

Our experiments with disruptive coloration and background pattern matching (Cuthill et al., 2005)

We tested two predictions pinpointed by Merilaita (1998) arising from the studies by Thayer (1909) and Cott (1940), which are as follows:

(i) patterns on the body's edge should render the animal less visible than equivalent patterns placed in other locations;
(ii) highly contrasting colors should also render the animal less visible as an animal than those of low contrast, *even though the pattern is more visible in the high-contrast condition.*

We developed artificial targets in which the wings were printed patterns on paper and the edible bodies consisted of dead mealworms. These were not designed to resemble any specific moth; rather, the triangular shapes of the bodies were designed to bear some generic similarity to moths, but no attempt was made to optimize or assess this. Thus, we can see these experiments as telling us about the detectability of certain kinds of object in the field as these objects are manipulated. This is thus an example of "field psychophysics" in which the dependent variable is the uptake rate of the edible "prey" and the independent variable the design of the printed targets.

In Experiment 1, we used brown targets with markings similar to those found on mature oak trees (from photographs of these). The five types of stimulus were as follows: (1) "Edge" in which the markings are coincident with the edge of the figure; (2) "Inside 1" in which the same markings are brought inward so that they do not overlap the edges; (3) "Inside 2" in which other randomly selected markings are placed inside the figure; (4) monochrome black; and (5) monochrome brown targets. Examples are shown in the top row of Fig. 9.

The first three moth types all contained similar elements derived from photographs of tree bark

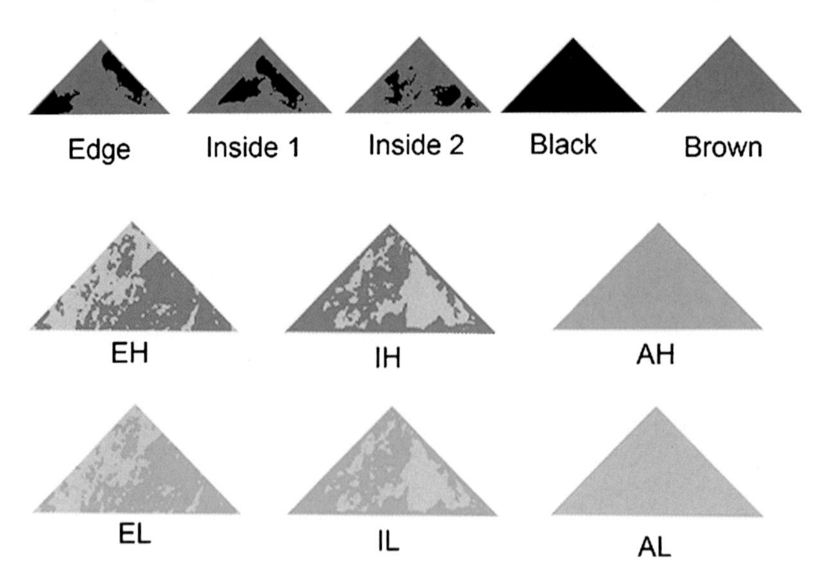

Fig. 9. Upper row: the five treatments of Cuthill et al., (2005) Experiment 1. Bottom two rows: the five treatments of Cuthill et al., (2005) Experiment 2. Adapted from Nature 434: 72–74. Nature Publishing Group.

and were thus expected to be equally cryptic in terms of background pattern matching (and less visible than plain Brown or Black). On the other hand, the theory of disruptive coloration predicts that the "Edge" moths will survive better than the others.

In Experiment 2, we varied the contrast of the patterns in the "Edge" and "Inside 2" conditions from Experiment 1, with a high-contrast and a low-contrast version of each. In addition, there were two monochrome conditions, which were the spatial average of the high- and low-contrast conditions, respectively. The theory of disruptive coloration predicts that the high-contrast targets will survive best in the "Edge" condition, with less survival in the low-contrast condition and no benefits from blobs in the "Inside 2" condition.

Each target consisted of a paper triangle and an edible body. The edible body consisted of dead (frozen overnight at $-80°C$, then thawed) mealworm (*Tenebrio molitor* larvae) pinned onto the 50 mm × 25 mm colored paper triangle. These targets were pinned onto oak trees in mixed deciduous woodland and their "survival" checked at ca. 2, 4, 6, and 24 h. Fig. 10 shows examples of the "moths" on their tree bark.

When the worm was eaten by a bird, most or all of it was removed. Sometimes, other predators took the worms: spiders sucked fluids out leaving a hollow exoskeleton, and slugs left slime trails; when we judged that nonbird predation had taken place, when the target disappeared completely, or when it survived intact for 24 h, the datum was treated as a "censored" value in survival analysis (Klein and Moeschberger, 2003). Both experiments had randomized block designs with 10 replicate blocks run in different areas of the wood on different dates between October 2003 and March 2004. Each block had 75 (Experiment 1; 15 per treatment) or 84 (Experiment 2; 14 per treatment) targets in a nonlinear transect of ca. 1.5 km × 20 m (targets on $<5\%$ of available trees in transect). Treatments were randomly allocated to trees, subject to the constraints that no lichen covered the trunk since lichen-free oak bark reflects negligible UV (Majerus et al., 2000; Cuthill et al., 2006, in press), and no young trees of trunk circumference less than 0.9 m were used. Color matches of treatments to natural bark were verified using spectrophotometry of stimuli and bark, followed by modeling of predicted photon catches (Maddocks et al., 2001) of a typical passerine bird, the blue tit's (*Parus caeruleus*) single cone photoreceptors (Hart et al., 2000) using irradiance spectra from overcast skies in the study site. Our acceptance criterion was simply that cone captures for the experimental stimuli fell within the measured range of those for oak bark.

Even a properly calibrated RGB image does not precisely simulate the avian-perceived color of many natural objects, due to differences in the spectral sensitivity of bird versus human, long-wave, medium-wave, and short-wave cones (Cuthill et al., 2000). However, as our treatments varied only in relative color contrast, any error associated with this method was considered minor; an assumption verified retrospectively using spectrophotometry and color-space modeling. We chose color pairs from the eight most frequent RGB triplets in the bark photos as follows: a "background" color, then a triplet that was similar

Fig. 10. Examples of types of moth used in Cuthill et al., (2005) experiments.

to the background (low contrast treatment), and one that differed markedly (high contrast). The major difference between colors was in overall brightness and not hue, but we could not systematically vary only one color dimension within the available common bark colors. Sample numbers of background and contrasting colors were balanced for which was darker/lighter, and so there were no significant differences between bicolored treatments in the brightest or darkest color or average color (ANOVAs on RGB sums and all possible ratios; $p > 0.9$). Monochrome treatments were also created as the means of the respective R, G, and B values of the two colors in bicolored high- and low-contrast treatments. Different color pairs and patterns, from different trees, were used for each replicate target.

Survival analysis was performed by Cox's regression (Cox, 1972) with the factors treatment and block. Cox's regression assumes that all survival functions have the same shape; this proportional hazards assumption was checked by plotting partial residuals against ranked survival times. There were significant block effects in both experiments (Experiment 1: Wald = 121.78, d.f. = 9, $p < 0.001$; Experiment 2: Wald = 271.50, d.f. = 9, $p < 0.001$), reflecting differences in average predation rates in different parts of the woods on different dates, but this was not relevant to our hypotheses.

Results

Experiment 1

The prediction that the "edge" condition would lead to longer survival than other treatments was fulfilled (Fig. 11).

We included the treatment "Inside 2" because of the possibility that moving the pattern elements present in treatment Edge from the periphery of the "wings" to form treatment "Inside 1" created pattern elements with straight lines that themselves could have enhanced conspicuousness. This indeed seemed to be the case, as treatment Inside 2 survived better than Inside 1, which lacked these straight edges to the pattern elements (Fig. 11). The inward displacement of pattern elements in Inside 1 also tended to enhance the outline of these targets, hence having the opposite effect to disruptive coloration. Nevertheless, all bicolored treatments survived better than monochrome Black or Brown, indicating that background pattern matching was, as expected, itself effective as camouflage.

Experiment 2 had six treatments: the 2×2 combination of bicoloured patterns with high or low contrast, placed as in Experiment 1 treatment "Edge" or "Inside 2," plus two monochrome treatments that were the average color of either the high- or the low-contrast color pairs. As uniquely predicted by the theory of disruptive coloration,

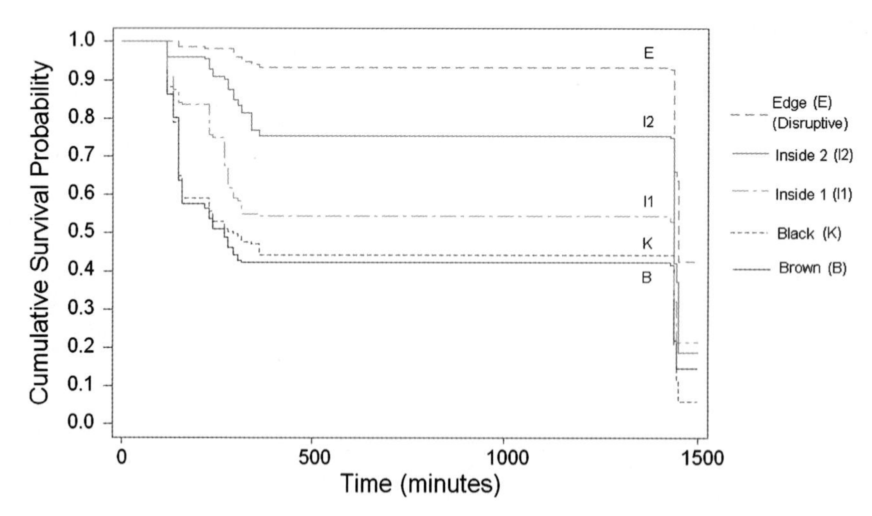

Fig. 11. Results of Cuthill et al., (2005) Experiment 1. Adapted from Nature 434: 72–74. Nature Publishing Group.

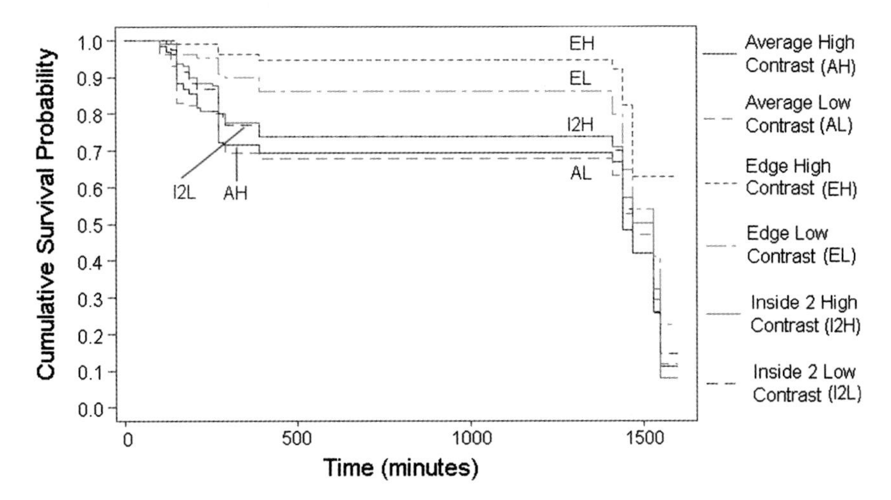

Fig. 12. Results of Cuthill et al., (2005) Experiment 2. Adapted from Nature 434: 72–74. Nature Publishing Group.

the high-contrast-edge treatment survived best (Fig. 12), with high contrast providing minimal benefit in nondisruptive "Inside" treatments.

Discussion

Taken together, our results provide the strongest support to date for the effectiveness of disruptive patterns against birds, the most commonly invoked visual predators shaping the evolution of protective coloration in insects. The extent to which disruptive patterns provide a general advantage over simple crypsis, with different background types (e.g., varying spatial and/or chromatic complexity) or different light environments (e.g., direct or diffuse lighting), awaits further experimentation. However, we have shown that the addition of high-visibility pattern information reduces the chances of predation of moth-like targets by birds. The results are not explicable on the basis of crypsis, which does not distinguish between the "Edge" and the "Inside" treatments in our experiments.

Many questions remain. We have not explored to what extent one can use disruptive coloration to render crypsis redundant — if the moths were bright red, for example, but with disruptive markings, how much predation would there be? Another major unknown with this method is that we cannot know at what distance the birds make their decision, and therefore a modeling of the results while, say, accounting for the visual acuity of a bird, is made difficult. Such issues could be addressed by a combination of further field study and laboratory experiments.

However, the results presented here are promising for two reasons. First, we present a powerful but simple technique for carrying our field psychophysics in natural conditions (including natural illumination), second, the results provide strong evidence of what was always assumed in textbooks — that disruptive coloration is a powerful concealment strategy.

References

Armbruster, J.W. and Page, L.M. (1996) Convergence of a cryptic saddle pattern in benthic freshwater fishes. Environ. Biol. Fish., 45: 249–257.

Beatson, R.R. (1976) Environmental and genetical correlates of disruptive coloration in the water snake, *Natrix s. sipedon*. Evolution, 30: 241–252.

Bennett, A.T.D., Cuthill, I.C. and Norris, K.J. (1994) Sexual selection and the mismeasure of color. Am. Nat., 144: 848–860.

Berhens, R.R. (2002) False Colors: Art, Design and Modern Camouflage. Bobolink Books Dysart, Iowa.

Burkhardt, D. (1989) UV vision: a bird's eye view of feathers. J. Comp. Physiol. A, 164: 787–796.

Cable, J., Harris, P.D. and Tinsley, R.C. (1997) Melanin deposition in the gut of the monogenean *Macrogyrodactylus polypteri* Malmberg 1957. Int. J. Parasitol., 27: 1323–1331.

Camin, J.H. and Ehrlich, P.R. (1958) Natural selection in water snakes (*Natrix sipedon* L.) on islands in Lake Erie. Evolution, 12: 504–511.

Chiao, C.-C. and Hanlon, R.T. (2001) Cuttlefish camouflage: visual perception of size, contrast and number of white squares on artificial checkerboard substrata initiates disruptive colouration. J. Exp. Biol., 204: 2119–2125.

Chiao, C., Kelman, E.J. and Hanlon, R.T. (2005) Disruptive body patterning of cuttlefish (*Sepia officinalis*) requires visual information regarding edges and contrast of objects in natural substrate backgrounds. Biol. Bull., 208: 7–11.

Cott, H.B. (1940) Adaptive Colouration in Animals. Methuen & Co. Ltd., London.

Cox, D.R. (1972) Regression models and life-tables. J. Roy. Stat. Soc. B, 34: 187–220.

Cuthill, I.C., Hiby, E. and Lloyd, E. (2006) The predation costs of symmetrical cryptic Coloration. Proc. Roy. Soc. B,, 273: 1267–1271.

Cuthill, I.C., Partridge, J.C., Bennett, A.T.D., Church, S.C., Hart, N.S. and Hunt, S. (2000) Ultraviolet vision in birds. Adv. Stud. Behav., 29: 159–214.

Cuthill, I.C., Stevens, S., Sheppard, J., Maddocks, T., Párraga, C.A. and Troscianko, T.S. (2005) Disruptive coloration and background pattern matching. Nature, 434: 72–74.

Edmunds, M. (1974) Defence in Animals: A Survey of Anti-Predator Defences. Longman Group Limited, Harlow, Essex.

Edmunds, M. and Dewhirst, R.A. (1994) The survival value of countershading with wild birds as predators. Biol. J. Linn. Soc., 51: 447–452.

Emberton, K.C. (1994) Morphology and aestivation behaviour in some Madagascan land snails. Biol. J. Linn. Soc., 53: 175–187.

Emmel, T.C. (1973) On the nature of the polymorphism and mate selection phenomena in *Anartia fatima* (Lepidoptera: Nymphalidae). Evolution, 27: 164–165.

Endler, J.A. (1978) A predator's view of animal color patterns. Evol. Biol., 11: 319–364.

Endler, J.A. (1984) Progressive background matching in moths, and a quantitative measure of crypsis. Biol. J. Linn. Soc., 22: 187–231.

Endler, J.A. (1991) Interactions between predators and prey. In: Krebs, J.R. and Davis, N.B. (Eds.), Behavioural Ecology: an Evolutionary Approach (3rd Edition). Blackwell, Oxford, pp. 169–196.

Götmark, F. and Hohlfält, A. (1995) Bright male plumages and predation risk in passerine birds: are males easier to detect than females? Oikos, 74: 475–484.

Hanlon, R.T. and Messenger, J.B. (1988) Adaptive coloration in young cuttlefish (*Sepia officinalis L.*): the morphology and development of body patterns and their relation to behavior. Philos. T. Roy. Soc. B, 320: 437–487.

Hanlon, R.T., Maxwell, M.R., Shashar, N., Loew, E.R. and Boyle, K.-L. (1999) An ethogram of body patterning behavior in the biomedically and commercially valuable squid *Loligo pealei* off Cape Cod, Massachusetts. Biol. Bull., 197: 49–62.

Hart, N.S., Partridge, J.C., Cuthill, I.C. and Bennett, A.T.D. (2000) Visual pigments, oil droplets, ocular media and cone photoreceptor distribution in two species of the passerine: the blue tit (*Parus caeruleus* L.) and the blackbird (*Turdus merula* L.). J. Comp. Physiol. A, 186: 357–387.

Harvey, P.H. and Pagel, M.D. (1991) The Comparative Method in Evolutionary Biology. Oxford University Press, Oxford.

Harvey, P.H. and Nee, S. (1997) The phylogenetic foundations of behavioural ecology. In: Krebs, J.R. and Davies, N.B. (Eds.), Behavioural Ecology. An Evolutionary Approach (4th edition). Blackwell, Oxford, pp. 334–349.

Holden, K.K. and Bruton, M.N. (1994) The early ontogeny of the southern mouthbrooder, *Pseudocrenilabrus philander* (Pisces, Cichlidae). Environ. Biol. Fish., 41: 311–329.

Klein, J.P. and Moeschberger, M.L. (2003) Survival Analysis: Techniques for Censored and Truncated Data. Springer, New York.

Kiltie, R.A. (1988) Countershading: universally deceptive or deceptively universal? TREE, 3: 21–23.

Majerus, M.E.N., Brunton, C.F.A. and Stalker, J. (2000) A bird's eye view of the peppered moth. J. Evol. Biol., 13: 155–159.

Marshall, N.J. and Messenger, J.B. (1996) Colour-blind camouflage. Nature, 382: 408–409.

Marr, D. (1982) Vision: A Computational Investigation into the Human Representation and Processing of Visual Information. W.H. Freeman, New York.

Merilaita, S. (1998) Crypsis through disruptive coloration in an isopod. Proc. Roy. Soc. B, 265: 1059–1064.

Merilaita, S. and Lind, J. (2005) Background-matching and disruptive coloration, and the evolution of cryptic coloration. Proc. Roy. Soc. B, 272: 665–670.

Platt, A.P. and Brower, L.P. (1968) Mimetic versus disruptive coloration in intergrading populations of *Limenitis arthemis* and *Astyanax* butterflies. Evolution, 22: 699–718.

Platt, A.P. (1975) Monomorphic mimicry in nearctic *Limenitis* butterflies: experimental hybridization of the *L. arthemis-astyanax* complex with *L. archippus*. Evolution, 29: 120–141.

Ruxton, G.D., Sherratt, T.M. and Speed, M.P. (2004) Avoiding Attack: The Evolutionary Ecology of Crypsis, Warning Signals & Mimicry. Oxford University Press, Oxford.

Scott, P. (1961) Eye of the Wind. Hodder & Stoughton, London.

Sherratt, T.N., Rashed, A. and Beatty, C. (2005) Hiding in plain sight. TREE, 20: 414–416.

Silberglied, R.E., Aiello, A. and Windsor, D.M. (1980) Disruptive colouration in butterflies: lack of support in *Anartia fatima*. Science, 209: 617–619.

Stoner, C.J., Caro, T.M. and Graham, C.M. (2003) Ecological and behavioural correlates of coloration in artiodactyls: systematic analyses of conventional hypotheses. Behav. Ecol., 14: 823–840.

Thayer, A.H. (1896) The law which underlies protective coloration. The Auk, 13: 477–482.

Thayer, G.H. (1909) Concealing-Coloration in the Animal Kingdom: An Exposition of the Laws of Disguise Through Color and Pattern: Being a Summary of Abbott H. Thayer's Discoveries. The Macmillan Co., New York.

Vorobyev, M. (2003) Coloured oil droplets enhance colour discrimination. Proc. Roy. Soc. B, 270: 1255–1261.

Vorobyev, M., Osorio, D., Bennett, A.T.D., Marshall, N.J. and Cuthill, I.C. (1998) Tetrachromacy, oil droplets and bird plumage colours. J. Comp. Physiol. A, 183: 621–633.

Waldbauer, G.P. and Sternburg, J.G. (1983) A pitfall in using painted insects in studies of protective coloration. Evolution, 37: 1085–1086.

SECTION II

From Perceptive Fields to Gestalt: A Tribute to Lothar Spillmann

Introduction

The 28th European Conference on Visual Perception hosted a special symposium to honor Lothar Spillmann. The symposium was entitled "From perceptive fields to Gestalt". It included a plenary lecture by Lothar Spillmann and three additional lectures by Michael Paradiso, Sabine Kastner, and Stuart Anstis, all of which are chapters in this section.

Lothar Spillmann, Dick Cavonius, John Mollon, and Ingo Rentschler founded the European Conference on Visual Perception in 1978 (Marburg). After 28 years, the conference keeps growing and bringing together new generations of visual scientists, not only from Europe, but also from all over the five continents.

Lothar Spillmann was instrumental in the recovery of the field of visual psychophysics in Germany after World War II. His chapter summarizes many of the numerous and important discoveries accomplished by his laboratory in Freiburg. As the title of this section indicates, Lothar Spillmann has studied perceptive fields, Gestalt processes, and almost everything in between. One defining characteristic of Lothar Spillmann's studies is the elegant use of psychophysical techniques to probe the neural mechanisms of perception in a noninvasive fashion. Stuart Anstis summarizes Lothar Spillmann's accomplishments in a definitive way: *"If he has not studied it, it is not psychophysics"*.

Susana Martinez-Conde

Martinez-Conde, Macknik, Martinez, Alonso & Tse (Eds.)
Progress in Brain Research, Vol. 155
ISSN 0079-6123

CHAPTER 5

From perceptive fields to Gestalt

Lothar Spillmann*

Dept. of Neurology, Neurozentrum, University Hospital, Breisacher Strasse 64, 79106 Freiburg, Germany

Abstract: Studies on visual psychophysics and perception conducted in the Freiburg psychophysics laboratory during the last 35 years are reviewed. Many of these were inspired by single-cell neurophysiology in cat and monkey. The aim was to correlate perceptual phenomena and their effects to possible neuronal mechanisms from retina to visual cortex and beyond. Topics discussed include perceptive field organization, figure-ground segregation and grouping, fading and filling-in, and long-range color interaction. While some of these studies succeeded in linking perception to neuronal response patterns, others require further investigation. The task of probing the human brain with perceptual phenomena continues to be a challenge for the future.

Keywords: perceptive fields; gestalt; neurophysiological correlates of perception; visual illusions; figure-ground segregation and grouping; fading and filling-in; long-range color interaction

When I was a student, Gestalt factors were hardly more than a set of phenomenological rules to describe figure-ground segregation and grouping. Nowadays, Gestalt factors have entered the fields of neurophysiology and neurocomputation (Spillmann and Ehrenstein, 1996, 2004; Ehrenstein et al., 2003). Rüdiger von der Heydt is studying them, Steve Grossberg incorporates them into his models, and Wolf Singer refers to them within the context of synchronization of oscillations. The common goal is to promote an understanding of Gestalt factors in terms of specified single-neuron activities and to find the neuronal correlates of perceptual organization.

In his 1923 classical paper, the founder of Gestalt psychology, Max Wertheimer, had proposed that what we see is the simplest, most balanced, and regular organization possible under the circumstances. He called this the *Prägnanz* principle and attributed it to the tendency of the brain towards equilibrium. *Gestalten* are distinguished by two main criteria:

(i) *Supra-additivity*, meaning that the whole is different from the sum of its parts. Michael Kubovy would call this a *preservative emergent property*, because the elements survive, while something new emerges.
(ii) *Transposition*, implying that a Gestalt maintains its perceptual properties regardless of figural transformations (e.g., distance, orientation, slant). This constancy is nowadays called *viewpoint invariance*.

The Gestalt approach challenged the view that vision can be understood from an analysis of stimulus elements. Instead, it proposed Gestalt factors according to which stimulus patterns are segregated into figure and ground and individual parts grouped into a whole. Gestalt factors include *proximity, similarity, symmetry, smooth continuation, closure, and common fate* and are described within the framework of "*good Gestalt*" or *Prägnanz*. Little was known at the time about the neuronal mechanisms underlying these factors.

*Corresponding author. Tel.: +49-761-270-5042;
E-mail: lothar.spillmann@zfn-brain.uni-freiburg.de

DOI: 10.1016/S0079-6123(06)55005-8

Recent psychophysical and neurophysiological studies have shed light on some of the processes that may be responsible for figure-ground segregation and grouping (Valberg and Lee, 1991; Spillmann, 1999). The filling-in of gaps by illusory contours, the formation of boundaries by texture contrast, and the binding by coherent motion are among the better understood of these processes.

Part A. Atmosphere

Vision scientists who visited Freiburg from 1971 to 1994 may remember the building depicted in Fig. 1 (left), which housed our laboratory during those years. It was an old villa in one of the nicest neighborhoods in town, not far from the Schlossberg Mountain. When I arrived from America there was nothing in it, just empty rooms. So I found myself some old furniture and used equipment, a telephone, and dedicated collaborators. Ken Fuld, who had already worked with me in Boston, was the first. Billy Wooten followed from Brown, and then Charles Stromeyer and Bruno Breitmeyer from Stanford. Next were Arne Valberg and Svein Magnussen from Oslo. John S. Werner (UC Boulder), Munehira Akita (Kyoto), and Ted Sharpe (Cambridge) came later. Over the years, coworkers arrived from as many as 10 different countries, several of them returning for a second and third time. On the German side Wolfgang Kurtenbach and Christa Neumeyer, both zoologists, were among the first generation members. In 1994, we moved into the former Neurological Clinic (Fig. 1, right), just 200 m away, where we stayed for another 11 years.

From 1971 to 2005, the laboratory supported some 80 people at different stages of their careers, half of them diploma or doctoral students from biology, medicine, psychology, and physics. All of them were paid by grant money. Three former laboratory members moved on to become professors at German universities. Three visiting scientists were Alexander-von-Humboldt Senior Prize winners, nine were Humboldt Research Fellows, and five were supported by the German Academic Exchange Program (DAAD). Two Heisenberg Professorships and two Hermann and Lilly Schilling Professorships were bestowed upon laboratory members. Even in our last year, we were fortunate to have a DFG-Mercator Guest Professor from the Netherlands. Altogether we published more than 200 research papers, 4 edited books, 1 book translation, 25 book chapters, and numerous conference contributions (http://www.lothar-spillmann.de/). It is fair to say that Freiburg became a spot on the (perception) map.

Because of the great variety of people, there was also a great diversity of research. Guy Orban once remarked during a visit: "Lothar, I see everybody working on a different topic. You will never get famous this way." He was right, but I always thought that people are best at what they like the best. So I let them do whatever they wanted.

Our villa was old, but cozy. It had been a physician's residence and I kept as many of the permanent fixtures as possible. We had a kitchen, bathtubs, and even beds. We did experiments on the effect of vodka, grew marihuana on the balcony, and had wild and multilingual parties. Twice a year, we would go to the *Kaiserstuhl* and enjoy the local specialties — asparagus, pheasant, and venison. The atmosphere in the laboratory was very conducive to creative research. It was informal and relaxed, with much interaction, both scientific and social, among ourselves. The laboratory was very much the center of everybody's life — not just a place to work.

While life in the laboratory was enjoyable, dealing with the University administration and the Medical faculty was not always easy. As a psychologist in a clinical setting, one had little status and virtually no power in the University hierarchy. To gain visibility and esteem we began organizing research seminars. Richard Jung, our director, once said: "When you can't travel, you bring the world to your doorstep." This is what we did, even though we traveled a lot. Professor Jung supported us generously and regularly attended our seminars. The *Freiburg School of Neurophysiology* (R. Jung, G. Baumgartner, O. Creutzfeldt, O.-J. Grüsser, and H.H. Kornhuber) had always attracted a good number of distinguished neurophysiologists from around the globe (for a historical review see Grüsser et al., 2005); now we added psychophysicists. There must have been some 300 such seminars

Fig. 1. This building on Stadtstr. 11 (left) was home to the visual psychophysics laboratory from 1971 to 1994. (Photo: Clemens Fach) Thereafter, the laboratory moved into the former Neurological Clinic on Hansastr. 9a (right). (Photo: Ralf Teichmann) It was closed on June 30, 2005 in its 34th year.

over the years; many co-organized with Michael Bach from the local Eye Clinic. Much of what I know came from listening to those invited speakers.

Freiburg is a beautiful town surrounded by the Black Forest, the gastronomy is among the best in Germany, and there is plenty of good wine. Sometimes wine proved mightier than words. The first *European Conference on Visual Perception* in Marburg in 1978 (which then was called *Workshop on Sensory and Perceptual Processes*) also owes its success to this kind of currency. The last evening session was supposed to end at 10 p.m. The janitor wanted us out, but three bottles of *Endinger Engelsberg* sent him to his bed and we stayed on until long after midnight. This was the evening when the Dutch delegation under the leadership of Maarten Bouman (Fig. 2) and Dirk van Norren decided that the next meeting should be held in the Netherlands, and a tradition was born.

The *Neurologische Klinik mit Abteilung für klinische Neurophysiologie* in Freiburg (Fig. 1, right) was unique in Germany as it combined excellent clinical studies with first-rate basic research in human and animal subjects. The clinic was housed in a former sanatorium surrounded by a beautiful park. Every spare corner of the building was used for research and bustled with activity. Neurophysiological experiments on the visual, vestibular, somatosensory, and nociceptive sense modalities — and their multimodal interactions — were done

Fig. 2. Professor Maarten Bouman, president and organizer with Hans Vos of the 1979 *European Conference on Visual Perception* in Noordwijkerhout (The Netherlands).

next to oculomotor, EEG, and sleep recordings. A well-stocked library, two workshops for instrument development, and generous funding provided ideal conditions for productive research, resulting in many hundreds of publications.[1]

To honor Richard Jung on the occasion of his 75th birthday, Jack Werner and I planned an international conference in Badenweiler. Progress was slow and in the summer of 1986, Professor Jung — while on a visit to Belgium — suffered a stroke and died. So we organized our conference in his memory and that of his friends and co-editors of the *Handbook of Sensory Physiology* — Donald M. MacKay (1922–1987) and Hans-Lukas Teuber (1916–1977). We were lucky: the German Research Council, the Airforce Office of Scientific Research, the Alexander von Humboldt-Foundation, and Heinz Wässle from the Max Planck Institute for Brain Research in Frankfurt supported us. In 1987, we took the participants to that wonderful old hotel, the *Römerbad*, where at the turn of the 19th century Friedrich Nietzsche, Richard Wagner, and Anton Chekhov had lodged, and had a great time. Conference participants first interacted in small groups and then presented a given topic for plenary discussion, with no chair assigned to a session. To our surprise it worked.

The book on *Visual Perception — The Neurophysiological Foundations* (Eds. Spillmann and Werner, 1990) came out of the Badenweiler conference. The individual chapters were written by some of the finest scientists in the field, all writing in their own style. This prompted Brian Wandell to say in his review in *Contemporary Psychology*, "The book jumped into my lap like an excited puppy." To judge from the number of sold copies (6500), the book appears to have served the vision community well. It is also one of the few that aimed primarily at correlating perceptual phenomena to their underlying neuronal mechanisms.

Phenomenology as a guide to brain research had always had a great tradition in Freiburg. Jung (1961, 1973) firmly believed that all percepts had physiological correlates. He had proposed B- and D-neurons for brightness and darkness perception even before they were called on- and off-neurons. He had read the writings of Purkinje, Mach, and Hering on subjective sensory physiology, and when I first arrived as a student in the spring of 1962, Hans Kornhuber asked me whether I wanted to do a doctoral thesis on the Hermann Grid illusion. The conference report on the *Neurophysiology and Psychophysics of the Visual System* (Eds. Jung and Kornhuber, 1961) had just appeared with a chapter by Baumgartner on the responses of neurons in the central visual system of the cat. In this chapter he presented his receptive, field model of the Hermann grid illusion (p. 309). To a young psychologist, the prospect of looking into the human brain without actually sticking an electrode into it was fascinating. This fascination has never left me throughout my entire life. In the following, I will describe some of the perceptual phenomena studied in our laboratory in conjunction with their possible neurophysiological correlates.

Part B. Science

Perceptive fields

Hermann grid illusion

The Hermann grid is characterized by the presence of dark illusory spots at the intersections of white bars. A physiological explanation of this illusion involves concentric center-surround receptive fields. A *receptive field* is the area on the retina from which the response of a ganglion cell or higher-level neuron can be modulated by light entering the eye. Take two on-center receptive fields, one superimposed on the intersection and one on the bar. While central excitation is the same for both, the receptive field on the intersection receives more lateral inhibition than the receptive field on the bar (Fig. 3A). As a result the intersection looks darker. On a black grid, the intersections look lighter due to less lateral activation in off-center fields.

To test his hypothesis, Baumgartner and collaborators (Schepelmann et al., 1967) recorded from neurons in the cat visual cortex and found

[1]Schriftenverzeichnis Richard Jung und Mitarbeiter, Freiburg im Breisgau, 1939–1971. Herausgegeben anläßlich des 60. Geburtstages von Richard Jung. Springer-Verlag Berlin-Heidelberg-New York 1971.

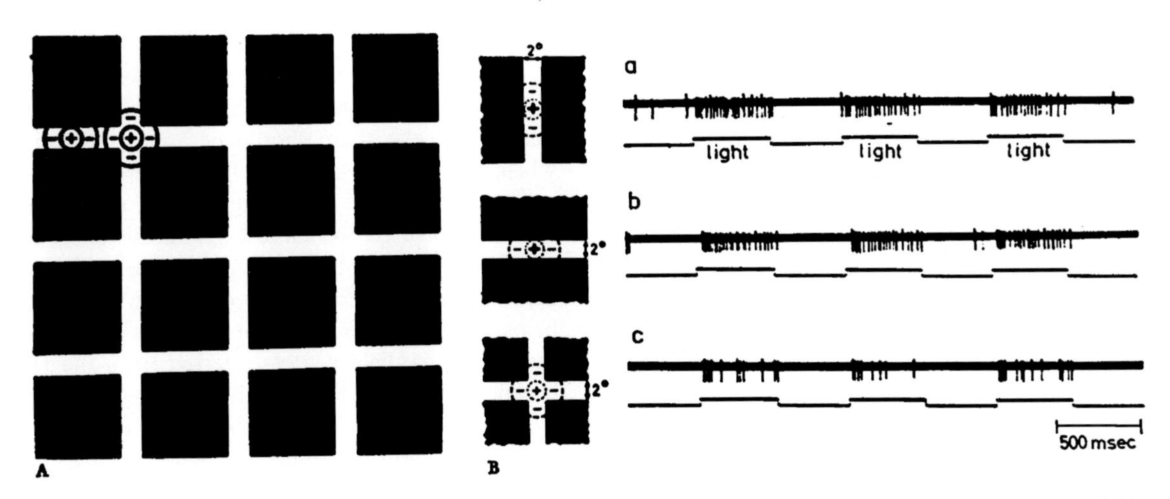

Fig. 3. Hermann grid illusion. (A) Dark illusory spots are attributed to more lateral inhibition of neurons whose receptive fields are stimulated by an intersection as compared to a bar. (B) Single-cell recording from first-order B-neuron in the cortex of the cat with one or two bars stimulating the receptive field. The firing rate is reduced when both bars are presented simultaneously, consistent with a darkening at the intersection. (Modified from Baumgartner, 1990, with kind permission from Springer.)

that each bar presented by itself on the receptive field of the neuron produced a strong response (Fig. 3B). However, when both bars were presented together as in the intersection of the grid, the neuronal response was greatly reduced. Baumgartner postulated that the illusion should be strongest when the width of the bar matched the receptive field center (Tom Troscianko would later say that a factor of 1.4 was more appropriate).

Here then was a psychophysical tool to study the receptive field organization in humans without invading the brain. All one needed to do was to find the grid that produced the strongest illusion. So I pasted a number of Hermann grids with different bar width on cardboard and presented them at various distances from the fixation point. The task of the subject was to select the grid that yielded the darkest illusory spots.

Foveal field centers turned out to be quite small, only 4–5 minarc (Spillmann, 1971). However, with increasing eccentricity, center size increased up to 3° in the outer periphery (Fig. 4). The small center size in the fovea is the reason why the Hermann grid illusion is typically not seen with direct fixation. The bars are just too wide (Baumgartner, 1960, 1961).

Jung called these centers *perceptive field centers* because they are revealed through our perception

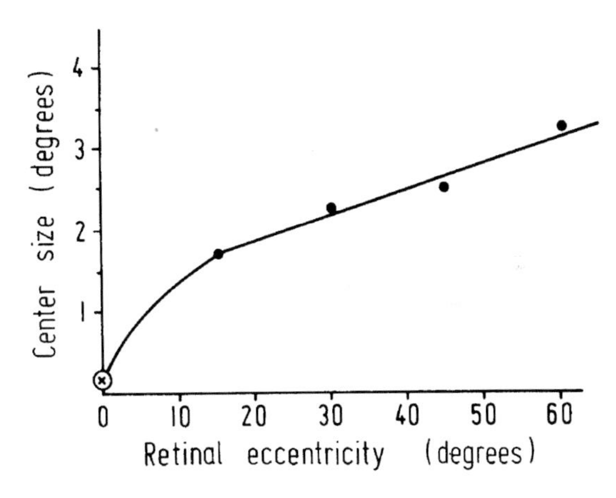

Fig. 4. Perceptive field center size derived from the bar width that elicited the maximum illusory effect in the Hermann grid illusion, plotted as a function of retinal eccentricity. Center size in the fovea is only 4–5 min of arc, which is the reason why the dark illusory spots are normally not seen in foveal vision. (Modified from Jung and Spillmann (1970), with kind permission from the National Academy of Sciences of the United States of America.)

(Jung and Spillmann, 1970). You may argue that a perceptive field reflects the activity of many neurons, not just one. This is undoubtedly true. Moreover, we do not know where these neurons reside in the visual pathway. So, it is difficult to assign a given percept to the retina, lateral geniculate body, or visual cortex.

However, there are ways to narrow down the possible brain loci. For example, if the Hermann grid illusion cannot be seen with dichoptic presentation, we would say that it is most likely of subcortical origin. On the other hand, if it exhibits a strong oblique effect, we would assume that it is cortical. Finally, if the illusion can be seen with isoluminant colors, it is likely mediated by the parvocellular pathway. All three statements apply to the Hermann grid illusion. We therefore tend to think that it is primarily a retinal effect with a cortical contribution (for a review see Spillmann, 1994).

As did Colin Blakemore, we call these and other techniques the *psychologist's microelectrode* (a term variously attributed to Bela Julesz, John Mollon, and John Frisby) because of the insights they can provide into the mechanisms of visual perception and their location in the visual pathway. Peter Schiller's (Schiller and Carvey, 2005) recent paper in *Perception* proposes a new kind of cortical neuron to explain the Hermann grid illusion. Yet his proposal is still awaiting neurophysiological confirmation in the trained monkey.

When I went to America in 1964, I thought I would continue my Freiburg work studying visual illusions. Hans-Lukas Teuber (at MIT) was supportive, but David Hubel on the other side of the River was reluctant and recommended that I do straightforward neurophysiology. Torsten Wiesel was more sympathetic. It took Margaret Livingstone (Livingstone and Hubel, 1987; Livingstone, 2002) to bridge the gap between neurophysiology and perception at Harvard Medical School. Perceptual labels were boldly attached to visual structures and functions, and even illusions became fashionable among former hardcore neuroscientists.

Phi-motion

After measuring perceptive fields and field centers in the Hermann grid, we wondered whether we could also measure perceptive field centers for motion. The obvious choice was the phi-phenomenon. In 1912, Max Wertheimer (1912) had published his landmark study on apparent motion, which he attributed to some kind of intracortical short circuit (*Querfunktionen*). Our idea was simple: when two successively presented stimuli fell within the same perceptive field, there should be apparent motion; when they fell into different fields, there should be no interaction and — consequently — no motion.

So I measured the largest spatial distance over which phi-motion could be seen. The results are again plotted against retinal eccentricity; perceptive fields for motion were about 20 times larger than the perceptive field centers inferred from the Hermann grid illusion (Fig. 5). From this discrepancy we concluded that there were different kinds of perceptive field organization depending on the response criterium. This finding anticipated neurophysiological measurements that show receptive fields of area MT-neurons much larger than those

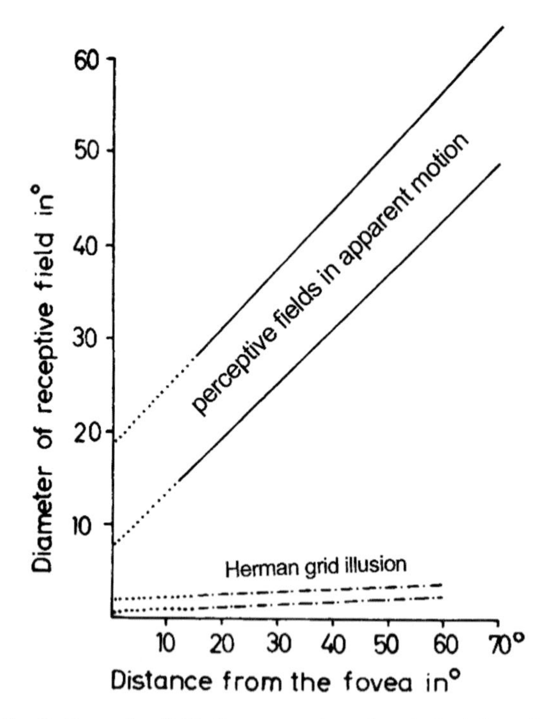

Fig. 5. Perceptive fields for apparent motion derived from the largest distance between two successively flashed stimuli across which phi-motion could still be seen, plotted as a function of retinal eccentricity. Regression lines refer to ascending and descending thresholds. Results obtained with the Hermann grid illusion are shown for comparison. (Modified from Jung and Spillmann, 1970), with kind permission from the National Academy of Sciences of the United States of America.)

of retinal ganglion cells or V1 neurons (Britten, 2004).

Westheimer paradigm

The 1960s and 1970s were the time of perceptual phenomena in search of neural mechanisms and neural mechanisms in search of perceptual phenomena. It was like a revelation; psychologists everywhere went wild. Colin Blakemore was the youthful leader of this group. Looking back, Baumgartner (1990) would later ask, "*Where do visual signals become a perception?*" The benefit was mutual; neurophysiologists looked for mechanisms that could not have been predicted from the physical stimulus alone. Vice versa, psychologists looked for percepts that may otherwise not have been discovered.

Naturally, we were not alone in our quest for psychophysical correlates of neuronal mechanisms. In 1965 and 1967, Gerald Westheimer (1965, 1967) published two influential papers in the *Journal of Physiology* on spatial interactions in the human retina (see also Westheimer, 2004). Westheimer used a small test spot centered on a variable background that in turn was superimposed on a large ambient field (Fig. 6A). With this kind of luminance hierarchy, he obtained the increment–threshold curve known as Westheimer function (Fig. 6B).

Threshold was plotted as a function of background diameter. When the background became larger, the threshold for the test spot first increased to a peak, then decreased, and finally leveled off. Westheimer attributed the initial increase to spatial summation within the perceptive field center (first arrow) and the subsequent decrease to lateral inhibition within the perceptive field surround (second arrow). In this way he derived the diameters of the center and the entire field.

I very much liked Westheimer's paradigm. So I asked Anne Ransom-Hogg (Ransom-Hogg and Spillmann, 1980), now Anne Kurtenbach, in my laboratory to measure perceptive fields and field centers in the light- and dark-adapted eye. To do so, we used an elaborate three-channel Maxwellian-view system, beautifully crafted by our master mechanics and wired up by our top electronics

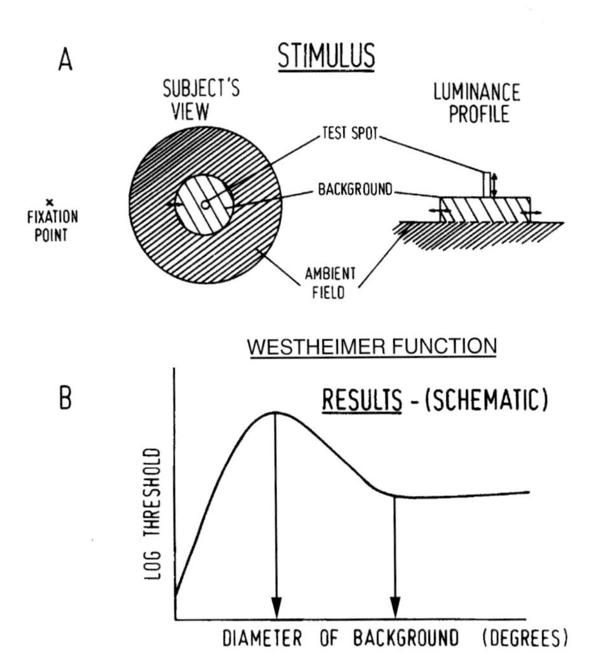

Fig. 6. Westheimer paradigm. (A) Stimulus configuration as seen by the subject (left) and corresponding luminance profile (right). A small test spot is flashed onto the center of a background of variable diameter that is presented on a diffusely illuminated ambient field. (Reprinted from Oehler, 1985, with kind permission from Springer Science and Business Media). (B) Increment threshold plotted as a function of background diameter (schematic). The first arrow marks the background diameter that corresponds to the size of the perceptive field center; the second marks the diameter corresponding to the entire perceptive field (center plus surround). (Modified from Ransom-Hogg and Spillmann, 1980, with kind permission from Elsevier.)

technician. It had a swivel support (adopted from Billy Wooten's laboratory), enabling us to do measurements out to 70° eccentricity without realigning the pupil. (We also had a four-channel Michelson interferometer, to which later were added two more Maxwellian-view systems, making our laboratory one of the best-equipped vision laboratories in Germany.)

With increasing retinal eccentricity, the position where the Westheimer curve peaked was displaced to the right, and so was the position of the point where the curve asymptoted. Therefore, when we plotted perceptive field center size as a function of eccentricity, both curves (for photopic and scotopic vision) increased from the fovea towards the periphery, just as for the Hermann grid illusion.

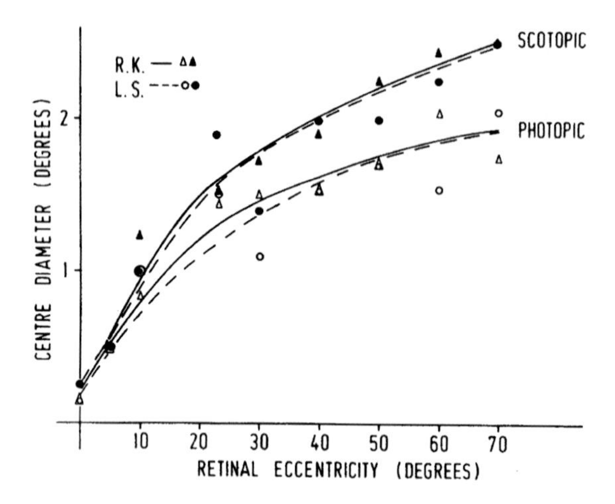

Fig. 7. Comparison of perceptive field centers for photopic and scotopic vision, plotted as a function of retinal eccentricity. Data are from two observers. (Modified from Ransom-Hogg and Spillmann, 1980, with kind permission from Elsevier.)

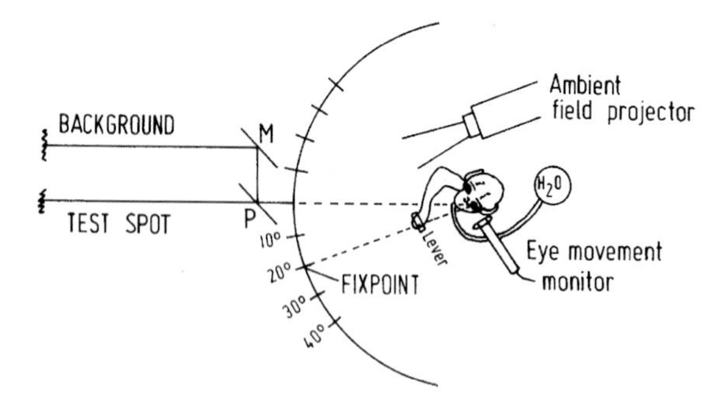

Fig. 8. Experimental setup (seen from above) for testing rhesus monkeys with the Westheimer paradigm. M = first surface mirror, P = pellicle. The same setup was used for testing human observers under identical conditions. (Reprinted from Oehler, 1985, with kind permission from Springer Science and Business Media.)

However, field centers were larger by approximately one fourth for scotopic than for photopic vision (Fig. 7). We attributed this difference to the peak shift caused by the decrease of lateral inhibition with dark adaptation and the resulting flattening of the curve. This finding agreed with Horace Barlow et al.'s (1957) discovery in the cat that at low-light levels the area for spatial summation increases when lateral inhibition gradually diminishes and disappears.

The study by Ransom-Hogg permitted a further conclusion. When we plotted perceptive field center size against the inverse of the cortical magnification factor (Drasdo, 1977), we obtained a straight line with a slope of 0.88. This finding suggested an almost constant size of the cortical representation of perceptive field centers; it also compared well with the slope of 0.81 found for cortical receptive field centers in the rhesus monkey (Hubel and Wiesel, 1974).

So far, we had tacitly assumed that the perceptive field organization in the human was similar to the receptive field organization in the monkey. But we had no evidence. Therefore, in a follow-up experiment, Regina Oehler (1985) in our laboratory used the Westheimer paradigm to measure perceptive fields and field centers in human and rhesus monkey. Fig. 8 shows her experimental setup.

The monkey and human curves were similar in shape and stacking order, but they differed in height (Fig. 9). However, when one derived the critical background diameters at which the curves peaked and leveled off, the resulting values were almost the same (Fig. 10). In fact, the match for perceptive *field centers* could not have been better. In comparison, perceptive *field sizes* were somewhat larger for the human observers, suggesting more extensive lateral inhibition.

After demonstrating that the perceptive field organization was equivalent in macaque monkeys and humans, the question remained: how do macaque *per*ceptive field centers compare to macaque *re*ceptive field centers obtained neurophysiologically?

To answer this question we plotted the diameters of macaque perceptive field centers (obtained with the Westheimer paradigm) and receptive field centers of retinal ganglion cells (from DeMonasterio and Gouras, 1975) against eccentricity. Again, the agreement between the two kinds of measurements was excellent (Fig. 11).

Now we had evidence that perceptive field centers and receptive field centers in the monkey were equivalent. And what holds for the monkey should also apply to the human observer. So whenever we measure perceptive fields and field centers in man using the Hermann grid illusion, the Westheimer paradigm, or another procedure, we can safely say that we are tapping the underlying receptive field

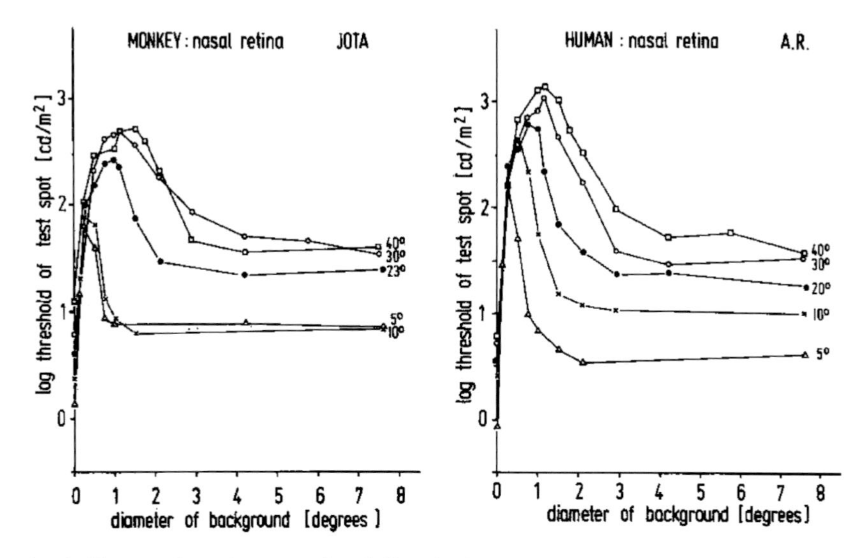

Fig. 9. Increment–threshold curves for a rhesus monkey (left) and a human observer (right). Curves refer to five retinal eccentricities in the nasal retina ranging from 5° to 40°. (Reprinted from Oehler, 1985, with kind permission from Springer Science and Business Media.)

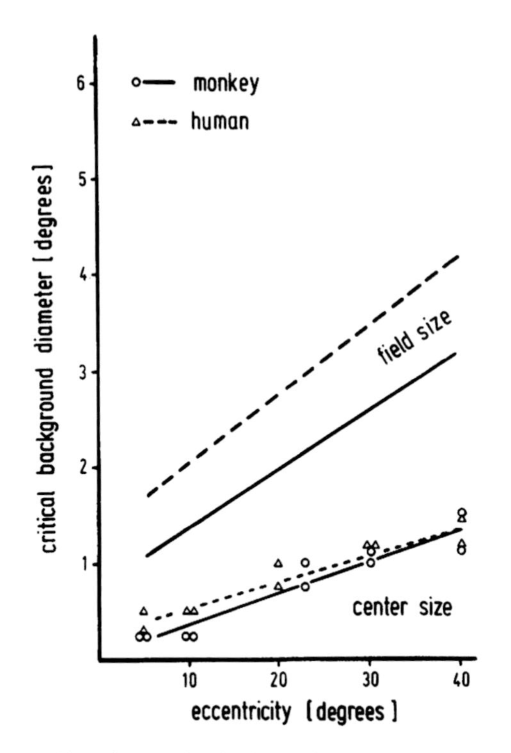

Fig. 10. Size of perceptive fields and field centers, plotted as a function of eccentricity on the horizontal meridian of the nasal retina. Continuous lines refer to monkey data and dashed lines to human data. Averages are from two monkeys and two humans. (Modified from Oehler, 1985, with kind permission from Springer Science and Business Media.)

organization (of ganglion cells) without using a microelectrode (Spillmann et al., 1987).

Beyond the classical receptive field

So far I have described center-surround organization in classical receptive fields. This section addresses neurons whose response is modulated by stimulus properties from *beyond the classically defined receptive field.* Following the finding by McIlwayn (1964) that retinal ganglion cells respond to stimuli in the far periphery, Bruno Breitmeyer and Arne Valberg in our laboratory embarked on a series of studies to identify related psychophysical responses. They found that the increment threshold for a foveal stimulus increased in the presence of a grating shift as far as 4° away — the *Jerk Effect* (Breitmeyer and Valberg, 1979).

The neurophysiological breakthrough for long-range interactions came from the Zurich group of Günter Baumgartner (Baumgartner et al., 1984). Rüdiger von der Heydt and Esther Peterhans showed that the response rate of neurons could be affected by stimuli that were clearly outside the classical receptive field (von der Heydt et al., 1984; von der Heydt, 1987; von der Heydt and Peterhans, 1989). They called this the *response field.* This

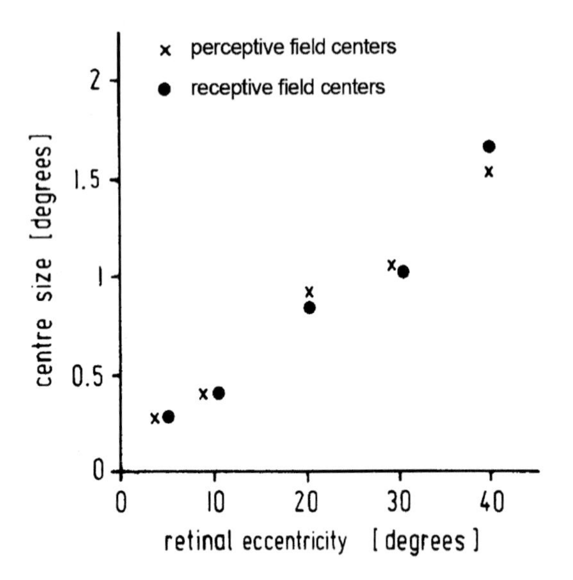

Fig. 11. Comparison of perceptive field centers measured with the Westheimer paradigm (crosses) and receptive field centers obtained with single-cell recording (dots), both in the rhesus monkey. Behavioral data are averages from the nasal and temporal hemiretinae of two monkeys. Neurophysiological data refer to broadband cells and are from DeMonasterio and Gouras (1975). (Reprinted from Oehler, 1985, with kind permission from Springer Science and Business Media.)

discovery opened up the study of perceptual completion across gaps and scotomata, surface filling-in, large-scale color effects, and context-dependent boundary formation. Our review paper on this topic, 12 years later, was requested as often as 400 times (Spillmann and Werner, 1996).

In the stimulus pattern shown in Fig. 12 (left), one can perceive a bright vertical bar delineated by illusory contours. Von der Heydt and Peterhans studied this illusion neurophysiologically in visual area V2 (Fig. 12, right). They first presented a solid bar moving back and forth across the receptive field. The response was vigorous in each direction (A). Then they presented the same bar, but with a large gap in the middle, to spare the classical receptive field. Under these conditions, one would predict the neuron to fall silent, as there is nothing to drive the cell. However, this was not the case. Instead, the neuron continued to respond, albeit less strongly (B). This can only be explained by assuming that it received input from the two short bars at the top and bottom. Finally, when the bars were closed off with thin lines, the response was essentially absent (C). In this condition, the perception of the illusory contour also breaks down.

This finding had enormous consequences. It meant that we can perceptually recover an object that is only partially given by virtue of filling-in. It also opened the possibility of explaining a number of illusions that are characterized by perceptual occlusion, such as the Kanizsa triangle and the Ehrenstein illusion.

Kanizsa triangle

The Kanizsa (1979) triangle exhibits a triangular surface that is brighter than the surround and delineated by illusory contours (Fig. 13). Although the illusion is typically elicited by black solid cues (a), it will also arise from concentric rings (b), and even small dots at the apices (c). Illusory contours may be straight or curved depending on the shape of the missing sectors (d). Supporting lines jutting in from the side enhance the illusion.

Von der Heydt and Peterhans (1989) suggested that neurons responding to discontinuous bars also mediate the perception of the Kanizsa triangle illusion. According to their model (Fig. 14), end-stopped neurons in area V1, whose receptive fields are activated by the corners at the edges of the missing sectors, feed their signals into a gating mechanism in V2 neurons. Signals from two aligned sectors will be multiplied ($\times$) and then sent to a higher order neuron, where they will be summed with the input from the straight edges of the missing sectors ($\sum$). The result is an illusory line delineating the bright triangle across the interspace.

This model is consistent with the observation that the Kanizsa triangle only emerges when the three cut-out sectors (pacmen) are properly aligned. When they are rotated just by a small amount, the illusion weakens and disappears. This need for collinearity is an example of the Gestalt *factor of good continuation*. Meanwhile, it has been shown that mammals, birds, and even insects behave as though they perceive the Kanizsa triangle (Nieder, 2002). This clearly speaks for a bottom-up mechanism.

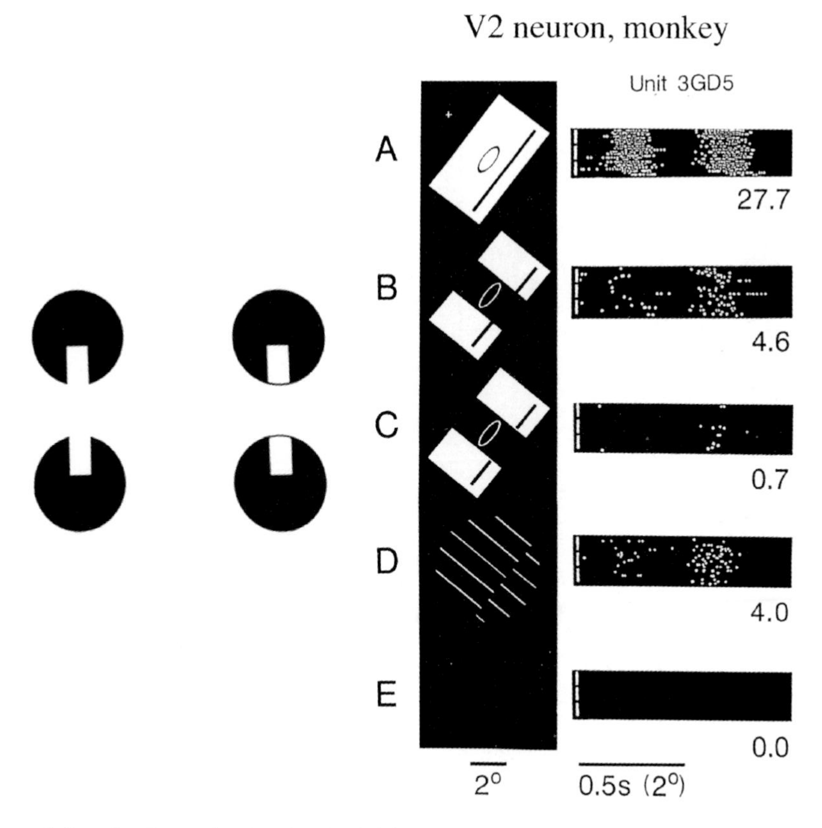

Fig. 12. Perceptual completion of an incomplete bar showing brightness enhancement and illusory contours (left) and macaque responses to variants of this stimulus (right). (A) A solid bar moved across the receptive field of a V2 neuron produces a vigorous response. (B) The same bar with its center section missing continues to produce a response, although the classical receptive field is no longer stimulated by this pattern. (C) When the top and bottom sections are closed off by thin orthogonal lines, they are no longer effective. Under these conditions, the illusion on the left is also abolished. (D) Two phase-shifted gratings opposing each other elicit a response as strong as stimulus B. (E) Control. (Reprinted from Peterhans and von der Heydt, 1991, with kind permission from Elsevier.)

Ehrenstein illusion

In the Ehrenstein (1954) brightness illusion, one perceives a bright disk in the center between the radial lines, delineated by an illusory ring that is orthogonal to the inducers (Fig. 15, left). This is known as line-end contrast. Ehrenstein pointed out that brightness enhancement disappears when a physical ring is superimposed onto the illusory contour. This observation suggests that one needs open gaps in order to have brightness enhancement. When the radial lines are laterally displaced or rotated out of alignment, the illusion becomes weaker and ultimately breaks down (for a review see Spillmann and Dresp, 1995). This again is evidence for the role of collinearity and the Gestalt factor of good continuation.

There is a fascinating property of the Ehrenstein illusion: the *neon color effect* (van Tuijl, 1975; for a review see Bressan et al., 1997). When a colored cross is used to connect the radial lines across the central gap, this region appears to be tinted with the color of the cross (Fig. 15, right). Christoph Redies (Redies and Spillmann, 1981; Redies, 1989) studied this effect in my laboratory and tentatively linked it to line-gap enhancement and end-stopped cells. Grossberg ("I like neon") proposed a computational model that interprets neon color in terms of diffusion (Grossberg and Mingolla, 1985; Pinna and Grossberg, 2005). Another well-known illusion that shows how illusory contours are formed at right angles to the inducing line ends is the abutting grating illusion.

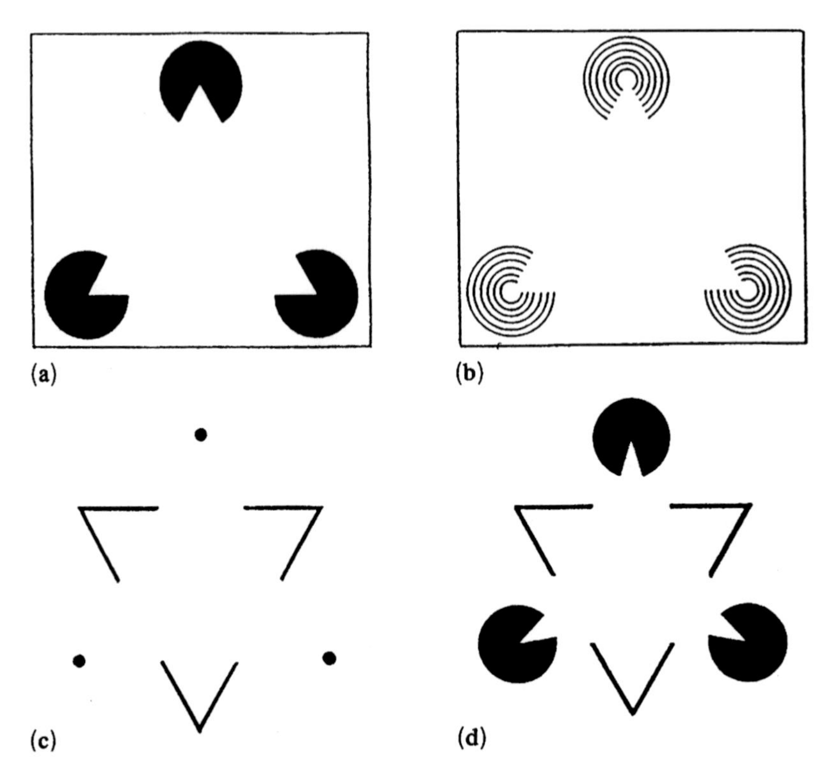

Fig. 13. Kanizsa triangle. The illusory triangle appears brighter than its surround and is delineated by an illusory edge. It also appears to lie slightly above the background. Various kinds of corner cues (a–d) elicit the same illusory percept. (Modified from Kanizsa, 1974, with kind permission from Il Mulino.)

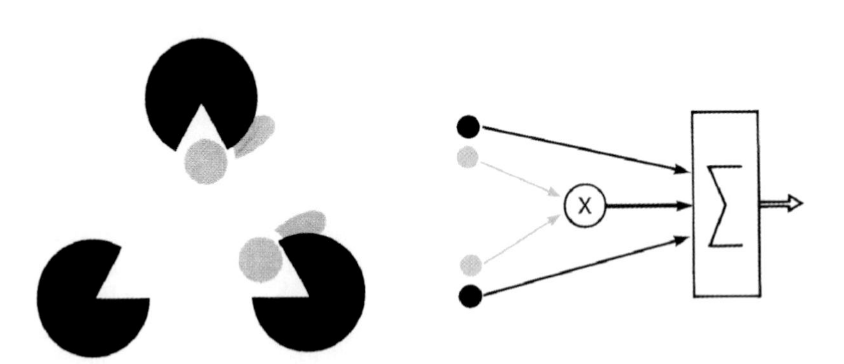

Fig. 14. Kanizsa triangle with receptive fields (gray patches) of end-stopped cells superimposed onto the corners of the pacmen (left). The model (right) distinguishes between two neuronal paths: an edge-detecting path (black arrows) that receives its input from the aligned edges of the cut-out sectors; and a grouping path (gray arrows) that receives its input from the end-stopped neurons. The latter signals are fed into a V2 neuron where they are multiplied ($\times$) and then sent to a higher order neuron, where they are summed with the input from the edge-detecting path (Σ). In this way, an illusory contour emerges at right angles to the inducing cues for which there is no physical equivalent. Receptive fields for the two paths are assumed to overlap on the same patch of retina. (Modified from Peterhans and von der Heydt, 1991, with kind permission from Elsevier.)

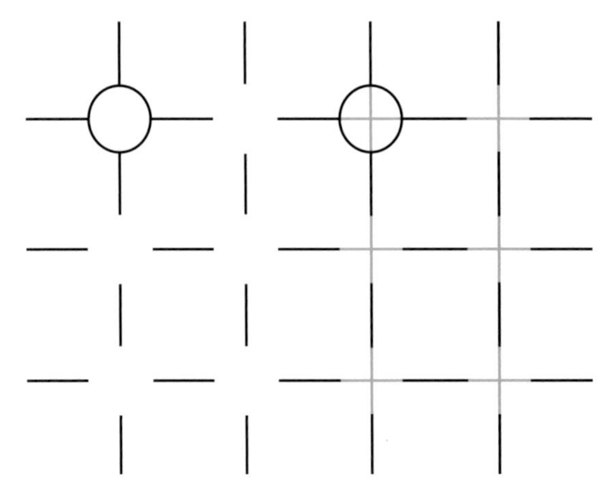

Fig. 15. Ehrenstein illusion. (Left half) The central area between the radial lines appears brighter than the surrounding background. A thin black ring abolishes the illusion. (Right half) Neon color spreading is observed when a colored cross bridges the gap between the radial inducing lines. (Modified from Redies and Spillmann, 1981, with kind permission from Pion.)

Abutting grating illusion

In the figure by Kanizsa (1974), a crisp line appears to run down the interface between two phase-shifted gratings (Fig. 16). Manuel Soriano et al. (1996) in my laboratory did a study on the abutting grating illusion and found that the illusory line depends crucially on the alignment of the tips of the horizontal lines along the same vertical. When the two gratings are slightly interleaved or pulled apart, the illusory line breaks down. For most stimulus parameters tested, such as number and spacing of lines, the psychophysical results paralleled the neurophysiological data obtained by Peterhans et al. (1986; Peterhans and von der Heydt, 1991) in the monkey.

These authors had demonstrated that an abutting grating line was almost as effective in eliciting a neuronal response as a real line (Fig. 12D). For an explanation, they suggested the same two-stage model as invoked for the Kanizsa triangle, except that in the abutting grating illusion, there are many more end-stopped neurons involved to support the illusory line between the two phase-shifted gratings.

Peterhans and von der Heydt (1989) also found neurons in area V2 of the monkey that may account for the high sensitivity of the illusory contour towards deviations from collinearity. They used a moving string of dots and found that for some cells a deviation of one of the dots by only 2 min arc from a straight line sufficed to offset the neuronal response. This finding points towards *alignment* detectors governing the perception of illusory contours according to the Gestalt factor of good continuation.

Fig. 16. Abutting grating illusion. A thin vertical line appears to separate the two phase-shifted gratings at the interface. For best result view from a distance. The illusion is highly sensitive towards misalignment of the terminators. (Modified from Kanizsa, 1974, with kind permission from Il Mulino.)

Figure-ground segregation

The Gestaltists already knew that a uniformly textured region would group together and become a surface. On the other hand, differences in texture would lead to segregation. The next section describes a study on figure-ground segregation by orientation contrast.

Orientation contrast

Victor Lamme (1995) asked whether a difference in orientation between the target and the background

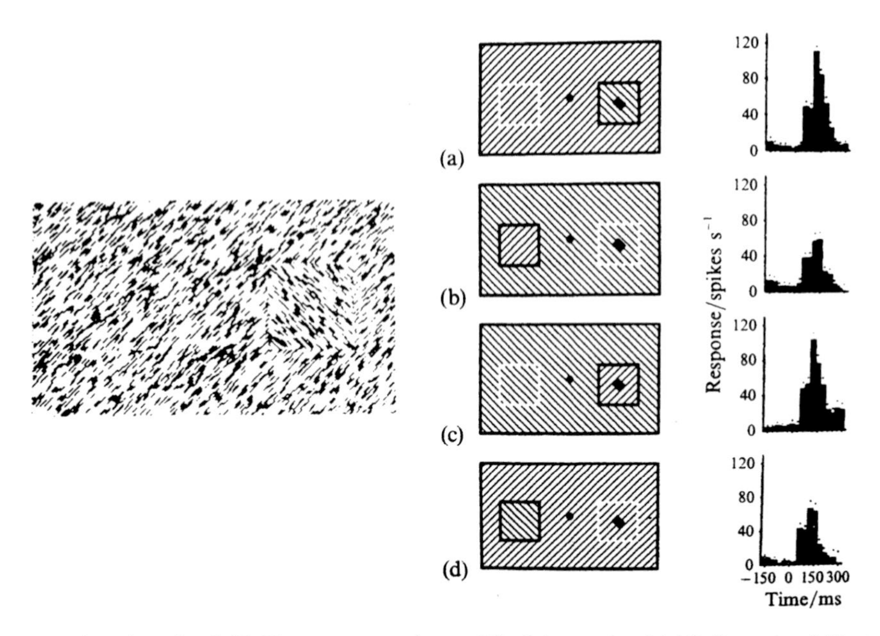

Fig. 17. Texture contrast by orientation (left). Neuron response in area V1 of the monkey (right). Boxes (a–d) illustrate schematically the stimulus relative to the background. The neuronal response is enhanced when the orientation of the target is orthogonal to that of the background. Note that the receptive field of the neuron (black rectangle) is entirely enclosed within the target (boundary not shown in the experiment). The difference in response must therefore be due to long-range interaction. (Modified from Lamme, 1995, with kind permission from the Society for Neuroscience.)

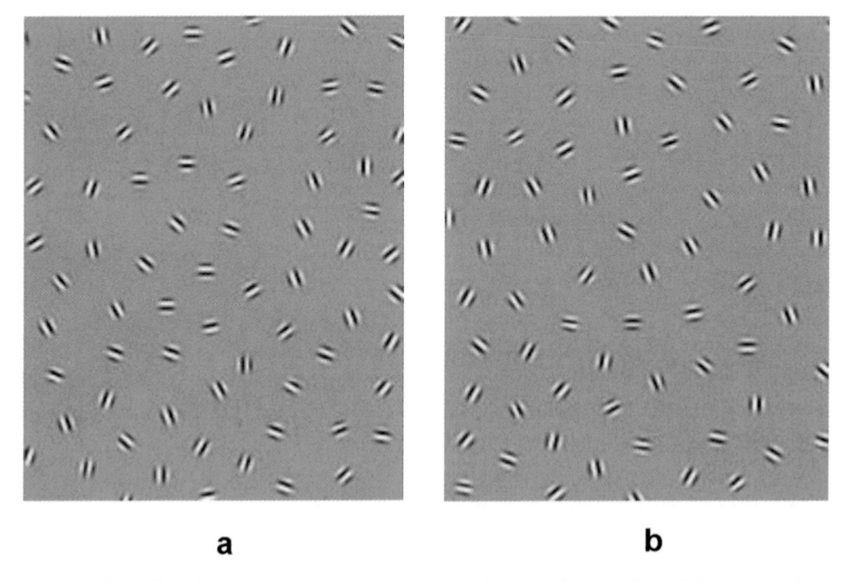

Fig. 18. Contour integration. (a) Aligned Gabor patches on a background of randomly oriented Gabor patches pop out perceptually when arranged as a semicircular curve, (b) but even more so when forming a complete circle. The number and overall distribution of Gabor patches is the same in both patterns, but the response of the brain is not. These percepts may be attributed to the Gestalt *factors of good continuation and closure*. (From Kovács and Julesz, 1993, with kind permission from the National Academy of Sciences of the United States of America.)

would affect the neuronal response, even if the receptive field of the neuron, onto which the target was superimposed, did not receive direct input from the surrounding background (Fig. 17). He found that when the orientation of the target was orthogonal to that of the background, the neuronal response was large (a) and (c). However, when the orientation of the target was the same as that of the background, the response was small (b) and (d). The difference in response between the two conditions suggested that the background had an effect on the target response through long-range interactions. This experiment shows that figure-ground segregation by orientation contrast can occur as early as area V1 (see also Lamme et al., 1992). In the cat, responses to motion contrast have also been found in striate cortex (Kastner et al., 1999).

Contour integration

While alignment plays a role in illusory contour formation, it is also essential for perceptual grouping. I had first seen chains of aligned Gabor patches in Robert Hess' laboratory in Montreal (Field et al., 1993), but at the time did not appreciate their importance for contour integration. Fig. 18 is from the work of Kovács and Julesz (1993). It shows an assembly of Gabor patches with different orientations. Within this pattern, there are six patches that line up to form a curvilinear contour (a). This is an example of the Gestalt *factor of good continuation*. However, the curved contour pops out much more easily when the ring is complete (b). This is an example of the Gestalt *factor of closure*. Obviously, the neuronal mechanism underlying this kind of contour integration must be effective over a rather large distance; otherwise, there would be no grouping (Spillmann and Werner, 1996).

Grouping by coherent motion

In the domain of motion, the Gestalt *factor of common fate* is probably the most important of all. This factor implies that coherently moving dots on a background of randomly moving dots will pop out as a group, even if the dots are fairly widely spaced.

Bill Uttal and Allison Sekuler (Uttal et al., 2000), both guest researchers in my laboratory, asked: How common must common fate be? They found that only 4 coherently moving dots, within a dynamic noise background of 100 dots, were sufficient in order to be seen as a group (Fig. 19). This is a very low signal-to-noise ratio. Frank Stürzel (Stürzel and Spillmann, 2004) further found that the time needed for grouping is only 430 ms. He also showed that coherent motion obeyed several of the constraints known from neurophysiological studies, such as speed and angular deviation from parallel trajectories. Finally, Gunnar Johansson's (1973) biological motion stimuli demonstrated that grouping by common fate occurs even when individual dots have different motion vectors.

Is there a neurophysiological correlate to support these psychophysical observations? The answer is yes. In a carefully designed experiment, Britten et al. (1992) showed that neurons in primate area MT respond strongly to coherently moving dots. Furthermore, they demonstrated that the neuronal threshold was comparable to the behavioral threshold measured simultaneously in the same animal. Neuroimaging in the human has confirmed area V5 as the brain locus responsible for mediating perception of motion coherence

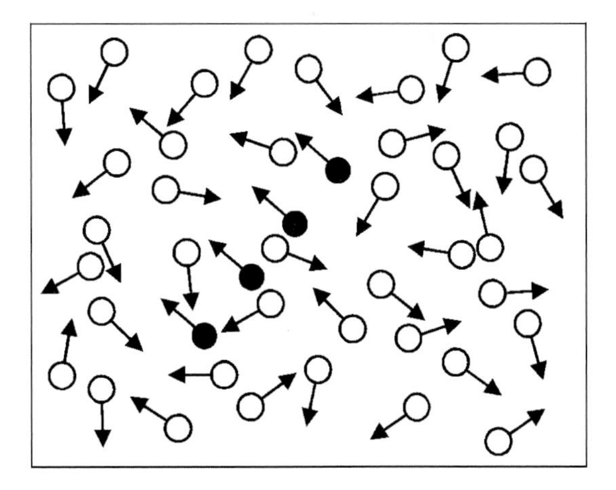

Fig. 19. Coherent motion. Four aligned dots moving in the same direction and at the same speed group perceptually together on a background of randomly moving dots. All dots were white on a black background. Parallel trajectories facilitate grouping, but are not necessary. This is an example of the Gestalt *factor of common fate*. (Reprinted from Stürzel and Spillmann, 2004, with kind permission from Elsevier.)

(Braddick et al., 2000). Evidence from a recent study by Peterhans et al. (2005) further shows strong responses to rows of coherently moving dots already in macaque areas V2 and V3. These results demonstrate that the Gestalt factor of common fate is a basic mechanism of our neuronal inventory.

Many of the Gestalt factors mentioned can be found in the animal kingdom in the interest of camouflage. In his book, *Laws of Seeing*, Wolfgang Metzger (1936) showed that it is not just mammals that break figure-ground segmentation to hide from predators, but also insects, fishes and birds. The purpose is to blend in with the ground. Ramachandran et al. (1996) demonstrated that a flounder displays on its skin the texture of a checkerboard on which it is placed. This is amazing as it occurs within minutes. Hiding through camouflage is particularly effective when an animal "freezes," although in a moving environment the absence of motion will likely reveal an animal.

Metzger argued that if predators get fooled by camouflage much in the same way as we do, their visual systems must be processing information in a way similar to ours. He therefore considered the Gestalt factors to be both innate and bottom-up. However, the early Gestaltists already knew that there were also top-down effects such as attention, set, motivation, and memory. Neurophysiologists, especially Mountcastle (1998) and Schiller (1998), have actively investigated the feedback loops, including guided eye movements, required for top-down modulation. It is now clear that visual perception uses both bottom-up and top-down processes.

Fading and filling-in

Troxler effect

The previous sections emphasized the spatial aspects of the perceptive field organization and grouping. This section will address the temporal aspects. Here we asked how a surface is sustained over time. Stuart Anstis' chapter in this volume mentions some of our earlier studies on the Troxler effect using static, rotating, and flickering targets. Christa Neumeyer in our laboratory was the first to study fading of large, centrally fixated disks (Neumeyer and Spillmann, 1977). She used various figure-ground contrasts and found that figures typically fade into the ground, not vice versa. Furthermore, when an oscillating grating surrounded the target, fading time was shorter. This observation is consistent with a later finding that a kinetic contour facilitates fading, rather than delaying it (Spillmann and Kurtenbach, 1992).

Research into fading picked up with the beautiful effects on color and texture filling-in demonstrated by Ramachandran and Gregory (1991). These findings involved grating patterns and page print, suggesting a postretinal origin. A few years later, Peter DeWeerd et al. (1995) showed neurophysiologically that texture fading occurred in area V3 of the visual cortex.

DeWeerd used a pattern with a white square on a dynamic background of vertical slashes (Fig. 20, top). The white target is called an *artificial scotoma* in analogy to a real scotoma. To make this target disappear, fixate at the small disk in the upper left corner for about 15 s. While fixating, you will see that the white target area becomes less distinct and eventually fades into the background. This is an example of *texture spreading or filling-in*. DeWeerd distinguished between two processes: a slow process for breaking down the border (cancellation) and a fast process for filling-in properties from the surround (substitution).

In Fig. 20 (bottom), the response rate of a V3 neuron in the monkey is plotted as a function of time. The continuous curve (hole) shows the response when the white target was present over the receptive field and the dotted curve the response when there was no hole in the background. The continuous curve first decreases, then gradually increases, and finally approaches the upper control curve. The interesting aspect here is that the firing rate for the "hole" condition recovers over time, although there is no change in the stimulus.

DeWeerd interpreted this "climbing activity" as the neuronal correlate of fading. When the two curves merged, the neuron could no longer distinguish between the two types of stimuli. He then asked human observers to look at the same pattern with steady fixation and report when the white target had faded into the background. The time needed for perceptual completion was quite similar

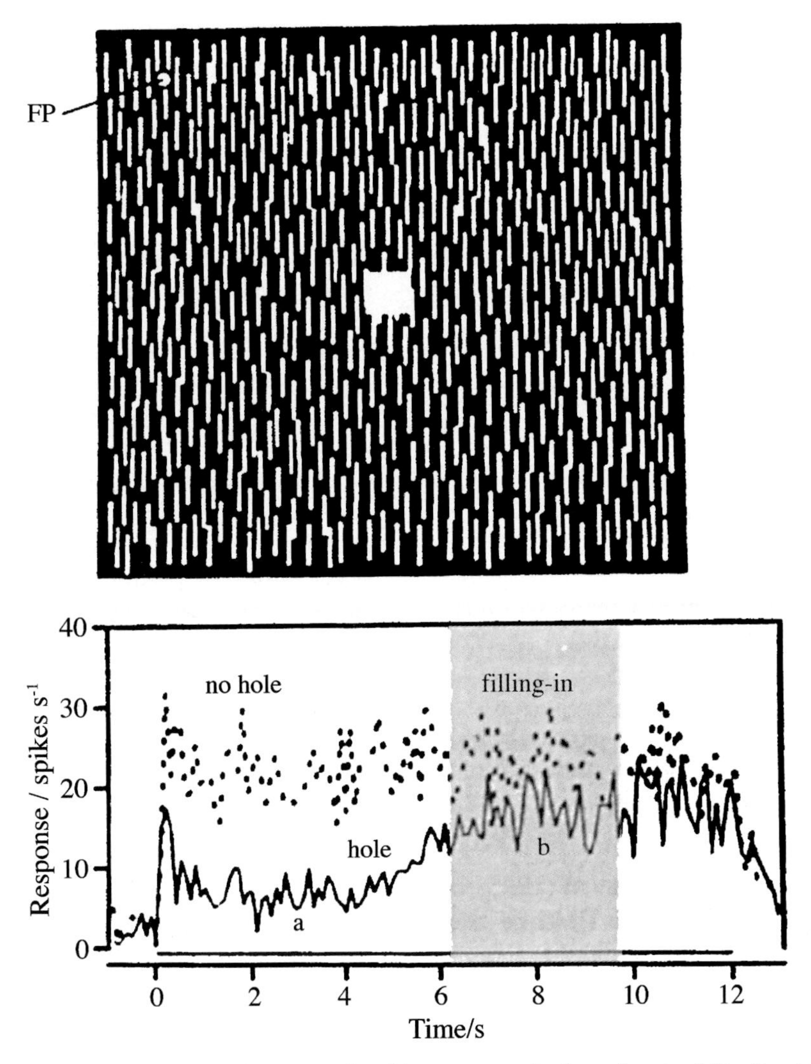

Fig. 20. Fading and filling-in. (Top) With fixation at FP, the white square on the dynamic noise field will quickly fade and become embedded in the background. (Bottom) Neuron response in area V3 of the monkey during fixation of the stimulus with the white square ("hole") and to a control stimulus presented without the white square (no hole). The receptive field of the neuron was located well within the square. The two curves converge at about the same time when human observers report (shaded area) that the white square has faded into the background. (Reprinted from De Weerd et al., 1995, with kind permission from Nature Publishing Group.)

to the time required for neuronal completion (shaded area). When a red square was used for a target, fading time increased. This suggests an effect of stimulus salience.

Texture fading

Ralf Teichmann in our laboratory studied the effect of salience on fading time, systematically (Teichmann and Spillmann, 1997). He presented a striped target disk within a grating background and varied the difference in angle between the target and the background (Fig. 21). Fading time was longest when target and background were oriented approximately orthogonally to each other, i.e., when the target was most salient.

Catherine Hindi Attar has recently taken this approach one step further. She used two patterns by Giovanni Vicario (1998), a randomly oriented center within a uniformly oriented surround (Fig. 22, left) and its converse, a uniformly oriented center within a randomly oriented surround (Fig. 22, right). Although the two patterns are made up from the same textures, they do not have the same perceptual salience. The center with the randomly oriented bars stands out much more clearly than its converse, and it takes several seconds longer to fade. This may be because this type of texture activates many orientation channels, thereby producing a stronger neuronal response than a uniform texture. (For a review see Spillmann and DeWeerd, 2003.)

Filling-in of the blind spot

While filling-in of an artificial scotoma requires a trained observer and good fixation, filling-in over the area of the blind spot is effortless and immediate (Ramachandran and Gregory, 1991). This is because in the first case, the hole is in the physical stimulus and must first be adapted to before it is filled-in with the color and pattern of the surround. In the case of the blind spot, however, the hole is on the retina and it has been there since birth. It is therefore not surprising that nature has provided us with a mechanism that replaces the hole with the stimulus properties of the surround without our doing. There are no photoreceptors in the retinal area corresponding to the blind spot and therefore no signals from there reach the brain. Nevertheless, we do not normally notice the blind spot. Even if we close one eye, we do not see it, although it is quite large ($6^\circ \times 9^\circ$). We asked, how much information at the edge is needed to fill-in the blind spot?

We started by plotting the blind spot of my left eye. Once the blind spot had been charted, we presented a large red blob, somewhat larger than the blind spot (Fig. 23, top). It looked uniform — as it should. Then we cut out the center (Fig. 23, bottom). It still looked uniformly red. Finally, we reduced the width of the frame, making it narrower and narrower until it no longer became filled-in. In this way we arrived at a critical frame width of 6 arcmin for the minimum information necessary to fill-in the blind spot.

The same procedure was used for a textured background (stripes, dots). Here the critical width of the surrounding frame was about three times greater than for color, implying that more information was needed. However, uniform filling-in often was short-lived due to unstable fixation. Slight deviations of the eye from the fixation point resulted in partial

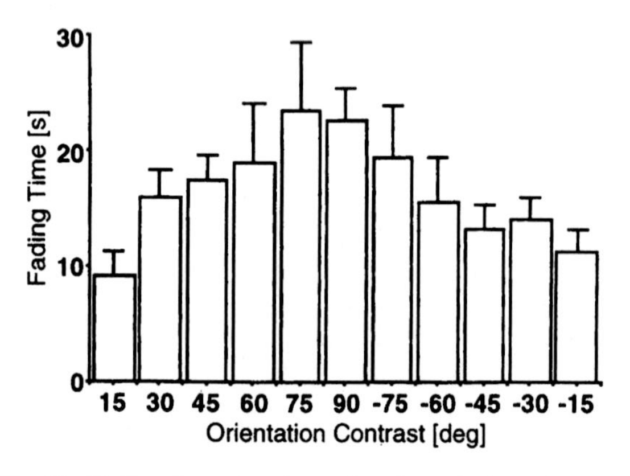

Fig. 21. Fading time plotted as a function of orientation contrast between target and background. The striped target disk was 2° in diameter and was centered at 15° from fixation. The spatial frequency of the target and background was 0.8 cpd. Results are averages for one presentation each in nine observers. (Reprinted from Teichmann and Spillmann, 1997, with kind permission from Thieme.)

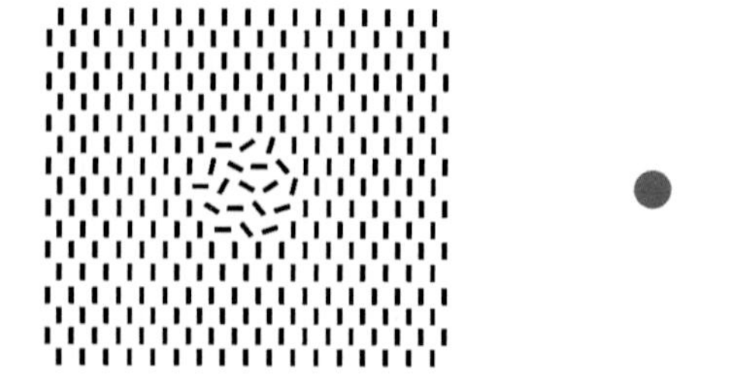

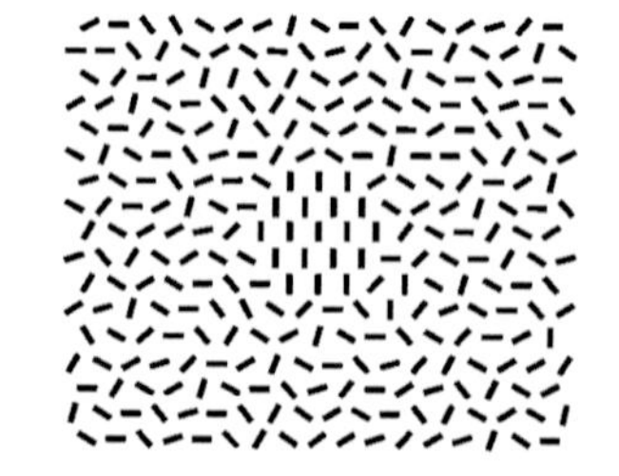

Fig. 22. "Order versus chaos." The left and right patterns are composed of the same two textures, however, with target and surround exchanged. The target on the left looks more salient than the one on the right and also takes longer to fade. Fixation is on the black dot in the middle. (Reprinted from Vicario, 1998, with kind permission from Springer Science and Business Media.)

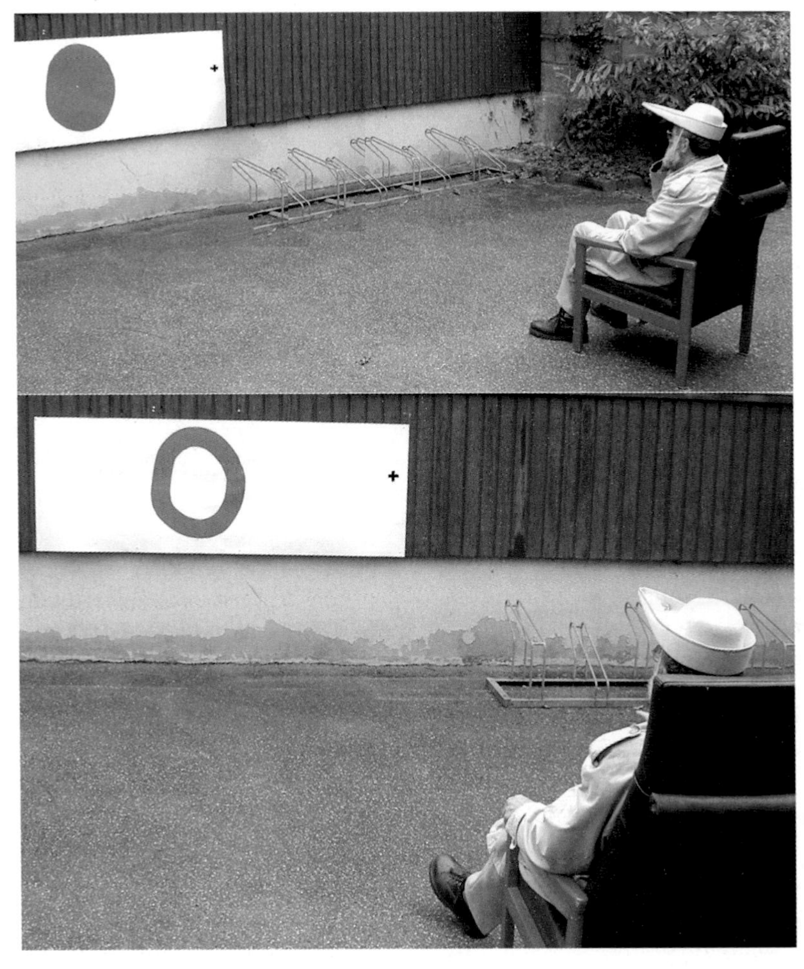

Fig. 23. Filling-in of the blind spot. (Top) A large red patch overlapping with the blind spot appeared uniformly colored, although there are no photoreceptors in the central area to signal its color to the brain. (Bottom) A narrow annulus at the edge of the blind spot had the same perceptual effect, suggesting that little information is required for filling-in. With strict fixation and under controlled stimulus conditions, a frame of only a few arcmin was found to suffice for uniform and complete filling-in. (Photo: Tobias Otte.)

filling-in, indicating that the frame was no longer spatially contiguous and in register to the border of the blind spot. To explain filling-in of the blind spot, we suggest a neuronal mechanism that detects the color at the edge and actively propagates it from there into the area of the blind spot (Spillmann et al., 2006).

Retinal scotomata

Next we asked whether acquired scotomata, such as those caused by a retinal lesion, also fill in. The answer is yes. Fig. 24 schematically presents a mechanism to account for filling-in. Neurons respond when light falls onto their receptive fields (left). However, when a patch of retina is destroyed by photocoagulation, the deafferented neuron falls silent and, as a consequence, there will be a scotoma in the visual field. Surprisingly, the silence lasts only for a short while. Charles Gilbert (1992) has shown that only a few minutes after deafferentation a neuron in area V1 will begin to fire again when light falls onto the area surrounding the lesion (see also Spillmann and Werner, 1996).

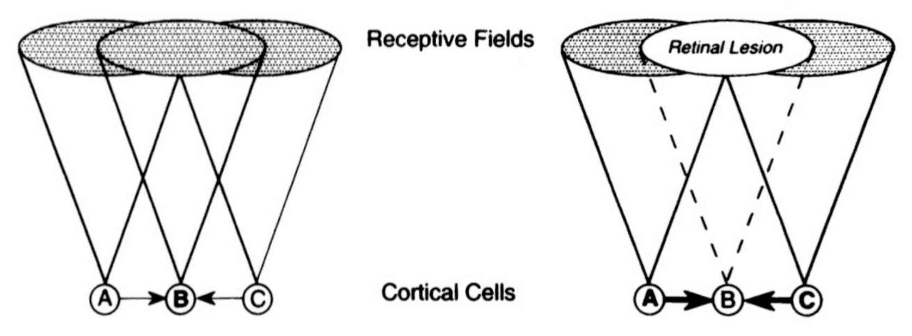

Fig. 24. Neural circuitry assumed to subserve filling-in of a real scotoma by intracortical horizontal connections (schematic). (Left) Uniform illumination of receptive fields of cortical neurons (A–C). (Right) After a retinal lesion, the associated neuron (B) in the visual cortex will stop firing. However, only a few minutes after deafferentation, it can be reactivated by illumination of the neighboring areas through lateral signals from neurons A and C (thick arrows). The resulting enlargement of the receptive field is assumed to underlie the perceptual filling-in of the scotoma with properties from the surround. Sustained brightness and color perception on large uniform surfaces (left) may be similarly explained by maintained edge signals from the surround. (Reprinted from Spillmann and Werner, 1996, with kind permission from Elsevier.)

One way to explain this reorganization is by collateral input activating the silenced neuron via horizontal fibers (thick arrows in Fig. 24, right). Such cortico-cortical connections are assumed to be functionally present all the time; however, in the case of local deafferentation, their influence may become potentiated owing to disinhibition. So one would expect that the receptive field of the neuron should become larger than it was before because of the inclusion of neighboring receptive fields. This is indeed the case. Gilbert found an enlargement of up to a factor 5. This enlargement may be responsible for the perceptual filling-in of the scotoma with stimulus properties from the surround.

I have asked a number of patients with diabetic retinopathy who had undergone retinal laser therapy how they perceived a uniformly white wall. If you consider that such patients have several hundred scars on their retinae, you would expect them to see a sieve with many dark holes in it. However, most of these patients said that their perception was largely unchanged. This is clear evidence for filling-in, although a control experiment using a textured wall remains to be done.

From perceptive field to Gestalt

Watercolor effect

How do we bridge the gap from perceptive fields to *Gestalten?* This is exemplified by the watercolor effect of Pinna et al. (2001, 2003). This effect is produced by a light-colored contour (e.g., orange) that runs alongside a darker chromatic contour (e.g., purple). In Fig. 25a, assimilative color (orange) is seen to spread from the chromatic double contour onto the enclosed surface area. The colored area is much larger than receptive fields of individual neurons. The watercolor effect may thus be thought of as an example of large-scale interaction from sparse stimulation, not unlike brightness and color perception on extended surfaces. Both percepts require transient edge signals and active propagation (i.e., filling-in) to sustain. Michael Paradiso (see his chapter in this volume) has presented psychophysical and neurophysiological evidence for such a mechanism.

We have shown that the watercolor effect exerts a strong effect on figure-ground organization. It thereby overrules the classical Gestalt factors such as *proximity, good continuation, closure, symmetry, and Convexity* (Fig. 25b). The asymmetric luminance profile of the stimulus defines what becomes figure and what ground. Invariably the side with the lower luminance contrast is seen as figure and that with the higher contrast as ground. By imparting illusory color the watercolor effect assigns unambiguous figure status to the perceptually tinted area.

This effect is consistent with Edgar Rubin's (1915) notion that the border belongs to the figure, not the ground. A neurophysiological correlate of

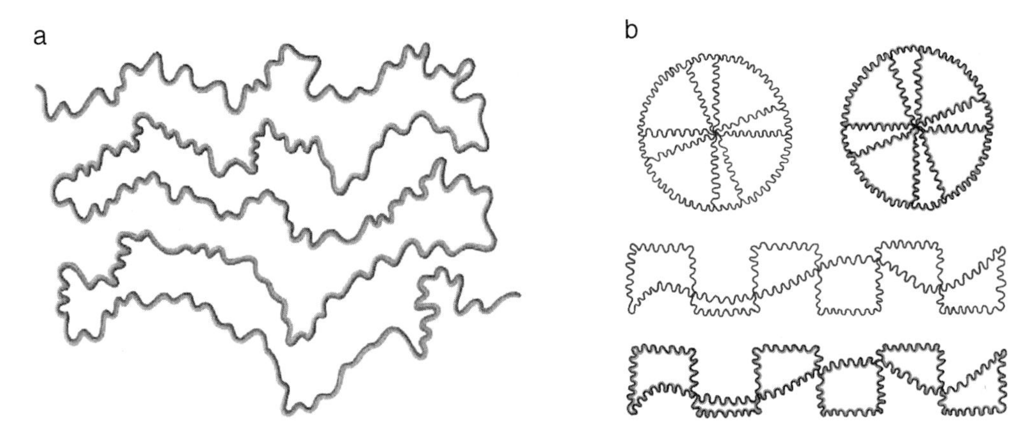

Fig. 25. Watercolor effect. Pinna's watercolor effect is an example of large-scale color assimilation arising from thin color boundaries. (a) A wiggly purple contour flanked by a lighter orange fringe elicits the perception of uniform orange coloration on the enclosed white surface area. The illusory surface appears to be slightly elevated relative to its surround. (Reprinted from Pinna, Brelstaff and Spillmann, 2001, with kind permission from Elsevier). (b) The watercolor effect overrules the classical Gestalt factors. In Rubin's (1915) Maltese cross (top) the narrow sectors are normally seen as figure. However, if pitted against the watercolor effect, the wider sectors are seen as figure, regardless of the *factor of proximity*. Similarly in the example of an undulating line superimposed onto a crenellated line or Greek fret (bottom), the *factor of good continuation* is overruled in favor of seeing closed cells. (Modified from Pinna et al., 2003, with kind permission from Elsevier.)

border ownership may be sought in area V2 neurons that respond to an edge — such as black and white — in one direction, but not in the other (Zhou et al., 2000; Qiu and von der Heydt, 2005). Neurocomputational models of form perception assume that the outflow of color depends on a weakening of the boundary between differentially activated edge neurons through lateral inhibition. The resultant assimilative color spreads through the enclosed surface area until it is stopped by boundary contours on the other side (Pinna and Grossberg, 2005).

Part C. Reminiscences and outlook

The studies mentioned so far cover only a fraction of what we did in Freiburg. Of the numerous other experiments, I will only mention the research on rod-monochromacy. We were fortunate to have Knut Nordby from Oslo, a former student and colleague of Svein Magnussen's (Fig. 26). Knut had no cones in his retinae, a very rare condition, making him an ideal subject for the study of rod vision.

Fig. 26. Knut Nordby (1942–2005). (Photo from K.N.'s website on the Internet.)

Mark Greenlee and Svein Magnussen (Greenlee et al., 1988) looked at Knut's spatial contrast sensitivity and orientation tuning and found that Knut had very low spatial frequency channels. They suggested that these had evolved by adapting to the lack of cones in his retinae. Arne Valberg says that he also had a much better contrast sensitivity, possibly due to reduced lateral inhibition.

For an account of his childhood and youth, see his autobiographic book chapter (Nordby, 1990).

Thereafter, Ted Sharpe (Sharpe and Nordby, 1990) in a series of sophisticated experiments looked at Knut's vision from every angle. Some of the very best threshold curves in the literature came from this research, although Knut could not fixate. This testifies to his patience and dedication. Invitations from Cambridge and other prestigious laboratories followed. In the end he was the world's best-researched rod monochromat. In 2004, we learned that Knut was very ill. Sadly, he died on April 25, of the following year.

There are a large number of experiments that I can only list by name: studies on Stiles' π-mechanisms by Charles Stromeyer and Charles Sternheim; studies on the nature of brown by Ken Fuld, Jack Werner, and Billy Wooten; masking and metacontrast studies by Bruno Breitmeyer; studies on the tilt effect and tilt aftereffect by Svein Magnussen and Wolfgang Kurtenbach; and studies on the Abney effect by Wolfgang Kurtenbach; experiments on the so-called Ouchi illusion by myself (Spillmann et al., 1986) and the motion aftereffect by Nick Wade; a study by Holger Knau on the Ganzfeld; and studies on the persistence of moving arcs by Adam Geremek; a beautiful collaboration with Barbara Heider and Esther Peterhans from Zurich on stereoscopic illusory contours in man and monkey. Studies on S-cones by Keizo Shinomori and on the foveal blue scotoma by Svein Magnussen and Frank Stürzel were done in collaboration with Jack Werner in Sacramento. On a different topic, Mark Greenlee and Svein Magnussen together with Jim Thomas and Rolf Müller conducted a whole series of experiments on grating adaptation and short-term memory. There were clinically oriented studies by Walter Ehrenstein on interocular time thresholds in MS-patients, and by myself and Dieter Schmidt on partial recovery in prosopagnosia (patient WL). Lately we have also become interested in functional magnetic resonance imaging of the Pinna–Brelstaff illusion in collaboration with the Freiburg Department of Radiology. I think we have gone a long way.

Looking back, I guess I was lucky. I had excellent teachers: Wolfgang Metzger in Münster, Richard Jung in Freiburg, and Hans-Lukas Teuber in Cambridge, MA (Fig. 27). From them I learned much about Gestalt psychology, neurophysiology, and neuropsychology. In 1966, I entered Ernst Wolf's laboratory in Boston, a German expatriate and utterly decent man, who introduced me to visual psychophysics, clinical testing of eye patients, and electrophysiology. I inherited from him my love for optical and mechanical apparatus and precise measurement. Ernst told me about Selig Hecht's laboratory at Columbia and his early years at the Harvard laboratories of Comparative Zoology. Nobody knew his age, but he was as active, enthusiastic, and untiring as anyone. All these people taught me a lot, but — most importantly — they made me aware that we stand on the shoulders of others that paved the way. Sadly, most of our heroes and heroines from that time are no longer with

W. Metzger R. Jung H.-L. Teuber

Fig. 27. Wolfgang Metzger (1899–1979), Richard Jung (1911–1986), and Hans-Lukas Teuber (1916–1976). (Sources unknown.)

us. But our admiration, respect, and affection for them continue.

It is the same with equipment. Who still remembers the cherished apparatus we used? Stimulus generators, oscilloscopes, Tektronix 602 displays, tachistoscopes, Maxwellian view systems? High-quality optics were necessary, precisely aligned components, stable light sources, IR and UV filters, prisms, collimators, achromatic lenses, first surface mirrors, narrow-band monochromators, electromagnetic shutters, neutral-density wedges, step filters, adjustable apertures, prisms, pellicles, and beamsplitters. All observations were done with subjects supporting themselves on a dentist-fitted, adjustable bite bar, so that the exit aperture was centered in the pupil. Highly sensitive radio-photometers were required for precise calibration. You had to be good with the soldering iron, too.

This is a bygone time. But I do remember how impressed I was when I visited Richard Gregory's laboratory in Cambridge in the late 1960s. It looked more like a mechanics workshop than a vision lab: Helen Ross was sitting in a swing testing for size constancy. In Freiburg, individual spikes were counted from a filmstrip (sometimes using an abacus), Jerry Lettvin listened to neuronal spike activity simply by ear, and Hubel and Wiesel used a stick and a screen to find orientation-specific neurons in the cat. Baumgartner may have missed his chance because it took too long to build an apparatus for precise stimulus presentation (Jung, 1975). There should be a museum to keep these memories alive. A whole generation of expertise in building instrumentation for the life sciences seems to have gone lost. Nowadays, computers are much faster, more convenient, and powerful. Sometimes, I feel like a man from the Stone Age. But not everything can be done using a monitor.

Epilogue

May I end by saying: It was wonderful. We had all the freedom in the world to do what we wanted, where, when, and with whom. We have precious memories of the many guests and visitors who came to Freiburg. It is great to see the international vision community growing. National borders no longer play a role. The East has opened up, so that we see more and more representatives of those countries. In fact, the 2006 *European Conference on Visual Perception* will take place in St. Petersburg. Scientists are so much better than politicians at striking friendly relationships.

It is also rewarding to see that in Germany there are many more psychophysics laboratories now than there were in 1971, when we started. For 25 years I sent out information on jobs and positions to several hundred addresses via D-CVnet, to keep the German vision community together. Vision research is now actively pursued in Mainz, Frankfurt, Düsseldorf, Dortmund, Münster, Bremen, Kiel, Potsdam, Giessen, Tübingen, Ulm, Regensburg, and Munich, among other places. The Freiburg laboratory, regrettably, was discontinued, although it was one of the few that enabled young German students to collaborate with established vision researchers from other countries. Wolfgang Kurtenbach and Frank Stürzel had as many as six papers each, jointly published with senior faculty from the US, Canada, and Scandinavia before receiving their doctorate.

Finally, I will always be grateful to my collaborators for their loyalty and help. None of the research that came out of our laboratory could have been done without them. Numerous publications owe their existence to the long-standing scientific exchange with the laboratories in Boulder, Sacramento, Oslo, Trondheim, Dortmund, Sassari, Padua, and New York. The social side of science always meant a great deal to me. Looking over the past 35 years, I will not forget the unfailing help received from my friends in the vision communities in Boston, Cambridge, MA, and Berkeley. I thank the University of Freiburg, the German Research Council, the Alexander von Humboldt-Foundation, the German Academic Exchange Program, and the other funding agencies for their most generous support. And I thank my family.

Summary

I have attempted to show how the study of simple perceptual phenomena enables us to learn more about the neuronal processing of visual stimuli in the human brain. Examples include the Hermann grid illusion, illusory contours, figure-ground segregation, coherent motion, fading and filling-in, and large-scale color assimilation.

The term *perceptive field* is heuristically valuable as it provides the bridge from the phenomenon to the underlying *receptive field* organization. The correlation between the two is not just qualitative; it also enables quantitative comparisons. Gestalt phenomena that were observed 80 years ago have not lost any of their meaning; to the contrary, they even have gained in importance. With today's knowledge of neuronal mechanisms, they serve as non-invasive tools to gain insight into the processes of how the visual system organizes the seemingly bewildering wealth of information from the outside world.

In his seminal article on visual perception and neurophysiology, Richard Jung (1973) published a table of neurophysiological correlates summarizing much of the Freiburg work. Given the speed of today's progress and the enthusiasm of researchers in the field of vision (Chalupa and Werner, 2004), we have good reasons to hope that in the next 30 years the neuronal mechanisms and processes underlying visual perception will be largely unveiled.

Acknowledgements

I thank S.C. Benzt for transcribing my talk at ECVP 2005 in A Coruna (Spain). The additional help by W.H. Ehrenstein, E. Peterhans, B. Heider, J.S. Werner, A. Kurtenbach, C. Stromeyer, and B. Breitmeyer is greatly appreciated. Tobias Otte kindly modified and assembled the figures.

References

Barlow, H.B., Fitzhugh, R. and Kuffler, W.S. (1957) Change of organisation in the receptive fields of the cat's retina during dark adaptation. J. Physiol. (Lond.), 137: 338–354.

Baumgartner, G. (1960) Indirekte Grössenbestimmung der receptiven Felder der Retina beim Menschen mittels der Hermannschen Gittertäuschung. Pflügers Arch., 272: 21–22.

Baumgartner, G. (1961) Die Reaktionen der Neurone des zentralen visuellen Systems der Katze im simultanen Helligkeitskontrast. In: Jung, R. and Kornhuber, H.-H. (Eds.), Neurophysiologie und Psychophysik des visuellen Systems. Springer, Berlin, pp. 296–313.

Baumgartner, G. (1990) Where do visual signals become a perception? In: Eccles J. and Creutzfeldt O. (Eds.), The Principles of Design and Operation of the Brain, Vol. 78. Pontificiae Academiae Scientiarum Scripta Varia, Vatican City, pp. 99–114.

Baumgartner, G., von der Heydt, R. and Peterhans, E. (1984) Anomalous contours: a tool in studying the neurophysiology of vision. Exp. Brain Res., (Suppl.), 9: 413–419.

Braddick, O.J., O'Brien, J.M.D., Wattam-Bell, J., Atkinson, J. and Turner, R. (2000) Form and motion coherence activate independent, but not dorsal/ventral segregated, networks in the human brain. Curr. Biol., 10: 731–734.

Breitmeyer, B. and Valberg, A. (1979) Local foveal inhibitory effects of global peripheral excitation. Science, 203: 463–464.

Bressan, P., Mingolla, E., Spillmann, L. and Watanabe, T. (1997) Neon color spreading. Perception, 26: 1353–1366.

Britten, K.H. (2004) The middle temporal area: motion processing and the link to perception. In: Chalupa, L.M. and Werner, J.S. (Eds.), The Visual Neurosciences. The MIT Press, Cambridge, MA, pp. 1203–1216.

Britten, K.H., Shadlen, M.N., Newsome, W.T. and Movshon, J.A. (1992) The analysis of visual motion: a comparison of neuronal and psychophysical performance. J. Neurosci., 12: 4745–4765.

Chalupa, L. and Werner, J.S. (Eds.). (2004) The Visual Neurosciences. The MIT Press, Cambridge, MA.

DeMonasterio, F.M. and Gouras, P. (1975) Functional properties of ganglion cells of the rhesus monkey retina. J. Physiol. (Lond.), 251: 167–195.

DeWeerd, P., Gattas, R., Desimone, R. and Ungerleider, L.G. (1995) Responses of cells in monkey visual cortex during perceptiual filling-in of an artifical scotoma. Nature, 377: 731–734.

Drasdo, N. (1977) The neural representation of visual space. Nature, 266: 554–556.

Ehrenstein, W.H. (1954) Probleme der ganzheitspsychologischen Wahrnehmungslehre (3rd edn). Barth, Leipzig.

Ehrenstein, W.H., Spillmann, L. and Sarris, V. (2003) Gestalt issues in modern neuroscience. Axiomathes, 13: 433–458.

Field, D.J., Hayes, A. and Hess, R.F. (1993) Contour integration by the human visual system — evidence for a local association field. Vision Res., 33: 173–193.

Gilbert, C.D. (1992) Horizontal integration and cortical dynamics. Neuron, 9: 1–13.

Greenlee, M.W., Magnussen, S. and Nordby, K. (1988) Spatial vision of the achromat: spatial frequency and orientation-specific adaptation. J. Physiol. (Lond.), 395: 661–678.

Grossberg, S. and Mingolla, E. (1985) Neural dynamics of form perception: boundary completion, illusory figures, and neon color spreading. Psychol. Rev., 92: 173–211.

Grüsser, O.-J., Kapp, H. and Grüsser-Cornehls, U. (2005) Microelectrode investigations of the visual system at the Department of Clinical Neurophysiology, Freiburg i.Br.: a historical account of the first 10 years, 1951–1960. J. Hist. Neurosci., 14: 257–280.

Hubel, D.H. and Wiesel, T.N. (1974) Uniformity of monkey striate cortex: a parallel relationship between field size, scatter and magnification factor. J. Comp. Neurol., 158: 295–306.

Johansson, G. (1973) Visual perception of biological motion and a model of its analysis. Percept. Psychophys., 14: 201–211.

Jung, R. (1961) Korrelation von Neuronentätigkeit und Sehen. In: Jung, R. and Kornhuber, H.-H. (Eds.), Neurophysiologie und Psychophysik des visuellen Systems. Springer, Berlin, pp. 410–435.

Jung, R. (1973) Visual perception and neurophysiology. In: Jung, R. (Ed.) Handbook of Sensory Physiology, Vol VII/3A, Central Processing of Visual Information. Springer, Berlin, N Y, pp. 1–152.

Jung, R. (1975) Some European scientists: a personal tribute. In: Worden, F.G., Swazey, J.P. and Adelman, G. (Eds.), The Neurosciences: Paths of Discovery. The MIT Press, Cambridge, MA, pp. 447–511.

Jung, R. and Kornhuber, H.-H. (Eds.). (1961) Neurophysiologie und Psychophysik des visuellen Systems. Springer, Berlin.

Jung, R., Spillmann, L. (1970) Receptive-field estimation and perceptual integration in human vision. In: Young, F.A. and Lindsley, D.B. (Eds.), Early Experience and Visual Information Processing in Perceptual and Reading Disorders. Proc. Natl. Acad. Sci., Washington DC, pp. 181 – 197.

Kanizsa, G. (1974) Contours without gradients or cognitive contours? Italian J. Psychol., 1: 93–112.

Kanizsa, G. (1979) Organization in Vision. Essays on Gestalt Perception. Praeger, New York.

Kastner, S., Nothdurft, H.-C. and Pigarev, I.N. (1999) Neuronal responses to orientation and motion contrast in cat striate cortex. Visual Neurosci., 15: 587–600.

Kovács, I. and Julesz, B. (1993) A closed curve is much more than an incomplete one: effect of closure in figure-ground segmentation. Proc. Natl. Acad. Sci. USA, 90: 7495–7497.

Lamme, V.A. (1995) The neurophysiology of figure-ground segregation in primary visual cortex. J. Neurosci., 15: 1605–1615 1995.

Lamme, V.A., van Dijk, B.W. and Spekreijse, H. (1992) Organization of contour from motion processing in primate visual cortex. Vision Res., 34: 721–735.

Livingstone, M.S. (2002) Vision and Art. The Biology of Seeing. Harry N. Abrams, New York.

Livingstone, M.S. and Hubel, D.H. (1987) Psychophysical evidence for separate channels for the perception of form, color, movement, and depth. J. Neurosci., 7: 3416–3468.

McIlwayn, J.T. (1964) Receptive fields of optic tract axons and lateral geniculate cells: peripheral extent and barbiturate sensitivity. J. Neurophysiol., 27: 1154–1174.

Metzger, W. (1936) Gesetze des Sehens (1st edn.). Kramer, Frankfurt/M. Engl. Transl. (2006) Laws of Seeing. MIT Press, Cambridge, MA.

Mountcastle, V.B. (1998) Perceptual Neuroscience: The Cerebral Cortex. Harvard University Press, Cambridge, MA.

Neumeyer, C. and Spillmann, L. (1977) Fading of steadily fixated large test field in extrafoveal vision. Pflügers Arch., 368: R40 (Abstract).

Nieder, A. (2002) Seeing more than meets the eye: processing of illusory contours in animals. J. Comp. Physiol. A, 188: 249–260.

Nordby, K. (1990) Vision in a complete achromat: a personal account. In: Hess, R.F., Sharpe, L.T. and Nordby, K. (Eds.), Night Vision: Basic, Clinical and Applied Aspects. Cambridge University Press, Cambridge, pp. 290–315 (Chapter 8).

Oehler, R. (1985) Spatial interactions in the rhesus monkey retina: a behavioural study using the Westheimer paradigm. Exp. Brain Res., 59: 217–225.

Peterhans, E., Heider, B. and Baumann, R. (2005) Neurons in monkey visual cortex detect lines defined by coherent motion of dots. Euro. J. Neurosci., 21: 1091–1100.

Peterhans, E. and von der Heydt, R. (1991) Subjective contours — bridging the gaps between psychophysics and physiology. Trends Neurosci., 14: 112–119.

Peterhans, E., von der Heydt, R. and Baumgartner, G. (1986) Neuronal responses of illusory contour stimuli reveal stages of visual cortical processing. In: Pettigrew, J.D., Sanderson, K.J. and Levick, W.R. (Eds.), Visual Neuroscience. Cambridge University Press, Cambridge, pp. 343–351.

Pinna, B., Brelstaff, G. and Spillmann, L. (2001) Surface color from boundaries: a new 'watercolor' illusion. Vision Res., 41: 2669–2676.

Pinna, P. and Grossberg, S. (2005) The watercolor illusion and neon color spreading: a unified analysis of new cases and neural mechanisms. J. Opt. Soc. Am. A, 22: 2207–2221.

Pinna, B., Werner, J.S. and Spillmann, L. (2003) The watercolor effect: a new principle of grouping and figure-ground organization. Vision Res., 42: 43–52.

Qiu, F.T. and von der Heydt, R. (2005) Figure and ground in the visual cortex: V2 combines stereoscopic cues with Gestalt rules. Neuron, 47: 155–166.

Ramachandran, V.S. and Gregory, R.L. (1991) Perceptual filling in of artificially induced scotomas in human vision. Nature, 350: 699–702.

Ramachandran, V.S., Tyler, C.W., Gregory, R.L., Rogers-Ramachandran, D., Duensing, S., Pillsbury, C. and Ramachandran, C. (1996) Rapid adaptive camouflage in tropical flounders. Nature, 379: 815–818.

Ransom-Hogg, A. and Spillmann, L. (1980) Perceptive field size in fovea and periphery of the light- and dark-adapted retina. Vision Res., 20: 221–228.

Redies, C. (1989) Discontinuities along lines. Psychophysics and neurophysiology. Neurosci. Biobehav. Rev., 13: 17–22.

Redies, C. and Spillmann, L. (1981) The neon color effect in the Ehrenstein illusion. Perception, 10: 667–681.

Rubin, E. (1915) Synsoplevede Figurer. Kopenhavn, Glydendalske.

Schepelmann, F., Aschayeri, H. and Baumgartner, G. (1967) Die Reaktionen der simple field — Neurone in Area 17 der Katze beim Hermann-Gitter-Kontrast. Pflügers Arch., 294: R57 (abstract).

Schiller, P.H. (1998) The neural control of visually guided eye movements. In: Richards, J.E. (Ed.), Cognitive Neuroscience of Attention: A Developmental Perspective. Erlbaum Assoc., Mahwah, NJ, pp. 3–50 (Chapter 1).

Schiller, P.H. and Carvey, C.E. (2005) The Hermann grid illusion revisited. Perception, 34: 1375–1397.
Sharpe, L.T. and Nordby, K. (1990) Total colour blindness: an introduction. In: Hess, R.F., Sharpe, L.T. and Nordby, K. (Eds.), Night Vision: Basic, Clinical and Applied Aspects. Cambridge University Press, London, pp. 253–289.
Soriano, M., Spillmann, L. and Bach, M. (1996) The abutting grating illusion. Vision Res., 36: 109–116.
Spillmann, L. (1971) Foveal perceptive fields in the human visual system measured with simultaneous contrast in grids and bars. Pflügers Arch. ges. Physiol., 326: 281–299.
Spillmann, L. (1994) The Hermann grid illusion: a tool for studying human perceptive field organization. Perception, 23: 691–708.
Spillmann, L. (1999) From elements to perception: local and global processing in visual neurons. Perception, 28: 1461–1492.
Spillmann, L. and DeWeerd, P. (2003) Mechanisms of surface completion: perceptual filling-in of texture. In: Pessoa, L. and DeWeerd, P. (Eds.), Filling-in: From Perceptual Completion to Cortical Reorganization. Oxford University Press, Oxford, pp. 81–105.
Spillmann, L. and Dresp, B. (1995) Can we bridge the gap between levels of explanation. Perception, 24: 1333–1364.
Spillmann, L. and Ehrenstein, W.H. (1996) From neuron to Gestalt: mechanisms of visual perception. In: Greger, R. and Windhorst, U. (Eds.) Comprehensive Human Physiology, Vol. 1. Springer, Berlin, pp. 861–893.
Spillmann, L. and Ehrenstein, W.H. (2004) Gestalt factors in the visual neurosciences. In: Chalupa, L. and Werner, J.S. (Eds.), The Visual Neurosciences. The MIT Press, Cambridge, MA, pp. 1573–1589.
Spillmann, L., Heitger, F. and Schüller, S. (1986) Apparent displacement and phase unlocking in checkerboard patterns. 9. Europ. Conf. Vis. Perception, Bad Nauheim (Poster).
Spillmann, L. and Kurtenbach, A. (1992) Dynamic noise backgrounds facilitate target fading. Vision Res., 32: 1941–1946.
Spillmann, L., Otte, T., Hamburger, K. and Magnussen, S. (2006) Perceptual filling-in from the edge of the blind spot. Vision Res. (Under revision).
Spillmann, L., Ransom-Hogg, A. and Oehler, R. (1987) A comparison of perceptive and receptive fields in man and monkey. Hum. Neurobiol., 6: 51–62.
Spillmann, L. and Werner, J.S. (Eds.). (1990) Visual Perception: The Neurophysiological Foundations. Academic Press, NY.
Spillmann, L. and Werner, J.S. (1996) Long-range interaction in visual perception. Trends Neurosci., 19: 428–434.
Stürzel, F. and Spillmann, L. (2004) Perceptual limits of common fate. Vision Res., 44: 1565–1573.
Teichmann, R. and Spillmann, L. (1997) Fading of textured targets on textured backgrounds. In: Elsner, N. and Wässle, H. (Eds.) Göttingen Neurobiology Report 1997, vol. II. Thieme, Stuttgart, p. 569 (Abstract).
Uttal, W.R., Spillmann, L., Stürzel, F. and Sekuler, A.B. (2000) Motion and shape in common fate. Vision Res., 40: 301–310.
Valberg, A. and Lee, B.B. (Eds.). (1991) From Pigments to Perception. Advances in Understanding Visual Processes. NATO ASI Series, Plenum Press, London.
Van Tuijl, H.F.J.M.van. (1975) A new visual illusion: neonlike color spreading and complementary color induction between subjective contours. Acta Psychol., 39: 441–445.
Vicario, G.B. (1998) On Wertheimer's principles of organization. In: Stemberg, G. (Ed.) Gestalt Theory, Vol. 20. Krammer, Vienna, pp. 256–270.
von der Heydt, R. (1987) Approaches to visual cortical function. Rev. Physiol. Biochem. Pharmacol., 108: 69–151.
von der Heydt, R. and Peterhans, E. (1989) Mechanisms of contour perception in monkey visual cortex. I. Lines of pattern discontinuity. J. Neurosci., 9: 1731–1748.
von der Heydt, R., Peterhans, E. and Baumgartner, G. (1984) Illusory contours and cortical neuron responses. Science, 224: 1260–1262.
Wertheimer, M. (1912) Experimentelle Studien über das Sehen von Bewegung. Z. Psychol., 61: 161–265.
Wertheimer, M. (1923) Untersuchungen zur Lehre von der Gestalt II. Psychol. Forsch., 4: 301.
Westheimer, G. (1965) Spatial interaction in the human retina during scotopic vision. J. Physiol. (Lond.), 181: 881–894.
Westheimer, G. (1967) Spatial interaction in human cone vision. J. Physiol (Lond.), 190: 139–154.
Westheimer, G. (2004) Center-surround antagonism in spatial vision: retinal or cortical locus? Vision Res., 44: 2457–2465.
Zhou, H., Friedman, H.S. and von der Heydt, R. (2000) Coding of border ownership in monkey visual cortex. J. Neurosci., 20: 6594–6611.

Martinez-Conde, Macknik, Martinez, Alonso & Tse (Eds.)
Progress in Brain Research, Vol. 155
ISSN 0079-6123

CHAPTER 6

In honour of Lothar Spillmann — Filling-in, wiggly lines, adaptation, and aftereffects

Stuart Anstis*

Department of Psychology, UCSD, 9500 Gilman Drive, La Jolla, CA 92093-0109, USA

Abstract: I have studied a number of visual phenomena that Lothar Spillmann has already elucidated. These include:

Neon spreading: when a small red cross is superimposed on intersecting black lines, the red cross seems to spread out into an illusory disk. Unlike the Hermann grid, neon spreading is relatively unaffected when the black lines are curved or wiggly. This suggests that the Hermann grid, but not neon spreading, involves long-range interactions. Neon spreading can be shown in random-dot patterns, even without intersections. It is strongest when the red crosses are equiluminous with the gray background.

Adaptation, aftereffects, and filling-in: direct and induced aftereffects of color, motion, and dimming. Artificial scotomata and filling-in: the "dam" theory is false. Staring at wiggly lines or irregularly scattered dots makes them gradually appear straighter, or more regularly spaced. I present evidence that irregularity is actually a visual dimension to which the visual system can adapt.

Conjectures on the nature of peripheral fading and of motion-induced blindness.

Some failed experiments on correlated visual inputs and cortical plasticity.

Keywords: adaptation; aftereffects; afterimages; color induction; filling-in; illusions

Introduction

It is impossible to summarize Lothar Spillmann's contributions to visual psychophysics because he has studied just about everything. If he has not studied it, it is not psychophysics. I shall just discuss some random samples taken from his formidable body of works on vision. The topics I have picked include the Hermann grid, neon spreading, filling-in and aftereffects, and visual plasticity. Note that many of the illusions described here are beautifully illustrated on the web page of Lothar's colleague Michael Bach at http://www.michaelbach.de/ot/

*Corresponding author. Tel.: +1-858-534-5456;
E-mail: sanstis@ucsd.edu

Long- and short-range interactions: Hermann's grid vs. neon spreading

Hermann grid. Spillmann has always been interested in the relationships between long- and short-range interactions in vision (Spillmann and Werner, 1996; Spillmann, 1999). A case in point is the Hermann-grid illusion (Hermann, 1870; Spillmann, 1971, 1994; Spillman and Levine, 1971; Oehler and Spillmann, 1981), which has long been regarded as a short-range process but has now been shown to require long-range processes as well (Geier et al., 2004). In the Hermann grid, illusory dark spots or blobs can be seen at every street crossing, except for the ones that are being directly fixated. A stronger version, known as the scintillating grid (Schrauf et al., 1997; Ninio and Stevens, 2000; Schrauf and Spillmann, 2000),

DOI: 10.1016/S0079-6123(06)55006-X

has a small disk at each intersection. This produces a smaller but much darker and more vivid illusory point. Both the Hermann grid and the scintillating grid work equally well in reversed contrast, with black stripes on a white ground.

The standard, short-range explanation comes from Baumgartner (1960). He suggested that an on-center retinal ganglion cell could be positioned by chance at an intersection, in which case it would have four bright regions in its inhibitory surround, one from each street, and these would reduce its response. A ganglion cell looking at a street would have only two inhibitory regions, so it would respond more strongly. A fixated intersection falls on the fovea, where the receptive fields are so small that it would make no difference whether or not it fell on an intersection. In fact, Spillmann (1994) and Ransom-Hogg and Spillmann (1980) measured the stripe widths that gave the maximum illusion at different eccentricities in order to determine the size of human "perceptive fields."

This explanation fails to explain why global factors are important. Wolfe (1984) pointed out that Baumgartner's model is local in nature, since it relies on cells with concentric on-off or off-on receptive fields. This model predicts that the magnitude of the illusion at a given intersection should be the same whether that intersection is viewed in isolation or in conjunction with other intersections in a grid. However, Wolfe showed that illusion magnitude grows with the number of intersections and that this growth is seen when the intersections are arranged in an orderly grid but not when they are placed irregularly. These results rule out any purely local model for the Hermann-grid illusion. Global factors must be involved. Geier et al. (2004) decisively overthrew the Baumgartner model by imparting a slight sinusoidal curvature to the lines. When the lines are straight the illusion is visible, but as soon as the lines become curved the illusion vanishes. The same distortions applied to the scintillating grid made the scintillations disappear. This implies that the Hermann grid and the scintillating grid both depend upon long-range interactions, probably operating along the length of the lines (see Fig. 1).

Neon spreading. Spillmann has also studied the neon spreading that can be seen at the intersection of two thin black lines (Bressan et al., 1997). A red + sign superimposed on the intersection appears to spread out into a pink disk, provided that the black lines are continuous with, and aligned with, the red lines (Redies and Spillmann, 1981; Spillmann and Redies, 1981; Redies et al., 1984; Kitaoka et al., 2001). Don Macleod and I wondered whether neon spreading, like the Hermann grid, would vanish for curved lines. If so, neon spreading would also depend upon long-range global interactions, and not merely upon local factors. Accordingly we (he) wrote a program that could apply any desired curvature to a neon-spreading lattice of black lines. *Result*: Curving the lines did *not* reduce the neon spreading, in sharp contrast to Geier's results with the Hermann grid. This suggests that neon spreading is a *local*, short-range affair.

Fig. 2 shows that neon spreading is strongest when the red crosses are equiluminous with the surround. In Fig. 2, the gray background is swept from dark on the left to light on the right, while the red crosses are swept from darkest at the bottom to lightest at the top. A glance at Fig. 2a shows that neon spreading is strongest along a positive diagonal where the luminances of the colored crosses and the gray background are equal.

In that case, what is the minimum stimulus that neon spreading requires? My own observations suggest, not much. It is well known that a square lattice of thin black lines on a white surround gives strong neon spreading when the intersections are replaced with red. But I also produced neon spreading in sparse, stationary random black dots scattered on a white surround, simply by coloring a ring-shaped subset of the black dots red (not illustrated). The ring was then moved around, but the red/black dots defining it remained stationary, merely turning red when they lay within the annular region that defined the moving ring and returning to black when they did not. Result: observers reported a pink neon annulus moving around across a stationary random-dot field. The neon effect was much stronger when the ring moved than when it was stationary. This shows that neon spreading is not necessarily dependent upon geometrical features such as intersections. It merely needs to replace black regions that lie on a white ground.

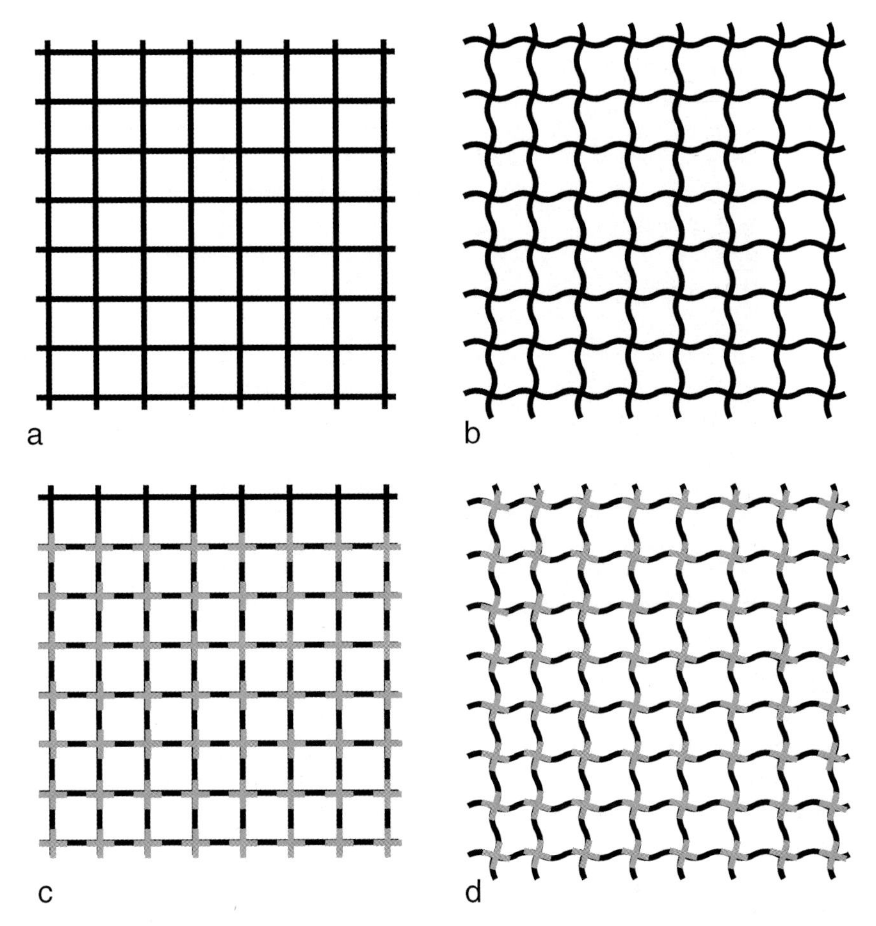

Fig. 1. (a) Hermann grid ...(b) is abolished by curving lines (Geier et al., 2004). (c) Neon spreading...(d) is unaffected by curving lines.

These results point to some low-spatial-frequency visual pathway that blurs the seen colors and spreads them outside the confines of the thin lines. Could neon spreading simply be a by-product of the famously low acuity of the chromatic pathways? (Kelly, 1983). This cannot be the whole story since neon spreading also works for gray (Bressan et al., 1997). It might be that the beautiful watercolor effect discovered by Pinna et al. (2001, 2003) is an extreme case of neon spreading.

Peripheral fading

Lothar Spillmann has always been fascinated by the fact that strict fixation can make peripherally viewed stimuli fade out and disappear from view. Here are some examples of peripheral fading.

1. Troxler fading of a luminance-defined object, such as a black or white disk on a gray surround (Troxler, 1804).
2. A window of drifting dense random dots embedded in a field of twinkling dynamic noise gradually fades from view and disappears (Anstis, 1989). When all motion is subsequently stopped, the window shows a negative aftereffect of motion (Figs. 3a, b).
3. A small, peripheral gray patch embedded in a field of twinkling dynamic noise also gradually fades from view (Ramachandran and Gregory, 1991; Spillmann and Kurtenbach,

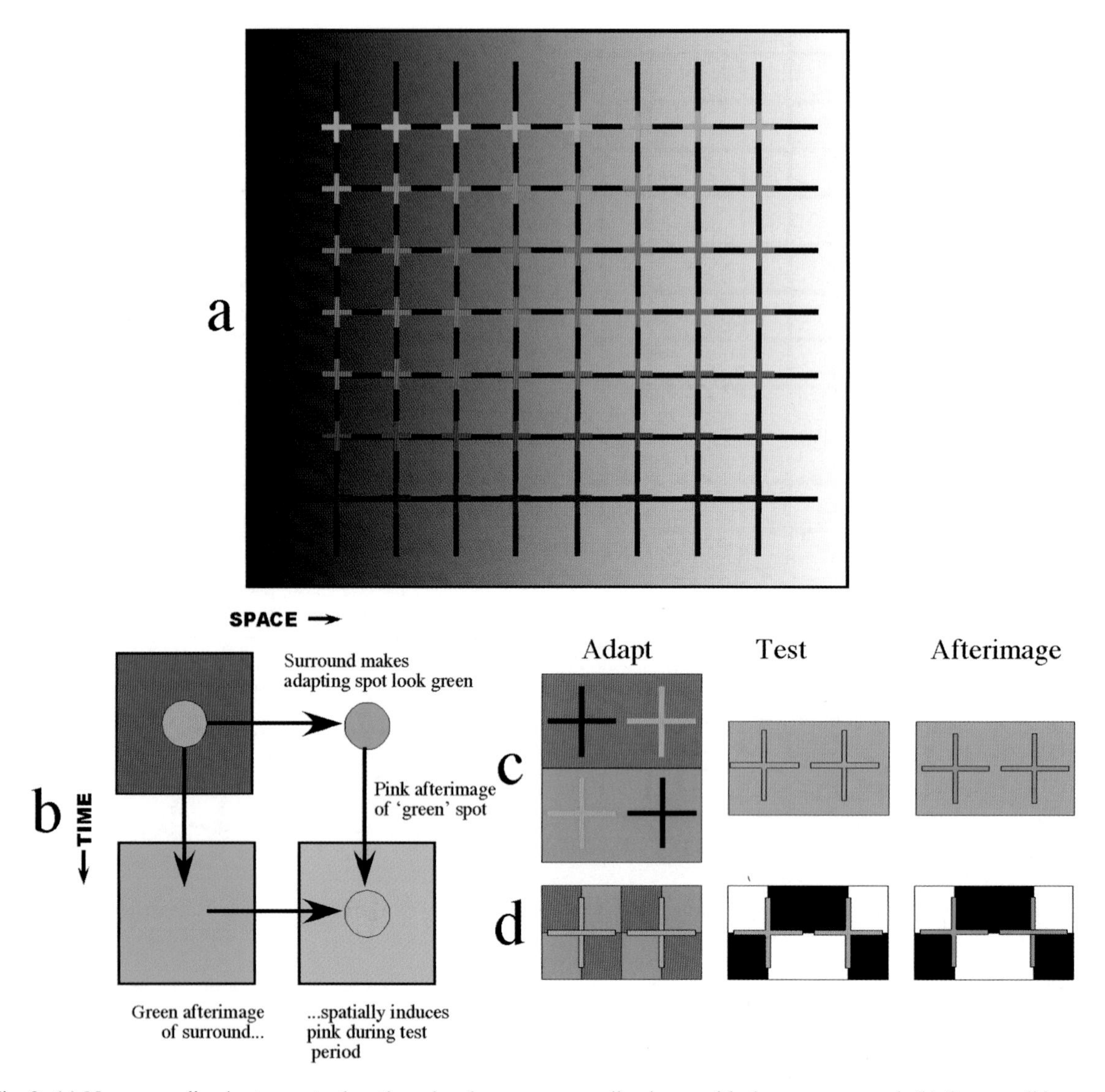

Fig. 2. (a) Neon spreading is strongest when the colored crosses are equiluminous with the gray surround. (b) Two possible neural routes for induced afterimages: (i, top right) surround spatially induces subjective green into during adaptation, then greenish spot has pink afterimage; (ii, bottom left) magenta surround has green afterimage, which spatially induces pink into the post during the test period. (c) shows existence of (i): following adaptation to alternating black and gray cross on magenta and green surrounds, a gray test field shows afterimages in which the left cross looks greenish and the right cross looks pinkish. (d) shows existence of (ii): following adaptation to gray crosses lying on green and magenta quadrants, the white test quadrants show afterimages which induce the left cross to look greenish and the right cross to look pinkish. (After Anstis et al., 1978.)

1992). When the field switches to a uniform gray test field, a twinkling aftereffect, resembling the original twinkling dots, fills the gray patch (Figs. 3c, d).

4. A gray patch, or a patch of one kind of static texture, embedded in another kind of static texture, will also fade out and disappear (Spillmann, 2003). The more salient the texture, the longer it takes to disappear (Sturzel and Spillmann, 2001).
5. A flickering peripherally viewed spot remains visible, but the perceived amplitude of its

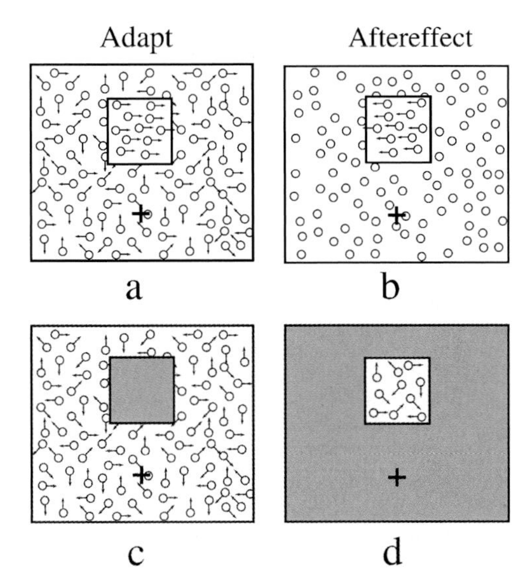

Fig. 3. (a) A window filled with drifting dots, embedded in dynamic noise, rapidly disappears from view and (b) gives a negative motion aftereffect on a stationary test field (Anstis, 1989). (c) A gray patch also disappears and (d) gives a twinkling aftereffect (Ramachandran and Gregory, 1991).

flicker falls steadily over time (Schieting and Spillmann, 1987; Anstis, 1996).

6. Peripherally viewed motion can appear to slow down and stop (Campbell and Maffei, 1979; Hunzelmann and Spillman, 1984).
7. A very blurred colored spot on an equiluminous gray surround gradually fades from view. On a gray test field an afterimage in the complementary color is visible. Jeremy Hinton has a nice demonstration on Michael Bach's web page. Twelve identical small blurred purple spots are arranged around a clock face, and the observer fixates in the middle. The spot at 1 o'clock is briefly turned off and then turned back on. Then the spot at 2 o'clock similarly disappears briefly, then the spot at 3 o'clock, and so on. At first one sees a "traveling gap," sometimes knows as omega movement (Tyler, 1973). But after a few seconds all the purple spots fade out from view and one simply perceives a bright green spot — an afterimage — running around clockwise.
8. A spectacular new illusion of peripheral fading is known as "motion-induced blindness" (Bonneh et al., 2001).

Filling-in and aftereffects: Spillmann's theory

Peripherally viewed stimuli often fade out from view during strict fixation, as listed above. For instance, if a large surround is filled with twinkling dynamic noise, a simple small gray square (Ramachandran and Gregory, 1991) or a static window filled with drifting random dots (Anstis, 1989) will disappear from view within $\sim$10 s. Spillmann and de Weerd (2003) have suggested a *two-stage theory* to explain all such examples of delayed peripheral fading. First, there is a slow adaptation of the figure's boundary representation that normally keeps the figure perceptually segregated from its surround, and then a fast filling-in or interpolation process takes place in which the area previously occupied by the figure becomes invaded by the surround. De Weerd et al. (1998) measured the time for filling-in of a gray square on a black background filled with small white vertical bars that moved dynamically, a form of anisotropic dynamic noise. They used various sizes and eccentricities of squares, and based upon estimates of cortical magnification, they found that the time for filling-in was linearly related to the total contour length of the square's projection upon the visual cortex, rather than its retinal image size. This is consistent with Spillmann and de Weerd's theory. I shall characterize (or caricature) Spillmann and de Weerd's theory as the "dam" theory. The boundary of a peripheral gray patch walls or dams it off from an ocean of twinkling dots. Adaptation slowly erodes the dam, and twinkling water then rushes in quickly. This fits the experimental facts. However, Ramachandran and Gregory applied a dam theory to the subsequent aftereffect, such that when the water is drained away, some twinkling water remains briefly trapped inside the dam. This story predicts that the water inside the dam — the aftereffect — should be the same color or texture as the adapting ocean. But I find that it is not. The twinkle aftereffect has a fixed grain size (spatial frequency), regardless of the adapting grain size!

Ramachandran and Gregory (1991) attribute their twinkling aftereffect to a process of interpolation that actively fills-in the gray test square with the twinkling dots from the adapting surround. However, Tyler and Hardage (1998) disagree, and my own observations also suggest otherwise. Like Ramachandran et al., (1991) I used a gray patch set in a field of twinkling dots, but I also systematically varied the grain size of the adapting random dots over an eightfold range, with dots ranging from 0.1° to 0.8° in diameter. When observers were then exposed to a uniform gray test field, they were able to match the apparent grain of the twinkling aftereffect by means of an adjustable random-dot field. If some active filling-in were responsible, the aftereffect grain size should match the adapting grain size. In fact, however, I found that the aftereffect was always matched to the *same* grain size of 0.1°, whatever the adapting grain be. Since the aftereffect remained constant even when the adapting field changed, it was clearly not the result of a simple filling-in process. But in that case, what was it? (see Fig. 4).

A clue comes from "induced afterimages of color" (Anstis et al., 1978). If you adapt to an equiluminous gray cross in a very large green surround and then switch to a uniform gray test field, the expected afterimage might be a gray cross in a pink surround. But in fact, the afterimage is of a strongly green cross in a neutral gray surround!

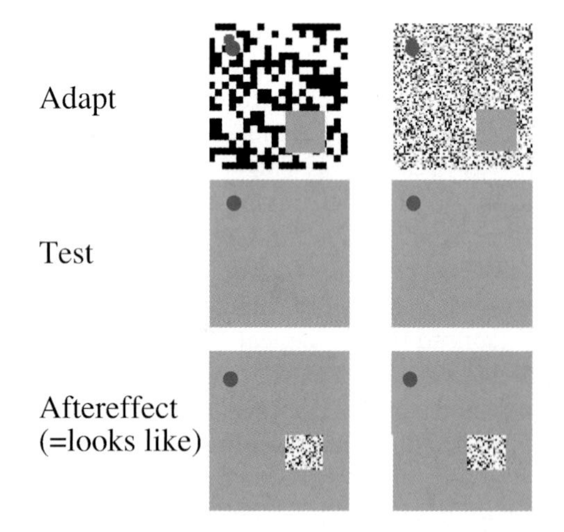

Fig. 4. Breakdown of the dam theory. If one fixates the black dot, the peripheral gray square soon disappears (top row), whatever the grain size of the noise. This fits the dam theory. When a gray test field is substituted (middle row) the previous location of the gray square is filled with a twinkling aftereffect (bottom row). However, this is not the simple fill-in that a dam theory might predict because the grain size of the aftereffect is constant and unrelated to the adapting grain size.

Thus, the cross afterimage matched the adapting surround. But this is not a simple fill-in. Instead, there are two possibilities (Fig. 2b). Either the adapting surround spatially induced an apparent pink into the cross during the adapting period, and this apparent pink was followed by its own afterimage in the cross, or alternatively, the adapting surround was followed by its own pink afterimage, which then spatially induced green into the cross during the test period. In the first case, simultaneous contrast (spatial induction) precedes successive contrast (afterimage). In the second case, successive contrast precedes simultaneous contrast. Many years ago (Anstis et al., 1978) we showed that both processes occur and can be elicited independently. We arranged for the background to switch every few seconds between two complementary colors, green and magenta. Two thin crosses lay side by side (Fig. 2c). When the surround was green, the left-hand cross was an equiluminous gray. By Grassman's (1853) third law, this induced a strong apparent pink into the cross. The right-hand cross was black, and only a minimum amount of pink was induced into it. Conversely, when the surround was magenta, the left-hand cross was black and the right-hand cross was an equiluminous gray, and looked apparently greenish. Following an adaptation period of 30–60 s, a uniform gray test field was presented. Two cross-shaped afterimages were visible. The left-hand afterimage looked green, and the right-hand afterimage looked pink. This shows that the subjective colors induced into the crosses during the adaptation period could generate their own afterimages. Note that the alternation of the complementary colors green and magenta in the adapting surround would cancel out and produce no net colored afterimage in the test surround.

In our second experiment, the adapting fields surrounding each cross were divided into green and magenta quadrants of equal sizes (Fig. 2d). The crosses always looked neutral gray because each was bordered by equal amounts of green and magenta, during the adapting period. But during the test period the surround quadrants were made black and white. The black quadrants showed little or no afterimage, but on the white quadrants strong negative afterimages were seen, which spatially induced secondary pink and green afterimages into the crosses. As before, the left-hand afterimage looked green and the right-hand afterimage looked pink, but now the reasons were different. This result shows that the surround afterimages could spatially induce subjective colors into the crosses during the test period, but not the adaptation period. Thus in our first experiment, simultaneous contrast preceded successive contrast, while in our second experiment, successive contrast preceded simultaneous contrast. An alternative formulation is that the visual system can adapt to color ratios that can be expressed as edge-redder-on-left and edge-greener-on-right. This description includes red/white, red/green, and white/green edges. Incidentally, in these experiments an outline of the crosses was included in the test field because afterimages are easier to see when they are outlined (Daw, 1962). I have also found similar interactions between simultaneous induction and successive aftereffects, both for motion (Anstis and Reinhardt-Rutland, 1976) and for adaptation to gradual change of luminance (Anstis, 1979), which I shall not describe here. For a useful review of color induction, see Zaidi (1999). For filling-in, see Pessoa and de Weerd, (2003).

I conclude that in these experiments on colored afterimages, and in the disappearance of small patches superimposed on texture (Anstis, 1989; Ramachandran and Gregory, 1991; Spillmann and Kurtenbach, 1992; Sturzel and Spillmann, 2001; Spillmann, 2003), there is a complex interplay between processes of simultaneous and successive contrast. The aftereffects were *not* a simple fill-in from the adapting surrounds.

Motion-induced blindness

Bonneh et al. (2001) have discovered a dramatic example of peripheral fading. Three small stationary yellow spots forming a triangle are displayed on a monitor screen, against a background of small dark blue spots that rotate around the center of the screen or else fly around randomly like a swarm of midges. If one gazes at the center of the triangle, the yellow spots dramatically disappear and reappear. Although the conditions are not too critical, the effect is strongest for small, high-contrast stationary yellow dots against numerous, high-contrast blue dots in rapid motion. The blue surround spots can be flickering instead of moving.

Fig. 5 shows a simplified version of their stimulus, comprising an array of stationary dark blue flickering spots. Each spot flickers independently between 100 different random luminance levels, retaining always the same blue hue. Three of the spots are static yellow instead of blue. When the observer looks at the middle of this triangle of yellow spots, the yellow spots seem to disappear

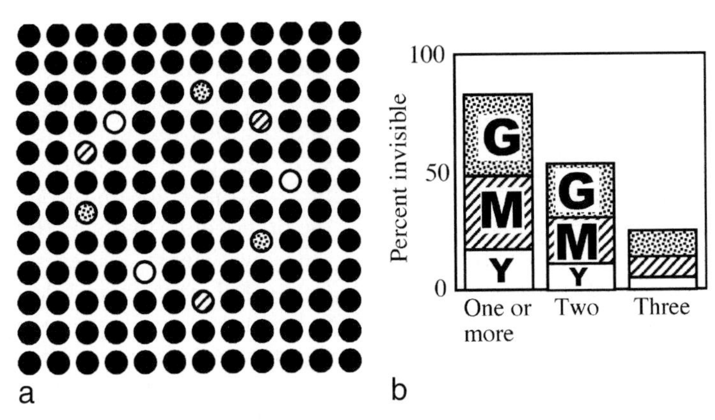

Fig. 5. (a) All spots flicker except the three yellow (unfilled) spots, which seem to disappear dramatically. If observers attend to the three flickering magenta (striped) or flickering green (spotted) dots, these also disappear. (b) Percentage of accumulated invisibility period for the disappearance of one or more spots, exactly two spots, and exactly three spots. The spots were invisible for about 50% of the time. The flickering green and magenta spots disappeared even more than the static yellow ones.

and reappear. Bonneh et al. refer to this as "motion-induced blindness." However, in my hypothesis motion does not actually *induce* anything. Instead, I conjecture that:

1. *All peripherally viewed targets tend to disappear over time.* This conjecture applies to all the peripheral stimuli listed above, not just to motion-induced blindness.
2. *Changing (*i.e., *flickering or moving) targets are more resistant than static targets to disappearing.* This does not mean that flickering or moving objects are a proof against disappearance: they are not (Campbell and Maffei, 1981; Hunzelmann and Spillmann, 1984; Schieting and Spillmann, 1987; Anstis, 1996). It merely means that statistically they are less likely to disappear. It is like the old joke about the married couple being pursued by a hungry grizzly bear. As the wife runs she (yellow spots) reflects that she need not outrun the bear (blue spots), she need only outrun her husband. Specifically, the moving blue dots resist disappearance more than the yellow dots, but they do not *induce* any blindness to the blue dots. The situation is like a horse race in which the winner does not slow down the losers, as opposed to a tug of war in which the winners do oppose and impede the efforts of the loser. The wife does not need to trip up her husband, only outrun him. (Incidentally, eye movements will shift all the spots across the retina and tend to keep them visible.)
3. *The disappearances are generally not noticed unless attention is specifically directed toward them.*

I call this process "fade blindness." In this model, *all* the spots disappear from time to time. The static yellow spots may disappear more often than the moving blue spots, but in addition only the yellow ones are salient enough, and receive enough attention, for us to notice their fading. Consider any peripherally viewed blue spot. If this were to fade, its fading would scarcely be noticed, since it is already randomly flickering and changing luminance and is harder to distinguish from the other blue spots in any case. However, the three yellow spots stand out very sharply from the surround, and if these fade, it is immediately noticed. I suggest that *all* the spots stand an almost equal chance of fading, but the observer only notices this when the spots happen to be yellow and hence pop out and seize his attention. Thus all spots are constantly subjectively disappearing and reappearing, but only those spots that capture the observer's attention are correctly seen to be fading. To demonstrate this, three flickering blue dots were painted magenta and three others were painted green. The nine dots (yellow, magenta, and green) were arranged in an irregular circle around the fixation point. Result: the yellow spots still showed dramatic disappearances. But when observers were asked to concentrate on the magenta spots, or on the green spots, these attended spots also disappeared at irregular times. The purpose of the magenta and green colors was simply to label some of the flickering dots to make it easier for the observer to attend to them. The fact that these disappeared supports our conclusion that any dot can fade out, and hence we conjecture that all of them do so at irregular intervals. The fading escapes our notice, rather as the temporal changes in change-blindness movies also escape our notice, until we specifically direct our attention toward them (see below).

This hypothesis predicts that selectively paying attention to some peripheral spots will paradoxically make them more likely, not less likely, to fade out. Lou (1999) reported exactly this effect. He presented a circular array of six disks — three green and three orange — in alternate positions, against a uniform gray background. Sixteen observers maintained steady fixation at the center of the array and were instructed to direct their attention to three disks of one color and to ignore the three disks of the other color. In about 10 s, some disks started to fade away from awareness. Of those starting to fade, 81% were those selected for attention. The faded disks remained out of awareness for 1–2 s during which time other disks were clearly visible. The fading increased with eccentricity, a defining characteristic of Troxler fading. Lou concluded from the selectivity of the fading that voluntary attention can have an inhibitory effect on early sensory processing. However, we take a rather different view; we suggest that attention did not alter the sensory processing in any way, but simply made the Ss aware of the fading that they (and indeed all of us) usually ignore.

It may seem bizarre to suggest that all our lives peripheral stimuli have been fading without our ever being aware of it. But an analogous situation has recently been discovered, in which our picture of the world is far more fragile and impoverished than everyday intuition would suggest. I refer to change blindness. Suppose that a photograph of a jet plane is flashed up, followed immediately by a doctored version of the same scene that has been changed in some obvious way, say by removal of one jet engine, and these two pictures cycle continuously. Observers immediately report the location and nature of the change. However, if this scenario is repeated with fresh observers, but now with a blank gray interstimulus interval (ISI) of 250 ms inserted after each picture (A-blank-B-blank-A-blank…), observers typically fail to notice the change and need to scan the alternating picture sequence carefully for up to 30 s before they notice the change. The conclusion is that in the first instance, the engine flickering on and off captures observer's attention and this cues the identification of the change. However, inserting the ISI generates flicker transients over the whole field. Without the local attention-grabbing transient, the observer fails to see the change. This remarkable "change blindness" forces us to abandon the traditional belief that the visual system builds up a detailed and complex picture of the world over time, and suggests instead that our visual representations are sparse and volatile (Dennett, 1991; O'Regan, 1992; Simons and Levin, 1997; Wolfe, 1998; O'Regan et al., 1999; Rensink, 2000; Simons, 2000). Careful parametric studies by Becker and colleagues (Becker et al., 2000; Becker and Pashler, 2002) confirm that our visual inputs are indeed far more impoverished than we usually imagine. I am conjecturing that "fade blindness" is a near-universal but hitherto unrecognized phenomenon that, together with change blindness, seriously restricts the amount of information that we actually take in from the world. Careful attentive scrutiny is necessary to detect both the vanishing jet engine and the vanishing peripheral spots.

Adapting to irregularity

Some of the peripherally viewed stimuli listed earlier actually disappear during strict fixation. Others remain visible but they lose some of their visual properties; for instance, moving or flickering objects do not vanish but they gradually seem to lose their motion (Campbell and Maffei, 1979; Hunzelmann and Spillmann, 1984) or flicker (Schieting and Spillmann, 1987; Anstis, 1996). In such cases the visual system is adapting to higher order properties such as motion, flicker, or texture (e.g., Anstis, 1983), rather than to the luminance contours that define the object's existence and location. Alan Ho and I have measured visual adaptation to an unusual and frequently overlooked visual property, namely geometrical *irregularity*. We found that peripherally viewed wiggly or curved lines, or irregularly arranged dots, remain visible but gradually look smoother and more regular. This hints at the way in which the visual system codes for spacing and curvature.

Figs. 6a, b show a set of wiggly vertical lines, wiggly circles, and irregularly arranged dots, each mirrored about a fixation point (Mackay, 1964a, b). In each case, cover the right-hand half with a piece of paper and gaze steadily at the fixation point for 30 s or so. Now remove the paper and you will notice that the adapted lines, circles, or dots on the left look much straighter and more regular than the freshly exposed ones on the right. Note that in all these cases the test and the adapting stimulus were identical — the stimulus changed its appearance during prolonged inspection. This is different from the commoner measure of an aftereffect in which (say) a tilted adapting line alters the appearance of a vertical test line. Our process was a form of normalization, comparable to the way in which a tilted line gradually normalizes toward the vertical (Gibson and Radner, 1937).

Wiggly lines. In our experiments, a fixed set of adapting wiggly vertical lines lay to the left of a fixation point. Once per second an adjustable set of wiggly test lines was flashed up to the right of the fixation point for 100 ms. These test lines were a mirror image of the adapting lines, except that their amplitude was under the observer's control, and she/he hit keys to adjust this amplitude in

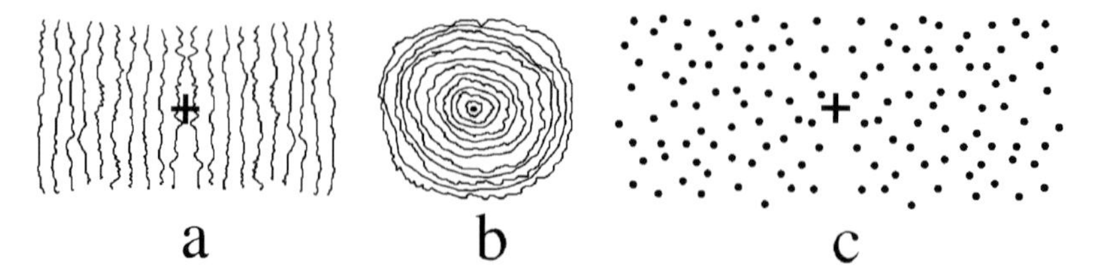

Fig. 6. (a)–(c) Adapt to irregularity (Mackay, 1964a, b). Cover one half, adapt for 30–60 s. Then uncover the half that should look more irregular.

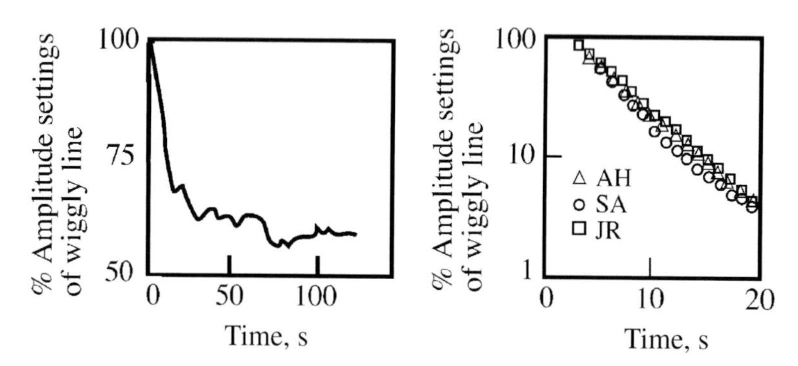

Fig. 7. (a) Illusory smoothing of wiggly lines over time. The *y*-axis shows the amplitude or gain of a matching wiggly line that subjectively matched a constant wiggly test line over time. (b) Illusory smoothing of sinusoidal curved lines over log time, averaged across spatial frequencies. All three observers had a time constant of 2.5 s. Note different scales on *y*-axis.

order to match subjectively the gradual decline of wiggliness in the adapting lines. Result: the illusory smoothing over time is shown in Fig. 7a.

Sinusoidal curves. Mathematically, every wiggly line is composed of a set of sinusoids of different amplitudes and spatial frequencies. So we also adapted to a single sinusoidal line, flashing up an adjustable comparison sinusoid next to it. Result: Fig. 7b shows that averaging across a range of different spatial frequencies, the perceived amplitude fell off exponentially with time, with a time constant of 2.5 s, the same for all three observers.

Irregular dot lattice. First, an irregular lattice was generated from a regular square lattice by sending each dot on a random walk, like a diffusion process. This adapting lattice was steadily fixated, and an adjacent, adjustable matching lattice was flashed up once every second for 100 ms, and again the observer continuously adjusted its irregularity to match the gradually changing appearance of the adapting irregular lattice. Result (not shown): the perceived regularity increased monotonically over time.

Interpretation. Instead of suggesting that these irregular stimuli became gradually straighter or more regular, I claim that they became less irregular. In other words, irregularity is a class of visual property that is explicitly coded by the visual system, analogous to color. Just as there are different hues, so there are different forms of irregularity, the most obvious being variance in curvature or in spacing. And just as adapting to colors tends to reduce their saturation, adaptation reduces the strength of irregularity. In support of this idea, we have also found that sinusoidally curved lines looked less curved (i.e., straighter) when they were masked with twinkling dynamic noise, or when they were defined by second-order texture or by twinkle instead of by luminance. We conclude that masking noise simulates the effects of prolonged adaptation, because both reduce the neural signal/noise ratio, in this case whatever it is that signals curvature. This was especially true of low-amplitude sinusoids. Thus, noise masking and prolonged adaptation both serve to reduce the signal/noise ratio of the neural signals of irregularity. So irregularity is a signal; added to a straight line, it gives curved or wiggly lines; added to a regular square lattice, it transforms it into an irregular one.

Visual search and added properties. Treisman and Souther (1985) showed that it is easier to find a Q hidden among O's than it is to find an O hidden among Q's. They argued that the tail of Q is an added visual feature or property, and it is easier to find a target that possesses the feature among distracters that lack it than vice versa. So if curvature is an added property, then curved lines should be easy to find among straight lines, whereas straight lines should be hard to find among curved lines. And they are (Treisman and Gormican, 1988).

Easy	Hard
OOQOO	QQOQQ
|||)|||	)))|)))

When searching arrays of dot lattices, it should be easier to find an irregular patch of dots hidden among regular dots than vice versa. Sturzel and Spillmann (2001) adapted to patches of irregular dots hidden among regular lattices, and found that they vanished after some seconds. This is similar to the adaptation effects that we found for lattices.

A differential-adaptation model. Prolonged inspection of a sinusoidal line adapts out the curvature signal so that the line gradually comes to look straighter. We conjecture that a single long receptive field running along the line codes its overall (mean) orientation, while the local curves are coded by small, end-stopped receptive fields parallel to the local curves. If these smaller units adapt out more rapidly than the larger long unit, the line will gradually come to look straighter.

Again, a tilted line is probably coded by overlapping, differently oriented receptive fields. Suppose that a line tilted 10° away from the vertical is coded by equal firing from one unit tuned to 0° (vertical) and a second unit tuned to 20° away from vertical. If the second more tilted unit adapts more rapidly, the tilted line will appear to become gradually more vertical. In general, we can model the normalization of any "self-adapting" stimulus by a firing-ratio model consisting of one unit tuned to a "norm," such as straight or vertical, and a second unit tuned somewhere away from the norm, such as oriented at some angle away from the vertical. If the norm unit adapts more slowly than the other unit, the stimulus will appear to normalize over time.

Glaucoma prediction. This leads to a clinical prediction. It has been shown that chronic glaucoma selectively damages large optic nerve fibers, which generally have large receptive fields. The loss of large cells tends to be greater in the lower than the upper retina, and to spare the fovea (Quigley et al., 1987, 1988; Glovinsky et al., 1991). Our model predicts that a sinusoidally curved line will look *more curved* to a glaucomatous patient because the straight-line signal from the long receptive field will be attenuated compared to the local curves signaled by the still intact smaller receptive fields. Projecting two identical sinusoidal lines on to more and less seriously affected retinal regions, such as the lower and upper retina of the same patient, will allow the observer to act as his own control; the more extensive the large cell loss in any retinal region, the more curved the line should look.

It might be possible to mimic the supposed visual effects of selective loss of large or small cells by adapting a normal observer to isotropic (nonoriented) dynamic noise that had been spatially filtered to contain only high or only low spatial frequencies. Preadaptation to coarse dynamic noise should temporarily attenuate large receptive fields and briefly simulate the visual experience of glaucoma. It should make a sinusoidally curved line look more curved. Adaptation to fine noise should have the opposite effect and make the line look apparently straighter.

A failure: do correlated sensory inputs teach the brain where things are?

Most of the experiments I do are rather dull, with results that would not surprise anybody. But at least they are very likely to work and lead to publications that keep my merit committee happy. But now and again it is worth attempting an experiment with a low probability of success but a very high payoff if it does succeed. I shall now describe such an attempt. The low probability won out over the high payoff and the experiment did not succeed, to put it mildly, but I am still glad that I tried it. The idea was to see whether the visual system can modify or rewire itself in response to experimentally correlated visual inputs.

Crick and Koch (2003, p. 120) asked:

> How are feature detectors formed? A broad answer is that neurons do this by detecting common and significant correlations in their inputs and by altering their synapses (and perhaps other properties) so that they can more easily respond to such inputs. In other words, the brain is very good at detecting apparent causation. Exactly how it does this is more controversial. The main mechanism is probably Hebbian, but Hebb's seminal suggestion needs to be expanded.

I narrowed down Crick and Koch's question to ask: how do we know where things are in space? Are we born with an innate knowledge of retinal sign, or do we acquire this by some learning process? One can imagine the brain to be like a TV technician sitting in a studio and receiving visual signals that arrive up the million cables of the optic nerve. At first he has no idea as to which signal comes from where in the retina, but by careful inspection he notices that some cables carry correlated signals. For instance, a pair of vertically aligned retinal receptors is likely to fire together when a vertical edge passes over them. Adjacent receptors are statistically likely to see the same colors, luminances, edge orientations, directions of motion, and so on. When he notices two cables with correlated inputs he ties them together. Gradually he can arrange the cables in a two-dimensional pattern that corresponds to the layout of the retina. Maloney and Ahumada (1989) offer some hints on how this might be done.

Much evidence from brain scans and neural recordings indicates that the brain may be an expert statistician that can draw inferences from correlated sensory inputs via Hebbian learning (reviewed by Cruikshank and Weinberger, 1996). Consider first the somatosensory system. Each body region sends signals to its own brain area. For instance, the five fingers on one hand all send signals to adjacent patches of the somatosensory cortex. Allard et al. (1991) glued together two fingers of an owl monkey's hand, and found that the two brain areas of the two fingers eventually coalesced into a single brain area. Were these brain changes produced by correlated inputs or correlated outputs? Gluing the two fingers together would correlate the motor outputs, since the two fingers were forced to move together, but it would also correlate the sensory inputs, since the two fingers would nearly always touch the same surfaces at the same time. Wang et al. (1995) showed that correlated inputs produced the brain changes. Without gluing the fingers together, they made a mechanical tapper that tapped both fingers simultaneously all day long at random times. Again the two brain areas coalesced. It occurred to me to do the same thing in vision, by applying correlated visual inputs to adjacent retinal areas.

The visual cortex also changes its firing patterns in response to correlated visual inputs (e.g., Eysel et al., 1998). Some computational models have suggested how the brain might wire up "local sign" on the retina by means of an unsupervised learning algorithm, based upon correlations of the retinal inputs that result from eye movements (Maloney and Ahumada, 1989).

Vision is obviously a more highly developed sense than skin sensitivity, so it seemed like a good idea to investigate correlated inputs in vision rather than on the skin. A prime example of correlated visual inputs is stereo vision, where the two eyes receive almost identical pictures which the brain can fuse or combine, and can use the small differences between the two pictures to calculate depth. Is this process hard wired or learned? There is much evidence that the two eyes compete for "brain space" during early development. Whereas acuity develops gradually over the first 6 months of life, stereo vision appears suddenly, usually within a few days, during the 20th week of life (reviewed by Atkinson, 2000). This suggests that stereo vision is hard wired, though it could in principle involve latent learning that shows no signs until a sudden winner-take-all decision process permits binocular fusion.

I correlated the visual inputs, not by putting near-identical pictures one into each eye but by putting both pictures side by side into the same eye(s), stimulating two nearby retinal areas with the same regime. Originally I thought of using two light-emitting diodes that would flash and flicker in step at the same rate, with matching brightness and color. But a much better idea is to expose two identical movies side by side on a split-screen television. With some difficulty I persuaded my university to buy me a new flat-panel TV with a built-in split-screen facility, which I installed in my living room. Whatever picture was exposed on the TV screen, it appeared as two identical pictures side by side and touching. Each picture was 7° wide at the viewing distance used, so that points that were horizontally separated spatially by 7° (one screen width) were always identical in color, flicker, motion, and brightness. Thus, the split-screen television set had the identical program on the two halves of the screen. Observers were instructed to watch the left-hand picture. There was now a point-to-point matching of brightness, color, edges, flicker, and movement over a considerable retinal area. Note that extremely strict fixation was not necessary; provided that the observer always looked somewhere in the middle of the left-hand picture, there was a large region ($<7°$) where points exactly 7° apart were correlated (see Fig. 8).

I invited three student research assistants over for the day, and all of us watched TV all day for six consecutive hours. We found that the correlation between the two screens felt "weird." This correlation was most obvious during periods of maximum motion and activity in the picture, and also whenever cuts between scenes, or camera zooms and pans, produced sudden massive changes in the picture. Of course the two pictures were always correlated 100%, but this was subjectively far more obvious for moving than for static pictures.

From time to time we ran some informal visual tests, looking for small perceptible changes. For instance, we thought it possible that correlated points would gradually begin to look closer together, so we ran a bisection task in which three dots were horizontally separated by $\sim 7°$. Observers adjusted the positions of the dots until the middle dot appeared to lie halfway between the two outer dots. We predicted that observers might underestimate the distance between the middle and the right-hand dots, corresponding to the separation between the fixated and the peripherally viewed screen.

We also thought that the visual system might start to predict the existence of the correlations, which might be visible as a form of contingent aftereffect (Shute, 1979). Accordingly, we occasionally covered the peripherally viewed screen with a black card to see whether observers could notice any ghostly duplicate of the centrally viewed picture. Also, we sometimes turned off the pair of pictures and substituted a single, foveally viewed stimulus such as a red cross rotating clockwise. Observers looked for any hint of an illusory duplicate shape 7° to the right, which might take the form of a ghostly red or green cross rotating clockwise or counterclockwise.

Unfortunately, all these tests came out negative. We never saw the slightest errors in bisection, nor

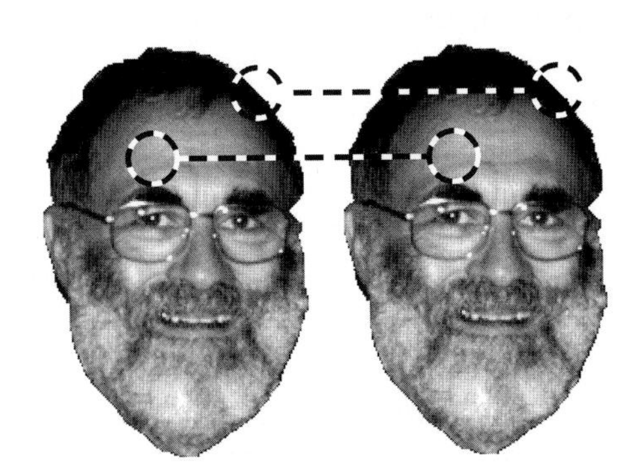

Fig. 8. (a) In normal vision, nearby points are statistically alike in orientation (hairline), color, and texture (forehead) ($0<p<1$). (b) In split-screen mode, points separated by 7° are precisely matched in orientation *and* color *and* texture and so on ($p = 1$).

ghostly illusions nor visual disturbances of any kind. So the experiment was a failure. However, it still looks like a promising line of enquiry, and given a better-designed experiment or a smarter experimenter it might still pay off. Perhaps I can null out the effects of eye movements if I spend more money.... My only problem is that my department may soon demand to get its TV back. When the TV goes, I shall have plenty of time to re-read Lothar Spillmann's huge collected works.

Conclusions

Among the new findings summarized here are the following:

1. Neon spreading is strongest at equiluminance. It does not necessarily require intersecting continuous lines, but can be seen even on scattered random dots.
2. The "dam" theory can explain the fading out of peripheral stimuli during adaptation, but not the subsequent aftereffects that are often misnamed as "filling-in."
3. We propose a speculative theory to explain phenomena of peripheral fading, in particular motion-induced blindness, which we suggest may not really be motion induced. In brief, all peripheral stimuli tend to disappear, especially during fixation, so we have a much sketchier picture of the peripheral scene than we generally believe. It is only when we attend to specific peripheral items that we become aware that they are fading. We regard this as a new cousin of change blindness.
4. During steady fixation, irregular lines, grids, and scattered dots gradually look more regular, or to be more accurate less irregular. We claim that irregularity or randomness is explicitly coded as a visual dimension, and adaptation reduces the strength of this coded neural signal.

References

Allard, T., Clark, S.A., Jenkins, W.M. and Merzenich, M.M. (1991) Reorganization of somatosensory area 3b representations in adult owl monkeys after digital syndactyly. J. Neurophysiol., 66: 1048–1058.

Anstis, S.M. (1979) Interactions between simultaneous contrast and adaptation to gradual change of luminance. Perception, 8: 487–495.

Anstis, S. (1983) Aftereffects of form, motion, and color. In: Spillmann, L. and Wooten, B.R. (Eds.), Sensory Experience, Adaptation, and Perception. Lawrence Erlbaum Associates, Hillsdale, NJ, pp. 583–601.

Anstis, S. (1989) Kinetic edges become displaced, segregated and invisible. In: Lam, D.M.K. and Gilbert, C.D. (Eds.), Neural Mechanisms of Visual Perception. Gulf Publishing Company, Houston, pp. 247–260.

Anstis, S. (1996) Adaptation to peripheral flicker. Vision Res., 36: 3479–3485.

Anstis, S.M. and Reinhardt-Rutland, A.H. (1976) Interactions between motion aftereffects and induced movement. Vision Res., 16: 1391–1394.

Anstis, S., Rogers, B. and Henry, J. (1978) Interactions between simultaneous contrast and coloured afterimages. Vision Res., 18: 899–911.

Atkinson, J. (2000) The Developing Visual Brain. Oxford University Press, Oxford.

Baumgartner, G. (1960) Indirekte Größenbestimmung der rezeptiven Felder der Retina beim Menschen mittels der Hermannschen Gittertäuschung. Pflügers Arch. gesamate Physiol., 272: 21–22.

Becker, M.W. and Pashler, H. (2002) Volatile visual representations: failing to detect changes in recently processed information. Psychonom. Bull. Rev., 9: 744–750.

Becker, M.W., Pashler, H. and Anstis, S. (2000) The role of iconic memory in change-detection tasks. Perception, 29: 273–286.

Bonneh, Y.S., Cooperman, A. and Sagi, D. (2001) Motion-induced blindness in normal observers. Nature, 411: 798–801.

Bressan, P., Mingolla, E., Spillmann, L. and Watanabe, T. (1997) Neon color spreading: a review. Perception, 26: 1353–1366.

Campbell, F.W. and Maffei, L. (1979) Stopped visual motion. Nature, 278: 192.

Crick, F. and Koch, C. (2003) A framework for consciousness. Nat. Neurosci., 6(2): 119–126.

Cruikshank, S.J. and Weinberger, N.M. (1996) Evidence for the Hebbian hypothesis in experience-dependent physiological plasticity of neocortex: a critical review. Brain Res. Rev., 22: 191–228.

Daw, N.W. (1962) Why after-images are not seen in normal circumstances. Nature, 196: 1143–1145.

De Weerd, P., Desimone, R. and Ungerleider, L.G. (1998) Perceptual filling-in: A parametric study. Vision Res., 38: 2721–2734.

Dennett, D.C. (1991) Consciousness Explained. Little, Brown and Co., Boston.

Eysel, U.T., Eyding, C.A.D. and Schweigart, G. (1998) Repetitive optical stimulation elicits fast receptive field changes in mature visual cortex. Neuroreport, 9: 949–954.

Geier, J., Sera, L. and Bernath, L. (2004) Stopping the Hermann grid illusion by simple sine distortion. Perception ECVP Abstracts.

Gibson, J.J. and Radner, M. (1937) Adaptation, after-effect, and contrast in the perception of tilted lines. I. Quantitative studies. J. Exp. Psychol., 20: 453–467.

Glovinsky, Y., Quigley, H.A. and Dunkelberger, G.R. (1991) Retinal ganglion cell loss is size dependent in experimental glaucoma. Invest. Ophthalmol. Vis. Sci., 32: 484–491.

Grassman, H. (1853) Zur Theorie der Farbenmischung. Poggendorfs Ann. Phys., 89: 69–84.

Hermann, L. (1870) Eine Erscheinung simultanen Contrastes. Pflügers Arch. gesamte Physiol., 3: 13–15.

Hunzelmann, N. and Spillmann, L. (1984) Movement adaptation in the peripheral retina. Vision Res., 24: 1765–1769.

Kelly, D.H. (1983) Spatiotemporal variation of chromatic and achromatic contrast thresholds. J. Opt. Soc. Am., 73: 742–750.

Kitaoka, A., Gyoba, J., Kawabata, H. and Sakurai, K. (2001) Two competing mechanisms underlying neon color spreading, visual phantoms and grating induction. Vision Res., 41: 2347–2354.

Lou, L. (1999) Selective peripheral fading: evidence for inhibitory sensory effect of attention. Perception, 28: 519–526.

Mackay, D.M. (1964a) Central adaptation in mechanisms of form vision. Nature, 203: 992–993.

Mackay, D.M. (1964b) Dynamic distortions of perceived form. Nature, 203: 1097.

Maloney, L.T. and Ahumada, A.J. (1989) Learning by assertion: two methods for calibrating a linear visual system. Neural Comput., 1: 392–401.

Ninio, J. and Stevens, K.A. (2000) Variations on the Hermann grid: an extinction illusion. Perception, 29: 1209–1217.

O' Regan, K.J. (1992) Solving the "real" mysteries of visual perception: The world as an outside memory. Can. J. Psychol., 46: 461–488.

O' Regan, K.J., Rensink, R.A. and Clark, J.J. (1999) Change blindness as a result of mudsplashes. Nature, 398: 34.

Oehler, R. and Spillmann, L. (1981) Illusory colour changes in Hermann grids varying only in hue. Vision Res., 21: 527–541.

Pessoa, L. and deWeerd, P. (2003) Filling-in: From Perceptual Completion to Cortical Reorganization. Oxford University Press, Oxford.

Pinna, B., Brelstaff, G. and Spillmann, L. (2001) Surface color from boundaries: a new 'watercolor' illlusion. Vision Res., 41: 2669–2676.

Pinna, B., Werner, J.S. and Spillman, L. (2003) The watercolor effect: a new principle of grouping and figure-ground organization. Vision Res., 43: 43–52.

Quigley, H.A., Dunkelberger, G.R. and Green, W.R. (1988) Chronic human glaucoma causing selectively greater loss of large optic nerve fibers. Ophthalmology, 95: 357–363.

Quigley, H.A., Sanchez, R.M., Dunkelberger, G.R., L'Hernault, N.L. and Baginski, T.A. (1987) Chronic glaucoma selectively damages large optic nerve fibers. Invest. Ophthalmol. Vis. Sci., 28: 913–920.

Ramachandran, V.S. and Gregory, R.L. (1991) Perceptual filling in of artificially induced scotomas in human vision. Nature, 350: 699–702.

Ransom-Hogg, A. and Spillmann, L. (1980) Perceptive field size in fovea and periphery of the light- and dark-adapted retina. Vision Res., 20: 221–228.

Redies, C. and Spillmann, L. (1981) The neon color effect in the Ehrenstein illusion. Perception, 10: 667–681.

Redies, C., Spillmann, L. and Kunz, K. (1984) Colored neon flanks and line gap enhancement. Vision Res., 24: 1301–1309.

Rensink, R.A. (2000) The dynamic representation of scenes. Vis. Cogn., 7: 17–42.

Schieting, S. and Spillmann, L. (1987) Flicker adaptation in the peripheral retina. Vision Res., 27: 277–284.

Schrauf, M., Lingelbach, B. and Wist, E.R. (1997) The scintillating grid illusion. Vision Res., 37: 1033–1038.

Schrauf, M. and Spillmann, L. (2000) The scintillating grid illusion in stereo-depth. Vision Res., 40: 717–721.

Shute, C.C.D. (1979) The McCollough effect. Cambridge University Press, Cambridge.

Simons, D.J. (2000) Current approaches to change blindness. Vis. Cogn., 7: 1–16.

Simons, D.J. and Levin, D.T. (1997) Change blindness. Trends Cogn. Sci., 1: 261–267.

Spillmann, L. (1971) Foveal perceptive fields in the human visual system measured with simultaneous contrast in grids and bars. Pflugers Archiv., 326: 281–299.

Spillmann, L. (1994) The Hermann grid illusion: a tool for studying human perspective field organization. Perception, 23: 691–708.

Spillmann, L. (1999) From elements to perception: local and global processing in visual neurons. Perception, 28: 1461–1492.

Spillmann, L. (2003) Re-viewing 25 years of ECVP — a personal view. Perception, 32: 777–791.

Spillmann, L. and de Weerd, P. (2003) Mechanisms of surface completion: perceptual filling-in of texture. In: Pessoa, L. and deWeerd, P. (Eds.), Filling-in: From Perceptual Completion to Cortical Reorganization. Oxford University Press, Oxford, pp. 81–105.

Spillmann, L. and Kurtenbach, A. (1992) Dynamic noise backgrounds facilitate target fading. Vision Res., 32: 1941–1946.

Spillmann, L. and Levine, J. (1971) Contrast enhancement in a Hermann grid with variable figure-ground ratio. Exp. Brain Res., 13: 547–559.

Spillmann, L. and Redies, C. (1981) Random-dot motion displaces Ehrenstein illusion. Perception, 10: 411–415.

Spillmann, L. and Werner, J.S. (1996) Long-range interactions in visual perception. Trends Neurosci., 19: 428–434.

Sturzel, F. and Spillmann, L. (2001) Texture fading correlates with stimulus salience. Vision Res., 41: 2969–2977.

Treisman, A. and Gormican, S. (1988) Feature analysis in early vision: Evidence from search asymmetries. Psychol. Rev., 95: 15–48.

Treisman, A. and Souther, J. (1985) Search asymmetry: a diagnostic for preattentive processing of separable features. J. Exp. Psychol. Gen., 114: 285–310.

Troxler, D. (1804) Ueber das Verschwinden gegebener Gegenstaende innerhalb unseres Gesichtskreises. In: Himly, K. and Schmidt, J.S. (Eds.), Ophthal. Bibliothek II.2. F Frommann, Jena, pp. 1–119.

Tyler, C.W. (1973) Temporal characteristics in apparent movement: omega movement vs. phi movement. Quart. J. Exp. Psychol., 25: 182–192.

Tyler, C.W. and Hardage, L. (1998) Long-range twinkle induction: An achromatic rebound effect in the magnocellular processing system. Perception, 27: 203–214.

Wang, X., Merzenich, M.M., Sameshima, K. and Jenkins, W.M. (1995) Remodelling of hand representation in adult cortex determined by timing of tactile stimulation. Nature, 378: 71–75.

Wolfe, J.M. (1984) Global factors in the Hermann grid illusion. Perception, 13: 33–40.

Wolfe, J.M. (1998) Visual memory: What do you know about what you saw. Curr. Biol., 8: R303–R304.

Zaidi, Q. (1999) Color and brightness induction: from Mach bands to three-dimensional configurations. In: Boynton, B.B., Gegenfurtner, K.R. and Sharpe, L.T. (Eds.), Color Vision: From Genes to Perception. Cambridge University Press, Cambridge, pp. 317–343.

Martinez-Conde, Macknik, Martinez, Alonso & Tse (Eds.)
Progress in Brain Research, Vol. 155
ISSN 0079-6123

CHAPTER 7

Lightness, filling-in, and the fundamental role of context in visual perception

Michael A. Paradiso*, Seth Blau, Xin Huang, Sean P. MacEvoy, Andrew F. Rossi and Gideon Shalev

Department of Neuroscience, Brown University, Providence, RI 02912, USA

Abstract: Visual perception is defined by the unique spatial interactions that distinguish it from the point-to-point precision of a photometer. Over several decades, Lothar Spillmann has made key observations about the nature of these interactions and the role of context in perception. Our lab has explored the perceptual properties of spatial interactions and more generally the importance of visual context for neuronal responses and perception. Our investigations into the spatiotemporal dynamics of lightness provide insight into underlying mechanisms. For example, backward masking and luminance modulation experiments suggest that the representation of a uniformly luminous object develops first at the borders and, in some manner, the center fills in. The temporal dynamics of lightness induction are also consistent with a filling-in process. There is a slow cutoff temporal frequency above which surround luminance modulation will not elicit perceptual induction of a central area. The larger the central area, the lower the cutoff frequency for induction, perhaps indicating that an edge-based process requires more time to "complete" the larger area. In recordings from primary visual cortex we find that neurons respond in a manner surprisingly consistent with lightness perception and the spatial and temporal properties of induction. For example, the activity of V1 neurons can be modulated by light outside the receptive field and as the modulation rate is increased response modulation falls off more rapidly for large uniform areas than smaller areas. The conclusion we draw from these experiments is that lightness appears to be computed slowly on the basis of edge and context information. A possible role for the spatial interactions is lightness constancy, which is thought to depend on extensive spatial integration. We find not only that V1 responses are strongly context dependent, but that this dependence makes V1 lightness constant on average. The dependence of constancy on surround interactions underscores the fundamental role that context plays in perception. In more recent studies, further support has been found for the importance of context in experiments using natural scene stimuli.

Keywords: lightness; filling-in; natural vision; context

Early in the 20th century, the founding fathers of Gestalt psychology powerfully demonstrated the fundamental role that context plays in visual perception. Some of the well-known laws, such as proximity and symmetry, emphasize the importance of spatial context. Other phenomena, such as apparent motion studied by Wertheimer, underscore the interplay between spatial and temporal context. Subsequent psychophysical research has extended our knowledge of the richness of visual interactions across space and time. While

*Corresponding author. Tel.: +1-401-863-1159; Fax: +1-401-863-1074; E-mail: michael_paradiso@brown.edu

DOI: 10.1016/S0079-6123(06)55007-1

contextual effects are often most dramatically demonstrated in illusions, the interactions are not mere curiosities; they get at the root of human visual perception. Over several decades, the importance of context has been powerfully demonstrated by Lothar Spillman and his colleagues. For example, they studied filling in of scotomas and using neon color filling they quantified the high sensitivity to contextual factors such as the continuity and orientation of inducers (Redies and Spillmann, 1981; Spillmann and Dresp, 1995). Some of the longest range influences of context ever found were reported by Pinna et al. in the watercolor effect (Pinna et al., 2001, 2003) in which border color and contrast appear to "paint" a large enclosed area.

Contextual interactions have also been explored with neurophysiological techniques. Lateral inhibition is the simplest example (Hartline et al., 1956). A considerable number of experiments have shown that the response of a neuron to a stimulus in its receptive field can be strongly modulated by other stimuli located outside the receptive field (Allman et al., 1985; MacEvoy et al., 1998; Albright and Stoner, 2002). However, aside from experiments specifically looking for larger-scale neural integration, a great many physiological studies are conducted with stimuli confined to a small receptive field.

The Gestalt psychologists argued that perception is made up first and foremost of Gestalten. The question from a neurophysiological perspective is whether an understanding of the neural basis of perception based on reduced stimuli is fundamentally flawed because of the lack of naturalistic context. Below we discuss aspects of our own research that explore the role of context and interactions in perception. From our studies, the answer to the question posed above appears to be both "no" and "yes". No, because most of what has been learned about neural representations in simple situations is still valid in more complex ones. However, there are basic aspects of neuronal coding that appear to depend critically on context. This means that a correct and full understanding of the relationship between vision and the brain requires that context be taken into account.

Perceptual studies of lightness, brightness, and filling-in

Brightness masking

A range of visual effects illustrate that the perception of surfaces is powerfully affected by both additive and subtractive influences from throughout the visual field (Hess and Pretori, 1894; Wallach, 1948; Helson, 1963; Cornsweet, 1970; White, 1979). In particularly dramatic situations, one area of the visual field can be perceptually filled in from information derived elsewhere. This is seen at the blind spot produced by the optic nerve and also with pathological and artificially created scotomas (Poppelreuter, 1917; Fuchs, 1921; Lashley, 1941; Bender and Teuber, 1946; Gassell and Williams, 1962; Gerrits and Timmerman, 1969). Experiments with images stabilized on the retina show that a stabilized patch is filled in by the lightness and color of the surrounding area (Riggs et al., 1953; Krauskopf, 1963; Gerrits et al., 1966; Yarbus, 1967; Larimer and Piantanida, 1988).

In experiments conducted with Dr. Ken Nakayama (Paradiso and Nakayama, 1991), we used a masking paradigm to explore the spatiotemporal dynamics of filling-in. A uniformly luminous disk was briefly flashed and, after a variable stimulus onset asynchrony (SOA), a masking stimulus was presented. The mask consisted of a bright circle on a black background with the circle internal to the boundaries of the large uniform disk. Subjects viewed multiple cycles of the disk followed by the mask. Their task was to indicate which element of a palette of gray tones was most similar to the brightness perceived at the center of the disk. As we expected, with a long SOA, the disk was perceived as filled-in before the mask was presented. However, as the SOA was decreased, brightness toward the center of the disk was diminished (Fig. 1A). Outside the masking circle, the disk appeared normal except for a small area of darkening just next to the outside of the circle. Inside the circular mask, there was a much more dramatic effect — at SOAs between 50—and 100 ms, the entire area of the disk inside the circular mask was significantly darker or black (Fig. 1B).

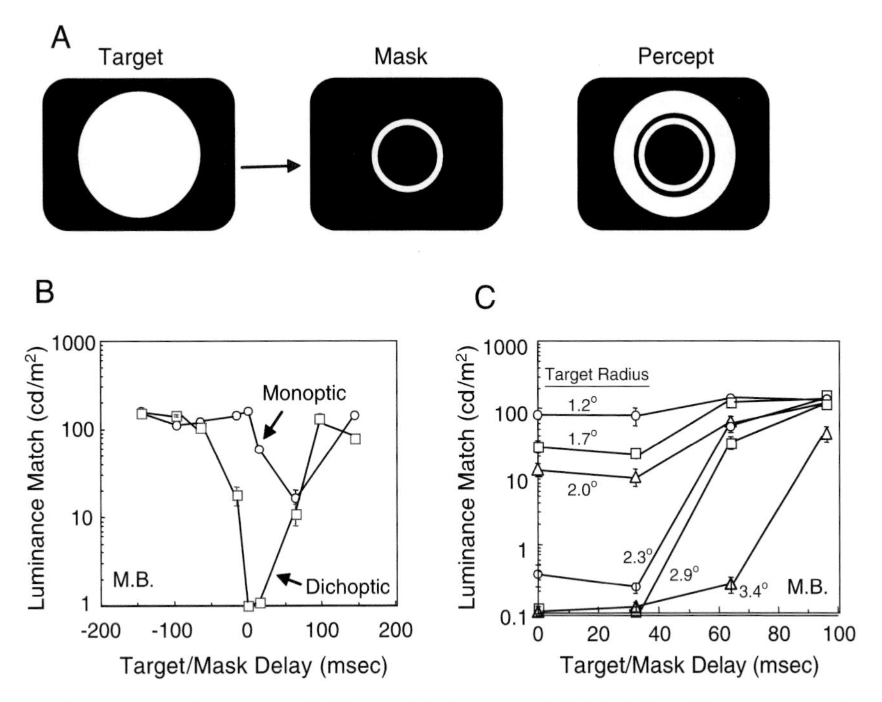

Fig. 1. Backward masking suppresses brightness. (A) When a briefly presented disk is followed by a ring-shaped mask, brightness in the interior of the disk is greatly diminished. (B) If the disk and ring are shown to the same eye (monoptic), maximal brightness suppression occurs with an SOA of 50–100 ms. With dichoptic masking, suppression is greater and maximal at zero SOA. (C) With a fixed-size mask, brightness suppression increases as the target increases in size (and the distance between the edges of target and mask increases). Masking is observed at longer delays as the target increases in size.

Under optimal conditions the brightness matches made in the center of the disk were reduced as much as 2 log units relative to the condition in which no masking was observed (i.e. at long target/mask intervals). Evidently, the circular shape of the mask was responsible for the large asymmetry in the masking strength inside and outside the circle. The normal appearance of the disk outside the area of the circular mask was consistent with the hypothesis that the outside edge of the disk played a role in determining its interior brightness and that the masking circle primarily interfered with this process inside its radius (Grossberg and Todorovic, 1988). However, if there is explicit filling-in, it appears to work at a coarse spatial scale: potent brightness masking was observed even if many small gaps were made in the masking contour; it was not the case that brightness "leaked through" the gaps. This finding is consistent with Spillmann and colleagues finding that the watercolor effect is relatively unaffected by introducing gaps into the inducing contours (Pinna et al., 2001).

The brightness masking effect was even stronger with dichoptic than monoptic presentation (square symbols in Fig. 1B). When the target and mask were presented to different eyes, the interior of the target disk was absolutely black. The fact that the masking was effective dichoptically suggests that the stimuli interact in visual cortex where there are binocular neurons.

If there is propagation of a signal related to surface brightness, one should be able to see masking at a later time if the masking contour is farther from the edge of the target disk. To test this prediction, targets with radii ranging from 1.2° to 3.4° were used with a 2.0° radius-masking ring. Consistent with the prediction, the suppressive effect of the masks was greater as the target disk increased in size (Fig. 1C). Also, masking remained effective at longer SOAs as the distance between the edges of the two stimuli increased. On the basis of the latest times at which masking was effective with different distances between the outer edge of the disk and circular mask, a velocity for brightness propagation was calculated. The estimated velocity was 110–150 deg/s. Using estimates of the human cortical magnification factor for primary visual cortex, this comes out to a roughly constant speed of 0.15–0.4 m/s for the propagation (if it occurred in V1).

Watching the filling-in process

If filling-in is a part of normal vision, we presumably do not notice it because it is rapid. But is it possible to make filling-in visible? We explored this question by gradually increasing or decreasing the luminance of a uniform disk, in an attempt to prolong the duration of time over which the edge signal is not filled into the disk center (Paradiso and Hahn, 1996). We found that in some situations the disk's brightness is noticeably inhomogeneous as the luminance changes. Most commonly it appears that brightness changes near the center of the uniform disk lag behind changes at the disk's edge. For example, if the disk's luminance is increased, the center is darker and brightness appears to move inward. The central lag in brightness is even more pronounced when the entire computer screen is bright and a disk's luminance begins bright and gradually decreases (Fig. 2A). There is a striking percept that the center of the disk is brighter than the edge and darkness sweeps into the center.

A critical determinant of these filling percepts is the rate at which the luminance changes: the disk appears uniform if the luminance is changed rapidly or slowly but nonuniform if it is changed at intermediate rates. In qualitatively exploring the phenomenon, we tried a variety of stimulus configurations (squares, disks, etc), stimulus sizes (0.5°–10°), and luminance modulation paradigms (linear, exponential, etc). Generally speaking, the qualitative results did not depend critically on these parameters. For example, if the stimulus consisted of several simultaneously presented uniform patches, each patch appeared to fill in independently from its own borders. The filling-in percepts were observed even when the modulated disk was viewed so that it encompassed the blind spot at the optic disk. This strongly suggests that the filling-in percept is based on a cortical process.

The most important variable for the perception of brightness filling-in was the dwell time at each luminance step. Spreading brightness was seen with short dwell times, but as the dwell increased above 50 ms, the darkness spreading effect decreased significantly (Fig. 2B). In other words, the perception of the modulated disk was uniform and filling-in was lost if the luminance was held longer than 50–100 ms at each luminance step.

One interpretation of the results is that the edge and center of the stimuli have different brightness when the luminance is swept up or down because the sweep speed exceeds the rate of an underlying brightness process. It is known that the brightness of an area is strongly dependent on the luminance contrast at the area's border (Hess and Pretori, 1894; Heinemann, 1972). Perhaps there is a spread of activity in visual cortex underlying perceptual filling-in. In order to account for the fact that inhomogeneities are not seen at fast luminance ramp speeds, one must postulate that the inhomogeneity exists for a period of time too short to be perceived. This would explain why we are not aware of any brightness nonuniformities in normal visual situations as we move our eyes about. Presumably, by stretching out the luminance ramp in time, an inhomogeneity can be maintained for a longer duration, making it perceptible. When the luminance ramp is very slow, each time the luminance is incremented, the filling-in process completes before the next increment and the inhomogeneity is preserved for too short a time to be perceived.

Dynamic lightness induction

DeValois et al. (1986) published a surprising observation about the temporal characteristics of brightness induction that provides another piece of evidence suggesting that brightness processes are slow and involve filling-in. DeValois et al. used a stimulus in which a static gray patch was surrounded by a larger area in which the luminance was modulated sinusoidally in time. The

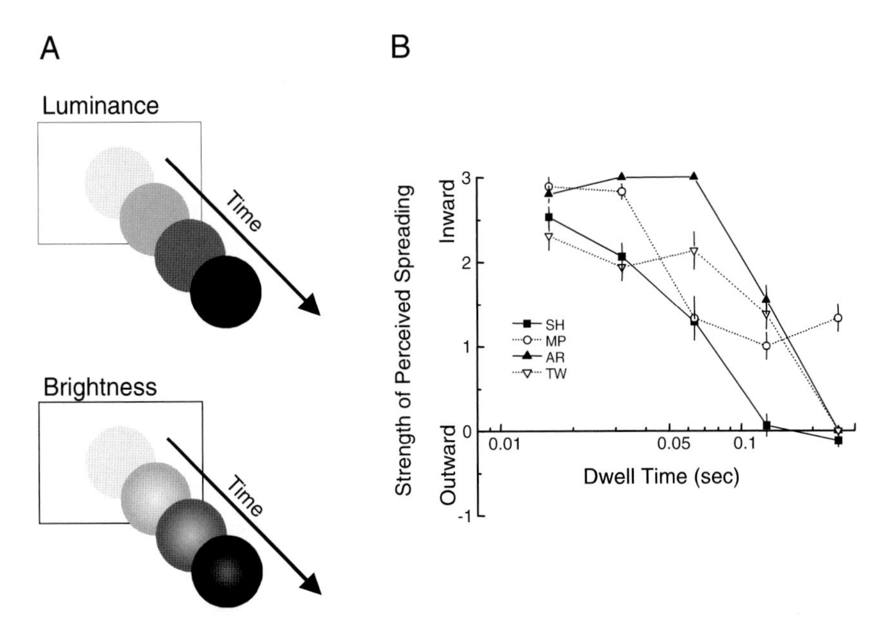

Fig. 2. Luminance sweeps produce filling-in percepts. (A) Brightness changes toward the center appear to lag behind changes at the border when the luminance of a uniform disk is progressively decreased. (B) The strength of the filling-in percept depends strongly on the dwell time spent at each luminance step. At short dwell times there are compelling filling-in percepts, but these are lost as the dwell time increases above about 50 ms.

luminance modulation of the surround produced powerful brightness induction in the gray patch, roughly in antiphase to the surround modulation. Surprisingly, brightness modulation was induced in the gray patch only when the surround was modulated at quite low temporal frequencies (i.e. below about 2.5 Hz). When the surround was modulated at higher rates the central patch appeared a static gray. This low cutoff for induced brightness modulation stands in stark contrast to the critical flicker fusion rate, which is an order of magnitude faster.

We extended the experiments of DeValois et al. to determine whether the properties of this dynamic form of brightness induction are consistent with the implications of the masking and luminance sweep experiments described above (Rossi and Paradiso, 1996). The stimulus we used was a temporally modulated squarewave grating (Fig. 3A). The grating was modulated in a manner such that the luminance of every other stripe varied sinusoidally in time and the intervening stripes had constant luminance. Perceptually, the modulation produced brightness induction in the constant stripes, roughly in antiphase to the brightness of the luminance-modulated stripes. In light of the masking and luminance-sweep results, we were particularly interested in any dependence the temporal properties of brightness induction might have on spatial scale. Using the method of adjustment, we had observers find the highest temporal modulation rate at which induction was perceived at different spatial frequencies. We found that the lower the spatial frequency (i.e. the larger the areas of uniform brightness), the lower was the cutoff temporal modulation rate (Fig. 3B). We also quantified the amplitude of the perceived brightness modulation that was induced. This showed that the amplitude of brightness induction was greatest at low temporal modulation rates. The luminance matches to the peak and trough of the brightness modulation approach each other as the temporal frequency is increased, eventually becoming equal when there is no perceived brightness modulation. These results make it clear that below the cutoff modulation rate, the amplitude of brightness induction is graded relative to temporal frequency and induction is not

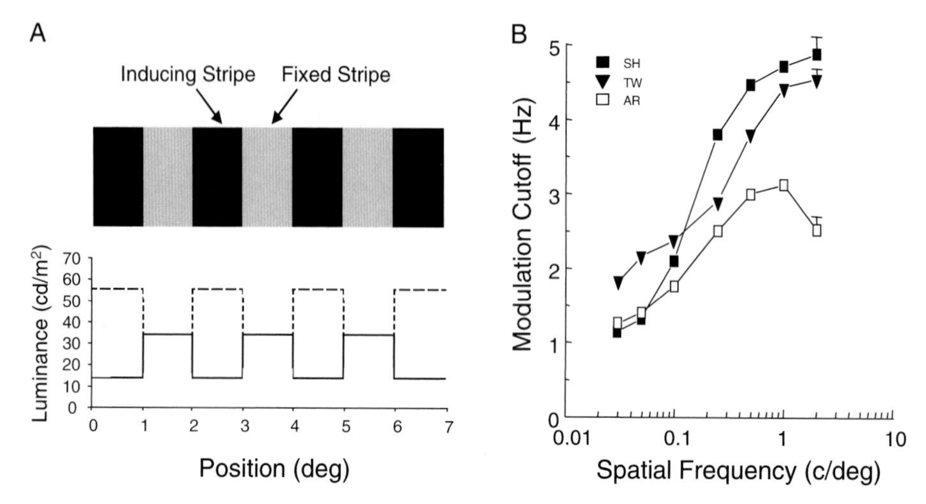

Fig. 3. Temporal properties of brightness induction. (A) Subjects viewed a fixed stripe with constant luminance while neighboring inducing stripes were luminance modulated. (B) Cutoff temporal frequencies for three observers above which induction was not observed.

simply "off" and "on" above and below the cutoff rate.

The results of the dynamic induction experiments make two important points about the mechanisms involved in brightness perception. First, the process responsible for brightness changes due to induction is considerably slower than the process responsible for brightness changes from direct luminance modulation. Second, the time course of induction is scale-dependent. The low cutoff frequency for induction and the effect of modulation rate on the amplitude of induction suggest that induction is a slow process that simply cannot "keep up" with fast modulation rates. In another way, it appears that larger spatial areas take more time to induce. Besides the fact that the cutoff rates for "real" and induced brightness changes differ by more than a factor of 10, they depend on spatial scale in opposite ways. While the critical flicker fusion rate increases with stimulus size, the cutoff frequency in our induction experiments decreases with size. The results indicate that there is a major difference between the mechanisms limiting perception of modulation with real and induced brightness. If induction were initiated at edges and propagated inward, this would explain why it takes longer to induce a larger area and why the cutoff frequency decreases with increasing size. We measured the spatial phase of induced brightness across spatial scale and this led to an estimate of 140–180°/s for the induction process. This estimate is in rough agreement with the filling-in velocities estimated in the brightness masking and luminance sweep experiments.

To reconcile filling-in with the high CFF, we have proposed that brightness involves two mechanisms — a fast process that is relatively unaffected by the size of a uniformly luminous area, and a slow filling-in process with a duration that increases with the size of a uniformly luminous area. A fast process largely based on luminance appears to be necessary to explain the high CFF for luminance modulation and the fact that this frequency does not decrease with the size of the modulated area. A slow process driven mainly by contrast appears to be required to account for the induction results as well as results in the masking and luminance-sweep experiments. We hypothesize that when the luminance of an area is modulated, both the fast and slow processes are involved in determining the final brightness percept of that area. Previous experiments suggest that slow filling-in occurs with luminance modulation (Paradiso and Nakayama, 1991; Paradiso and Hahn, 1996), but the fast process presumably determines the CFF. The situation is different when brightness modulation occurs solely because of induction. In this case, we suggest that only the slow filling-in process is responsible for the perceived brightness modulation of the induced area. Thus, the velocity of the filling-in process would determine the cutoff frequency for induced modulation.

Implications for mechanisms

On the basis of the psychophysical results discussed above, we can hypothesize how a surface representation and filling-in might be physiologically implemented. First, it seems that there must be some form of explicit surface representation. If surfaces were represented entirely in terms of edge signals, it is not clear why filling-in would be perceived. Second, the representation of brightness should have a scale-dependence in accord with the psychophysical results. Third, the filling-in process should be slow compared to the CFF. The psychophysical data give a measure of the expected speed. The conclusion we cannot draw from the psychophysical results is that filling-in is an isomorphic process akin to painting the interior of a surface. While this is conceivable, there is little evidence in favor of it. More likely, filling-in involves a rapid temporal process of neural interactions leading to the final surface percept; the brain representation is probably not an isomorphic representation of the stimulus. An edge-dominated process involving filling-in is a way to reconcile the psychophysical and physiological results presented here with studies showing coding of border ownership (Zhou et al., 2000).

Physiological studies of lightness and filling-in

Brightness induction and the cortical representation of surfaces

For there to be a form of neural filling-in underlying perceptual filling-in of a surface, there needs to be an explicit representation of the surface. If surfaces were represented entirely by the responses to their boundaries, nothing would need to fill-in. We investigated whether there is an explicit representation of surface brightness using the dynamic induction stimulus previously used in psychophysical experiments. The question was whether the response of a neuron is modulated when its receptive field is positioned on the induced surface.

We recorded from single cells in the retina, lateral geniculate nucleus (LGN), and primary visual cortex of anesthetized cats (Rossi and Paradiso, 1996; Rossi et al., 1996; Rossi and Paradiso, 1999). After mapping the RF, the dynamic induction stimulus was positioned such that a large central gray patch covered the RF (middle left icon in Fig. 4A). The size of the patch was always much larger than the RF, extending 3–5° beyond the RF boundary on each side. Flanks, the same size as the central patch, were positioned to each side. The luminance of the flanking patches was modulated from light to dark sinusoidally in time. Two principal control stimuli were used (top and bottom icons to left in Fig. 4A). In one, the luminance of the area covering the RF was modulated, rather than the flanks (top row). This tested whether responses were better correlated with the luminance or brightness of the central patch. The second control had a large black patch covering the RF, instead of a gray patch, while the flanks were modulated (bottom row). If light from the flanks directly modulated the response of the neuron (e.g. scattered light), this should be revealed with this control stimulus. However, as there is no perceptual induction when the central patch is black, the neuron's response should not be affected by the flank modulation if the response represents brightness.

To study the responses of retinal ganglion cells, we recorded from their axons in the optic tract. It is important to keep in mind that the stimuli employed were sized such that the central patch encompassed both the center and surround of the RF — any effects of the flanks were from beyond the RF. As would be expected from the low-pass frequency response of retinal neurons, many cells were somewhat activated by a gray patch of light covering their RF. Furthermore, some cells responded in a phase-locked manner to luminance modulation within the RF. When the dynamic induction stimulus was used, cell responses were generally constant along with the luminance of the patch covering the RF, rather than modulated.

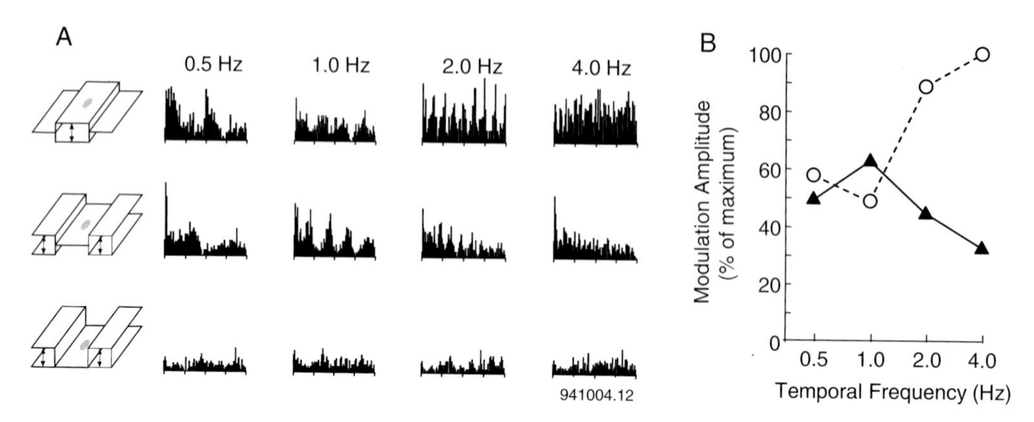

Fig. 4. Responses in cat V1 to brightness induction stimuli. (A) Responses of one cell to three stimuli (left) were recorded at four temporal modulation rates. In the stimulus icons, the small gray patch is the receptive field. The stimuli were: luminance modulation control (top), induction (center), and surround modulation control with central area black (bottom). (B) As temporal frequency increased, V1 responses on average were more modulated in the luminance modulation control (dashed line) but less modulated in the induction condition (solid line).

The optic tract recordings showed that retinal neurons are responsive to luminance in the absence of contrast, but the responses did not correlate with perceived brightness.

In many regards, the results were similar when recordings were made in lateral geniculate nucleus layers A and A1. A static gray patch in the RF often elicited a neuronal response, and the response varied as the luminance of the central patch was modulated. There were also neurons that showed modulated responses when the flank luminance was varied in time. However, the fact that the response modulation to the flanks was generally greater when the central patch covering the receptive field was black rather than gray suggests that light scattering was probably involved rather than a brightness representation. Very occasionally, cells were encountered that did appear to have responses correlated with brightness.

Many cortical neurons were found to respond only in stimulus conditions that produced perceptual changes in brightness in the area corresponding to the RF. An example of this is shown in Fig. 4A. The middle row of this figure shows the response of a neuron to the presentation of a constant gray field flanked on either side by luminance-varying fields of equal size. The RF was 4° wide and was centered on the central gray area that was 14° across. The response of the neuron was phase-locked to the frequency of the luminance modulation in the stimulus flanks. There was no such response when the central portion of the stimulus was black (bottom row). In our sample of 160 striate neurons, 120 (75%) had responses that were modulated and phase locked to the luminance changes outside the RF.

A comparison of responses in the different conditions suggests that the firing of the neuron in Fig. 4A (and others) was more correlated with the brightness in the area covering the RF than with the luminance of any particular portion of the stimulus. This was seen in the strong response modulation when the stimulus center was gray (a condition that yields perceptual induction) compared to the weak response when the center was black (a condition that does not cause induction). A correlation with brightness is also found when comparing the responses to luminance modulation within the receptive field and luminance modulation in the flanks. When light covering the receptive field was modulated, the response was maximal when the central area was brightest (and the luminance of the modulated light was highest). When the flanks were modulated, the response was again greatest when the central area was brightest, but in this case the luminance of the modulated flanks was lowest. The response was clearly not determined by the overall amount of light present in the stimulus.

As described in the previous section, one of the hallmarks of the dynamic brightness induction effect is that brightness changes are only perceived at low modulation rates in comparison to the rates of direct luminance modulation that elicit brightness variations. In other words, at higher modulation rates there is a significant difference in the degree of perceived brightness modulation with real and induced brightness. Physiologically, we found differences in the amplitudes of response modulation in the induction and center modulation conditions. In the induction condition (middle row), the response of the neuron in Fig. 4A was largest at low temporal frequencies and *decreased* as the rate of flank modulation was increased above 1.0 Hz. However, when the luminance of the central area was modulated (top row), the response amplitude progressively *increased* with increasing temporal frequency. A significant difference in the induction and center modulation conditions is clearly evident in the averaged data (Fig. 4B).

The responses of neurons in striate cortex correlate with perceived brightness in four ways: the neural responses were modulated at the frequency of the surrounding luminance modulation; the response modulation occurred in conditions that elicited brightness induction but not in similar conditions that did not produce induction; in the induction conditions, the response modulation greatly decreased as temporal frequency increased; there was a complete phase shift between induction and center modulation conditions. However, cells that appeared to follow brightness based on one criterion did not always do so according to other criteria. We estimate that about 30% of V1 neurons have responses correlated with brightness in all conditions.

Lateral interactions in visual cortex

The induction studies described above suggest that at least for a subgroup of neurons in striate cortex, there is an explicit representation of surface qualities such as brightness. The logical hypothesis to explain the induction physiology results is that outside cortical-receptive fields, there are areas that modulate the response to a stimulus in the RF. While there have been many studies of interactions from outside striate-receptive fields, virtually all have used lines or gratings as stimuli (i.e. contrast within the RF). However, when a uniform surface covers the RF, interactions from areas beyond the receptive field are not reliably the same as those that would be found with line or grating stimuli (MacEvoy et al., 1998). From one cell to another, there is considerable diversity in the modulatory effects that light outside the RF has on the response to a surface covering the RF. The most common effect is surround suppression. The prevalence of surround suppression and its large spatial range, offers an explanation for the induction effect noted above. The modulatory areas outside the RF cannot drive the cortical neuron alone (i.e. no response when the area covering the RF is black), but when the cell is excited by a central stimulus, such as the gray patch, the flanks can alter the response. The "in phase" and "out of phase" responses recorded with luminance modulation of the patch covering the RF versus the flanks can be accounted for by areas beyond the RF that are either facilitatory or inhibitory.

Lightness constancy in visual cortex

Presumably, the interactions from beyond the receptive fields of V1 neurons serve some valuable purpose(s) and it appears that lightness constancy may be one such purpose. Over the course of a day and across the seasons of the year, the illumination coming from the Sun varies considerably. Humans and animals that have been tested, perceive the lightness of objects to be stable despite large variations in illumination. This perceptual constancy for lightness, and related color constancy, were presumably of great evolutionary value. For example, in the absence of perceptual constancies, there might not be reliable cues upon which to select ripe fruit to consume.

We conducted experiments in cat V1 to test responses for lightness constancy (Mac Evoy and Paradiso, 2001). Stimuli consisted of monochromatic patches (i.e. a monochromatic "Mondrian" stimulus) on a computer monitor simulating surfaces with a wide range of

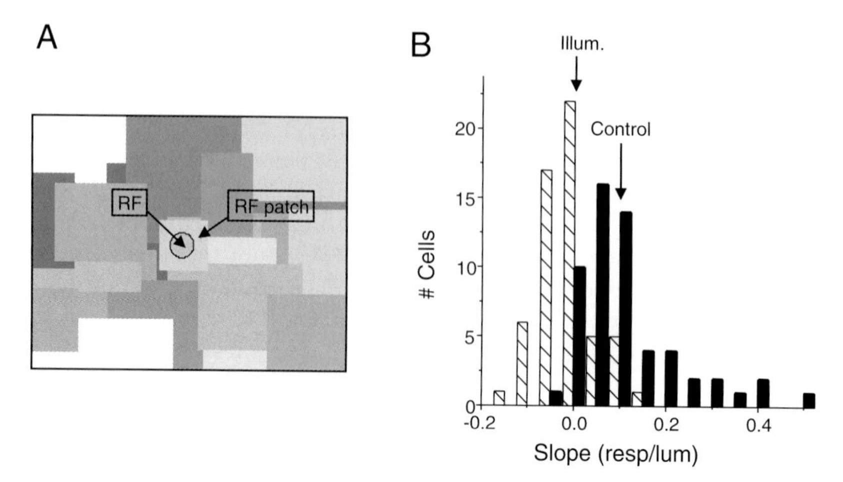

Fig. 5. Lightness constancy in cat V1. (A) On a computer display, a Mondrian of patches with different reflectances was simulated. The receptive field was centered on one of the patches (the RF patch). (B) In control conditions, only the luminance of the RF patch was varied. As shown by the solid bars and arrow, the average response was positively correlated with luminance (and lightness). In illumination conditions (hatched bars), the average response was constant as RF patch luminance changed along with the average background luminance.

reflectances (Fig. 5A). One patch of the stimulus encompassed the receptive field and the rest composed the background. Changes in the luminance values of the patches were made in a manner either consistent (illumination conditions) or inconsistent (control conditions) with overall changes in illumination. Comparisons were made between the two situations when the patches covering the receptive field in each case were identical.

In control conditions only the luminance of the patch covering the receptive field increased, a situation in which the perceived lightness of the patch increases with luminance. In contrast, in the illumination conditions the lightness percept was stable. While there was considerable cell-to-cell variability, on average the responses of V1 neurons to identical stimuli in their receptive field were significantly different in the two conditions (Fig. 5B). In the control condition the average V1 response correlated with the luminance of the RF patch. Since perceived lightness correlated with the luminance, the neural response also correlated with lightness. In marked contrast, in the illumination conditions there was essentially no change in the average V1 response as the luminance of the RF patch increased. This mirrored the perceptual constancy of the patch. These results are reminiscent of the constancy demonstrations of Land (Land and McCann, 1971; Land, 1986). In these demonstrations, the color of patches in a Mondrian were shown to be constant when the overall illumination was varied, but viewing any single patch in isolation (for instance, through a tube) revealed that the patch luminance changed dramatically. The context provided by the changes in the larger area is essential to normalize the local percept. In our physiological studies, in both the illumination and control conditions the response correlated with lightness, but only in the illumination conditions did the response exhibit lightness constancy.

A likely basis for the response invariance in the illumination conditions is the predominant surround suppression seen with uniform patches of light (Schein and Desimone, 1990; MacEvoy et al., 1998; Wachtler et al., 2003). Evidently in the illumination conditions the increased response of the neuron that comes from more light in the RF is counterbalanced by increased suppression from the surround. What is somewhat surprising is that on average the net input to V1 neurons balances the added RF drive with increased surround suppression.

The importance of natural visual and behavioral context

The effects of natural scenes and saccades on V1 activity

The results described above suggest that in the domain of lightness, modulatory inputs from lateral or feedback connections play an important role in the V1 representation of lightness and in lightness constancy. Although Mondrian's paintings can be viewed in museums, flashing similar pictures to a fixating animal can hardly be considered a natural visual situation. Visual stimulation in the real world typically involves complex arrangements of light, color, and contrast, quite unlike the simple stimuli usually used in the laboratory. Moreover, natural stimuli fill the visual field whereas many experiments are conducted with small RF stimuli isolated on a large blank display. Another obvious difference between typical experiments and natural vision is behavioral. In a natural setting, the eyes move to bring new stimuli into view; typical fixations are about 300 ms with brief intervening saccades. In most experiments, animals are trained to hold fixation (or anesthetized) while stimuli are flashed into the receptive field.

To explore the significance of stimulus complexity and saccades on V1 responses, we conducted an experiment in alert macaques in which the same stimulus was presented to a neuron under four different conditions, varying in how natural the visual situation was. In the first condition, the animal fixated a point on an otherwise gray visual display and a small bar was flashed into the receptive field. In the second condition the animal fixated the same point and the bar was flashed, but in this case the background was a grayscale outdoor scene (van der Schaaf and van Hateren, 1996) with the same mean luminance as the gray background. In the third and fourth conditions the background image was a uniform gray or a grayscale photo, respectively. However, in these conditions the fixation of the animal was guided such that a saccade brought the bar stimulus into the receptive field rather than it being flashed on.

We find that the response to a bar stimulus in the receptive field is influenced by both stimulus complexity and the method by which the stimulus comes into the RF (flash or saccade). Examples are shown in Fig. 6. The cell illustrated in Fig. 6A shows the difference in response associated with the gray and natural-scene backgrounds. The response to a small bar is significantly higher when the background is a uniform gray than when it is a grayscale picture. In this case the response was roughly 50% greater when the bar was presented on a gray background compared to a

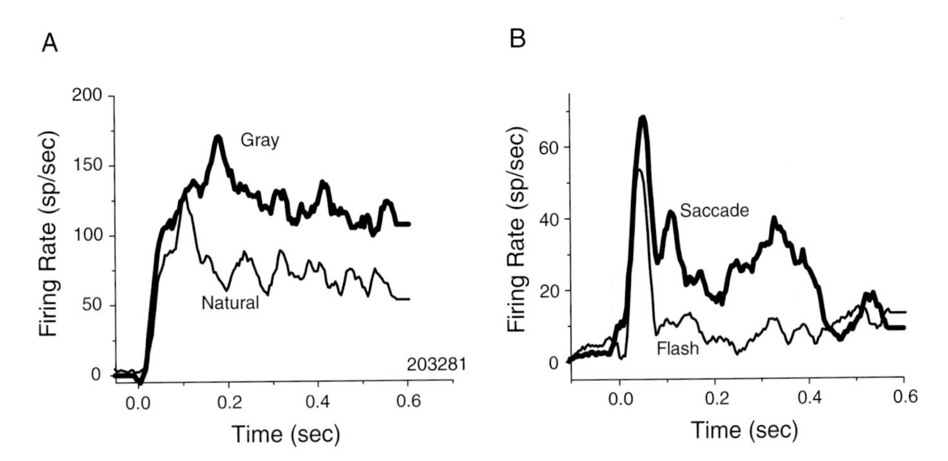

Fig. 6. Macaque V1 responses influenced by background and saccades. (A) In this neuron, the response to the same stimulus bar was significantly higher on a gray background than a natural background. (B) The response to the identical stimulus was higher when it entered the receptive field via saccade rather then being flashed.

natural scene. This response difference is found regardless of whether a saccade or flash introduced the bar into the receptive field. Also interesting is the delay in the separation of the two curves. The initial response with gray and natural backgrounds is similar but after about 50 ms there is a reduction in the natural-scene response. It is not possible to say why the response differs between the two conditions, but the natural scene obviously has contrast and structure not present with the gray background. Previous studies have reported the suppressive effect of contrast outside the receptive field (Allman et al., 1985) and our result can be interpreted in that context. The delay in the background effect suggests that different circuitry may be involved.

More surprising is the influence of presentation method shown in Fig. 6B. When a saccade brought the bar stimulus into the receptive field, the initial response was similar to a flashed stimulus, but after about 50 ms the response in the saccade condition was much larger (more than 100% greater). The neuron shown in this figure has a particularly pronounced difference, but even in the population average, the saccade response with the natural-scene background was 15% higher than the flash response on the same background. Several factors were considered to account for the response difference when a stimulus appeared in the receptive field via saccade versus flash. For example, we considered the possibility that the stimulus present in the receptive field prior to the saccade might affect response magnitude. While there was a hint of this in some cells, it could not account for most of the response difference. We also considered the possibility that stimuli swept across the receptive field during the saccade might make the saccade response greater. Again, this was not able to account for the response difference. These and other factors were considered (MacEvoy et al., 2002), and while several factors have small effects on response rate, no single factor has yet been found that can account for the bulk of the response difference.

The combined (and opposed) effects of scene complexity and saccades suggest that it is impossible to predict responses in natural situations from responses to small stimuli flashed into the receptive field.

Background changes and delayed form information

The experiments in which saccades on a natural scene brought stimuli into receptive fields represent a more natural visual situation than flashing a bar isolated to a receptive field. However, the use of complex scenes and saccades complicates interpretation of the results. For example, natural scenes had a suppressive effect on V1 activity relative to a uniform gray background, but the scene complexity made it difficult to ascertain what aspect(s) of the picture was responsible for the suppression.

In a parallel series of experiments we used a somewhat less natural visual paradigm to gain greater control over the effects of image complexity and saccades (Huang and Paradiso, 2005). When an animal makes a saccade while viewing a natural scene, the "contents" of a V1 receptive field change. Perhaps a rock was initially in the receptive field and afterwards the branch of a tree. At the same time that a new local feature is introduced, the background or context usually changes. When the branch of the tree sweeps into the receptive field, adjacent areas in the RF and outside the RF might "see" other vegetation. In our experiments we simulated this natural sort of visual stimulation in the context of a well-controlled fixation paradigm. Macaques fixated a point on the visual display and bars of light or Gabor patches were presented in the receptive field. On some trials the background was static as in most visual physiology experiments. On other trials the background luminance or pattern changed simulating what would occur with a saccade. Comparisons of neural responses were made when identical stimuli were within and beyond the RF in the static and changing background trials. The only difference between the conditions was the stimulus in the receptive field *before* the bar and background used for response measurements.

We found that when context changes with the introduction of a local feature (the changing background condition), the response pattern is qualitatively and quantitatively different than the standard static-background situation. An example of this is shown in Fig. 7. When a bar was introduced on a static background, the response

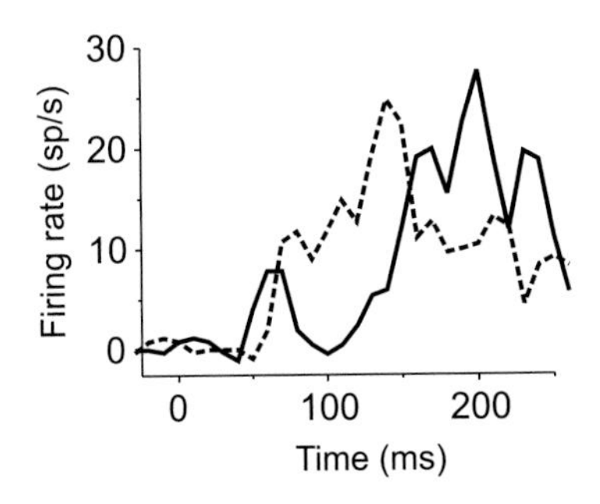

Fig. 7. Temporal context effects in Macaque V1. When a bar stimulus was flashed on a static background, the response had a single peak (dashed curve). To mimic the effect of a saccade, the same stimulus and background were shown, but preceded by a different background. In the changing background condition the response had two peaks (solid curve). With the static background, orientation was reflected in the amplitude of the initial response and contrast was anticorrelated with response latency. With the changing background the initial response carried little orientation or contrast information. Instead, these attributes were represented in the amplitude and latency of the second peak.

typically showed one peak. In this case, orientation selectivity and contrast sensitivity were represented in the amplitude and latency of the earliest response, respectively. The response is quite different in the changing background situation. The data shown in Fig. 7 were obtained when a bar stimulus appeared simultaneously with a change in background luminance, but similar results were obtained with an isoluminant background pattern change. The early response in the changing background condition appears similar to that in the static condition. However, there tended to be response suppression after the initial transient followed by rebound to a higher firing rate. In other words, the average response in the changing background condition showed two peaks rather than one. Of particular interest is the observation that form information is no longer represented in the initial transient as in the static background condition. Instead, orientation and contrast sensitivity are reflected in the amplitude and latency of the delayed second response peak. This peak occurred about 50 ms after the initial peak, suggesting that in the changing background condition, form information was delayed by this amount of time. Other neurons showed only this late response component.

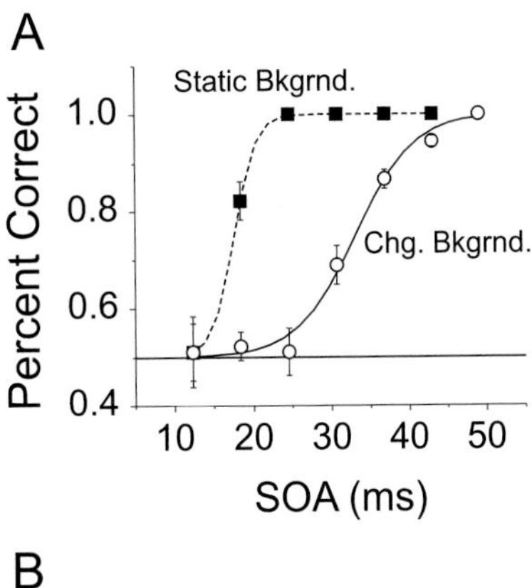

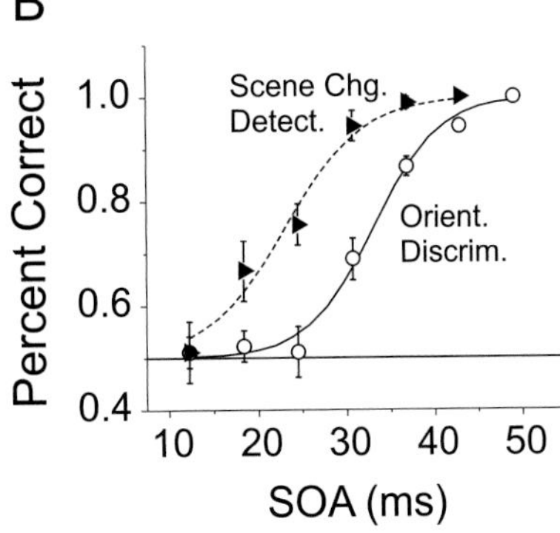

Fig. 8. Human discrimination and detection with a changing background. (A) As SOA between the stimulus and mask increased, performance in static and changing background conditions improved. Performance increased significantly faster with a static background and, at intermediate SOAs, performance was optimal when performance with a changing background was still at chance levels. (B) Subjects were able to detect a change in the scene at significantly shorter SOAs than those at which they could discriminate bar orientation.

The response differences recorded in the static and changing background conditions suggest that there might be a temporal difference in the brain's access to form information. We tested this hypothesis in a series of human psychophysics experiments (Huang et al., 2005). Bar stimuli were briefly presented and followed by a masking stimulus at various SOAs to limit the duration of visual processing (Breitmeyer, 1984). As predicted by the physiology data, it appears that perceptual access to form information (orientation and contrast) is delayed when a background change (luminance or pattern) accompanies the presentation of a bar of light. In Fig. 8A, orientation discrimination improves significantly as SOA increases with either static or changing background. However, performance saturates at much shorter SOAs with the static background. This suggests that the orientation information is present earlier in that condition. The psychophysics experiments also showed that in the changing background situation, observers are able to detect that the scene has changed well before (about 20 ms) they are able to discriminate orientation (Fig. 8B). This suggests that in the changing background condition the early response signals that the scene has changed but does not carry the bulk of the information about the details of the stimulus.

Abbreviations

CFF	critical flicker frequency
RF	receptive field
SOA	stimulus onset asynchrony

Acknowledgments

The authors wish to thank Aaron Gregoire, Lisa Hurlburt Kinsella, and Amber Pierce for technical assistance. This research was supported by the National Eye Institute and the National Science Foundation.

References

Albright, T.D. and Stoner, G.R. (2002) Contextual influences on visual processing. Annu. Rev. Neurosci., 25: 339–379.

Allman, J., Miezin, F. and McGuinness, E. (1985) Stimulus specific responses from beyond the classical receptive field. Annu. Rev. Neurosci., 8: 407–430.

Bender, M.B. and Teuber, H.L. (1946) Phenomena of fluctuation, extinction and completion in visual perception. Arch. Neurol. Psychiatry, 55: 627–658.

Breitmeyer, B.G. (1984) Visual Masking: An Integrative Approach. Oxford University Press, New York.

Cornsweet, T.N. (1970) Visual Perception. Academic Press, New York.

DeValois, R.L., Webster, M.A., DeValois, K.K. and Lingelbach, B.L. (1986) Temporal properties of brightness and color induction. Vision Res., 26: 887–897.

Fuchs, W. (1921) Completion phenomena in hemianopic vision. In: A Source Book of Gestalt Psychology. Kegan Paul, London.

Gassell, M.M. and Williams, D. (1962) Visual function in patients with homonymous hemianopia. Part III. The completion phenomenon; insight and attitude to the defect; and visual functional efficiency. Brain, 85: 229–260.

Gerrits, H.J.M. and Timmerman, G.J. (1969) The filling-in process in patients with retinal scotomata. Vision Res., 9: 439–442.

Gerrits, H.J.M., de Haan, B. and Vendrik, A.J.H. (1966) Experiments with retinal stabilized images. Relations between the observations and neural data. Vision Res., 6: 427–440.

Grossberg, S. and Todorovic, D. (1988) Neural dynamics of 1-d and 2-d brightness perception: a unified model of classical and recent phenomena. Percept. Psychophys., 43: 241–277.

Hartline, H.K., Wagner, H.G. and Ratliff, F. (1956) Inhibition in the eye of limulus. J. Gen. Physiol., 39: 651–673.

Heinemann, E. (1972) Simultaneous brightness induction. In: Jameson, D. and Hurvich, L.M. (Eds.), Handbook of Sensory Physiology. Springer, New York, pp. 146–169.

Helson, H. (1963) Studies of anomalous contrast and assimilation. J. Opt. Soc. Am., 53: 179–183.

Hess, C. and Pretori, H. (1894) Messende Untersuchungen uber die Gesetzmassigkeit des simultanen Helligkeitscontrastes. Arch. Ophthalmal., 40: 1–24.

Huang, X. and Paradiso, M.A. (2005) Background changes delay information represented in Macaque V1 neurons. J. Neurophysiol., 94: 4314–4330.

Huang, X., Blau, S. and Paradiso, M.A. (2005) Background changes delay the perceptual availability of form information. J. Neurophysiol., 94: 4331–4343.

Krauskopf, J. (1963) Effect of retinal image stabilization on the appearance of heterochromatic targets. J. Opt. Soc. Am., 53: 741–744.

Land, E.H. (1986) Recent advances in retinex theory. Vision Res., 26: 7–21.

Land, E.H. and McCann, J.J. (1971) Lightness and retinex theory. J. Opt. Soc. Am., 61: 1–11.

Larimer, J. and Piantanida, T. (1988) The impact of boundaries on color: stabilized image studies. Soc. Photo-Opt. Instr. Eng., 901: 241–247.

Lashley, K.S. (1941) Patterns of cerebral integration indicated by the scotomas of migraine. Arch. Neurol. Psychiatry, 46: 331–339.

MacEvoy, S.P. and Paradiso, M.A. (2001) Lightness constancy in primary visual cortex. Proc. Natl. Acad. Sci. USA, 98: 8827–8831.

MacEvoy, S.P., Kim, W. and Paradiso, M.A. (1998) Integration of surface information in primary visual cortex. Nat. Neurosci., 1: 616–620.

MacEvoy, S.P., Hanks, T.D. and Paradiso, M.A. (2002) Responses of macaque V1 neurons with natural scenes and saccades. Soc. Neurosci., Abstract 622.4.

Paradiso, M.A. and Nakayama, K. (1991) Brightness perception and filling-in. Vision Res., 31: 1221–1236.

Paradiso, M.A. and Hahn, S. (1996) Filling-in percepts produced by luminance modulation. Vision Res., 36: 2657–2663.

Pinna, B., Brelstaff, G. and Spillmann, L. (2001) Surface color from boundaries: a new 'watercolor' illusion. Vision Res., 41: 2669–2676.

Pinna, B., Werner, J.S. and Spillmann, L. (2003) The watercolor effect: a new principle of grouping and figure-ground organization. Vision Res., 43: 43–52.

Poppelreuter, W. (1917) Die Psychischen Schadigungen durch Kopfschuss im Kriege 1914–1916: Vol. 1, Die Storunge n der niederen und hoheren Sehleistungen durch Verletzungen des Okzipitalhirns, Band I. Leopold Voss, Leipzig.

Redies, C. and Spillmann, L. (1981) The neon color effect in the Ehrenstein illusion. Perception, 10: 667–681.

Riggs, L.A., Ratliff, F., Cornsweet, C. and Cornsweet, T.N. (1953) The disappearance of steadily fixated visual test objects. J. Opt. Soc. Am., 43: 495–501.

Rossi, A.F. and Paradiso, M.A. (1996) Temporal limits of brightness induction and mechanisms of brightness perception. Vision Res., 36: 1391–1398.

Rossi, A.F. and Paradiso, M.A. (1999) Neural correlates of perceived brightness in the retina, lateral geniculate nucleus, and striate cortex. J. Neurosci., 19: 6145–6156.

Rossi, A.F., Rittenhouse, C.D. and Paradiso, M.A. (1996) The representation of brightness in primary visual cortex [see comments]. Science, 273: 1104–1107.

Schein, S.J. and Desimone, R. (1990) Spectral properties of V4 neurons in the macaque. J. Neurosci., 10: 3369–3389.

Spillmann, L. and Dresp, B. (1995) Phenomena of illusory form: can we bridge the gap between levels of explanation? Perception, 24: 1333–1364.

van der Schaaf, A. and van Hateren, J.H. (1996) Modelling the power spectra of natural images: statistics and information. Vision Res., 36: 2759–2770.

Wachtler, T., Sejnowski, T.J. and Albright, T.D. (2003) Representation of color stimuli in awake macaque primary visual cortex. Neuron, 37: 681–691.

Wallach, H. (1948) Brightness constancy and the nature of achromatic colors. J. Exp. Psychol., 38: 310–324.

White, M. (1979) A new effect on perceived lightness. Perception, 8: 413–416.

Yarbus, A.L. (1967) Eye Movements and Vision. Plenum Press, New York.

Zhou, H., Friedman, H.S. and von der Heydt, R. (2000) Coding of border ownership in monkey visual cortex. J. Neurosci., 20: 6594–6611.

Martinez-Conde, Macknik, Martinez, Alonso & Tse (Eds.)
Progress in Brain Research, Vol. 155
ISSN 0079-6123

CHAPTER 8

Beyond a relay nucleus: neuroimaging views on the human LGN

Sabine Kastner*, Keith A. Schneider and Klaus Wunderlich

Department of Psychology, Center for the Study of Brain, Mind and Behavior, Princeton University, Green Hall, Princeton, NJ 08544, USA

Abstract: The lateral geniculate nucleus (LGN) is the thalamic station in the retinocortical projection and has traditionally been viewed as the gateway for sensory information to enter the cortex. Here, we review recent studies of the human LGN that have investigated the retinotopic organization, physiologic response properties, and modulation of neural activity by selective attention and by visual awareness in a binocular rivalry paradigm. In the retinotopy studies, we found that the contralateral visual field was represented with the lower field in the medial-superior portion and the upper field in the lateral-inferior portion of each LGN. The fovea was represented in posterior and superior portions, with increasing eccentricities represented more anteriorly. Functional MRI responses increased monotonically with stimulus contrast in the LGN and in visual cortical areas. In the LGN, the dynamic response range of the contrast function was larger and contrast gain was lower than in the cortex. In our attention studies, we found that directed attention to a spatial location modulated neural activity in the LGN in several ways: it enhanced neural responses to attended stimuli, attenuated responses to ignored stimuli, and increased baseline activity in the absence of visual stimulation. Furthermore, we showed in a binocular rivalry paradigm that neural activity in the LGN correlated strongly with the subjects' reported percepts. The overall view that emerges from these studies is that the human LGN plays a role in perception and cognition far beyond that of a relay nucleus and, rather, needs to be considered as an early gatekeeper in the control of visual attention and awareness.

Keywords: fMRI; retinotopy; magno- and parvocellular LGN; contrast response; flicker response; selective attention; binocular rivalry

Introduction

The lateral geniculate nucleus (LGN) is the thalamic station in the retinocortical projection and has traditionally been viewed as the gateway for sensory information to enter the visual cortex (Jones, 1985; Sherman and Guillery, 2001). Its topographic organization and the response properties of its neurons have been extensively studied in nonhuman primates (e.g., Polyak, 1953; Kaas et al., 1972; Malpeli and Baker, 1975; Connolly and Van Essen, 1984). The LGN is typically organized into six main layers, each of which receives input from either the contra- or ipsilateral eye and contains a retinotopic map of the contralateral hemifield. The four dorsal layers contain small (parvocellular, P) neurons characterized by sustained discharge patterns and low-contrast gain, and the two ventral layers contain large (magnocellular, M) neurons characterized by transient discharge patterns and high-contrast gain (Wiesel

*Corresponding author. Tel.: +1-609-258-0479; Fax: +1-609-258-1113; E-mail: skastner@princeton.edu

DOI: 10.1016/S0079-6123(06)55008-3

and Hubel, 1966; Dreher et al., 1976; Creutzfeldt et al., 1979; Shapley et al., 1981; Derrington and Lennie, 1984; Merigan and Maunsell, 1993). In addition to retinal afferents, which constitute only 10% of its overall afferent input, the LGN receives modulatory inputs from multiple sources including striate cortex, the thalamic reticular nucleus (TRN), and the brainstem. The LGN therefore represents the first stage in the visual pathway at which cortical top-down feedback signals could affect visual processing. The functional role of these top-down inputs to the LGN is, however, not well understood (Guillery and Sherman, 2002).

In the human brain, it has proven difficult to study subcortical nuclei because of spatial resolution and signal-to-noise limitations of brain-mapping techniques. Thus, surprisingly little is known about the functional anatomy, physiological response properties, and functional role in perception and cognition of the human LGN. Here, we review a series of studies from our laboratory that utilized optimized neuroimaging techniques at conventional or high resolution to scan the human thalamus. First, we will describe high-resolution fMRI studies on the topographic organization of the LGN and its functional subdivisions into magno- and parvocellular parts. Second, we will report its basic response properties to stimulus contrast and flicker rate. Third, we will describe our advances in understanding the role of the LGN in attentional processing, one of the better understood cognitive operations in primates. And fourth, we will report recent studies on neural correlates related to perceptual experiences in binocular rivalry, suggesting an important role of the human LGN in visual awareness and conscious perception. The overall view that emerges from these studies is that the human LGN plays a role in perception and cognition far beyond that of a relay nucleus and, rather, needs to be considered as an early gatekeeper in the control of visual attention and awareness.

We chose to provide an update on the human LGN as a tribute to Lothar Spillmann, because many of the perceptual phenomena that were at the heart of Lothar's interest and study such as the Hermann grid and the "perceptive fields" (e.g., Spillman, 1971, 1994) can be explained by physiological response properties of retinal and geniculate neurons. Indeed, the lead author of this chapter first met Lothar in Freiburg over a discussion on neurophysiological correlates of simultaneous color contrast in LGN neurons, her graduate work at the time conducted with Otto Creutzfeldt in Goettingen (Creutzfeldt et al., 1991; Kastner et al., 1992).

Retinotopic organization

The topographic organization of the LGN has been studied extensively in macaques, using anatomical (Brouwer and Zeemann, 1926), physiological (Kaas et al., 1972; Malpeli and Baker, 1975; Connolly and Van Essen, 1984; Malpeli et al., 1996; Erwin et al., 1999), and lesion techniques (Clark and Penman, 1934). These studies have shown that the contralateral visual hemifield is represented in the LGN with the horizontal meridian dividing the structure into a superior and medial half representing the lower visual field and an inferior and lateral half representing the upper visual field. The fovea is represented medially in the posterior pole of the nucleus, whereas more peripheral visual field representations are located more anteriorly and laterally (Malpeli and Baker, 1975).

In the human LGN, anatomical studies have revealed a similar organization compared to the macaque LGN in terms of laminar patterns. The layout of the representation of the visual field, however, is less well understood because its study has been restricted to postmortem anatomical analyses of degeneration patterns following retinal or cortical lesions (Rönne, 1910; Juba and Szatmári, 1937; Kupfer, 1962; Hickey and Guillery, 1979). In one neuroimaging study, a retinotopic organization was suggested by demonstrating distinct and inverted activations associated with stimulation of the upper and lower visual hemifields in the inferior and superior parts of the LGN, respectively (Chen et al., 1999). We used high-resolution ($1.5 \times 1.5 \times 2\,mm^3$) fMRI at 3 T to derive a detailed account of the retinotopic organization of the human LGN, including estimates of the eccentricity magnification factor (Schneider et al., 2004). Representations of polar angle and eccentricity were measured within the central 15° of the visual field.

Polar-angle maps

The polar-angle component of the retinotopic map in the LGN was determined by using a smoothly rotating, flickering hemifield checkerboard stimulus. The checkerboard stimulus rotated counterclockwise about a central fixation point, at which subjects were instructed to maintain fixation throughout the presentation, and swept through the visual field with a period of 32 s, thereby evoking waves of activation in neurons whose receptive fields (RFs) they passed. With this stimulus, bilateral activations were found in the posterior thalamus, in the anatomical location of the human LGN, in all the seven subjects tested. These activations were strictly confined to stimulation of the contralateral hemifield in each LGN. Individual activation maps are shown for two representative subjects (S1 and S2) in the left and right columns of Fig. 1. The activation maps, overlaid on structural scans (shown in the central panel), are displayed in $15 \times 12\,\mathrm{mm}^2$ windows for five contiguous brain slices. The color given to each voxel was determined by the phase of its response and represents the region of the visual field to which the voxel was most responsive, as indicated in the color legend at the top of each column. Regions of the upper visual field are indicated in red-yellow, regions along the horizontal meridian in green, and regions of the lower visual field in blue. In the coronal plane, the representation of the horizontal meridian was oriented at an approximately 45° angle, dividing the lower visual field,

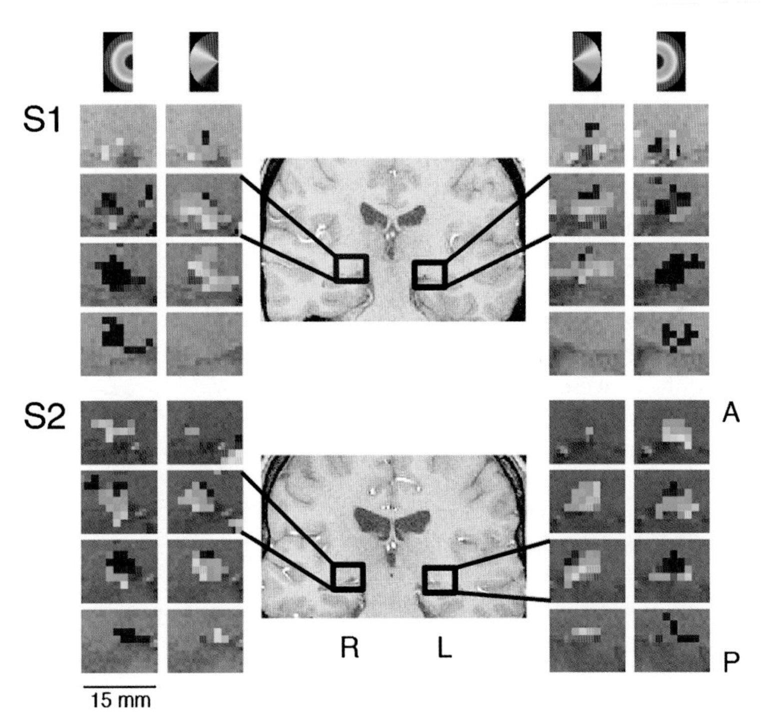

Fig. 1. Retinotopic maps in the LGN. Polar-angle and eccentricity maps are shown for two representative subjects (S1 and S2). The central panel shows an anatomical image in the coronal plane through the posterior thalamus. The boxes indicate the locations of the panels to the left and right. Details of the polar-angle maps in the right (R) and left (L) LGNs are shown in the near left and right columns, arranged in several sequential slices from anterior (A) to posterior (P). The eccentricity maps are shown in the far left and right columns and have been spatially registered with the polar-angle maps. The color code (shown for voxels whose responses were correlated with the fundamental frequency of the stimulus, ($r \geq 0.25$) indicates the phase of the response and labels the region of the visual field to which the voxel is most responsive, as depicted in the visual field color legend at the top of each column. (From Schneider et al., 2004, with permission.)

represented in the medial-superior section of the LGN, and the upper visual field, represented in the lateral-inferior section. Although the extent of activations varied somewhat among subjects, the overall pattern of retinotopic polar-angle organization was consistent among them.

Eccentricity maps

Eccentricity maps were measured in response to an expanding or contracting flickering checkerboard ring stimulus. The expanding ring stimulus consisted of an annulus with thickness equal to half of the radius of the visual display that expanded from the fixation point. The annulus increased in eccentricity (i.e., the distance from fixation) and wrapped around to the center once it reached the outer edge of the display, while subjects maintained fixation throughout the presentation. This stimulus activated the LGN bilaterally in all subjects ($N = 7$). Eccentricity maps are shown for the same subjects and in register with the polar-angle retinotopic maps in Fig. 1 (far away columns to the right and left). The color code, as indicated by the legend at the top of each column, indicates the region of the visual field to which each voxel was most responsive. Voxels representing the central 5° are indicated in dark to light blue; those representing 5°–10° in cyan to green to yellow; and those representing the peripheral 10°–15° in orange to red. The central 5° were represented mainly in the posterior portion of the LGN; in more anterior planes, the representation of the central 5° was confined to superior sections. More peripheral representations of the visual field were systematically arranged in anterior and inferior regions of the nucleus. As with the polar angle maps, the organization of the eccentricity maps was consistent across subjects.

Taken together, polar angle and eccentricity maps found in the human LGN indicate striking similarities in topographic organization compared to that reported in the macaque, as outlined above.

Eccentricity magnification factor

As in other visual areas, more LGN neurons are devoted to the representation of the fovea than to an equivalent area of the visual periphery. This distortion can be parameterized by an eccentricity magnification factor (Talbot and Marshall, 1941; Daniel and Whitteridge, 1961), which has been measured using a number of techniques in the macaque LGN and in both macaque and human V1. In the human brain, the eccentricity magnification factor for the LGN is unknown. We estimated this magnification factor on the basis of a volumetric analysis of the data obtained with eccentricity mapping. The cumulative volumes of the right and left LGNs were plotted as a function of eccentricity, as shown for three representative subjects (S1, S2, and S6) and the group (S1–S7) in Fig. 2. The cumulative volume functions were similarly shaped among the subjects, though differing in slope and extent. For many of the subjects, the cumulative volume function was steep and nearly linear for the initial 2°–5° of the eccentricity, after which the slope abruptly became shallower. A similar broken function has been observed in mac-

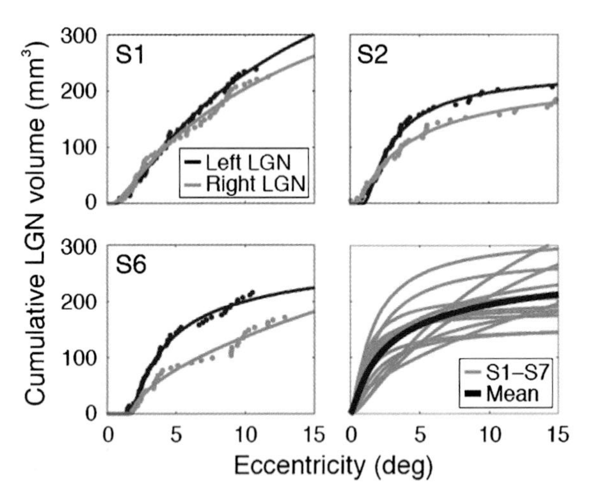

Fig. 2. Eccentricity magnification factor in the LGN. Eccentricity magnification factors for three representative subjects (S1, S2, S6) in each LGN are shown, computed on the basis of the phase responses of voxels activated by the expanding ring stimulus (see Fig. 1). The cumulative volume representing the area of the visual field from the fixation point to an eccentricity of r or less was fit to the integral of the magnification function $M(r) = A(r + B)^{-C}$ over the area of the visual hemifield (see Schneider et al., 2004 for more details). The fits for each LGN in each of the seven subjects are superimposed in the bottom right panel, along with the mean cumulative volume function fit for all subjects, which was $M(r) = 46.6(r + 0.52)^{-2.43}$. (From Schneider et al., 2004, with permission.)

aque visual cortex, where the RFs are nearly constant within the foveal 5° and begin to increase rapidly thereafter (Van Essen et al., 1984).

We then compared our estimate of the eccentricity magnification factor found in the group of seven subjects (see Fig. 2) with those obtained in macaque LGN and V1 reported in the literature. This comparison indicated a relative over-representation of the fovea in the human LGN as compared to the macaque LGN (see Fig. 6 in Schneider et al., 2004). We also compared our estimates for the human LGN with measurements of the magnification factor in human V1 that were obtained with different techniques including visually evoked potentials, fMRI, and phosphenes evoked by migraines or electrical stimulation. This comparison revealed that our estimates were similar to those obtained for human V1 (see Fig. 6 in Schneider et al., 2004). In the macaque, it is an open question whether the relative representation of the fovea expands progressively from the retina through the LGN and V1 (Malpeli and Baker, 1975; Myerson et al., 1977; Connolly and Van Essen, 1984; Van Essen et al., 1984; Perry and Cowey, 1985; Azzopardi and Cowey, 1996), or is preserved throughout the visual hierarchy with no additional magnification present at the level of the LGN or V1 (Webb and Kaas, 1976; Schein and de Monasterio, 1987; Wässle et al., 1989, 1990; Malpeli et al., 1996). In humans, our results of similar estimates of the magnification factor in LGN and V1 support the latter notion.

Magno- and parvocellular subdivisions

The spatial resolution of our imaging technique did not permit a dissociation of the parvocellular (P) and magnocellular (M) layers of the LGN. However, we attempted to dissociate the P- and M- subdivisions based upon two criteria: their anatomical locations and differences in functional properties, particularly in response sensitivity to stimulus contrast. On the basis of the anatomy, we expected M parts of the LGN to be located medially, inferiorly, and posteriorly. Typically, the M layers are flat and located on the inferior surface of the LGN, but particularly in the posterior planes, the LGN is oriented at an angle such that the M layers are located medially. Further, the LGN often exhibits folding such that the M layers would be located in the interior of the structure (Hickey and Guillery, 1979; Andrews et al., 1997). On the basis of the physiology, we expected that M cells should respond more sensitively to stimulus contrast than P cells. In single-cell recording studies, it has been shown that P cells are typically not responsive to contrast stimuli lower than 10% and have a 10-fold lower contrast gain than M cells, which typically respond to contrast stimuli as low as 2% (Shapley et al., 1981; Lee et al., 1989; Sclar, 1990). Therefore, we assumed that the M subdivision could be identified by two functional criteria: high sensitivity in response to a low-contrast stimulus, and small or no differences in responses to a low- and a high-contrast stimulus.

To identify P and M subdivisions of the LGN, flickering checkerboard stimuli of low (10%) and high (100%) luminance contrast were used presented in alternation to the right and left hemifields, while subjects maintained fixation at a central fixation point. We assumed that the high-contrast stimulus activated both the P and the M parts of the LGN, and its evoked activity was used to define a region of interest (ROI) encompassing both. Activation maps from two representative subjects (S1 and S6) are shown in Fig. 3A. The activations in the right and left LGNs evoked by the low- and high-contrast stimuli are displayed, similar to the format used in Fig. 1, on four sequential brain slices for each LGN. Activations evoked by the high-contrast stimulus are shown in the left column of each pair. Next, we identified the regions within this ROI that were most responsive to the low-contrast stimulus, and therefore were candidate areas to contain the M subdivision. The voxels activated by the low-contrast stimulus, shown in the right column of each pair in Fig. 3A, constituted a subset of the voxels activated by the high-contrast stimulus. The voxels most responsive to the low-contrast stimulus formed clusters that varied among the subjects in location relative to the activations evoked by the high-contrast stimulus.

In addition to response sensitivity to low-contrast stimuli, the second criterion that we employed to

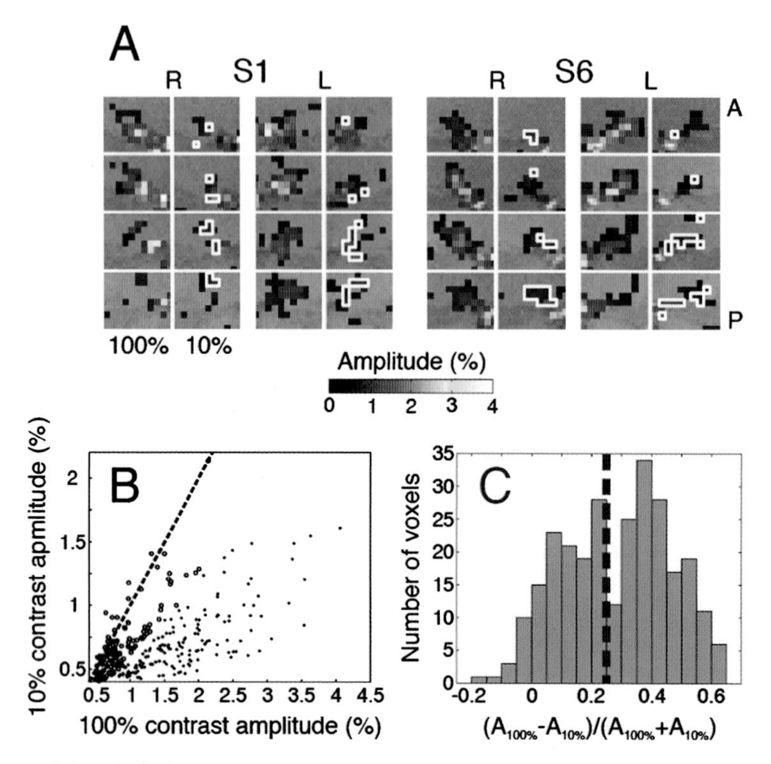

Fig. 3. Magno- and parvocellular subdivisions. Magno- and parvocellular subdivisions were functionally identified on the basis of the contrast sensitivity maps. (A) Activity evoked by an alternating hemifield stimulus in the anatomical location of the left (L) and right (R) LGNs is shown for two representative subjects. Four sequential slices are shown, ordered anterior (A) to posterior (P). The left column of each pair indicates the amplitude of the response to the high-contrast (100%) stimulus, and the right column indicates the response to the low-contrast (10%) stimulus. Only those voxels with correlations to the fundamental frequency of the stimulus, $r \geq 0.25$, are shown. Voxels surrounded by white lines responded similarly to the low- and high-contrast stimuli (see text and (B) below). On the basis of their high-contrast sensitivity, these voxels are likely dominated by magnocellular neurons. (B) For each subject ($N = 5$) and each voxel activated by both the low- and high-contrast stimuli ($r \geq 0.25$), the amplitudes of the mean fMRI time series evoked by the high- ($A_{100\%}$) and low- ($A_{10\%}$) contrast alternating hemifield stimuli are plotted against each other. The dashed diagonal line indicates equality between the amplitudes. The open circle symbols represent the voxels whose contrast modulation indices (CMI), defined as $(A_{100\%}-A_{10\%})/(A_{100\%}+A_{10\%})$, were less than 0.25. These voxels are bordered with solid white lines in (A). (C) The distribution of CMI. Voxels predominately containing M neurons are expected to be similarly activated by both the low- and high-contrast stimuli, and hence have a small CMI. P voxels are expected to have a strong differential response to the low- and high-contrast stimuli and therefore will have a large CMI. The dotted vertical line marks the 0.25 threshold used to select the voxels in panels (A) and (B). (From Schneider et al., 2004, with permission.)

identify the M subdivisions was that the responses of M voxels evoked by the low-contrast stimulus should be nearly saturated and marginally different from the responses evoked by the high-contrast stimulus, whereas P voxels should exhibit larger differences in response to the two contrast stimuli. Therefore, we analyzed the contrast modulation for those voxels that were reliably activated by both the low- and the high-contrast stimuli and plotted the averaged response amplitudes evoked by the two stimuli (Fig. 3B). A high correlation is evident, such that for each voxel, the larger the amplitude evoked by the high-contrast stimulus, the larger is the amplitude tended to be evoked by the low-contrast stimulus ($r = 0.59$, $p = 7.3 \times 10^{-27}$). The linear regression line has a slope of 0.22, but the population is distributed, including voxels clustered around the unity slope line, which indicates equality in the amplitudes evoked by the two contrast stimuli (see Fig. 3B). To quantify the response modulation, we calculated a contrast modulation index (CMI) for each voxel, defined as $(A_{100\%}-A_{10\%})/(A_{100\%}+A_{10\%})$, where $A_{100\%}$ and $A_{10\%}$ are the response amplitudes evoked by the 100% and 10% contrast stimuli, respectively. Voxels with CMI values near 0 were weakly modulated by the increase from low to high stimulus contrast, and those voxels with CMI near 1 were strongly modulated. The distribution of the CMIs is shown in Fig. 3C. The proportion of voxels with CMIs <0.25 are indicated by open circle symbols in Fig. 3B and are bordered with white lines in Fig. 3A. We found that 16.7% of all voxels activated by the high-contrast stimulus fulfilled both criteria, exhibiting significant responses to the low-contrast stimulus and exhibiting contrast saturation (CMI <0.25). These are the most likely candidates for voxels dominated by M responses. Although the anatomical locations of these voxels varied, when clustered, they tended to be located medially and/or posteriorly, as expected from the anatomical location of the M layers. This is also the case for the two subjects shown in Fig. 3A. In human anatomical studies, it has been shown that 19–28% of the LGN volume is occupied by the M layers (Andrews et al., 1997), which is similar to the proportion of LGN voxels identified as potential M voxel candidates using our functional criteria. It should be noted that our estimate depended on the choice of the activation and contrast modulation thresholds. Future studies using additional functional criteria will be necessary to further characterize the functional subdivisions within the human LGN.

Basic physiological response properties

Physiological response properties of LGN neurons have been extensively studied in nonhuman primates (for reviews see Jones, 1985; Sherman and Guillery, 2001). For example, P cells are characterized by sustained discharge patterns, sensitivity to color, and low-contrast gain, and M cells are characterized by transient discharge patterns and high-contrast gain (Wiesel and Hubel, 1966; Dreher et al., 1976; Creutzfeldt et al., 1979; Shapley et al., 1981; Merigan and Maunsell, 1993). In the series of studies reviewed in this section (Kastner et al., 2004), we investigated basic physiological response properties of the human LGN, specifically responses as a function of stimulus contrast and flicker reversal rate. Collective responses of neural populations in the LGN including both P and M parts were compared with population responses obtained in visual cortical areas.

Responses to stimulus contrast

To measure responses to stimulus contrast in the LGN and visual cortex, checkerboard stimuli with a constant flicker reversal rate of 7.5 Hz encompassing the central 12° of the visual field were presented in alternation to either the left or the right visual hemifield at six different contrast levels ranging from 4 to 100%. Subjects were instructed to maintain fixation at a central cross throughout the presentations. Time series of fMRI signals evoked by checkerboard stimuli presented at 4, 9, 35, and 100% contrast, averaged across scans and subjects ($N = 6$), are presented for the LGN, V1, V4, and medial temporal area (MT) in Fig. 4A.

In the LGN and visual cortical areas except MT, fMRI responses increased monotonically but nonlinearly as a function of stimulus contrast. In the LGN, responses to stimulus contrast less than 10% amounted to 41% of the maximum response. In visual cortex, an even greater sensitivity to low-contrast stimulus was seen. In areas V1 and V4, responses to the lowest contrast stimulus tested (4%) evoked 62% of the maximum response. In area MT, responses were saturated at the lowest contrast level (Fig. 4A). These findings confirmed previous single-cell physiology and neuroimaging studies (Dean, 1981; Tolhurst et al., 1981; Albrecht and Hamilton, 1982; Sclar et al., 1990; Cheng et al., 1994; Tootell et al., 1995; Boynton et al., 1996; Carandini and Ferster, 1997; Logothetis et al., 2001; Avidan et al., 2002). In the LGN, populations of neurons with different contrast sensitivities contributed to the collective responses measured with fMRI. As discussed in the last section, P cells are typically not responsive to contrast stimuli lower than 10% and have a 10-fold lower contrast gain than M cells, which typically respond to contrast stimuli as low as 2% (Shapley et al., 1981; Lee et al., 1989; Sclar, 1990). Our previous results (Schneider et al., 2004) demonstrated response saturation in the M subdivision of the LGN with contrast stimuli of 10%, suggesting

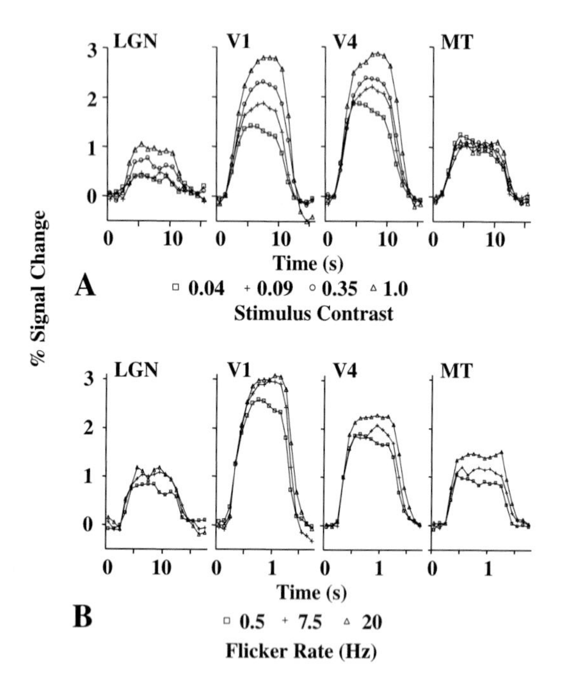

Fig. 4. Modulation by stimulus contrast (A) and flicker reversal rate (B): fMRI signals in LGN, V1, V4, and MT. Time series of fMRI signals in response to varying contrast (A) and flicker reversal rate (B) averaged over all subjects ($N = 6$) and scans. Data were combined across left and right hemispheres. (A) In the LGN, V1, and V4, responses increased monotonically with stimulus contrast. In MT, responses were saturated at the lowest contrast tested when stimuli were presented at increasing contrast levels. (B) In all areas, the 0.5 Hz stimulus evoked significantly smaller responses than the 20 Hz stimulus. In the LGN and in V1 responses evoked by the 7.5 Hz stimulus and the 20 Hz stimulus were similar, whereas in V4 and MT responses evoked by the 0.5 Hz stimulus and 7.5 Hz stimulus were similar. (From Kastner et al., 2004 with permission.)

high-contrast sensitivity for the magnocellular stream in the human visual system. Therefore, the relatively small LGN responses in the low-contrast range (< 10%) may be attributed to a dominant influence from P cells, which outnumber M cells several times in the LGN (Dreher et al., 1976; Perry et al., 1984; Andrews et al., 1997).

Responses of the LGN as a function of stimulus contrast differed in several respects from cortical contrast response functions (CRFs). First, responses in the LGN were evoked by a wider range of contrast stimuli, i.e., the dynamic range of the CRF was larger (Fig. 4A). In cortical areas, CRFs were steeper and saturated more readily, thereby reducing the dynamic range of the contrast functions. These results are in agreement with single-cell physiology studies (Sclar, 1990) and suggest that neural populations in the LGN can provide information about changes in contrast over a wider range than in cortex. Second, the contrast gain in LGN was lower than in cortical areas, as indicated by a steeper slope and a leftward shift of cortical CRFs along the contrast axis (Fig. 4A; see Fig. 4 in Kastner et al., 2004). In inactivation studies, it has been shown that cooling of V1 leads to decreases of contrast gain in LGN neurons suggesting that contrast gain in the LGN is controlled by cortical mechanisms that are mediated via corticofugal pathways (Przybyszewski et al., 2000). And third, LGN and V1 were significantly less sensitive to low luminance contrast than extrastriate cortex. A gradual increase of sensitivity to low luminance contrast was obtained from early to intermediate processing levels of the visual system (see also Avidan et al., 2002). These differences in contrast sensitivity may be attributed to the increasing receptive field size of neurons across visual cortex. For example, a neuron in area MT may receive inputs from as many as 10,000 M cells, which would increase its contrast sensitivity due to summation of inputs (Sclar, 1990). Similarly, the larger contrast sensitivity of M cells relative to P cells has been attributed to the larger receptive field sizes of M cells (Lennie et al., 1990).

Responses to flicker reversal rate

To measure responses to flicker reversal rate in the LGN and visual cortex, checkerboard stimuli with a constant contrast of 100% encompassing the central 12° of the visual field were presented in alternation to either the left or the right visual hemifield at three different rates: 0.5, 7.5, and 20 Hz. Subjects were instructed to maintain fixation at a central cross throughout the presentations. Time series of fMRI signals evoked by the stimuli presented at different flicker rates, averaged across sessions and subjects, are shown for the LGN, V1, V4, and MT in Fig. 4B. Differences in flicker rate modulated fMRI signals evoked by the checkerboard stimuli in the LGN and in cortical areas. In all areas, the 0.5 Hz stimulus evoked a significantly smaller response than the 20 Hz stimulus (Fig. 4B). However, the response evoked by the 0.5 Hz stimulus was surprisingly strong and totaled about 80% of the response elicited by the 20 Hz stimulus in the LGN and in cortical areas other than MT (Fig. 4B). In MT, the 0.5 Hz stimulus was only 62% ($\pm$7% S.E.M.) of the response evoked by the 20 Hz stimulus and elicited a significantly smaller response than in the other areas. In the LGN and in V1, the 7.5 and 20 Hz stimuli evoked similar responses that were significantly stronger than the response to the 0.5 Hz stimulus (Fig. 4B). In extrastriate areas V4 and MT, on the other hand, the 0.5 and 7.5 Hz stimuli evoked similar responses that were significantly smaller than the ones evoked by the 20 Hz stimulus (Fig. 4B). These results suggest that the LGN and V1 respond most sensitively to changes in flicker rate in the 0.5–7.5 Hz range. Extrastriate areas V4 and MT, on the other hand, appear to respond most sensitively within the frequency range of 7.5–20 Hz.

In the macaque monkey, P-LGN neurons have been found to respond most to stimuli at temporal frequencies close to 10 Hz, and M-LGN neurons to stimuli at frequencies close to 20 Hz (Hicks et al., 1983; Derrington et al., 1984; Merigan and Maunsell, 1990, 1993). Further, it was shown that P cells still responded to stimuli lower than 1 Hz, whereas such stimuli did not evoke responses in M cells (Hicks et al., 1983). Our results suggest that LGN responses evoked by the lowest frequency stimulus may be attributed to a predominant parvocellular influence. The low spatial frequency of the checkerboard stimulus presumably favored the activation of P cells, which, unlike M cells, do not show response attenuation at low spatial frequency (Enroth-Cugell et al., 1983; Hicks et al., 1983). In area MT, the relatively small responses evoked by the lowest frequency stimulus and the response preference in the high-frequency range are consistent with the notion that this area receives a dominant magnocellular input. Neurons in areas V1, V2, and V3 have been shown to respond optimally to temporal frequencies between 3 and 6 Hz (Foster et al., 1985; Levitt et al., 1994; Gegenfurtner et al., 1997). Despite significant differences in visual stimuli and methods to estimate neural activity, our finding of peak responses at temporal frequencies around 4 Hz (i.e., 7.5 Hz reversal rate) in these early cortical areas is in remarkable agreement with the results from single-cell physiology.

Finally, these results can also be related to psychophysical data. At spatial frequencies around 1 cycle/deg, contrast detection curves peak at temporal frequencies of about 3 Hz (Kelly, 1979). Thus, neural responses in the LGN and V1 with peak sensitivity around 4 Hz might predict psychophysical temporal frequency functions better than neural responses in extrastriate cortex with peak sensitivity at higher frequencies. However, studies using a combination of fMRI and psychophysics in the same subjects will be needed to test this idea further.

Attentional response modulation

Thus far, we have reported evidence that fMRI can be effectively used to study the functional topography and basic response properties of thalamic nuclei such as the LGN. Because the LGN represents the first stage in the visual pathway at which cortical top-down feedback signals could affect information processing, we took another step and investigated the functional role of the human LGN in a cognitive operation, which has been well defined at the neural level in visual cortex, selective visual attention.

At the cortical level, selective attention has been shown to affect visual processing in (at least) three different ways. First, neural responses to attended visual stimuli are enhanced relative to the same stimuli when unattended (attentional enhancement; e.g., Moran and Desimone, 1985; Corbetta et al., 1990). Second, neural responses to unattended stimuli are attenuated depending on the load of attentional resources engaged elsewhere (attentional suppression; Rees et al., 1997). And third, directing attention to a location in the absence of visual stimulation and in anticipation of the stimulus onset increases neural baseline activity (attention-related baseline increases; Luck et al., 1997; Kastner et al., 1999).

It has been proven difficult to study attentional response modulation in the LGN using single-cell physiology due to the small RF sizes of LGN neurons and the possible confound of small eye movements. Several single-cell physiology studies have failed to demonstrate attentional modulation in the LGN supporting a notion that selective attention affects neural processing only at the cortical level (e.g., Mehta et al., 2000). We revisited the role of the LGN in attentional processing using fMRI in humans (O'Connor et al., 2002; Kastner, 2004a, b). Functional MRI measures neural activity at a population level that might be better suited to uncover large-scale modulatory activity. Small modulatory effects that cannot be reliably found by measuring neural activity at the single- or multi-unit level may be revealed when summed across large populations of neurons. We investigated the three effects of selective attention demonstrated previously at the cortical level in a series of three experiments, which were designed to optimally activate the human LGN. Flickering checkerboard stimuli of high or low contrast were used in all experiments, which activated the LGN (Chen et al., 1999) and areas in visual cortex, including V1, V2, ventral and dorsal V3, V4, TEO, V3A, and MT/MST (referred to as MT), as determined on the basis of retinotopic mapping (Sereno et al., 1995; Kastner et al., 2001).

Attention effects of target enhancement, distracter suppression, and increases of baseline activity

To investigate attentional response enhancement in the LGN, checkerboard stimuli were presented to the left or right hemifield, while subjects directed attention to the stimulus (attended condition) or away from the stimulus (unattended condition). In the unattended condition, attention was directed away from the stimulus by having subjects count letters at fixation. The letter counting task ensured proper fixation and prevented subjects from covertly attending to the checkerboard stimuli (Kastner et al., 1998). In the attended condition, subjects were instructed to covertly direct attention to the checkerboard stimulus and to detect luminance changes that occurred randomly in time at 10° eccentricity. In our statistical model, stimulation of the left visual hemifield was contrasted with stimulation of the right visual hemifield. Thereby, the analysis was restricted to voxels activated by the peripheral checkerboard stimuli and excluded foveal stimulus representations. Relative to the unattended condition, the neural activity evoked by both the high-contrast stimulus and the low-contrast stimulus increased significantly in the attended condition (Fig. 5A). The attentional response enhancement was shown to be spatially specific. These results suggest that attention facilitates visual processing in the LGN by enhancing neural responses to an attended stimulus relative to those evoked by the same stimulus when ignored.

To investigate attentional-load-dependent suppression in the LGN, high- and low-contrast checkerboard stimuli were presented to the left or right hemifield while subjects performed either an easy attention task or a hard attention task at fixation and ignored the peripheral checkerboard stimuli. During the easy attention task, subjects counted infrequent, brief color changes of the fixation cross. During the hard attention task, subjects counted letters at fixation. Behavioral performance was 99% correct on average in the easy attention task and 54% in the hard attention task, thus indicating the differences in attentional demands. Relative to the easy task condition, neural activity evoked by the high- and low-contrast stimuli decreased significantly in the hard task condition (Fig. 5B). This finding suggests that neural activity evoked by ignored stimuli was attenuated in the LGN depending on the load of attentional resources engaged elsewhere.

To investigate attention-related baseline increases in the LGN, subjects were cued to covertly direct attention to the periphery of the left or right visual hemifield and to expect the onset of the stimulus. The expectation period was followed by attended presentations of a high-contrast checkerboard stimulus during which subjects counted the occurrence of luminance changes. During the expectation period, fMRI signals increased significantly relative to the preceding blank period in which subjects were fixating but not directing attention to the periphery. Because the visual input, a gray blank screen, was identical in both conditions, the increase in baseline activity appeared to be related to directed attention and may be interpreted as a bias in favor of the attended location.

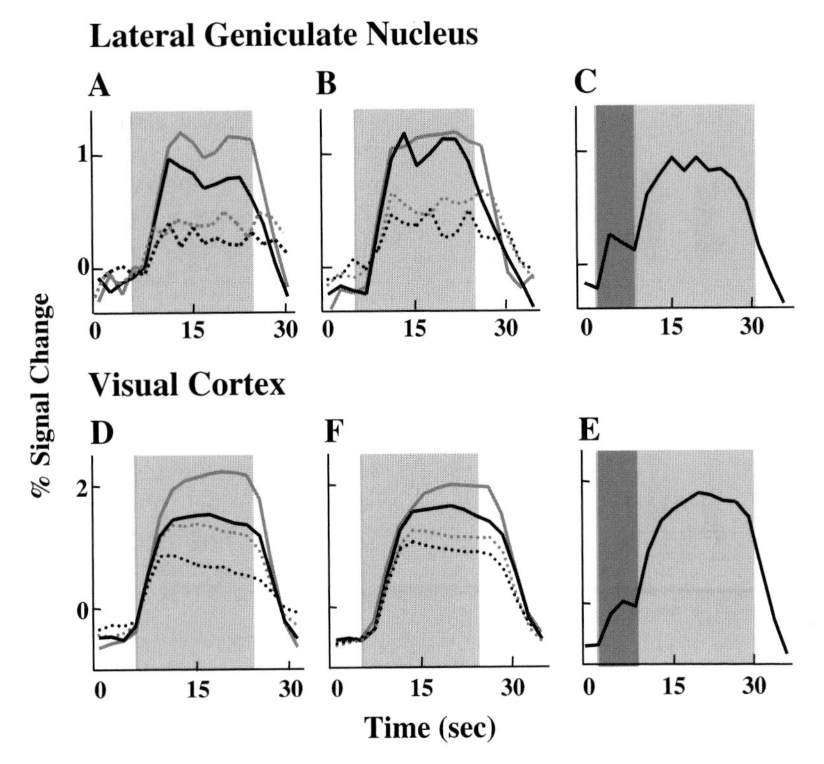

Fig. 5. Time series of fMRI signals in the LGN and in visual cortex. Group analysis ($n = 4$). Data from the LGN and visual cortex were combined across left and right hemispheres. Activity in visual cortex was pooled across areas V1, V2, V3/VP, V4, TEO, V3A, and MT/MST. (A), (D): Attentional enhancement. During directed attention to the stimuli (gray curves), responses to both the high-contrast stimulus (100%, solid curves) and low-contrast stimulus (5%, dashed curves) were enhanced relative to an unattended condition (black curves). (B), (E): Attentional suppression. During an attentionally demanding "hard" fixation task (black curves), responses evoked by both the high-contrast stimulus (100%, solid curves) and low-contrast stimulus (10%, dashed curves) were attenuated relative to an easy attention task at fixation (gray curves). (C), (F): Baseline increases. Baseline activity was elevated during directed attention to the periphery of the visual hemifield in expectation of the stimulus onset (darker gray shaded areas). The lighter gray shaded area indicates the beginning of checkerboard presentation periods. (From O'Connor et al., 2002, with permission.)

The baseline increase was followed by a further response increase evoked by the visual stimuli (Fig. 5C). It is important to note that, because of our statistical model, the increase in baseline activity was not related to the cue, which was presented at fixation. This finding suggests that neural activity in the LGN can be affected by attention-related top-down signals even in the absence of any visual stimulation whatsoever.

In summary, these studies indicate that selective attention modulates neural activity in the LGN by enhancing neural responses to attended stimuli, by attenuating those to ignored stimuli, and by increasing baseline activity in the absence of visual stimulation.

Comparison of attention effects in the LGN and the visual cortex

At the cortical level, qualitatively similar effects of attention were found, as shown in the time series of fMRI signals averaged across all activated areas in visual cortex, i.e., areas V1, V2, V3, V4, TEO, V3A, and MT (Figs. 5D–F). The attention effects found at the thalamic and at the cortical level were compared by normalizing the mean fMRI signals evoked in the LGN and in each activated cortical area and by computing index values for each attention effect and each area, which are measures of the magnitude of a given attention effect. This analysis is shown in Fig. 6; larger index values

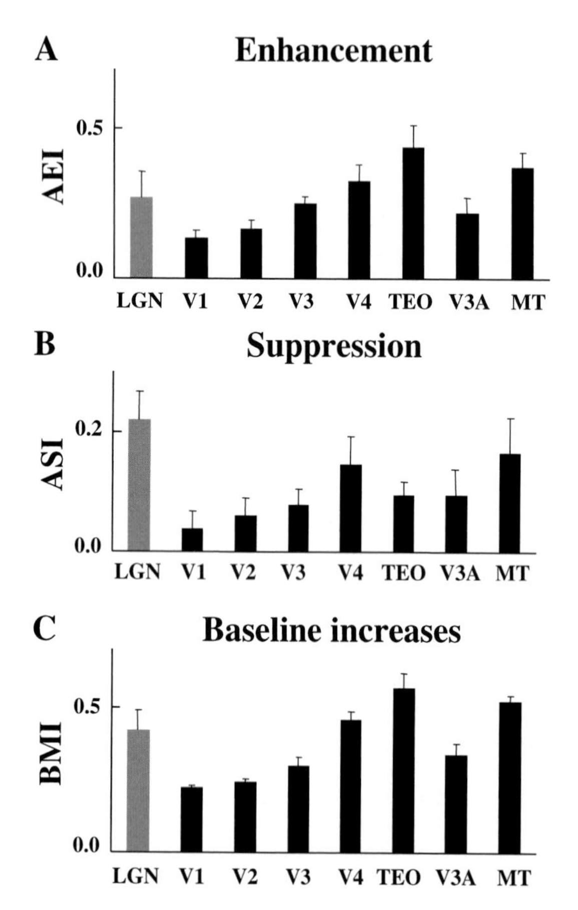

Fig. 6. Attentional response modulation in the LGN and in visual cortex. Attention effects that were obtained in the experiments presented in Fig. 1 were quantified by defining several indices: (A) attentional enhancement index (AEI), (B) attentional suppression index (ASI), (C) baseline modulation index (BMI). For all indices, larger values indicate larger effects of attention. Index values were computed for each subject based on normalized and averaged signals obtained in the different attention conditions and are presented as averaged index values from four subjects (for index definitions, see O'Connor et al., 2002). In visual cortex, attention effects increased from early to later processing stages. Attention effects in the LGN were larger than in V1. Vertical bars indicate S.E.M. across subjects. (From O'Connor et al., 2002, with permission.)

indicate larger effects of attention. It should be noted that index values cannot be easily compared across attention effects due to differences in index definitions and attention tasks. In accordance with previous findings (Kastner et al., 1998; Martinez et al., 1999; Mehta et al., 2000; Cook and Maunsell, 2002), the magnitude of all attention effects increased from early to more advanced processing levels along both the ventral and dorsal pathways of visual cortex (Figs. 6A–C). This is consistent with the idea that attention operates through top-down signals that are transmitted via corticocortical feedback connections in a hierarchical fashion. Thereby, areas at advanced levels of visual cortical processing are more strongly controlled by attention mechanisms than are early processing levels. This idea is supported by single-cell recording studies, which have shown that attention effects in area TE of inferior temporal cortex have a latency of approximately 150 ms (Chelazzi et al., 1998), whereas attention effects in V1 have a longer latency of approximately 230 ms (Roelfsema et al., 1998). According to this account, one would predict smaller attention effects in the LGN than in striate cortex. Surprisingly, it was found that all attention effects tended to be larger in the LGN than in striate cortex (Figs. 6A–C). This finding suggests that attentional response modulation in the LGN is unlikely to be due solely to corticothalamic feedback from striate cortex, but may be further influenced by additional sources of input (see below). Other possibilities that may explain the differences in magnitude of the modulation between the LGN and V1 include regional disparities underlying the blood oxygenation level-dependent signal or nonlinearities in thalamocortical signal transmission. Further, it is possible that differences in strength of attention effects at different processing stages may reflect the degree to which multiple parallel inputs converge on a given area rather than a feedback mechanism that reverses the processing hierarchy.

Sources of modulatory influences on the LGN

The findings reviewed thus far challenge the classical notion that attention effects are confined to cortical processing. Further, they suggest the need to revise the traditional view of the LGN as a mere gateway to the visual cortex. In fact, due to its afferent input, the LGN may be in an ideal strategic position to serve as an early "gatekeeper" in attentional gain control. In addition to corticothalamic feedback projections from V1, which comprise about 30% of its modulatory input, the

LGN receives another 30% of modulatory inputs from the TRN (Sherman and Guillery, 2002). For several reasons, the TRN has long been implicated in theoretical accounts of selective attention (Crick, 1984). First, all feed-forward projections from the thalamus to the cortex as well as their reverse projections pass through the TRN. Second, the TRN receives not only inputs from the LGN and V1, but also from several extrastriate areas and the pulvinar. Thereby, it may serve as a node where several cortical areas and thalamic nuclei of the visual system can interact to modulate thalamocortical transmission through inhibitory connections to LGN neurons (Guillery et al., 1998). And third, the TRN contains topographically organized representations of the visual field and can thereby modulate thalamocortical or corticothalamic transmission in spatially specific ways. Similarly, all corticofugal projections are organized in topographic order. Other modulatory influences on the LGN stem from the parabrachial nucleus of the brainstem. These cholinergic projections, another 30% of the modulatory input to the LGN, are more diffusely organized (Erisir et al., 1997), which makes a possible role in spatially selective attention more difficult to account for.

In summary, the LGN appears to be the first stage in the processing of visual information that is modulated by attentional top-down signals. Much remains to be learnt about the complex thalamic circuitry that may subserve attentional functions related to the control of neural gain in the LGN.

Neural correlates of visual awareness

Given the modulation of LGN activity by selective attention, we considered the possibility that the LGN may play a functional role in visual awareness. An ideal paradigm to study the neural basis underlying visual awareness is binocular rivalry. In binocular rivalry, the input from the two eyes cannot be fused to a single, coherent percept. Rivalry can be induced experimentally by simultaneously presenting dissimilar stimuli to the two eyes, such as a vertical and a horizontal grating. Rather than being perceived as a merged plaid, the two stimuli compete for perceptual dominance such that subjects perceive only one stimulus at a time while the other is suppressed from visual awareness (Levelt, 1965; Blake, 1989). Since the subjects' perceptual experiences change over time while the retinal stimulus remains constant, binocular rivalry provides an intriguing paradigm to study the neural basis of visual awareness (Crick and Koch, 1998).

Neural correlates of binocular rivalry in striate and extrastriate cortex

The neural mechanisms underlying binocular rivalry have been much debated. Single-cell physiology studies in monkeys trained to report their perceptual experiences during rivalry have identified neural correlates of binocular rivalry mainly in higher order visual areas (Sheinberg and Logothetis, 1997). Responses of about 90% of neurons in inferior temporal cortex increased when the neuron's preferred stimulus was perceived during rivalry, whereas only about 40% of neurons in areas V4 and MT showed such response enhancement, and even fewer in early visual areas V1 and V2 (Logothetis and Schall, 1989; Leopold and Logothetis, 1996). From these findings, it was concluded that binocular rivalry is mediated by competitive interactions between binocular neural populations representing the two stimuli at multiple stages of visual processing subsequent to the convergence of the input from the two eyes in V1 (pattern competition account). Alternatively, it has been suggested that binocular rivalry reflects competition between monocular channels either at the level of V1 or the LGN and is mediated by mutual inhibition and reciprocal feedback suppressing the input from one eye (Blake, 1989; Lehky and Blake, 1991). This interocular competition account has recently been supported by fMRI studies showing signal fluctuations correlated with subjects' perceptual experiences in area V1 (Polonsky et al., 2000) and more importantly in the monocular V1 neurons representing the blind spot (Tong and Engel, 2001). Given its anatomical organization and afferent projections, the LGN has often been considered as a possible site of suppression in accounts of interocular competition (Lehky, 1988; Lehky and Blake, 1991). However, single-cell recording studies in the LGN of awake monkeys viewing rivalrous stimuli did not find evidence to support this hypothesis (Lehky and Maunsell, 1996). We recently investigated the functional role of the human LGN in binocular rivalry using fMRI in subjects viewing dichoptically presented contrast-modulated grating stimuli (Wunderlich et al., 2005).

Modulation of LGN activity during binocular rivalry

In the rivalry experiment, superimposed sinusoidal gratings were viewed through red/green filter glasses; a high-contrast, green, horizontal grating was presented to one eye and a low-contrast, red, vertical grating was presented to the other eye. The gratings filled an annular aperture centered at a fixation point and reversed contrast to minimize adaptation. Their orthogonal orientations prevented the two gratings from being fused and induced rivalrous perceptual oscillations between them. Subjects ($N = 5$) maintained fixation and reported which grating was perceived by pressing a button; periods of mixed "piecemeal" percepts of the two stimuli were indicated with a third button. The same subjects also participated in a consecutive scanning session, the physical alternation experiment, in which sequential monocular presentations of the same grating stimuli were used to produce similar perceptual but different physical stimulation than during rivalry. The low- or high-contrast gratings were presented to one eye while a uniform field was presented to the other eye using the identical temporal sequence of stimulus alternations reported by the same subject in the rivalry experiment. During these physical alternations, subjects maintained fixation and pressed buttons to indicate which grating they viewed.

In the rivalry experiment, subjects experienced vigorous perceptual alternations between the horizontal high-contrast and the vertical low-contrast gratings. The perceptual durations were random and distributed according to a gamma-shaped function for both stimuli, as typically found in rivalry studies (Levelt, 1965). In accordance with classical findings (Levelt, 1965), the perceptually more salient high-contrast grating was perceived significantly longer than the low-contrast grating. In the group of subjects, the high-contrast stimulus was perceived on average for 5.1 ± 0.09 s (mean $\pm$ S.E.M.) compared to 3.1 ± 0.09 s for the low-contrast stimulus ($p \leq 0.001$).

In the LGN and V1, fMRI signals increase monotonically with stimulus contrast. Reliable fMRI signals are typically evoked by stimuli of more than 10% contrast, and signal saturation occurs with stimuli of more than 35% contrast (Boynton et al., 1996; Kastner et al., 2004; Schneider and Kastner, 2005). Therefore, the different fMRI signal amplitudes evoked by low- and high-contrast stimuli can be used as a "neural signature" of the LGN and V1 populations representing these stimuli, as previously shown for physical and rivalrous alternations of contrast-modulated gratings in V1 (Polonsky et al., 2000). In the physical alternation experiment, we expected fMRI signals to increase when the high-contrast gratings were shown monocularly and to decrease when the low-contrast gratings were presented. Further, we reasoned that, if the subjects' perceptual experiences during rivalry were reflected in fMRI signals, signal fluctuations similar to those obtained during physical alternations should occur in relation to the reported percepts despite the unchanging retinal stimulation.

Functional MRI signals in the LGN and V1 fluctuated while subjects perceived the rivalrous grating stimuli. The signals increased when subjects reported perceiving the high-contrast grating and decreased when they reported perceiving the low-contrast stimulus. To analyze the fMRI time series obtained in the rivalry experiment in relation to subjects' behavioral responses, an event-related analysis was performed for the LGN and V1 of each subject. Mean fMRI signals were derived by averaging the fMRI time series across all events of a reported switch to the high-contrast grating and, separately, across all events of a reported switch to the low-contrast grating. The events were time-locked to the subjects' manual responses and spanned a period of 4 s before and 9 s after each response, and the amplitude at the time of the response was subtracted to align the time series to 0. The mean fMRI signals were then averaged across subjects and are presented as group data ($N = 5$)

for the LGN and V1 in Fig. 7. Although both gratings were constantly present during rivalry, the fMRI signals in both LGN and V1 increased shortly after switches to the percept of the high-contrast grating (black lines) and decreased when the percept changed to the low-contrast grating

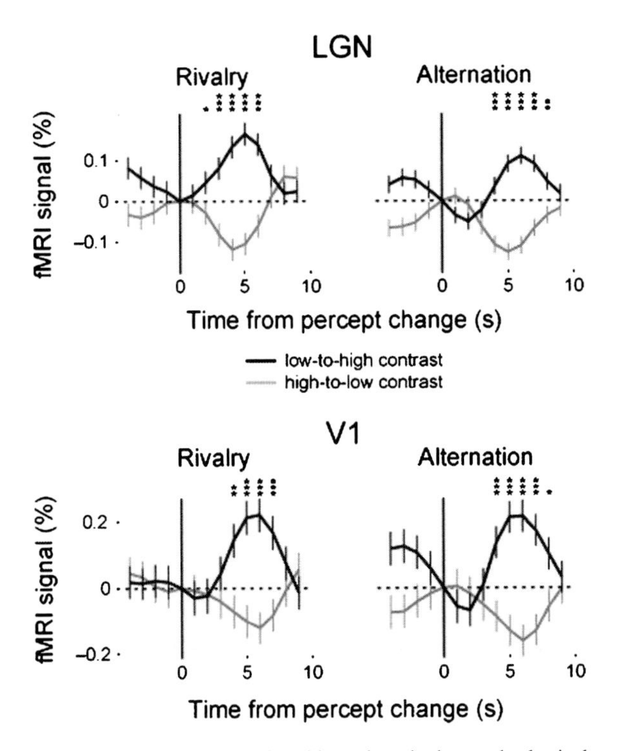

Fig. 7. fMRI signals during binocular rivalry and physical stimulus alternations in the LGN and V1 (group analysis). Data from the LGN and V1 of five subjects were combined across left and right hemispheres. Neural activity was averaged across all occurrences of perceptual switches from the low-contrast to the high-contrast grating (black curve) and across those from the high-contrast to the low-contrast grating (gray curve). The responses were time-locked to each subject's manual response, as indicated by the black vertical line at time point 0, and are shown within a relative time window of -4 to $+9$ s. All events were normalized, so that responses at time point 0 started at a value of 0% signal change. The vertical bar on each curve indicates one standard error of the mean. Asterisks indicate significant differences between data points of the two curves (one-tailed t-test, $*p<0.05$; $**p<0.01$; $***p<0.001$). Left: results from rivalry scans. Right: results from physical stimulus alternation scans. Neural activity increased when subjects perceived the high-contrast stimulus and decreased when they perceived the low-contrast stimulus during rivalry conditions. A similar response pattern was found when subjects viewed physical alternations of the same gratings. (From Wunderlich et al., 2005, with permission.)

(gray lines). Activity related to the percept of the high-contrast grating was significantly different from that related to the percept of the low-contrast grating in the LGN panel (Fig. 7) and in the V1 panel (Fig. 7) (one-tailed t-test, $*p\leq.05$; $**\ p\leq.01$; $***\ p\leq.001$ in Fig. 7). A strikingly similar pattern of responses was found in the LGN and V1 in the physical alternation condition (Figs. 7). Importantly, in both the LGN and V1, the magnitude and dispersion of fMRI signals evoked during rivalry were correlated with the duration of the subjects' perceptual experience suggesting that neural activity at the earliest stages of visual processing reflect both the content and duration of the percept and is therefore closely linked to visual awareness during binocular rivalry.

Possible neural mechanisms underlying binocular rivalry in the LGN

Advocates of interocular competition accounts have considered the LGN as a possible site at which the invisible stimulus is suppressed during binocular rivalry. Neurons in the LGN are exclusively monocular, with inputs from each eye segregated into separate layers. These adjacent laminae form an ideal substrate for inhibitory interactions between the two eyes, which would allow the signal from one eye to be selectively suppressed. Binocular interactions, predominantly inhibitory, have been widely reported in both monkey (Rodieck and Dreher, 1979; Marrocco and McClurkin, 1979; Schroeder et al., 1990) and cat LGN (Singer, 1970; Sanderson et al., 1971; Schmielau and Singer, 1977; Pape and Eysel, 1986; Varela and Singer, 1987; Sengpiel et al., 1995) and might provide a neural substrate in producing rivalry. These inhibitory interactions may be mediated by several anatomical pathways including interneurons extending between LGN layers, corticogeniculate feedback from striate cortex, or modulatory input from the TRN, as discussed above. One possibility is that feedback from binocular neurons in layer 6 of V1 (Lund and Boothe, 1975; Livingstone and Hubel, 1987) to monocular LGN layers could provide a descending control signal indicating whether stimuli are binocularly fused and regulating the strength of the inhibitory network (Lehky, 1988; Lehky and Blake, 1991). The importance of feedback from V1 in controlling the observed LGN activity cannot be overstressed. With our current temporal resolution, it is not possible to determine whether the LGN is controlling V1 activity or merely inherits the binocular resolution that might take place in a higher cortical area. Another possibility is that the TRN exerts additional control in modulating thalamocortical transmission through inhibitory connections to LGN neurons. It should be noted that these possibilities are not mutually exclusive. Taken together, based on its anatomy and the organization of its retinal and cortical feedback input, the LGN appears to be in an ideal position to play an important functional role in binocular rivalry, as suggested by our findings.

We conclude from our studies that the LGN appears to be the first stage in the processing of visual information at which neural correlates of visual awareness during binocular rivalry can be found. Our findings further suggest the need to revise the traditional view of the LGN as a mere gateway to the visual cortex. The LGN may, in fact, serve among a network of widely distributed cortical and subcortical brain systems as an early gatekeeper of visual awareness.

Conclusions

The studies reviewed in this chapter shed light on the retinotopic organization, basic response properties, and functional roles in perception and cognition of the human LGN. The LGN has been traditionally viewed as a relay nucleus, which conveys neural signals faithfully from the retina to primary visual cortex. Our studies have begun to cast doubt on this notion by demonstrating modulation of neural responses by cognitive operations such as selective attention and by perceptual content related to binocular rivalry. For example, the attentional response modulation in the LGN was on the order of that observed in extrastriate cortical areas, which have thus far been assumed to be the major sites of response modulation. We conclude from our studies thus far that the LGN plays an important role as an early gatekeeper in controlling attentional response gain and visual awareness. Neuroimaging of the human thalamus provides an important opportunity to revisit some of the classical notions on thalamic function that were derived mainly from studies in anesthetized animals and to ultimately define the functional role of thalamic nuclei such as the LGN in cognition and perception.

Acknowledgments

We thank K. Weiner for help with manuscript preparation. The research reviewed in this chapter was supported by the National Institute of Mental Health (RO1 MH-64043, P50 MH-62196, T32 MH-065214) and the Whitehall Foundation.

References

Albrecht, D. and Hamilton, D. (1982) Striate cortex of monkey and cat: contrast response function. J. Neurophysiol., 48: 217–237.

Andrews, T.J., Halpern, S.D. and Purves, D. (1997) Correlated size variations in human visual cortex, lateral geniculate nucleus, and optic tract. J. Neurosci., 17: 2859–2868.

Avidan, G., Harel, M., Hendler, T., Ben Bashat, D., Zohary, E. and Malach, R. (2002) Contrast sensitivity in human visual areas and its relationship to object recognition. J. Neurophysiol., 87: 3102–3216.

Azzopardi, P. and Cowey, A. (1996) The overrepresentation of the fovea and adjacent retina in the striate cortex and dorsal lateral geniculate nucleus of the macaque monkey. Neuroscience, 72: 627–639.

Blake, R. (1989) A neural theory of binocular rivalry. Psychol. Rev., 96: 145–167.

Boynton, G.M., Engel, S.A., Glover, G.H. and Heeger, D.J. (1996) Linear systems analysis of functional magnetic resonance imaging in human V1. J. Neurosci., 16: 4207–4221.

Brouwer, B. and Zeemann, W.B.C. (1926) The projection of the retina in the primary optic neuron in monkeys. Brain, 49: 1–35.

Carandini, M. and Ferster, D. (1997) A tonic hyperpolarization underlying contrast adaptation in cat visual cortex. Science, 276: 949–952.

Chelazzi, L., Duncan, J., Miller, E.K. and Desimone, R. (1998) Responses of neurons in inferior temporal cortex during memory-guided visual search. J. Neurophysiol., 80: 2918–2940.

Chen, W., Zhu, X., Thulborn, K.R. and Ugurbil, K. (1999) Retinotopic mapping of lateral geniculate nucleus in humans using functional magnetic resonance imaging. Proc. Natl. Acad. Sci. USA, 96: 2430–2434.

Cheng, K., Hasegawa, T., Saleem, K.S. and Tanaka, K. (1994) Comparison of neuronal selectivity for stimulus speed, length, and contrast in the prestriate cortical areas V4 and MT of the macaque monkey. J. Neurophysiol., 71: 2269–2280.

Clark, W.E.L.G and Penman, G.G. (1934) The projection of the retina in the lateral geniculate body. Proc. R. Soc. Lond. B Biol. Sci., 114: 291–313.

Connolly, M. and Van Essen, D. (1984) The representation of the visual field in parvicellular and magnocellular layers of the lateral geniculate nucleus in the macaque monkey. J. Comp. Neurol., 226: 544–564.

Cook, E.P. and Maunsell, J.H. (2002) Attentional modulation of behavioral performance and neuronal responses in middle temporal and ventral intraparietal areas of macaque monkey. J. Neurosci., 22: 1994–2004.

Corbetta, M., Miezin, F.M., Dobmeyer, S., Shulman, G.L. and Petersen, S.E. (1990) Attentional modulation of neural processing of shape, color, and velocity in humans. Science, 248: 1556–1559.

Creutzfeldt, O., Crook, J.M., Kastner, S., Li, C.Y. and Pei, X. (1991) The neurophysiological correlates of colour and brightness contrast in lateral geniculate neurons. Exp. Brain Res., 87: 3–21.

Creutzfeldt, O.D., Lee, B.B. and Elepfandt, A. (1979) A quantitative study of chromatic organisation and receptive fields of cells in the lateral geniculate body of the rhesus monkey. Exp. Brain Res., 35: 527–545.

Crick, F. (1984) Function of the thalamic reticular complex: the searchlight hypothesis. Proc. Natl. Acad. Sci. USA, 81: 4586–4590.

Crick, F. and Koch, C. (1998) Consciousness and neuroscience. Cereb. Cortex, 8: 97–107.

Daniel, P.M. and Whitteridge, D. (1961) The representation of the visual field on the cerebral cortex in monkeys. J. Physiol., 159: 203–221.

Dean, A. (1981) The relationship between response amplitude and contrast for cat striate Cortical neurons. J. Physiol., 318: 413–427.

Derrington, A.M. and Lennie, P. (1984) Spatial and temporal contrast sensitivities of neurones in lateral geniculate nucleus of macaque. J. Physiol., 357: 219–240.

Dreher, B., Fukada, Y. and Rodieck, R.W. (1976) Identification, classification and anatomical segregation of cells with X-like and Y-like properties in the lateral geniculate nucleus of old-world primates. J. Physiol., 258: 433–452.

Enroth-Cugell, C., Robson, J.G., Schweitzer-Tong, D.E. and Watson, A.B. (1983) Interactions in cat retinal ganglion cells showing linear spatial summation. J. Physiol., 341: 279–307.

Erisir, A., Van Horn, S.C. and Sherman, S.M. (1997) Relative numbers of cortical and brainstem inputs to the lateral geniculate nucleus. Proc. Natl. Acad. Sci. USA, 94: 1517–1520.

Erwin, E., Baker, F.H., Busen, W.F. and Malpeli, J.G. (1999) Relationship between laminar topology and retinotopy in the rhesus lateral geniculate nucleus: results from a functional atlas. J. Comp. Neurol., 407: 92–102.

Foster, K.H., Gaska, J.P., Nagler, M. and Pollen, D.A. (1985) Spatial and temporal frequency selectivity of neurons in visual cortical areas V1 and V2 of the macaquemonkey. J. Neurophysiol., 365: 331–363.

Gegenfurtner, K.R., Kiper, D.C. and Levitt, J.B. (1997) Functional properties of neurons in macaque area V3. J. Neurophysiol., 77: 1906–1923.

Guillery, R.W., Feig, S.L. and Lozsadi, D.A. (1998) Paying attention to the thalamic reticular nucleus. Trends Neurosci., 21: 28–32.

Guillery, R.W. and Sherman, S.M. (2002) Thalamic relay functions and their role in corticocortical communication: generalizations from the visual system. Neuron, 33: 163–175.

Hickey, T.L. and Guillery, R.W. (1979) Variability of laminar patterns in the human lateral geniculate nucleus. J. Comp. Neurol., 183: 221–246.

Hicks, T.P., Lee, B.B. and Vidyasagar, T.R. (1983) The response of cells in macaque lateral geniculate nucleus to sinusoidal gratings. J. Physiol., 337: 183–200.

Jones, E.G. (1985) The Thalamus. Plenum Press, New York.

Juba, A. and Szatmári, A. (1937) Ueber seltene hirnanatomische Befunde in Fällen von einseitiger peripherer. Blindheit. Klin. Monatsbl. Augenheilkd., 99: 173–188.

Kaas, J.H., Guillery, W.R. and Allman, J.M. (1972) Some principles of organization in the dorsal lateral geniculate nucleus. Brain Behav. Evol., 6: 253–299.

Kastner, S. (2004a) Towards a neural basis of human visual attention: Evidence from functional brain imaging. In: Kanwisher, N. and Duncan, J. (Eds.), Attention & Performance XX. Oxford University Press, Oxford, pp. 299–318.

Kastner, S. (2004b) Attentional response modulation in the human visual system. In: Posner, M. (Ed.), Attention. Guilford Press, New York, pp. 144–156.

Kastner, S., Crook, J.M., Pei, X. and Creutzfeldt, O. (1992) Neurophysiological studies of colour induction on white surfaces. Eur. J. Neurosci., 4: 1079–1086.

Kastner, S., De Weerd, P., Desimone, R. and Ungerleider, L.G. (1998) Mechanisms of directed attention in the human extrastriate cortex as revealed by functional MRI. Science, 282: 108–111.

Kastner, S., De Weerd, P., Pinsk, M.A., Elizondo, M.I., Desimone, R. and Ungerleider, L.G. (2001) Modulation of sensory suppression: implications for receptive field sizes in the human visual cortex. J. Neurophysiol., 86: 1398–1411.

Kastner, S., O'Connor, D.H., Fukui, M.M., Fehd, H.M., Herwig, U. and Pinsk, M.A. (2004) Functional imaging of the human lateral geniculate nucleus and pulvinar. J. Neurophysiol., 91: 438–448.

Kastner, S., Pinsk, M.A., De Weerd, P., Desimone, R. and Ungerleider, L.G. (1999) Increased activity in human visual cortex during directed attention in the absence of visual stimulation. Neuron, 22: 751–761.

Kelly, D.H. (1979) Motion and vision. II. Stabilized spatio-temporal threshold surface. J. Opt. Soc. Am., 69: 1340–1349.

Kupfer, C. (1962) The projection of the macula in the lateral geniculate nucleus of man. Am. J. Ophthalmol., 54: 597–609.

Lee, B.B., Martin, P.R. and Valberg, A. (1989) Sensitivity of macaque retinal ganglion cells to chromatic and luminance flicker. J. Physiol., 414: 223–243.
Lehky, S.R. (1988) An astable multivibrator model of binocular rivalry. Perception, 17: 215–228.
Lehky, S.R. and Blake, R. (1991) Organization of binocular pathways: Modeling and data related to rivalry. Neural Comput., 3: 44–53.
Lehky, S.R. and Maunsell, J.H. (1996) No binocular rivalry in the LGN of the alert macaque monkey. Vision Res., 36: 1225–1234.
Lennie, P., Krauskopf, J. and Sclar, G. (1990) Chromatic mechanisms in striate cortex of macaque. J. Neurosci., 10: 649–669.
Leopold, D.A. and Logothetis, N.A. (1996) Activity changes in early visual cortex reflect monkeys' percepts during binocular rivalry. Nature, 379: 549–553.
Levelt, W.J. (1965) Binocular brightness averaging and contour information. Br. J. Psychol., 56: 1–13.
Levitt, J.B., Kiper, D.C. and Movshon, J.A. (1994) Receptive fields and functional architecture of macaque V2. J. Neurophysiol., 71: 2517–2542.
Livingstone, M.S. and Hubel, D.H. (1987) Psychophysical evidence for separate channels for the perception of form, color, movement, and depth. J. Neurosci., 7: 3416–3468.
Logothetis, N., Pauls, J., Augath, M., Trinath, T. and Oeltermann, A. (2001) Neurophysiological investigation of the basis of the fMRI signal. Nature, 412: 150–157.
Logothetis, N.K. and Schall, J.D. (1989) Neuronal correlates of subjective visual perception. Science, 245: 761–763.
Luck, S.J., Chelazzi, L., Hillyard, S.A. and Desimone, R. (1997) Neural mechanisms of spatial selective attention in areas V1, V2, and V4 of macaque visual cortex. J. Neurophysiol., 77: 24–42.
Lund, J.S. and Boothe, R.G. (1975) Interlaminar connections and pyramidal neuron organization in the visual cortex, area 17, of the Macaque monkey. J. Comp. Neurol., 159: 305–334.
Malpeli, J.G. and Baker, F.H. (1975) The representation of the visual field in the lateral geniculate nucleus of Macaca mulatta. J. Comp. Neurol., 161: 569–594.
Malpeli, J.G., Lee, D. and Baker, F.H. (1996) Laminar and retinotopic organization of the macaque lateral geniculate nucleus: magnocellular and parvocellular magnification functions. J. Comp. Neurol., 375: 363–377.
Marrocco, R.T. and McClurkin, J.W. (1979) Binocular interaction in the lateral geniculate nucleus of the monkey. Brain Res., 168: 633–637.
Martinez, A., Anllo-Vento, L., Sereno, M.I., Frank, L.R., Buxton, R.B., Dubowitz, D.J., Wong, E.C., Hinrichs, H., Heinze, H.J. and Hillyard, S.A. (1999) Involvement of striate and extrastriate visual cortical areas in spatial attention. Nat. Neurosci., 2: 364–369.
Mehta, A.D., Ulbert, I. and Schroeder, C.E. (2000) Intermodal selective attention in monkeys. I: distribution and timing of effects across visual areas. Cereb. Cortex, 10: 343–358.
Merigan, W.H. and Maunsell, J.H. (1990) Macaque vision after magnocellular lateral geniculate lesions. Visual Neurosci., 5: 347–352.
Merigan, W.H. and Maunsell, J.H. (1993) How parallel are the primate visual pathways? Annu. Rev. Neurosci., 16: 369–402.
Moran, J. and Desimone, R. (1985) Selective attention gates visual processing in the extrastriate cortex. Science, 229: 782–784.
Myerson, J., Manis, P.B., Miezin, F.M. and Allman, J.M. (1977) Magnification in striate cortex and retinal ganglion cell layer of owl monkey: a quantitative comparison. Science, 198: 855–857.
O'Connor, D.H., Fukui, M.M., Pinsk, M.A. and Kastner, S. (2002) Attention modulates responses in the human lateral geniculate nucleus. Nat. Neurosci., 5: 1203–1209.
Pape, H.C. and Eysel, U.T. (1986) Binocular interactions in the lateral geniculate nucleus of the cat: GABAergic inhibition reduced by dominant afferent activity. Exp. Brain Res., 61: 265–271.
Perry, V.H. and Cowey, A. (1985) The ganglion cell and cone distributions in the monkey's retina: implications for central magnification factors. Vision Res., 25: 1795–1810.
Perry, V.H., Oehler, R. and Cowey, A. (1984) Retinal ganglion cells that project to the dorsal lateral geniculate nucleus in the macaque monkey. Neuroscience, 12: 1101–1123.
Polonsky, A., Blake, R., Braun, J. and Heeger, D.J. (2000) Neuronal activity in human primary visual cortex correlates with perception during binocular rivalry. Nat. Neurosci., 3: 1153–1159.
Polyak, S. (1953) Santiago Ramon y Cajal and his investigation of the nervous system. J. Comp. Neurol., 98: 3–8.
Przybyszewski, A.W., Gaska, J.P., Foote, W. and Pollen, D.A. (2000) Striate cortex increases contrast gain of macaque LGN neurons. Visual Neurosci., 17: 485–494.
Rees, G., Frith, C.D. and Lavie, N. (1997) Modulating irrelevant motion perception by varying attentional load in an unrelated task. Science, 278: 1616–1619.
Rodieck, R.W. and Dreher, B. (1979) Visual suppression from nondominant eye in the lateral geniculate nucleus: a comparison of cat and monkey. Exp. Brain Res., 35: 465–477.
Roelfsema, P.R., Lamme, V.A. and Spekreijse, H. (1998) Object-based attention in the primary visual cortex of the macaque monkey. Nature, 395: 376–381.
Rönne, H. (1910) Pathologisch-anatomische Untersuchungen über alkoloische Intoxikationsamblyopie. Arch. Ophthalmol., 77: 1–95.
Sanderson, K.J., Bishop, P.O. and Darian-Smith, I. (1971) The properties of the binocular receptive fields of lateral geniculate neurons. Exp. Brain Res., 13: 178–207.
Schein, S.J. and de Monasterio, F.M. (1987) Mapping of retinal and geniculate neurons onto striate cortex of macaque. J. Neurosci., 7: 996–1009.
Schmielau, F. and Singer, W. (1977) The role of visual cortex for binocular interactions in the cat lateral geniculate nucleus. Brain Res., 120: 354–361.
Schneider, K.A. and Kastner, S. (2005) Visual responses of the human superior colliculus: a high-resolution fMRI study. J. Neurophys., 94: 2491–2503.

Schneider, K., Richter, M. and Kastner, S. (2004) Retinotopic organization and functional subdivisions of the human lateral geniculate nucleus: A high- resolution fMRI study. J. Neurosci., 24: 8975–8985.

Schroeder, C.E., Tenke, C.E., Arezzo, J.C. and Vaughan Jr., H.G. (1990) Binocularity in The lateral geniculate nucleus of the alert macaque. Brain Res., 521: 303–310.

Sclar, G., Maunsell, J.H. and Lennie, P. (1990) Coding of image contrast in central visual pathways of the macaque monkey. Vision Res., 30: 1–10.

Sengpiel, F., Blakemore, C. and Harrad, R. (1995) Interocular suppression in the primary visual cortex: a possible neural basis of binocular rivalry. Vision Res., 35: 179–195.

Sereno, M.I., Dale, A.M., Reppas, J.B., Kwong, K.K., Belliveau, J.W., Brady, T.J., Rosen, B.R. and Tootell, R.B. (1995) Borders of multiple visual areas in humans revealed by functional magnetic resonance imaging. Science, 268: 889–893.

Shapley, R., Kaplan, E. and Soodak, R. (1981) Spatial summation and contrast sensitivity of X and Y cells in the lateral geniculate nucleus of the macaque. Nature, 292: 543–545.

Sheinberg, D.L. and Logothetis, N.K. (1997) The role of temporal cortical areas in perceptual organization. Proc. Natl. Acad. Sci. USA, 94: 3408–3413.

Sherman, S.M. and Guillery, R.W. (2001) Exploring the Thalamus. Academic Press, San Diego.

Sherman, S.M. and Guillery, R.W. (2002) The role of the thalamus in the flow of information to the cortex. Philos. Trans. R. Soc. Lond. Ser. B: Biol. Sci., 357: 1695–1708.

Singer, W. (1970) Inhibitory binocular interaction in the lateral geniculate body of the cat. Brain Res., 18: 165–170.

Spillman, L. (1971) Foveal perceptive fields in the human visual system measured with simultaneous contrast in grids and bars. Pflugers Arch. Ges. Physiol., 326: 281–299.

Spillman, L. (1994) The Hermann grid illusion: a tool for studying human perceptive field organization. Perception, 23: 691–708.

Talbot, S.A. and Marshall, W.H. (1941) Physiological studies on neuronal mechanisms of visual localization and discrimination. Am. J. Ophthalmol., 24: 1255–1263.

Tolhurst, D., Movshon, J. and Thompson, I. (1981) The dependence of response amplitude and variance of cat visual cortical neurons on stimulus contrast. Exp. Brain Res., 41: 414–419.

Tong, F. and Engel, S.A. (2001) Interocular rivalry revealed in the human cortical blind-spot representation. Nature, 411: 195–199.

Tootell, R.B.H., Reppas, J.B., Kwong, K.K., Malach, R., Born, R.T., Brady, T.J., Rosen, B.R. and Belliveau, J.W. (1995) Functional analysis of human MT and related visual cortical areas using magnetic resonance imaging. J. Neurosci., 15: 3215–3261.

Van Essen, D.C., Newsome, W.T. and Maunsell, J.H. (1984) The visual field representation in striate cortex of the macaque monkey: asymmetries, anisotropies, and individual variability. Vision Res., 24: 429–448.

Varela, F.J. and Singer, W. (1987) Neuronal dynamics in the visual corticothalamic pathway revealed through binocular rivalry. Exp. Brain Res., 66: 10–20.

Wässle, H., Grunert, U., Rohrenbeck, J. and Boycott, B.B. (1989) Cortical magnification factor and the ganglion cell density of the primate retina. Nature, 341: 643–646.

Wässle, H., Grunert, U., Rohrenbeck, J. and Boycott, B.B. (1990) Retinal ganglion cell density and cortical magnification factor in the primate. Vision Res., 30: 1897–1911.

Webb, S.V. and Kaas, J.H. (1976) The sizes and distribution of ganglion cells in the retina of the owl monkey *Aotus trivirgatus*. Vision Res., 16: 1247–1254.

Wiesel, T.N. and Hubel, D.H. (1966) Spatial and chromatic interactions in the lateral geniculate body of the rhesus monkey. J. Neurophysiol., 29: 1115–1156.

Wunderlich, K., Schneider, K.A. and Kastner, S. (2005) Neural correlates of binocular rivalry in the human lateral geniculate nucleus. Nat. Neurosci., 8: 1595–1602.

SECTION III

The Neural Bases for Visual Awareness and Attention

Introduction

Establishing the neural underpinnings of visual awareness and attention is one of the ultimate goals of basic and clinical neuroscience. And no wonder: these functions are central to human cognition. However, despite being critical to normal life, the study of attention and awareness is among the most elusive of all scientific endeavors. Scientists have not yet even achieved an agreed-upon definition of the terms. Nevertheless, awareness and attention are at the very core of what makes inanimate groups of neurons become living, thinking creatures. This transformation is perhaps the greatest and most awesome scientific mystery of all time, which does not make the scientific process any easier.

Many laboratories have begun to chip away at these questions bit-by-bit from a number of different directions. The following four chapters represent a microcosm of this diversified effort.

Bichot and Desimone begin the section by discussing the parallel and serial neural mechanisms used for visual selection of relevant data. Goldberg, Bisley, Powell, and Gottlieb discuss the role of the lateral intraparietal cortex (LIP) in attending to a location in space and in planning an eye movement. My own chapter follows with a description of the neurophysiological underlying our awareness of visible stimuli. Stoerig presents data from blindsight patients, who are not conscious of visual stimuli presented in their blind field and yet have the ability to somehow use this unconscious information. Valle-Inclán and Gallego wrap up the section with a study of frontal lobe function in binocular rivalry by testing of one of the last surviving human frontal lobotomy patients.

Stephen L. Macknik

Martinez-Conde, Macknik, Martinez, Alonso & Tse (Eds.)
Progress in Brain Research, Vol. 155
ISSN 0079-6123

CHAPTER 9

Finding a face in the crowd: parallel and serial neural mechanisms of visual selection

Narcisse P. Bichot and Robert Desimone*

McGovern Institute for Brain Research, Massachusetts Institute of Technology, Bldg. 46–6121, Cambridge, MA 02139, USA

Abstract: At any given moment, our visual system is confronted with more information than it can process. Thus, attention is needed to select behaviorally relevant information in a visual scene for further processing. Behavioral studies of attention during visual search have led to the distinction between serial and parallel mechanisms of selection. To find a target object in a crowded scene, for example a "face in a crowd", the visual system might turn on and off the neural representation of each object in a serial fashion, testing each representation against a template of the target object. Alternatively, it might allow the processing of all objects in parallel, but bias activity in favor of those neurons representing critical features of the target, until the target emerges from the background. Recent neurophysiological evidence shows that both serial and parallel selections take place in neurons of the ventral "object-recognition pathway" during visual search tasks in which monkeys freely scan complex displays to find a target object. Furthermore, attentional selection appears to be mediated by changes in the synchrony of responses of neuronal populations in addition to the modulation of the firing rate of individual neurons.

Keywords: attention; selection; saccades; visual search; serial; parallel; area V4; synchrony

Introduction

Vision is of primary importance in gathering information about the surrounding world, and we spend much of our time engaged in visual search to find and process behaviorally relevant information in crowded scenes. When viewers know the location of the relevant object, the brain mechanisms that guide spatial attention to the object are largely overlapping with those for selecting the targets for eye movements (Nobre, 2001; Corbetta and Shulman, 2002), consistent with behavioral studies showing a strong functional link between spatial attention and eye movements (Kowler et al., 1995; Deubel and Schneider, 1996; Liversedge and Findlay, 2000). Selection for attention or eye movements lead to an enhancement of the responses of visual cortex neurons to the relevant object, at the expense of distractors (Luck et al., 1997; Colby and Goldberg, 1999; Seidemann and Newsome, 1999; Treue and Maunsell, 1999; Andersen and Buneo, 2002), leaving object recognition mechanisms in the temporal cortex with only a single relevant stimulus at a time (Desimone and Duncan, 1995). However, in most common visual scenes, viewers rarely know the specific location of the relevant object in advance — instead, they must search for it, based on its distinguishing features such as color or shape, which is commonly termed visual search.

For decades, psychologists have debated on how the brain filters out irrelevant information and focuses attention on information that matters, with

*Corresponding author. Tel.: +1-617-324-0141; Fax: +1-617-452-4119; E-mail: desimone@mit.edu

DOI: 10.1016/S0079-6123(06)55009-5

many debates centered around the roles of serial and parallel mechanisms in selection (Shiffrin and Schneider, 1977; Treisman and Gelade, 1980; Nakayama and Silverman, 1986; Wolfe et al., 1989; Townsend, 1990). This distinction can be illustrated by considering a complex visual search such as finding Waldo in a crowded page of a "Where's Waldo?" book. When searching for Waldo, one possibility is that the brain scans the page spatially (serial processing) like a mental spotlight (Posner et al., 1980) moving across an otherwise dark page. In this model, the attentional spotlight would track across the page, checking each object within its "field of illumination" against a mental image of Waldo. Another possibility is that the brain takes in the entire page at once and gradually zooms in on relevant features such as color and shape (parallel processing). In this model, based for example on a bias towards Waldo's red-striped shirt, objects with the color red would gradually stand out from the cluttered background. Here, we review recent studies from our lab, as well as others, that have investigated the brain mechanisms underlying top-down, feature-based selection during visual search guided by the knowledge of the target's visual properties.

Parallel selection during visual search

The key element of parallel search models is a neural bias in favor of stimuli containing features (e.g., color or shape) of the searched-for target that occurs throughout the visual field and throughout the time period of the search, long before a target is identified. We recently investigated the presence of such a bias in the activity of visual cortical neurons in monkeys that freely scanned complex visual search arrays to find a target defined by color (Fig. 1A), shape, or both (Fig. 1B) (Bichot et al., 2005). Recordings were conducted in area V4, a key area of the ventral stream for object recognition (Mishkin et al., 1983), where neurons have smaller receptive fields (RFs) compared with the extremely large RFs in inferior temporal (IT) cortex and are selective for basic stimulus features such as color, orientation, and simple shapes (Desimone et al., 1985; Desimone and Schein,

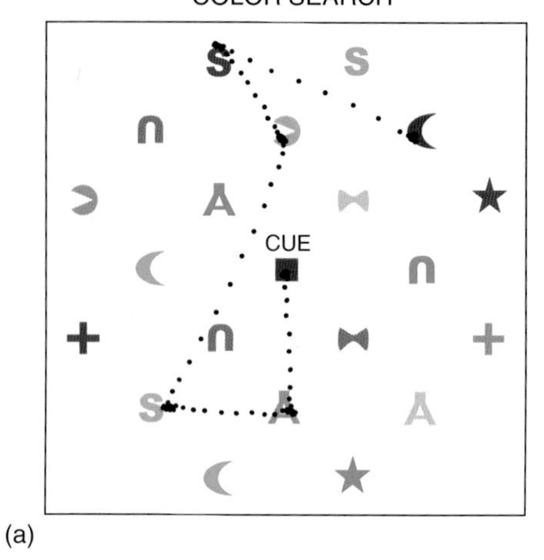

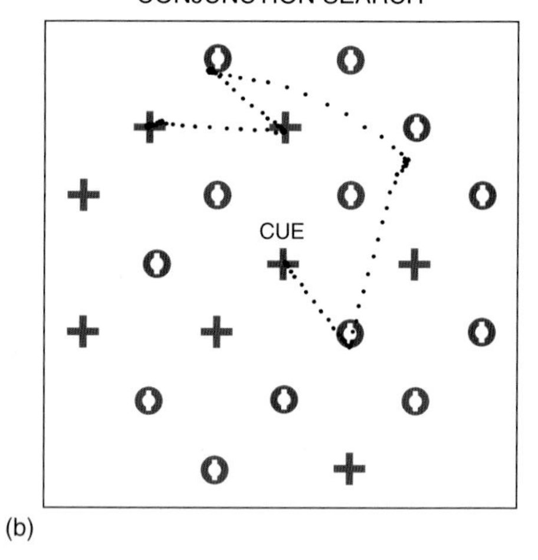

Fig. 1. Free-viewing visual search tasks. (a) Examples of a color search trial and a (b) conjunction search trial. The black dots show the eye position of the monkey during representative correctly performed trials. In all searches, monkeys fixated a central fixation spot that was then replaced by a central cue. The cues at the center of the screen are shown for illustration purposes only as they were extinguished before the onset of the search array in the experiment. During color search, the cue was a colored square that instructed monkeys with the color of the target, and during shape search, the cue was a gray shape that instructed monkeys with the shape of the target. When color was relevant, shape was irrelevant, and vice versa. During conjunction search, the cue (and target) was one combination of two colors and two shapes, and the distractors either shared the color, shape or no feature with the target. Monkeys were trained to find and fixate the target stimulus for a prescribed duration to receive reward. As shown in the examples, monkeys made several saccades (~5.8 per trial across all search types) before finding the target.

1987; Schein and Desimone, 1990). Furthermore, previous studies have shown that attentional modulation in area V4 is larger than in areas V1 and V2 and therefore easier to study (Luck et al., 1997).

To test for parallel, feature-based biasing, Bichot et al. (2005) specifically considered responses to stimuli in the RF of neurons at times when the monkey was actively "attending" elsewhere, when the monkey was preparing to make a saccade to a stimulus outside the RF. More specifically, the response to an unselected RF stimulus with the neuron's preferred or nonpreferred color was compared on trials during which the cue was of the preferred or nonpreferred color for the neuron (Fig. 2A). When a stimulus of the preferred color was in the RF, neurons gave enhanced responses when it matched the cue color. Responses to a RF stimulus of the nonpreferred color, on the other hand, were not modulated by cue color. Similar results were found during shape search taking into account the neurons' selectivity for stimulus shapes. Thus, neurons responded most strongly when an unselected RF stimulus with the preferred feature was the search target, even though the stimulus was not selected for a saccadic eye movement. For example, if the animal was searching for red, the cells preferred red, and there happened to be a red stimulus inside the RF; this is when firing rate was enhanced, even though the animal was preparing an eye movement to a different stimulus. Furthermore, when the cue was of the preferred color of the neurons, responses to stimuli with colors similar to the cue color also showed some enhancement, explaining why the animals tended to fixate distractors similar to the target.

The feature-based bias observed by Bichot et al. (2005) is consistent with the findings of another study examining neural activity in area V4 during free-viewing visual search (Mazer and Gallant, 2003). In that study, monkeys were trained to search for a target grayscale natural image patch among distractor ones. Across all fixations, they found significant differences in the activity of about one in four V4 neurons across the different search targets. However, it is not clear whether the overall enhancement of neural responses in that study for a particular search target affected all stimuli regardless of features and their similarity to the target, and whether the results were confounded by differences in gaze patterns with different search targets.

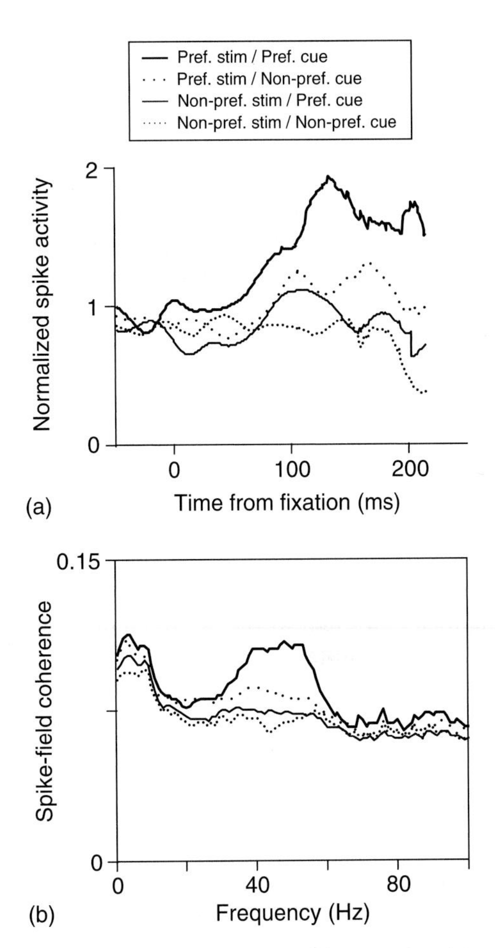

Fig. 2. Feature-related enhancement of neuronal activity and spike-field synchronization in area V4 during color search. (a) Normalized firing rates averaged over a population of V4 neurons during color search trials during fixations at the end of which the monkey made a saccade away from the RF. Thick lines show responses when the stimulus in the RF was of the preferred color for the recorded neurons; thin lines show responses when the stimulus was of the nonpreferred color; solid lines show responses on trials in which the cue was the preferred color; and dotted lines show responses on trials in which the cue was the nonpreferred color. (b) Spike field coherence for the same conditions.

As with most studies of the neural substrates of attention (Desimone and Duncan, 1995; Maunsell, 1995), the findings described above show a bias for stimuli that are likely to be the search targets resulting from the increase in the firing rate of individual neurons in response to those stimuli. However, another potential "amplifier" of the effects of one population of neurons on another is the synchronization of activity in the input population (Gray et al., 1989; Salinas and Sejnowski, 2000; Kara and Reid, 2003). Small changes in high frequency synchronization of spike trains with attention at one stage might lead to pronounced firing rate changes at subsequent stages (Niebur et al., 1993; Salinas and Sejnowski, 2000) because cells generally have short synaptic integration times. Indeed, V4 neurons synchronize their activity when attention is directed to their RFs (Fries et al., 2001), as do neurons in parietal cortex during a memory-saccade task (Pesaran et al., 2002). Also, in monkey somatosensory cortex, it has been reported that cells synchronize their activity when monkeys perform a tactile task compared to a visual task, presumably due to an increase in "tactile attention" in that task (Steinmetz et al., 2000).

To investigate potential changes in neuronal synchronization with feature-based attention, Bichot et al. (2005) measured the coherence between spikes and the local field potential (LFP) (Fries et al., 1997; Jarvis and Mitra, 2001). Spike-field coherence measures phase synchronization between the LFP and spike times as a function of frequency, and is independent of any changes in the firing rate of the spikes and in the power spectrum of the LFP. Coherence for a given frequency ranges from 0 (when the spikes do not have any systematic phase relation to the LFP component at this frequency) to 1 (when all spikes appear at exactly the same phase relation relative to this frequency component). During color search, similar to the effects found on firing rates, neurons increased their synchronization in the gamma-range (30–60 Hz) when an unselected RF stimulus with the preferred color was the target that the animal was searching for (Fig. 2B). There was also a smaller increase in gamma-band synchrony when the preferred color was cued and distractors with similar colors fell in the RF along with a marginal increase for distractors with the nonpreferred color. Similar effects on synchrony were found during shape search. Thus, these results suggest that when the animal is searching for a particular feature, the neurons that prefer that feature begin to synchronize their activity, reaching maximal synchronization when a stimulus with that feature falls within their RF (e.g., when the animal is searching for red, the neurons prefer red, and a red stimulus falls within the RF).

The results described so far show that neurons gave enhanced responses and synchronized their activity in the gamma-range whenever a preferred stimulus in their RF was the target the animal was looking for, but had not found as yet. However, it is not clear from these results whether a distractor with a target feature would share in the bias for target features as proposed by parallel models of visual search (Cave and Wolfe, 1990; Desimone and Duncan, 1995). For example, when searching for a red cross among red circles, green crosses, and green circles, we should see evidence for enhancement of responses and/or synchrony when the RF stimulus contains a single feature of the target (i.e., red circle or green cross) but is not, itself, a target as it lacks the other target feature. Indeed, we found that during a conjunction search, neurons increased and synchronized their activity for unselected distractors in the RF with the preferred color when that color was shared with the target (Fig. 3). Interestingly, although sharing in the bias for the target shape also led to neural enhancement, the magnitude of the effect was much smaller, consistent with behavioral evidence that the monkey used the color information more than the shape information in guiding its search to the color–shape conjunction target (i.e., fixated distractors with the target color more often than distractors with the target shape). Altogether, these results suggest that responses are enhanced whenever a RF stimulus contains a preferred feature of the neurons, and that feature is used in guiding the search for the target.

The source of the top-down bias on V4 activity most likely originates, at least in part, in prefrontal cortex, which has been shown to play an important role in working memory and executive control (Miller and Cohen, 2001). Accordingly, during a match-to-sample task, sample-selective delay activity in prefrontal cortex is maintained throughout the trial even when other test stimuli intervene during the delay, whereas delay activity in IT cortex is disrupted by intervening stimuli (Miller

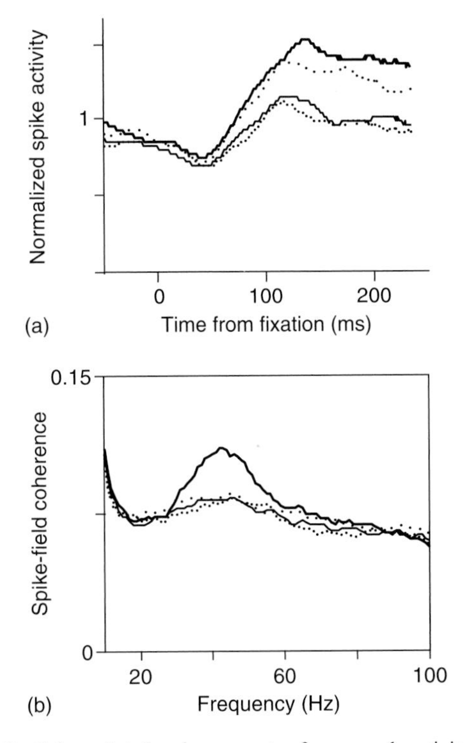

Fig. 3. Color-related enhancement of neuronal activity and spike-field synchronization in area V4 during conjunction search. Conventions are as in Fig. 2.

et al., 1996). Consistent with the idea that feedback inputs from prefrontal cortex to visual cortex bias activity in favor of behaviorally relevant stimuli, recent preliminary evidence shows that prefrontal lesions impair monkeys' ability to attend to stimuli based on color cues, but not to stimuli that are salient and pop-out (Rossi et al., 2001).

Serial selection during visual search

Although both the behavioral and the neural evidence for parallel processing during the visual search tasks described above is compelling, it is equally clear that these searches have a serial component in that the monkeys make several saccades to find the target (Fig. 1). To test for spatial attention effects on responses, Bichot et al. (2005) compared neural responses to any stimulus in the RF when it was either selected for a saccade or the saccade was made to a stimulus outside the RF during color and shape feature searches. Neurons

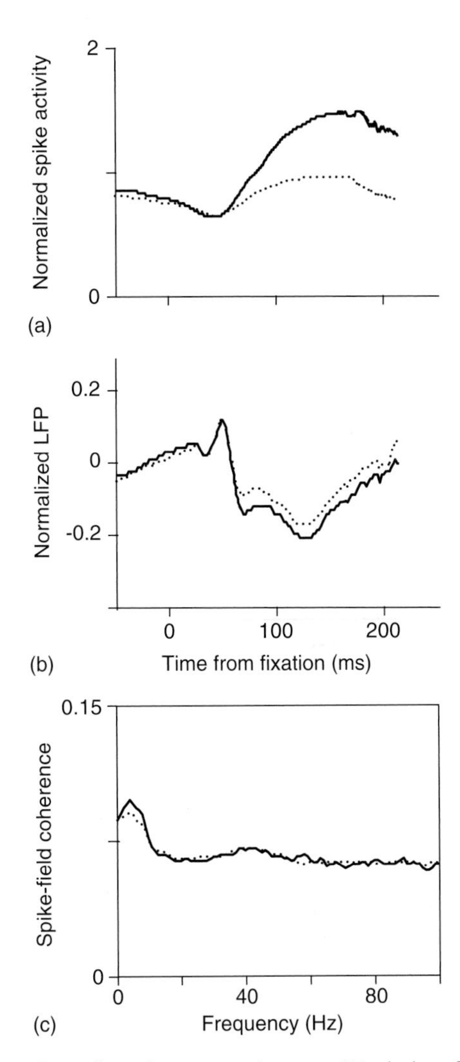

Fig. 4. Saccadic enhancement in area V4 during feature searches. (a) Normalized firing rates for the population of neurons when a saccade was made to a stimulus inside the RF (solid line) and when a saccade was made to a stimulus outside the RF (dotted line) across all saccades. Data from color and shape searches were combined. (b) and (c) Average normalized LFP and spike-field coherence for the same conditions.

responded more strongly to a stimulus in the RF when it was the goal of the impending saccade (Fig. 4A), consistent with the findings of Mazer and Gallant (2003), using a similar free-viewing visual search task. Furthermore, both the magnitude (Fig. 4B) and the spectral power of the LFP in the gamma-frequency range were significantly modulated by monkeys' decision to make a saccade to the RF stimulus. However, despite such strong modulation of neuronal firing rates and LFPs, spike-field coherence was unaffected by spatial selection (Fig. 4C).

The lack of modulation of neuronal synchrony with spatial selection during visual search stands in contrast to changes in synchrony that were observed with feature selection in the same search. Furthermore, superficially at least, this result seems at odds with a previous finding by Fries et al. (2001) that gamma-frequency synchronization increases and beta-frequency synchronization decreases when a monkey attends to a stimulus inside the RF of neurons (Fig. 5). A critical factor appears to be the length of time for which attention is maintained for a given feature or location. In the Fries et al. study, monkeys monitored and attended to the target location for up to several seconds, whereas during visual search the time that the animal takes to attend to the location of the next stimulus that will be the target of a saccade is only about 250 ms. The effect of feature-based attention on synchrony during visual search is consistent with this explanation as the animal maintains a state of attention to stimulus features lasting several seconds. Thus, it seems plausible that attentional effects on synchrony take longer to develop than attentional effects on firing rate.

The modulation of responses by the locus of spatial attention that we (Bichot et al., 2005) and others (Mazer and Gallant, 2003) have found in V4 likely involve feedback from structures involved in spatial attention and saccade production. One such structure, the frontal eye field (FEF), is reciprocally connected with areas of both the dorsal and the ventral visual processing streams, and these connections are topographically organized (Schall et al., 1995). Consistent with the idea that feedback from FEF to visual cortex plays a role in spatial selection, recent studies by Moore and colleagues have shown that subthreshold stimulation of FEF improves perceptual ability (Moore and Fallah, 2004) and enhances the visual responses of V4 neurons (Moore and Armstrong, 2003), similar to the effects of spatial attention on perception and neuronal responses.

Studies of neural selection in FEF during visual search have led to the proposal that this area represents the behavioral significance of stimuli

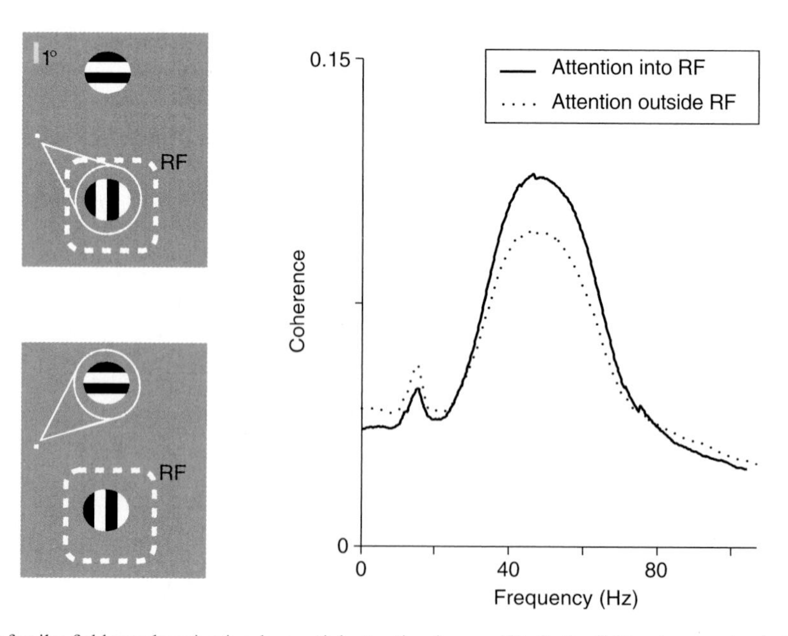

Fig. 5. Enhancement of spike-field synchronization by spatial attention in area V4. Spike-field coherence is plotted when the monkey attended to a drifting grating inside the RF (solid line) and when the monkey attended to one outside the RF (dotted line). Both gratings were presented on each trial, and the monkey was instructed to attend to one of them in blocks of trials. In the task displays, the RF is illustrated by the dotted square, and the location of attention is illustrated by the cone.

regardless of their visual features (Thompson and Bichot, 2005), much like the concept of a "salience map" found in many models of visual search (Koch and Ullman, 1985; Treisman, 1988; Cave and Wolfe, 1990; Olshausen et al., 1993; Itti and Koch, 2001). During conjunction search for example, visually responsive FEF neurons initially respond indiscriminately to the target and distractors of the search array (Bichot and Schall, 1999), consistent with the fact that FEF neurons are not selective for visual features such as color (Mohler et al., 1973). However, over time, these neurons not only discriminate the target from distractors, but also discriminate among distractors on the basis of their similarity to the target, even though a saccade is only made to the target location (Fig. 6). In other words, while the highest activation was observed when the target was in the neurons' RF, the activity in response to RF distractors that shared the target color or shape was also relatively enhanced. Such a spatial map of potential targets would result from topographically organized, convergent input from visually selective neurons with activity biased for target features, and in turn, spatial selection within this map would modulate the activity of visual neurons through feedback connections enhancing the representation of a stimulus selected for an eye movement (Hamker, 2005). Other structures that likely encode a map of behavioral relevance for spatial selection include the lateral intraparietal area (Gottlieb et al., 1998) and the superior colliculus (Findlay and Walker, 1999; McPeek and Keller, 2002).

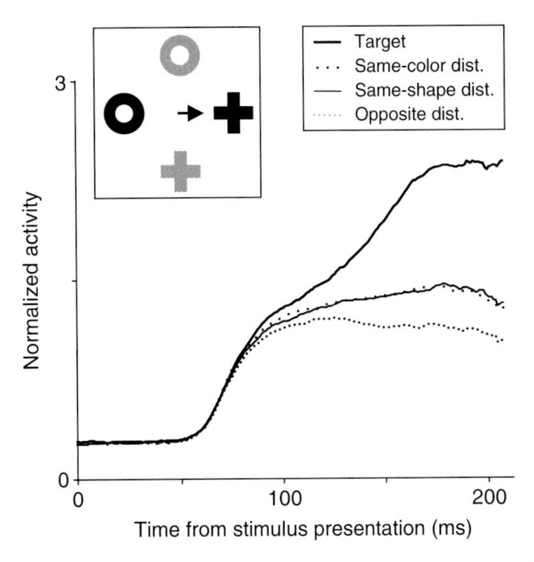

Fig. 6. Visual selection in FEF during conjunction search. Normalized firing rates averaged over a population of visually responsive FEF neurons on correctly performed trials (i.e., first and only saccade was made to the target illustrated by the arrow in the example display). Activity aligned on the time of search array presentation is shown when the target was in the RF (thick solid line); when a distractor that shared the target color (i.e., same-color distractor) was in the RF (thick dotted); when a distractor that shared the target shape (i.e., same-shape distractor) was in the RF (thin solid); and when a distractor that did not share any features with the target (i.e., opposite distractor) was in the RF (thin dotted). In the example search display, stimuli of one color are shown in black and stimuli of the other color are shown in gray.

Conclusion

We have reviewed recent neurophysiological studies showing that both serial and parallel processing of visual information takes places in the brain during visual search, consistent with hybrid models of visual selection (Cave and Wolfe, 1990; Hamker, 2005). Furthermore, these processes are observed in the same brain area and in the same neurons (e.g., area V4) (Mazer and Gallant, 2003; Bichot et al., 2005).

The search for a target based on its features appears to enhance and synchronize the activity of populations of V4 neurons that are selective for and respond preferentially to those features. As a result, stimuli that are similar to the target or that share target features are better represented in the cortex, leading to serial selection among candidate stimuli (Fig. 7). These results offer an explanation for why some visual search tasks are difficult, including some, where targets are defined by the conjunction of different features. Of course, if the results of parallel processing were as clean as depicted in Fig. 7, and the target selection mechanism picked the stimulus location with the highest activation (i.e., winner-take-all), finding the search target would require only one attentional shift or saccadic eye movement. However, evidence from the reviewed studies clearly shows a serial component to the visual search in that several stimuli are fixated before the target is found. Several sources of variability in the activation map from which a target is selected

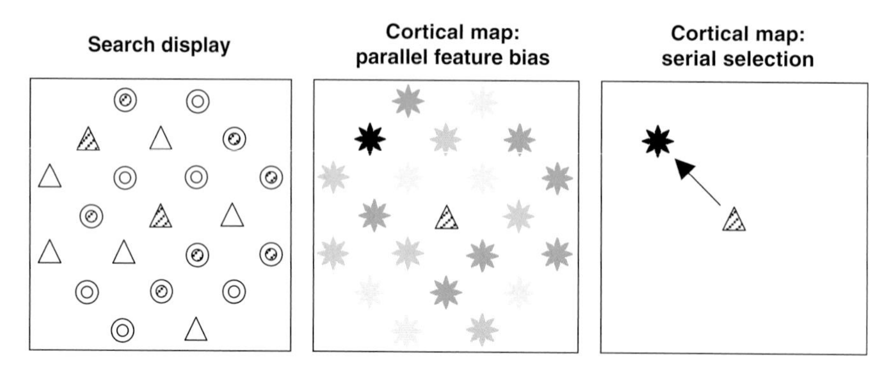

Fig. 7. Schematic illustration of selection mechanisms during a conjunction search task. Left: example stimulus display. Stimuli of one color are shown as filled with diagonal lines, and the ones of the other color are shown as not filled. Middle: representation of stimulus display in the cortex with a bias in favor of potential targets, illustrated by the intensity of the stimulus representation, as a result of parallel, feature-related selection throughout the visual field. Right: the bias for potential targets triggers spatial selection mechanisms resulting in eye movements (illustrated by the arrow towards the target location).

would account for the inability of the visuomotor system to detect the target immediately. For example, the guided-search model of visual selection has attributed this variability to simple "noisiness" in the firing rates of neurons (Cave and Wolfe, 1990), consistent with the known variability of neuronal responses (Henry et al., 1973; Tolhurst et al., 1983; Shadlen et al., 1996; Bichot et al., 2001). The feature-gate model of selection, on the other hand, suggests the possibility that bottom-up interactions between stimuli are a source of variability in their neural representation (Cave et al., 2005). Furthermore, decreased acuity for peripheral stimuli and cortical magnification also affect the neural representation of stimuli, and it has been shown that high probability detection of targets occurs only within a restricted area surrounding the fixation point (Motter and Belky, 1998).

Finally, several studies have shown that neuronal synchrony (especially in the gamma-frequency range) plays an important role in spatial selection (Fries et al., 2001; Pesaran et al., 2002) as well as featural selection (Bichot et al., 2005), suggesting that synchronizing signals could be a general way the brain focuses on important information. Gamma-frequency synchronization may also contribute to the increased "activation" found in functional magnetic resonance imaging (fMRI) studies of attention, based on findings from a recent study suggesting that the BOLD signal measured in fMRI is closely correlated with synchronous activity in the gamma-range (Kayser et al., 2004). Furthermore, the contribution of neural synchrony to feature-based selection during visual search (Bichot et al., 2005) lends additional support to the idea that synchronized activity has an amplifying role in relaying certain types of behaviorally relevant information from one neural population to the next. Although it has been argued that gamma-frequency synchronization solves the binding problem in visual perception (Engel et al., 1997), the results described here suggest that synchrony may play a much broader role in neural processing.

References

Andersen, R.A. and Buneo, C.A. (2002) Intentional maps in posterior parietal cortex. Annu. Rev. Neurosci., 25: 189–220.

Bichot, N.P., Rossi, A.F. and Desimone, R. (2005) Parallel and serial neural mechanisms for visual search in macaque area V4. Science, 308: 529–534.

Bichot, N.P. and Schall, J.D. (1999) Effects of similarity and history on neural mechanisms of visual selection. Nat. Neurosci., 2: 549–554.

Bichot, N.P., Thompson, K.G., Chenchal Rao, S. and Schall, J.D. (2001) Reliability of macaque frontal eye field neurons signaling saccade targets during visual search. J. Neurosci., 21: 713–725.

Cave, K.R., Kim, M.S., Bichot, N.P. and Sobel, K.V. (2005) The FeatureGate model of visual selection. In: Itti, L., Rees, G. and Tsotsos, J.K. (Eds.), Neurobiology of Attention. Elsevier Academic Press, San Diego, CA, pp. 547–552.

Cave, K.R. and Wolfe, J.M. (1990) Modeling the role of parallel processing in visual search. Cogn. Psychol., 22: 225–271.

Colby, C.L. and Goldberg, M.E. (1999) Space and attention in parietal cortex. Annu. Rev. Neurosci., 22: 319–349.

Corbetta, M. and Shulman, G.L. (2002) Control of goal-directed and stimulus-driven attention in the brain. Nat. Rev. Neurosci., 3: 201–215.

Desimone, R. and Duncan, J. (1995) Neural mechanisms of selective visual attention. Annu. Rev. Neurosci., 18: 193–222.

Desimone, R. and Schein, S.J. (1987) Visual properties of neurons in area V4 of the macaque: sensitivity to stimulus form. J. Neurophysiol., 57: 835–868.

Desimone, R., Schein, S.J., Moran, J. and Ungerleider, L.G. (1985) Contour, color and shape analysis beyond the striate cortex. Vision Res., 25: 441–452.

Deubel, H. and Schneider, W.X. (1996) Saccade target selection and object recognition: evidence for a common attentional mechanism. Vision Res., 36: 1827–1837.

Engel, A.K., Roelfsema, P.R., Fries, P., Brecht, M. and Singer, W. (1997) Role of the temporal domain for response selection and perceptual binding. Cereb. Cortex, 7: 571–582.

Findlay, J.M. and Walker, R. (1999) A model of saccade generation based on parallel processing and competitive inhibition. Behav. Brain Sci., 22: 661–674., discussion 674–721.

Fries, P., Reynolds, J.H., Rorie, A.E. and Desimone, R. (2001) Modulation of oscillatory neuronal synchronization by selective visual attention. Science, 291: 1560–1563.

Fries, P., Roelfsema, P.R., Engel, A.K., Konig, P. and Singer, W. (1997) Synchronization of oscillatory responses in visual cortex correlates with perception in interocular rivalry. Proc. Natl. Acad. Sci. USA, 94: 12699–12704.

Gottlieb, J.P., Kusunoki, M. and Goldberg, M.E. (1998) The representation of visual salience in monkey parietal cortex. Nature, 391: 481–484.

Gray, C.M., Konig, P., Engel, A.K. and Singer, W. (1989) Oscillatory responses in cat visual cortex exhibit inter-columnar synchronization which reflects global stimulus properties. Nature, 338: 334–337.

Hamker, F.H. (2005) The reentry hypothesis: the putative interaction of the frontal eye field, ventrolateral prefrontal cortex, and areas V4, IT for attention and eye movement. Cereb. Cortex, 15: 431–447.

Henry, G.H., Bishop, P.O., Tupper, R.M. and Dreher, B. (1973) Orientation specificity and response variability of cells in the striate cortex. Vision Res., 13: 1771–1779.

Itti, L. and Koch, C. (2001) Computational modelling of visual attention. Nat. Rev. Neurosci., 2: 194–203.

Jarvis, M.R. and Mitra, P.P. (2001) Sampling properties of the spectrum and coherency of sequences of action potentials. Neural Comput, 13: 717–749.

Kara, P. and Reid, R.C. (2003) Efficacy of retinal spikes in driving cortical responses. J. Neurosci., 23: 8547–8557.

Kayser, C., Kim, M., Ugurbil, K., Kim, D.S. and Konig, P. (2004) A comparison of hemodynamic and neural responses in cat visual cortex using complex stimuli. Cereb. Cortex, 14: 881–891.

Koch, C. and Ullman, S. (1985) Shifts in selective visual attention: towards the underlying neural circuitry. Hum. Neurobiol., 4: 219–227.

Kowler, E., Anderson, E., Dosher, B. and Blaser, E. (1995) The role of attention in the programming of saccades. Vision Res., 35: 1897–1916.

Liversedge, S.P. and Findlay, J.M. (2000) Saccadic eye movements and cognition. Trends Cogn. Sci., 4: 6–14.

Luck, S.J., Chelazzi, L., Hillyard, S.A. and Desimone, R. (1997) Neural mechanisms of spatial selective attention in areas V1, V2, and V4 of macaque visual cortex. J. Neurophysiol., 77: 24–42.

Maunsell, J.H. (1995) The brain's visual world: representation of visual targets in cerebral cortex. Science, 270: 764–769.

Mazer, J.A. and Gallant, J.L. (2003) Goal-related activity in V4 during free viewing visual search. Evidence for a ventral stream visual salience map. Neuron, 40: 1241–1250.

McPeek, R.M. and Keller, E.L. (2002) Saccade target selection in the superior colliculus during a visual search task. J. Neurophysiol., 88: 2019–2034.

Miller, E.K. and Cohen, J.D. (2001) An integrative theory of prefrontal cortex function. Annu. Rev. Neurosci., 24: 167–202.

Miller, E.K., Erickson, C.A. and Desimone, R. (1996) Neural mechanisms of visual working memory in prefrontal cortex of the macaque. J. Neurosci., 16: 5154–5167.

Mishkin, M., Ungerleider, L.G. and Macko, K.A. (1983) Object vision and spatial vision: two cortical pathways. Trends Neurosci., 6: 414–417.

Mohler, C.W., Goldberg, M.E. and Wurtz, R.H. (1973) Visual receptive fields of frontal eye field neurons. Brain Res., 61: 385–389.

Moore, T. and Armstrong, K.M. (2003) Selective gating of visual signals by microstimulation of frontal cortex. Nature, 421: 370–373.

Moore, T. and Fallah, M. (2004) Microstimulation of the frontal eye field and its effects on covert spatial attention. J. Neu rophysiol., 91: 152–162.

Motter, B.C. and Belky, E.J. (1998) The zone of focal attention during active visual search. Vision Res., 38: 1007–1022.

Nakayama, K. and Silverman, G.H. (1986) Serial and parallel processing of visual feature conjunctions. Nature, 320: 264–265.

Niebur, E., Koch, C. and Rosin, C. (1993) An oscillation-based model for the neuronal basis of attention. Vision Res., 33: 2789–2802.

Nobre, A.C. (2001) The attentive homunculus: now you see it, now you don't. Neurosci. Biobehav. Rev., 25: 477–496.

Olshausen, B.A., Anderson, C.H. and Van Essen, D.C. (1993) A neurobiological model of visual attention and invariant pattern recognition based on dynamic routing of information. J. Neurosci., 13: 4700–4719.

Pesaran, B., Pezaris, J.S., Sahani, M., Mitra, P.P. and Andersen, R.A. (2002) Temporal structure in neuronal activity during working memory in macaque parietal cortex. Nat. Neurosci., 5: 805–811.

Posner, M.I., Snyder, C.R. and Davidson, B.J. (1980) Attention and the detection of signals. J. Exp. Psychol., 109: 160–174.

Rossi, A.F., Bichot, N.P., Desimone, R. and Ungerleider, L.G. (2001) Top-down, but not bottom-up: deficits in target selection in monkeys with prefrontal lesions. J. Vis., 1: 18a.

Salinas, E. and Sejnowski, T.J. (2000) Impact of correlated synaptic input on output firing rate and variability in simple neuronal models. J. Neurosci., 20: 6193–6209.

Schall, J.D., Morel, A., King, D.J. and Bullier, J. (1995) Topography of visual cortex connections with frontal eye field in macaque: convergence and segregation of processing streams. J. Neurosci., 15: 4464–4487.

Schein, S.J. and Desimone, R. (1990) Spectral properties of V4 neurons in the macaque. J. Neurosci., 10: 3369–3389.

Seidemann, E. and Newsome, W.T. (1999) Effect of spatial attention on the responses of area MT neurons. J. Neurophysiol., 81: 1783–1794.

Shadlen, M.N., Britten, K.H., Newsome, W.T. and Movshon, J.A. (1996) A computational analysis of the relationship between neuronal and behavioral responses to visual motion. J. Neurosci., 16: 1486–1510.

Shiffrin, R.M. and Schneider, W. (1977) Controlled and automatic human information processing: II. Perceptual learning, automatic attending, and a general theory. Psychol. Rev., 84: 127–190.

Steinmetz, P.N., Roy, A., Fitzgerald, P.J., Hsiao, S.S., Johnson, K.O. and Niebur, E. (2000) Attention modulates synchronized neuronal firing in primate somatosensory cortex. Nature, 404: 187–190.

Thompson, K.G. and Bichot, N.P. (2005) A visual salience map in the primate frontal eye field. Prog. Brain Res., 147: 251–262.

Tolhurst, D.J., Movshon, J.A. and Dean, A.F. (1983) The statistical reliability of signals in single neurons in cat and monkey visual cortex. Vision Res., 23: 775–785.

Townsend, J.T. (1990) Serial vs. parallel processing: sometimes they look like tweedledum and tweedledee but they can (and should) be distinguished. Psychol. Sci., 1: 46–54.

Treisman, A. (1988) Features and objects: the fourteenth Bartlett memorial lecture. Q. J. Exp. Psychol. A, 40: 201–237.

Treisman, A.M. and Gelade, G. (1980) A feature-integration theory of attention. Cogn. Psychol., 12: 97–136.

Treue, S. and Maunsell, J.H. (1999) Effects of attention on the processing of motion in macaque middle temporal and medial superior temporal visual cortical areas. J. Neurosci., 19: 7591–7602.

Wolfe, J.M., Cave, K.R. and Franzel, S.L. (1989) Guided search: an alternative to the feature integration model for visual search. J. Exp. Psychol. Hum. Percept. Perform., 15: 419–433.

Martinez-Conde, Macknik, Martinez, Alonso & Tse (Eds.)
Progress in Brain Research, Vol. 155
ISSN 0079-6123

CHAPTER 10

Saccades, salience and attention: the role of the lateral intraparietal area in visual behavior

Michael E. Goldberg[1,2,3,*], James W. Bisley[1,3], Keith D. Powell[3] and Jacqueline Gottlieb[1,3]

[1]*Mahoney Center for Brain and Behavior, Center for Neurobiology and Behavior, Columbia University College of Physicians and Surgeons, and the New York State Psychiatric Institute, New York, NY 10032, USA*
[2]*Departments of Neurology and Psychiatry, Columbia University College of Physicians and Surgeons, New York, NY 10032, USA*
[3]*Laboratory of Sensorimotor Research, National Eye Institute, National Institutes of Health, Bethesda, MD 20892, USA*

Abstract: Neural activity in the lateral intraparietal area (LIP) has been associated with attention to a location in visual space, and with the intention to make saccadic eye movement. In this study we show that neurons in LIP respond to recently flashed task-irrelevant stimuli and saccade targets brought into the receptive field by a saccade, although they respond much to the same stimuli when they are stable in the environment. LIP neurons respond to the appearance of a flashed distractor even when a monkey is planning a memory-guided delayed saccade elsewhere. We then show that a monkey's attention, as defined by an increase in contrast sensitivity, is pinned to the goal of a memory-guided saccade throughout the delay period, unless a distractor appears, in which case attention transiently moves to the site of the distractor and then returns to the goal of the saccade. LIP neurons respond to both the saccade goal and the distractor, and this activity correlates with the monkey's locus of attention. In particular, the activity of LIP neurons predicts when attention migrates from the distractor back to the saccade goal. We suggest that the activity in LIP provides a salience map that is interpreted by the oculomotor system as a saccade goal when a saccade is appropriate, and simultaneously is used by the visual system to determine the locus of attention.

Keywords: lateral intraparietal area; saccade; attention; contrast sensitivity; monkey

Introduction

We live in a world of sensory overload. Sights, sounds, smells and touches bombard our sensory apparatus constantly, and the primate brain cannot possibly deal with all of them simultaneously. Instead, it chooses the objects most relevant to its behavior for further processing. This act of selection is called attention. William James described attention as "the taking possession by the mind in clear and vivid form, of one out of what seem several simultaneously possible objects or trains of thought … .It implies withdrawal from some things in order to deal effectively with others … (James, 1890)". He then described two different kinds of attention: "It is either passive, reflex, non-voluntary, effortless or active and voluntary. In passive immediate sensorial attention the stimulus is a sense-impression, either very intense, voluminous, or sudden … big things, bright things, moving things … blood." More recently, these two

*Corresponding author. Tel.: +1-212-543-0920;
Fax: +1-212-543-5816; E-mail: meg2008@columbia.edu

DOI: 10.1016/S0079-6123(06)55010-1

kinds of attention have been described as exogenous and endogenous attention or bottom-up and top-down. Attention is clinically important: the syndrome formerly known as 'hyperactivity' is now known as 'attention deficit and hyperactivity disorder (ADHD).' Patients with parietal (Critchley, 1949, 1953) and occasionally frontal (Heilman and Valenstein, 1972) deficits show neglect, the inability to attend to a part of the visual field.

The parietal cortex has long thought to be important in the neural mechanisms underlying spatial attention. One parietal area in particular, the lateral intraparietal area (LIP), has been implicated in attentional and oculomotor processes. Although it is clear that LIP has a visual representation, it is not clear whether this visual representation is dedicated to the processing of saccadic eye movements or has a more general attentional function independent of the generation of any specific movement. In this review, we describe three different experiments that examine the role of attention in LIP and its relation to the generation of saccadic eye movements. The first deals with the nature of the visual representation in LIP; the second deals with the independence of LIP activity from saccade planning; and the third with the nature and determinants of visual attention in the monkey.

Behavioral modulation of visual activity in LIP

The neurophysiological basis of attention is not clearly understood, although much evidence suggests that it arises from a gain control on the sensory activity of neurons in various association areas of the brain. This observation, first made in the superior colliculus (Goldberg and Wurtz, 1972), has been extended to a number of other areas in both the dorsal and ventral streams. The standard method for determining the visual properties of a neuron has been, since the development of the fixation task by Wurtz (1969), the response of the neuron to a stimulus that appears suddenly in its receptive field. This definition has a problem. Abruptly appearing stimuli are not only associated with photons exciting rods and cones on the retina, but are also, as James noted, attentional attractors, and the abrupt appearance of a task-irrelevant distractor decreases manual reaction time (Posner, 1980) and increases contrast sensitivity at the site of the distractor (Yantis and Jonides, 1984). Stimuli can enter receptive fields in several ways: one is when a light appears suddenly in the receptive field and a second is when a saccade brings a stable object into the receptive field. Since activity in parietal cortex is associated with attention as well as with vision, the question arises as to whether the 'visual responses' of parietal neurons are predominantly visual, i.e. the result of the appearance of photons on the retina, like a retinal ganglion cell, or attentional, the result of extraretinal modulation of the retinal signal.

To distinguish between these alternatives we devised a number of tasks in which the stimulus, rather than appearing de novo in the receptive field, entered the receptive field by virtue of a saccadic eye movement (Gottlieb et al., 1998). This enabled us to stimulate the receptive field using stimuli that did not have the attentional tag of abrupt onset. In these stable array tasks, the monkeys were presented with an array of eight stimuli arranged uniformly in a circular array. These stimuli did not appear or disappear from trial to trial. Instead, they were constant for a block of trials (Fig. 1). The stimuli were roughly 2° in diameter, and varied in shape and color. They were not equated for luminance. They were positioned so that when the monkey fixated the center of the array at least one stimulus appeared in the receptive field of the neuron under study. In the simplest of these tasks the monkey fixated at a position outside the array so that no stimulus was in the receptive field of the neuron being studied, and then, when the saccade brought one of the stable stimuli into the receptive field.

The typical neuron had a brisk response to the sudden appearance of a stimulus in its receptive field during a fixation task (Fig. 2A), and a much smaller response when the same stimulus as a member of the stable array entered the receptive field following the saccade (Fig. 2B). The decrement of response could have been related to the behavioral irrelevance of the stable target, or, it could have been due to a series of other confounds. For example, the movement of the stimulus into the receptive field by the saccade is not exactly the same as its appearance from the flash; the other members of the array might exert some purely visual local inhibition that suppresses the response. To test if these other factors could be responsible for the diminished response to the stable target, we developed the recently flashed stimulus task. In this task, the stable array contained only seven stimuli, but not the one that would be brought into the receptive field by the saccade. This eighth stimulus appeared while the monkey was fixating at the initial position, and remained on for the remainder of the trial. The monkey then made a saccade that brought this recently appeared stimulus into the receptive field. The neuron responded almost as briskly as it did to the abrupt appearance of the stimulus in the receptive field (Fig. 2C, cf. Fig. 2A). Therefore, the difference between the fixation case and the stable target case was not due to the visual or oculomotor differences between the tasks, but to the lack of salience of a stable component of the visual environment. Note that the neuron began to respond at or before the end of the saccade. This was a much lower latency than when the stimulus appeared in the receptive field abruptly (cf. Fig. 2A with Fig. 2C). Presumably this occurred because of the predictive response described previously: neurons in LIP may respond to stimuli that will be brought into their receptive field by saccades earlier than they do to the abrupt appearance of the same stimulus in their receptive fields (Duhamel et al., 1992). The recently appeared stimulus evoked a greater response across

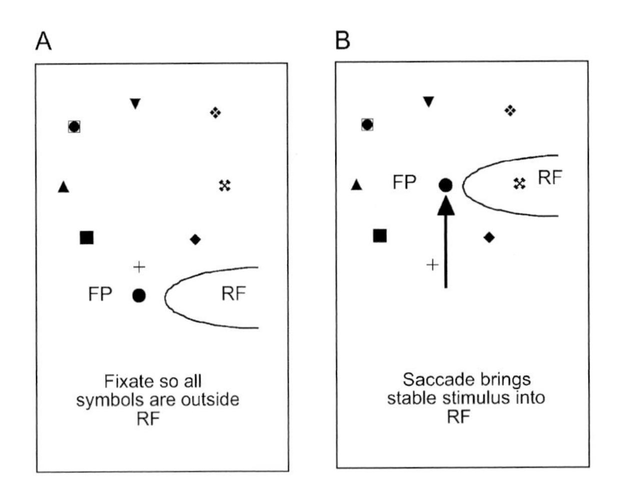

Fig. 1. Stable array task. An array of symbols remains on the screen unchanging throughout the task. (A) The monkey looks at a fixation point (black dot, marked FP) situated so no member of the array is in the receptive field (parabolic solid line, RF) of the neuron. (B) The fixation point jumps and monkey makes a saccade (arrow) to follow it, bringing the receptive field onto the spatial location of a symbol (in this case the X). Adapted from Gottlieb et al. (1998).

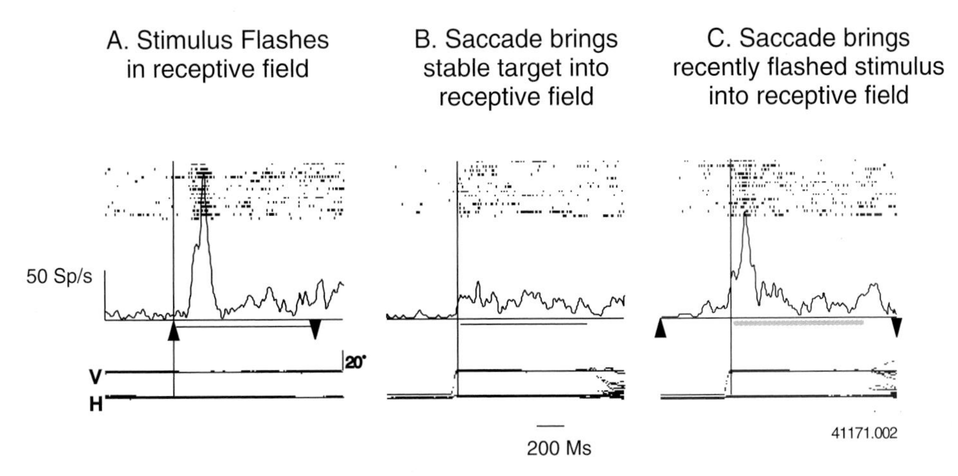

Fig. 2. Activity of a neuron in LIP to a stimulus brought into its receptive field by different strategies. (A) The stimulus flashes in the receptive field during a fixation task. Raster diagrams, spike density plot ($\sigma = 10\,\text{ms}$), and eye position (V: vertical, H: horizontal) are shown. Raster and spike density plots synchronized on the stimulus appearance. (B) Stable stimulus brought into receptive field by a saccade. Data synchronized on saccade onset. (C) Recently flashed stimulus brought into receptive field by the saccade. Data synchronized on saccade onset. Adapted from Gottlieb et al. (1998).

the population than did the stable stimulus (using an average response in an interval 200 ms after the end of the saccade: $p < 0.001$ by Wilcoxon signed-rank test; 31 neurons), and evoked a statistically significantly greater response in a majority (23/31, $p < 0.05$ by two-tailed t-test) of single neurons.

The original studies of behavioral modulation emphasized enhanced responses to stimuli that were the subjects of various behavioral tasks (Goldberg and Wurtz, 1972; Robinson et al., 1978; Moran and Desimone, 1985). In this experiment, however, the enhanced activity arises from an intrinsically salient stimulus that has no task relevance: the abruptly appearing stimulus. The intrinsic properties of the stimulus were, perforce, salient. However, it was the sudden appearance of the stimulus in the environment, and not merely the sudden appearance of the stimulus in the receptive field that endowed the stimulus with salience.

Salience does not arise only from intrinsic properties of the stimulus. Stable objects can become important by virtue of their relevance to current behavior, and under these circumstances a member of a stable array can evoke a response from a neuron in LIP. We can show this using the stable target task, a more complicated version of the stable array task (Fig. 3). In this task, the monkey fixated so that the stimulus was not in the receptive field, and a cue appeared during the first fixation. This cue matched one of the symbols in the stable array. The fixation point then jumped to the center of the array and the monkey tracked it with a saccade. Finally, when the fixation point disappeared, the monkey made a saccade to the member of the array that matched the cue. The target indicated by the cue was randomly chosen on each trial among the members of the array. Neurons responded strongly to stable stimuli brought into their receptive field if these were designated as the target of the next saccade (Fig. 4A). The neuron discharged from the first saccade to the second. In contrast, if the identical stable stimulus entered the receptive field but was not designated as the saccade target (the monkey was instructed to saccade elsewhere), neurons responded minimally (Fig. 4B).

It is possible that LIP neurons respond in the stable target task because the monkey is planning a purposive saccade, and the activity is less related to the salience of the stable target than it is to the processes underlying saccade planning. To see how much activity in LIP can be allocated to the planning and generation of a saccade itself we used a task in which the monkey had to make a saccade to a spatial location that had no visual stimulus (the 'black hole' task). This task began as a simple version of the stable target task (Fig. 5A). The monkey fixated a point in the center of the array, a cue appeared outside the receptive field matching the symbol in the receptive field, and the monkey made a saccade to that symbol. This trial was repeated a number of times, with the cue always matching the symbol in the receptive field. Then

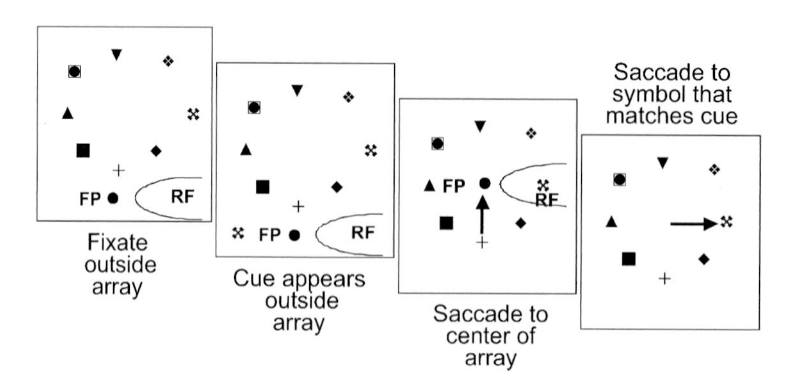

Fig. 3. Stable target task. First panel: the monkey fixates so that all symbols in the array are outside the receptive field. Second panel: a cue appears, also outside the receptive field. Third panel: the fixation point jumps and the monkey makes a saccade that brings a symbol into the receptive field. In this example, the symbol in the receptive field matches the cue. Fourth panel: the fixation point disappears and the monkey makes a saccade to the symbol that matched the cue. Adapted from Gottlieb et al. (1998).

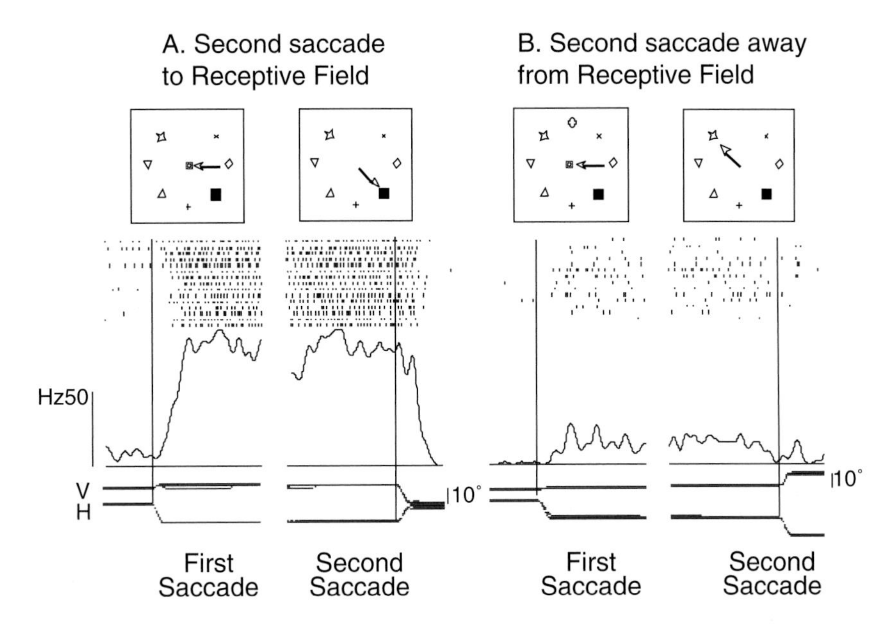

Fig. 4. Activity of LIP neuron in the stable target task. Each subfigure consists of two panels. The cartoon above each panel shows the saccade upon which the underlying raster and spike density figures are synchronized on the saccade beginning. The first saccade brings a stable stimulus into the receptive field; the second saccade is made to the member of the stable array that matched the cue. (A) The monkey makes the second saccade to the receptive field. (B) The monkey makes the second saccade away from the receptive field. Adapted from Gottlieb et al. (1998).

the symbol in the receptive field was removed, leaving a gap, the 'black hole' in the array. The monkey, nonetheless, had learned to make a saccade to the center of the hole in the array. Although the cell in Fig. 5 had a robust response when there was a stimulus in the receptive field, it failed to respond when the monkey made the same saccade when the stimulus was absent. Many LIP neurons active in the stable target task frequently failed to respond in this task, and no neurons responded more in this task than in the stable target task when target was in the receptive field. These results illustrated that the visual activity of neurons in LIP are under behaviorally modulated control, and suggested that this visual activity is related to the salience, or attention-worthiness of the stimuli that drove it.

The effect of saccade planning on the visual response of LIP neurons

The stable target and stable array task show that LIP neurons respond to the sorts of stimuli that usually attract attention, but much less to stable stimuli that usually do not attract attention. If LIP were primarily to have a saccade-planning function, then neurons in LIP should filter out irrelevant stimuli that are not the targets for saccades. Once the monkey was committed to a given saccade a stimulus appearing far away from the saccade goal should evoke a weaker response than a stimulus at the saccade goal. When we tested this hypothesis, we found the contrary result: neurons in LIP gave *enhanced* responses to task-irrelevant distractors that appeared away from the goal of a memory-guided saccade (Powell and Goldberg, 2000). In this experiment we first studied neurons in the delayed saccade task (Fig. 6A), and only chose those that had delay period and/or presaccadic activity (Figs. 7A and B). The neuron illustrated in Figs. 7A and B had a visual response, a striking delay period response but a lesser presaccadic increase.

We then studied 27 neurons with visual and delay and/or presaccadic activity in the distractor task. In this task the monkey (Figs. 6B and C) performs a delayed saccade task, but 300 ms before

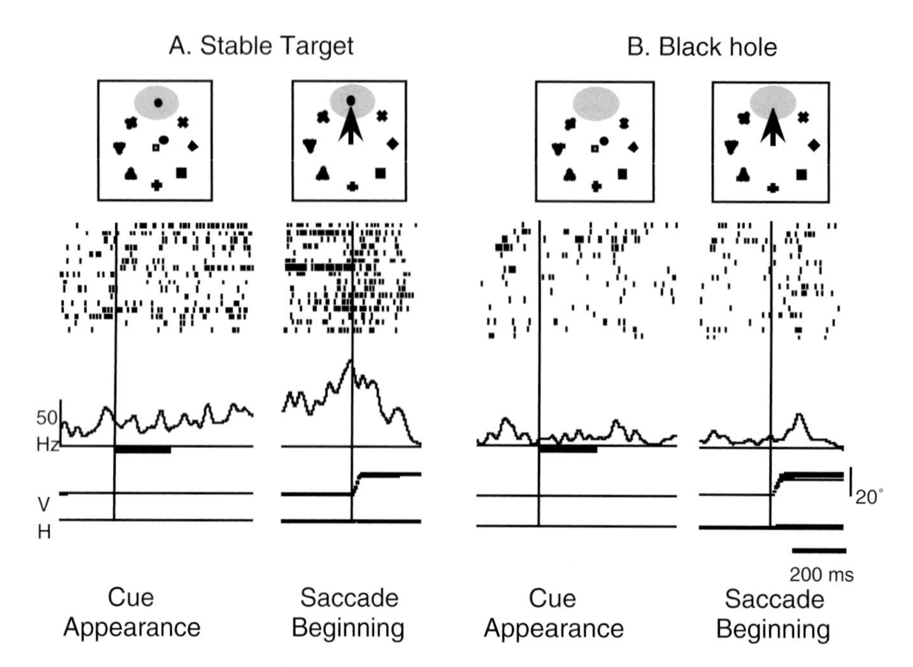

Fig. 5. Activity of LIP neuron in the black hole task. (A) Activity of a neuron in the stable target task. Left panel: activity synchronized on the appearance of the cue. Rasters, spike densities, horizontal (H) and vertical (V) eye position traces are shown. The heavy black line denotes the presence of the cue. Note that the cue is outside the receptive field of the neuron, and the activity develops slowly. Right panel: activity synchronized on the saccade beginning. (B) Activity of the same cell in the black hole task. Adapted from Gottlieb and Goldberg (1998).

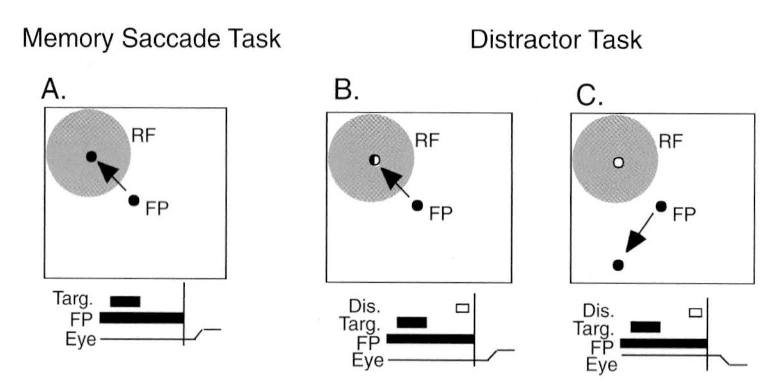

Fig. 6. Memory-guided saccade and distractor tasks. (A) Memory-guided saccade. The monkey fixates the fixation point (FP), and a stimulus (the target) appears and disappears in the receptive field of the neuron. The monkey maintains fixation until the fixation point disappears, at which point the monkey makes a saccade to the remembered location of the vanished target. (B) Distractor task. The monkey fixates, the target appears in the receptive field, and in the interval 300–100 ms before the disappearance of the fixation point a task-irrelevant distractor appears in the receptive field. When the fixation point disappears, the monkey makes a saccade to the spatial location of the vanished stimulus. (C) Distractor task. Same as B, except the distractor is in the receptive field and the saccade target appears elsewhere. Adapted from Powell and Goldberg (2000).

the fixation disappeared we flashed a distractor identical to the saccade target for 200 ms. The distractor could appear at the saccade goal (Fig. 6B) or away from it (Fig. 6C). The saccade goal could be in the receptive field or away from it. The second point was symmetric across the vertical and

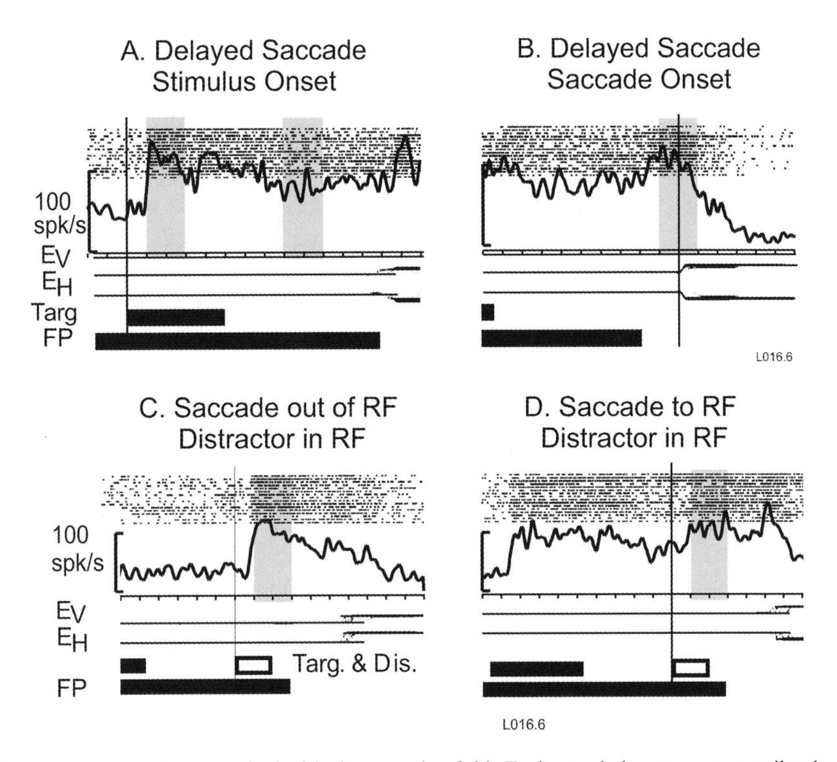

Fig. 7. Response of a neuron to a distractor flashed in its receptive field. Each panel shows a raster, spike density histogram and stimulus traces for fixation point, saccade target and distractor, and horizontal (Eh) and vertical (Ev) eye position. Shaded areas in the raster show stimulus, delay and distractor-activity periods used for quantitative analyses. (A) Response in the delayed saccade task synchronized on the stimulus appearance. (B) Same activity synchronized on the saccade onset. (C) Response of the same neuron to the stimulus flashed in the receptive field during the delay period of a saccade made outside the receptive field. (D) Response of the same neuron during the delay period of a saccade made to the receptive field. Adapted from Powell and Goldberg (2000).

horizontal meridians from the receptive field, and could appear in the same or the opposite hemifield. We randomly interleaved six types of trials: simple memory-guided delayed saccade tasks to the receptive field or to a second point that was outside the receptive field; trials with memory-guided saccades to the receptive field and distractor either in the receptive field or away; and trials with memory-guided saccades away from the receptive field and the distractor either in the receptive field or at the saccade goal.

We found that when a stimulus flashed in the receptive field of an LIP neuron while the monkey was planning a memory-guided saccade away from the receptive field, not only was the response of the neuron not suppressed, it was also slightly enhanced relative to the case when the stimulus appeared at the saccade goal. Figs. 7C and D illustrate this for the same neuron whose activity in the memory-guided saccade task was shown in Figs. 7A and B. This was true across the population and for individual neurons as well. Fig. 8 shows the population response in this experiment, comparing the activity of each cell when to a stimulus appearing away from the saccade goal — i.e. the saccade was made away from the receptive field (ordinate) to a stimulus appearing at the saccade goal (abscissa). If one looked at the activity of the neurons in the 50 ms before the beginning of the saccade, there was no difference between the case when the monkey made a saccade to the receptive field and the distractor was elsewhere, and the case when the monkey made a saccade away from the receptive field and the distractor was in the receptive field. Although the distractor evoked significant activity before the saccade, it had no effect on any measure of saccadic performance: velocity, reaction time, accuracy, or even the early trajectory. Thus LIP, which does filter out

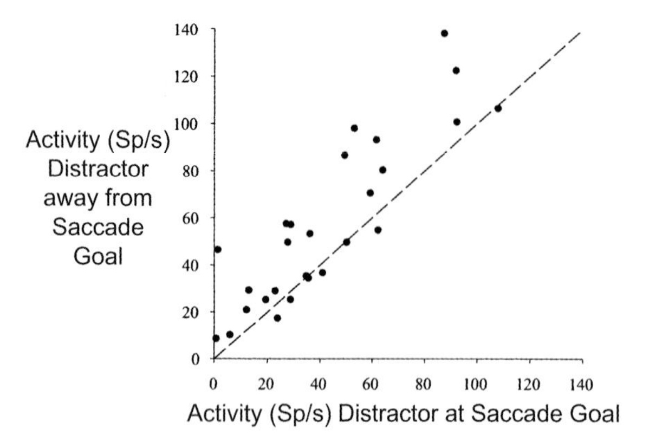

Fig. 8. Effect of saccade planning plotted across the sample. Each symbol represents a single cell, with the response to the distractor when the monkey plans a saccade away from the RF, plotted on the abscissa against the response to the distractor when the monkey plans a saccade into the RF, plotted on the ordinate. The response to stimulus appearance when the monkey makes a saccade away from the stimulus is significantly greater across the population of neurons than the response to the same stimulus when the monkey plans a saccade to the receptive field ($p = 0.001$, Wilcoxon paired signed-rank test). The dashed black line is the unity line. Figure reproduced with permission from Powell and Goldberg (2000).

responses from behaviorally irrelevant stable stimuli, does not filter out responses from salient stimuli that are known, a priori, not to be saccade goals.

It is well known that in humans there is an attentional advantage at the goal of a saccade (Kowler et al., 1995; Deubel and Schneider, 1996). There is also an attentional advantage at the site of an abruptly appearing stimulus (Yantis and Jonides, 1984). One logical way of interpreting our demonstration of an enhanced response to a stimulus that appears away from the goal of a planned memory-guided saccade is that the activity represents an attentionally salient object in the visual field. Both the saccade goal and the distractor should evoke attention, the former for endogenous reasons and the latter for exogenous reasons. The problem with the results described up to this point, however, is that we did not explicitly measure attention.

The psychophysics of attention in the monkey

In a subsequent study we actually measured attention, and were not only able to correlate the activity of neurons with the monkeys' attention but make predictions about the monkeys' performance on the basis of the temporal properties of the neuronal activity (Bisley and Goldberg, 2003). Three methods have been used to describe the locus of attention: a post hoc method which says that if a subject responds to a stimulus it must have attended to it (Goldberg and Wurtz, 1972); a reaction time method, defining the locus of spatial attention as the area of the visual field in which the response to a discriminandum has the lowest latency (Posner, 1980; Bowman et al., 1993); and a contrast sensitivity method, which defines the spatial locus of attention as the area of the visual field with enhanced visual sensitivity (Bashinski and Bacharach, 1980). We chose to use the latter, since it allows us to examine how attention changes over time and under different visual conditions. In addition, it allows us to rule out the possibility that any attentional advantage may be on the motor side of the response, a problem present when defining attention by changes in reaction time.

Our first problem was to establish that in the monkey, like in the humans, attention is pinned at the goal of a saccade. Our task had two components: the monkeys had to plan a saccade to a remembered location and later had to discriminate the orientation of a probe stimulus. On any given day we used four possible saccade targets, which were symmetric across the horizontal and vertical meridians. The probe stimulus consisted of three circle distractors and a Landolt ring whose gap could be on the right or on the left. The orientation of the Landolt ring instructed the monkey either to cancel (gap on right) or to execute (gap on left) the planned saccade when the fixation spot disappeared. We used the contrast sensitivity of the probe to measure attention, by comparing the sensitivity at different spatial locations during the task. We measured the animal's GO/NOGO discrimination performance at a number of contrasts and calculated the contrast threshold, which we defined as the contrast at which the animal could correctly discriminate the probe in 75% of the trials. The animal's performance was better when the probe appeared at the saccade goal than when it appeared elsewhere (Fig. 9A).

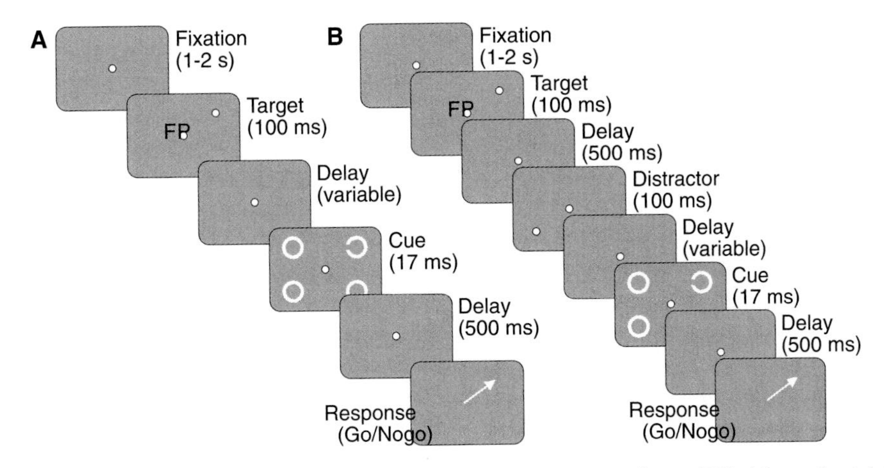

Fig. 9. Psychophysical attention task. (A) The monkeys began the trial by fixating a small spot (FP). After a short delay a second spot (the target) appeared for 100 ms at one of four possible positions equidistant from the fovea and evenly distributed throughout the four visual quadrants. The exact target locations varied from day to day, to prevent long-term perceptual learning. This target specified the goal for the memory-guided saccade that the monkey would have to make unless the probe told it otherwise. At some time after the target disappeared, a Landolt ring (the probe) and three complete rings of identical luminance to the probe flashed for one video frame (~17 ms) at the four possible saccade target positions. Five hundred milliseconds after the probe the fixation point disappeared, and the animals had to indicate the orientation of the Landolt ring by either maintaining fixation for 1000 ms (when the gap was on the right — a NOGO trial) or making a saccade to the goal and remaining there for 1000 ms (when the gap was on the left — a GO trial). The Landolt ring could appear at any of the four positions. The luminance of the rings varied from trial to trial, changing the contrast between the probe and the background. (B) In half of the trials a task-irrelevant distractor, identical to the target, was flashed 500 ms after the target either at or opposite the saccade goal. Reproduced with permission from Bisley and Goldberg (2003).

Because the performance was not statistically different at each of the three locations away from the saccade goal (mean normalized threshold $\pm$ SEM: 1.03 ± 0.05, 1.02 ± 0.06 and 0.95 ± 0.03 for monkey B; 1.01 ± 0.08, 1.04 ± 0.08 and 0.95 ± 0.06 for monkey I), we pooled the data from these locations (hollow circles in Fig. 10A). To enable us to compare the thresholds and enhancements measured on different days, we normalized the data from each day, although all statistical comparisons were performed on the prenormalized data. The normalizing factor for each delay from each session was the threshold from trials in which the probe was not at the saccade goal (i.e. the threshold from the right-hand function in Fig. 10A). The attentional advantage at the saccade goal, illustrated by the enhanced sensitivity, was significant throughout the task ($p < 0.05$ by paired t-test): we used stimulus onset asynchronies (SOAs) from the saccade target to probe of 800, 1300 and 1800 ms, and found enhanced performance in both animals for all SOAs that we studied. Fig. 10B shows the normalized contrast thresholds for each monkey at each SOA. In keeping with the human studies we define the attentional advantage as this lowering of threshold at the saccade goal. We assume that the equally high thresholds for the probe at other locations represent the monkey's performance at loci to which attention has not been allocated by the endogenous process of the saccade plan.

We then had to establish that the monkey's attention could be drawn to the spatial location of an abruptly appearing distractor. We flashed a task-irrelevant distractor during the delay to see if it could draw attention away from the goal of the planned saccade (Fig. 9B). The distractor appeared on half of the trials and was presented either at the saccade goal or opposite the saccade goal (as in Fig. 9B). The distractor was identical to the target in size, brightness and duration, but appeared 500 ms after the target. It remained on the screen for a duration of 100 ms.

When the distractor appeared in the opposite location of the target, and the probe appeared 200 ms after the distractor, the perceptual

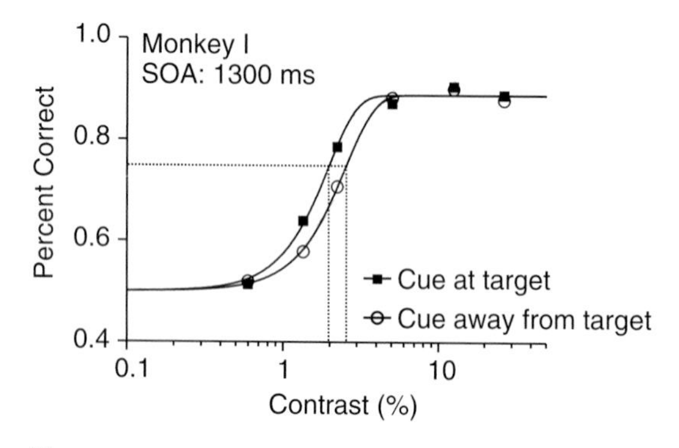

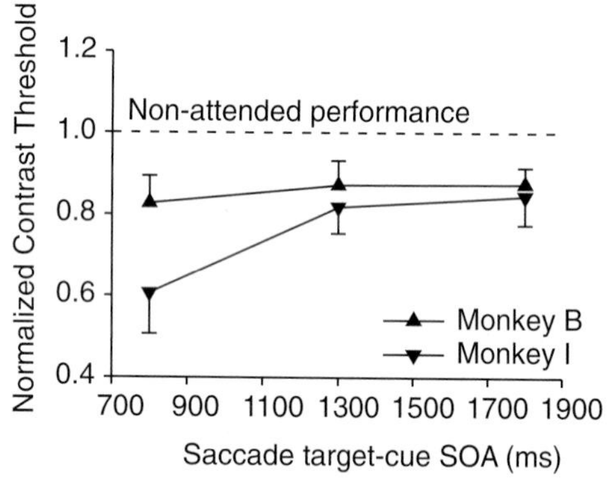

Fig. 10. Effect of saccade planning on perceptual threshold. (A) Psychometric functions from Monkey I from trials with a target-probe SOA of 1300 ms. The solid squares are from trials in which the probe was at the location of the target, the hollow circles are from trials in which the probe was not at the saccade goal. The data are pooled results from 22 sessions (approximately 800 trials per point). The performance from the two conditions was significantly different on the slopes of the functions ($p<0.01$, Chi-squared test at each contrast). The solid lines were fitted to the data with a Weibull function, weighted by the number of trials at each point, using the maximum likelihood method programmed in Matlab. (B) Normalized contrast thresholds for the three SOAs from the two monkeys when the probe was at the location of the saccade goal (solid triangles). Data for each delay were normalized by the performance at that delay when the probe was not at the saccade goal (illustrated by the dashed line). Points significantly beneath the dashed line show attentional enhancement, and all points were significant when tested with paired t-test comparing the prenormalized performance when the probe was at the saccade goal with the prenormalized performance when the probe was away from the saccade goal. No distractor appeared in any of these trials. Reproduced with permission from Bisley and Goldberg (2003).

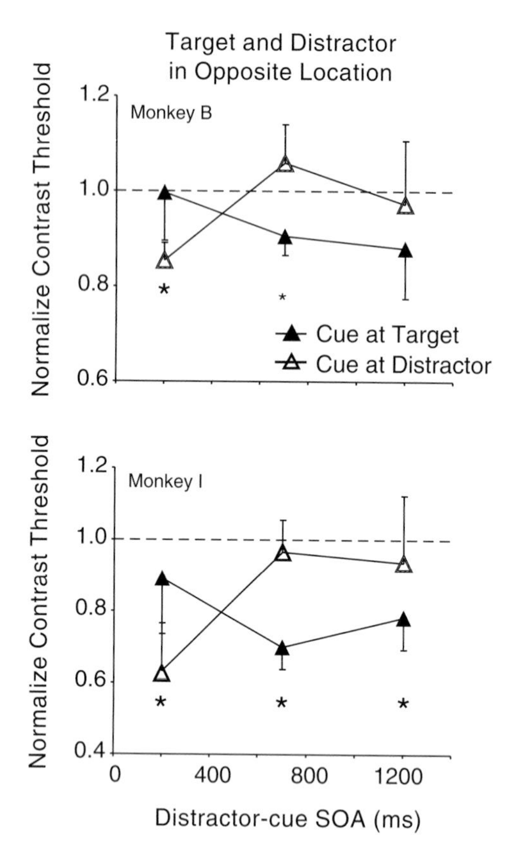

Fig. 11. Effect of the distractor on perceptual threshold. Data for each delay were normalized by the performance at that delay when the probe was not placed at the saccade goal in trials without a distractor (the same normalizing factor from Fig. 12). In the trials shown here the distractor appeared opposite the saccade goal. Normalized contrast threshold was plotted against stimulus onset asynchrony (SOA) for trials in which the probe appeared at the saccade goal (blue symbols) or at the distractor site (red symbols). Performance for both animals was recorded at SOAs of 200, 700 and 1200 ms. Points significantly beneath the dashed line show attentional enhancement (*$p<0.05$, paired t-test on prenormalized data). Reproduced with permission from Bisley and Goldberg (2003).

threshold went down to the attentionally advantaged level at the site of the distractor (Fig. 11, open points) and rose to the baseline level at the saccade goal (Fig. 11, filled points). However, 700 ms after the distractor had appeared, performance was once again enhanced at the saccade goal and not at the distractor location, and this was also the case with the 1200 ms SOA in monkey I, with a trend towards that result in monkey B. Thus, as in humans the abrupt onset of a distractor in the visual field draws attention. In the monkey this occurs even while the animal is planning a saccade, but the attentional effect of the distractor lasts for less than 700 ms, by which time attention has returned to the saccade goal. The distractor and the saccade plan had the same effect on the monkey's attention, lowering the contrast sensitivity threshold by the same amount.

The activity of LIP in the attention task

Having established that a saccade plan and a flashed distractor could attract attention in the monkey, we were then able to assess how neuronal activity in LIP correlated with the monkey's attentional performance. We found that the activity of LIP neurons in the 100 ms interval before the appearance of the probe predicted the monkey's attention to the site of the probe (Bisley and Goldberg, 2003). We recorded the activity of 41 neurons in LIP with peripheral receptive fields in two hemispheres of the two monkeys from whom we gathered the psychophysical data. All neurons were visually responsive, and the majority had delay period or perisaccadic activity as well. Because we were interested in the neuronal response to the saccade plan or distractor, and because these were constant throughout a given experiment (as opposed to the probe, whose contrast changed from trial to trial), we usually recorded cell activity only with the probe at the highest contrast; for four cells we also recorded activity throughout the range of probe contrasts. Figs. 12A and B show the response of a single neuron, measured at threshold, during the trials in which the target appeared in the receptive field of the neuron while the distractor flashed elsewhere (blue trace) and during the trials in which the distractor appeared in the receptive field and the target elsewhere (red trace). There was no difference between the responses to the distractor and during the saccade plan delay activity measured at threshold or suprathreshold probe contrasts. In this analysis we were not interested in the response to the probe itself, and the illustration shows trials with the longest interval between distractor and probe. This interval is so long that the response to the probe is not present in the time window of the illustration. This neuron displays the typical visual and memory responses of an LIP neuron. In particular, it replicates the earlier finding described above that the events underlying the generation of a memory-guided saccade do not suppress the response to a task-irrelevant distractor away from the saccade goal.

Rather than trying to relate the response of a single LIP neuron to the locus of attention, we normalized the responses of all the neurons by the mean value of all the points for each trial type and then calculated the average normalized activity for each animal (Figs. 12C and D). Presented like this, the data represent a population response to two different events: the appearance of the target and the subsequent generation of the memory-guided saccade; and the appearance of the distractor. Although we recorded the response of each of our neurons to those two events, one could as easily reinterpret the activity as that simultaneously seen in two different populations of neurons with the same overall properties, one with receptive fields at the saccade goal (the 'target population' in blue) and the other with receptive fields at the distractor site (the 'distractor population' in red).

The activity of the neuronal population in LIP parallels the attentional performance of the monkey. There is a consistent relationship between the activity in LIP (Figs. 12E and F; lower plot) and the behavioral results from the three SOAs measured previously (Figs. 12E and F; triangles). At any given time throughout the trial the attentionally advantaged part of the visual field was that which lay in the receptive fields of the most active neurons. For example, 200 ms after the appearance of the distractor the greatest activity in LIP was in the distractor population (red traces), and the attentional advantage lay at the distractor site. Then 500 ms later the target population again had the greatest activity (blue traces), and attention had returned to the saccade goal.

The appearance of the distractor outside of the receptive field had no significant effect on the delay period activity in the target population. The distractor evoked a brief transient response, which decayed rapidly and soon crossed the level of activity in the target population. We were curious if this crossing point had behavioral significance; if the site with greatest activity in LIP showed the locus of attention, then what happens when

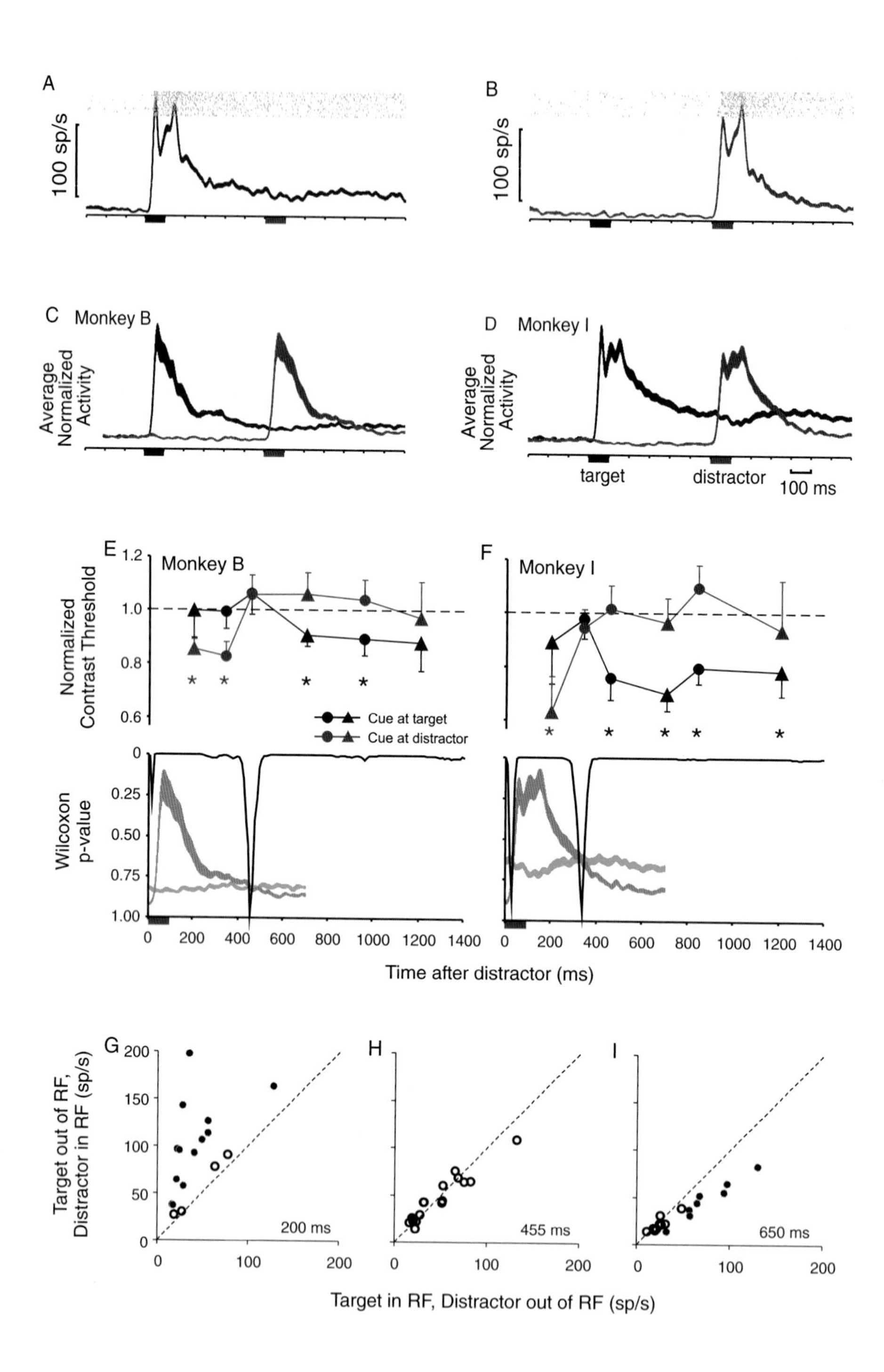
A
100 sp/s
B
100 sp/s
C Monkey B
Average Normalized Activity
D Monkey I
Average Normalized Activity
target
distractor
100 ms
E
Monkey B
Normalized Contrast Threshold
F
Monkey I
Cue at target
Cue at distractor
Wilcoxon p-value
Time after distractor (ms)
G
H
I
Target out of RF, Distractor in RF (sp/s)
200 ms
455 ms
650 ms
Target in RF, Distractor out of RF (sp/s)

the activity at both sites was transiently equal. To determine the period in which there was no single significant preponderance of activity in LIP, we compared the activity at the two sites in a 100 ms window that slid across the entire period in 5 ms steps (Figs. 12E and F; black traces). For each monkey there was a window of 80–90 ms in which there was no significant difference between the activity evoked by the distractor and the activity related to the saccade plan ($p > 0.05$ by Wilcoxon signed-rank test). We hypothesized that during this window of neuronal ambiguity, when activity at both sites did not differ, the monkey would not show a psychophysical attentional advantage at either site.

To test this hypothesis we then went back and measured the contrast sensitivities at the saccade goal and the distractor site at the crossing point in each monkey (455 ms for monkey B and 340 ms for monkey I) and 500 ms later. These times were the center of the window of neuronal ambiguity for each monkey. The top sections of Figs. 12E and F show the behavioral data from the original sessions (triangles) and from the sessions recorded after the physiological experiments (circles). At the crossing point we found no spatial region of enhanced sensitivity in either monkey, but within 500 ms attention had shifted back to the site of the target in both monkeys, with normalized thresholds similar to those seen in the earlier experiment. This is in stark contrast to the effect on saccades: when there was equal activity in the two populations, even in the 50 ms epoch immediately before the saccade, there was no measurable effect on the planned saccade (Powell and Goldberg, 2000). It is possible that there is a period of time following the distractor when attention is shifting, and this period just happens to coincide with the change in activity, while not being related to it. On the other hand, if the activity in LIP were related to attention, then we would expect the behavior to be different in the two animals, because the windows of neuronal ambiguity did not overlap between the two monkeys (see the troughs of the black traces in Figs. 12E and F). We presented the probe to monkey I at the crossing point for monkey B (455 ms) and to monkey B at the crossing point for monkey I (340 ms). We found that the location of the attentional advantage in monkey I had already returned to the saccade goal at the crossing point for monkey B, and that for monkey B the attentional advantage was still at the location of the distractor

Fig. 12. Neural activity and behavior in the attention task. (A) Raster diagram of response to the target appearing in the receptive field and to the distractor appearing outside of the receptive field (blue traces) and to the distractor appearing in the receptive field after the target had appeared outside of the receptive field (red traces). The thickness of the traces represents the standard error of the mean, and the solid blue and red bars show the time and duration of the target and distractor, respectively. (B) Spike density function calculated with a sigma of 10 ms from the same activity. These data were recorded while the monkey was performing the task on threshold. (C) Averaged normalized spike density functions from 18 cells from Monkey B. (D) Averaged normalized spike density functions from 23 cells from Monkey I. (E) and (F) Comparison of neural activity with behavior for each monkey. The top sections show the behavioral performance of the monkeys when the probe was placed at the saccade goal (blue data) or at the location of the distractor (red data in trials in which the target and distractor were in opposite locations. The triangles are data collected before the recording, and the circles are from data collected after recording the activity of LIP neurons in the same monkeys (red and blue traces in bottom section) The circle data were recorded at the crossing point in each monkey (455 ms for monkey B, 340 ms for monkey I) 500 ms later. Data were also collected from both animals at the crossing point recorded in the other animal. Statistical significance was confirmed with a paired t-test on the prenormalized data ($*p < 0.05$). The black traces in the bottom section show the p-values from Wilcoxon paired signed-rank tests performed on the activity of all the neurons for a monkey over a 100-ms bin, measured every 5 ms. A low p-value (high on the axis) represents a significant difference in the activity from the two conditions. The gray column signifies when there is no statistical difference between the activity in both populations. The normalized spike density functions from Figs. 4C and D have been superimposed to show the time course of activity in LIP following the onset of the distractor for the two monkeys. The thickness of the traces represents the standard error of the mean. (G–I) A comparison of the activity when the distractor, but not the saccade goal, was in the receptive field to the activity when the saccade goal, but not the distractor, was in the receptive field for one monkey. Solid circles represent cells with significant differences in response (t-test, $p < 0.05$). (G) Mean activity 150–250 ms following the onset of the distractor for Monkey B. The responses were different across the population ($p < 0.001$, Wilcoxon paired signed-rank test). (H) Mean activity during a 100-ms epoch centered at the crossing point for Monkey B (455 ms after the onset of the distractor). The responses were not different across the population ($p > 0.95$). (I) Mean activity 600–700 ms following the onset of the distractor for Monkey B. The responses were different across the population ($p < 0.01$).

at the crossing point for monkey I. These data support the hypothesis that there is indeed a correlation between activity in LIP and the locus of attention. Note that both the attentional performance of the monkey and the neuronal properties of LIP were extraordinarily stable. In each monkey we ran the psychophysical experiments for several months after the monkeys learned the task. We then recorded neuronal activity for several months. After we had collected enough neurons to determine the crossing point for each monkey, we returned to the psychophysics for another several months. The psychophysics before and after the recording session had the same characteristics as can be seen by comparing, in Figs. 12E and F, the triangles, which were data points collected before the neuronal recording, and the circles, which were data points collected after the neuronal recording. Similarly, the recordings, from roughly 20 cells per monkey, recorded over a period of months, were able to predict how the monkey would behave months after the recordings.

The absolute level of neural activity did not determine the locus of attention. Instead the locus of attention lay at the part of the visual field associated with the greatest neural activity in LIP. Thus, the delay period activity, which determined the locus of attentional advantage when it was the greatest activity in LIP, could not sustain that advantage when it was swamped by the huge transient response to the distractor. Although at times there was only a small difference in the normalized activity of neurons representing the attentionally advantaged and disadvantaged spatial locations, this difference was extraordinarily robust across the population. This is clear from the examples shown in Figs. 12G–I. These plots compare the mean activity of each neuron measured in monkey B when the saccade goal was in the receptive field with the mean activity during the same 100 ms epoch when the distractor was in the receptive field, at three different times during the paradigm: 200 ms after the appearance of the distractor (Fig. 12G); the crossing point (Fig. 12(H); and 650 ms after the appearance of the distractor (Fig. 12I). In our analysis we included all classes of neurons that we encountered, since the major outputs from LIP contain all the classes of neurons found in LIP (Pare and Wurtz, 1997), and have separately illustrated those with (filled circles) and without (open circles) statistically significant differences in their responses ($p<0.05$ by t-test). Generally those neurons without significant differences in delay activity during the task had no memory activity based on their responses to a regular memory-guided saccade task.

Although we recorded the bulk of the neurons while the monkey performed the task in a suprathreshold fashion, this had no effect on the neural responses. To test this we recorded the activity of four neurons while the animal performed the task with the full range of contrasts to see if there was any difference in the activity of the neurons during these periods in a more demanding situation. We calculated the mean rates of activity over the same epochs in Figs. 12G–I for both the stimulus configurations (i.e. target in receptive field, distractor out, and distractor in receptive field, target out) when the monkey was working with all contrasts and with only the suprathreshold contrast. We found that the activity was the same regardless of the difficulty level — the regression coefficient for a line fitted through the 24 points was 1.04 with a shift of 1.34 spikes per second (R^2 of 0.91).

There was a relationship between the performance of the animal and the activity in LIP, manifest in the responses during the 100 ms before the probe appeared in correct and incorrect trials for the two stimulus configurations (Figs. 13A, B). We found that the activity evoked by the saccade plan was lower on error trials than on correct trials, but the activity evoked by the distractor was higher on error trials than on correct trials. However, this activity did not vary with probe location. This suggests that while LIP activity predicts the locus of a perceptual advantage in our experiment, it does not predict on a given trial what the monkey will decide. However, when the monkey is performing the task well, even for suprathreshold probes, there is increased activity at the saccade goal and decreased activity at the distractor site. This is not merely arousal, because the response to the distractor is less than it is when the monkey performs poorly. Arousal would be expected to raise all activity levels.

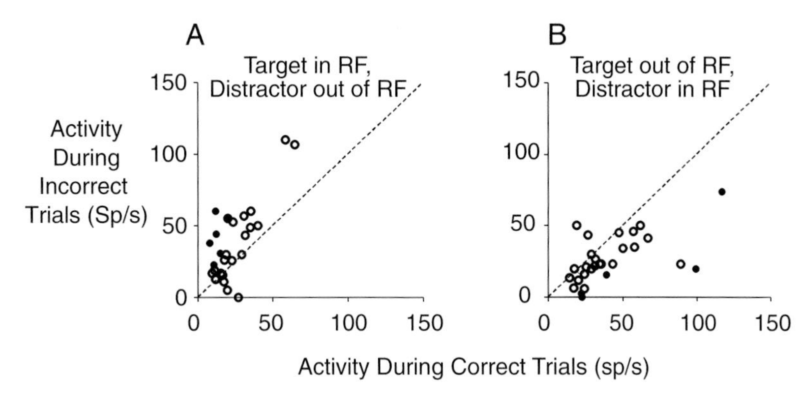

Fig. 13. Activity during error trials. (A) Comparison of activity in correct and incorrect trials. The mean activity of individual neurons in the 100 ms before the appearance of the probe in trials in which the target, but not the distractor, had appeared in the receptive field shown for incorrect (ordinate) and correct (abscissa) trials. Solid circles represent cells which had a significant difference by themselves ($p<0.05$ by t-test). Across the population there were significant response differences between correct and incorrect trials ($p<0.001$, Wilcoxon paired signed-rank test). (B) The mean activity in the 100 ms before the appearance of the probe in trials in which the distractor, but not the target, had appeared in the receptive field. Across the population there were significant response differences between correct and incorrect trials ($p<0.001$). Data in (A) and (B) are shown from the 30 neurons that had errors in both stimulus configurations. Adapted with permission from Bisley and Goldberg (2003).

The results we have discussed up to here show that the attention to be paid to a probe flashed for one video frame is predicted by the activity present in LIP at the time that the probe appears. This is in contradistinction to all previous studies of attentional modulation, which have suggested that the enhanced parietal response to an attended object is responsible for the attention to that object (Bushnell et al., 1981; Colby et al., 1996; Cook and Maunsell, 2002). We found, instead, that the responses evoked by the probe itself did not correlate with our measure of attention. When the probe was in the receptive field, the initial on-responses were identical whether the cue dictated GO to the receptive field, GO elsewhere, or NOGO (Fig. 14A shows the responses of a single cell; Fig. 14B shows the mean responses for every cell in the sample). After 100 ms these responses diverge. When the probe signals GO elsewhere, the response falls rapidly (dashed trace in Fig. 14A). When the probe signals GO to the receptive field, the response falls slightly more slowly and resumes the pre-probe delay period level (black trace in Fig. 14A). When the probe signals NOGO and the monkey was planning a saccade to the receptive field, the response falls far less rapidly, as if the stimulus requiring a cancellation of a saccade plan evokes attention longer than one confirming the saccade plan (gray trace in Fig. 14A). Across the sample the response to this cancellation of a saccade plan is significantly greater than the response to the continuation signal both when the saccade plan is to the receptive field (Fig. 14C) and even more so when the saccade plan and its associated attentional advantage is directed away from the receptive field (Fig. 14D). When the response finally falls, however, it falls to the level of the GO-elsewhere response. Remember that on every trial there was either a probe (the Landolt ring) or a complete ring in the receptive field. We found no difference between the response to the GO probe or the ring in trials in which the saccade plan was directed to the receptive field (Fig. 14E), or away from it ($p>0.2$, Wilcoxon paired signed-rank test), nor was there any difference in the responses to the probe in correct and incorrect trials. However, the enhanced cancellation response was only seen for the actual NOGO probe and not for a ring in the receptive field when the NOGO probe appeared outside of the receptive field (Fig. 14F).

Discussion

We have described three different experiments that illuminate the role of LIP in eye movements and

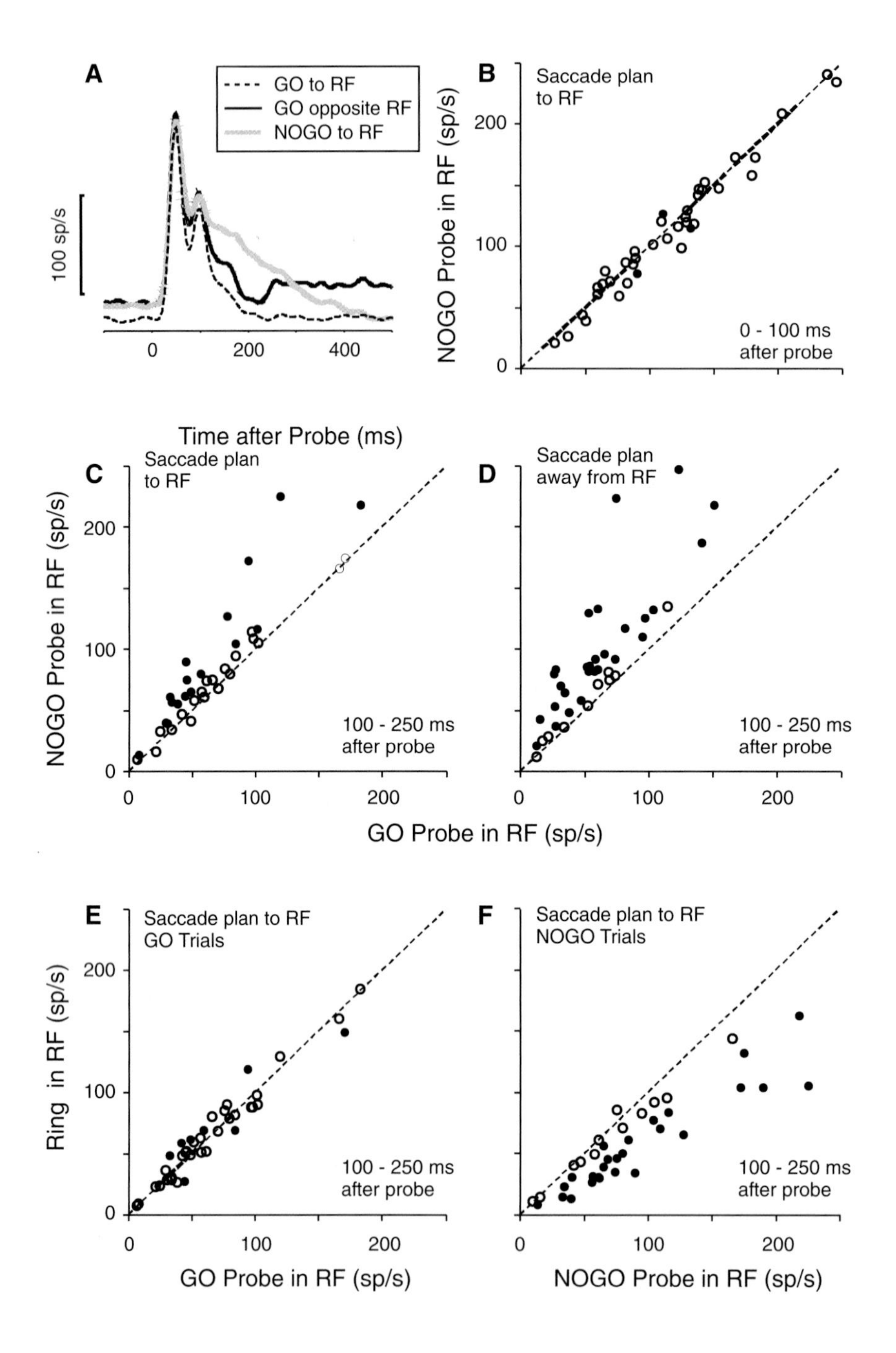
A
GO to RF
GO opposite RF
NOGO to RF
100 sp/s
0
200
400
Time after Probe (ms)
B
Saccade plan
to RF
NOGO Probe in RF (sp/s)
0 - 100 ms
after probe
C
Saccade plan
to RF
100 - 250 ms
after probe
D
Saccade plan
away from RF
100 - 250 ms
after probe
GO Probe in RF (sp/s)
E
Saccade plan to RF
GO Trials
Ring in RF (sp/s)
100 - 250 ms
after probe
F
Saccade plan to RF
NOGO Trials
100 - 250 ms
after probe
NOGO Probe in RF (sp/s)

visual attention. In the first, we show that LIP has representation of the visual world from which static behaviorally unimportant objects have been filtered. Two sorts of objects pass through the filter: those which are important to the animal because they are important to its behavior, in our case the goal of a saccade to a stable object in the environment that matches a cue, and an abruptly appearing stimulus which, although irrelevant to the monkey's task, is the sort of stimulus that evokes attention. These two different sorts of stimuli evoke the two different sorts of attention described by William James as voluntary and involuntary, which have been more recently described as endogenous and exogenous, or top-down and bottom-up. LIP describes both, and it is impossible to discern from the activity in LIP into which category the stimulus that evoked a response falls.

In the second we show that when a monkey plans a saccade, an abruptly appearing stimulus flashed elsewhere in the visual is not filtered out. Instead, it evokes an even greater response than a similar stimulus flashed at the goal of the saccade. These data are not consistent with LIP having a role exclusively dedicated to the planning of saccadic eye movements, but they are consistent with LIPs providing a representation of significant objects in the visual field, acting as a salience map. Of course, the goal of a saccade is usually the object of attention (Yantis and Jonides, 1984, 1996), and in fact Rizzolatti has postulated that attention in the primate is merely a readout of a possible saccade plan (Rizzolatti et al., 1987).

In the third we provide psychophysical evidence that a monkey's attention can be influenced by a saccade plan and by the appearance of a task-irrelevant distractor elsewhere in the visual field. Activity in LIP parallels the monkey's psychophysical performance: when one location in the visual field evokes significantly greater activity in LIP, the monkey exhibits a perceptual advantage for objects in that location.

One surprising result of these experiments is that the absolute value of activity in LIP does not necessarily predict the locus of a monkey's attention. Instead, the locus of attention is determined by the part of the visual field that evokes the greatest activity in LIP at a given time. For example, during the delay period, before the distractor appeared, the activity in the neurons whose receptive field was at the saccade goal was the highest, and the attentional advantage lay at the saccade goal. A few hundred milliseconds after the distractor appeared, however, the locus of attention had shifted to the distractor site. The activity at the saccade goal was unchanged. It is therefore impossible to know in these cases where the monkey's attention is by merely looking at one neuron or the representation of one part of space in LIP. The

Fig. 14. The response to the probe in the receptive field. (A) Spike density functions from the same neuron illustrated in Figs. 12A and B. Data are from trials in which the monkey was instructed to plan a saccade into the receptive field and either the GO stimulus (black) or the NOGO stimulus (gray trace) appeared in the receptive field and from trials in which the saccade goal was opposite the receptive field and the GO probe appeared in the receptive field (dashed trace). The timing of the stimulus presentation is represented by the black bar starting at 0 ms. (B) The response to the NOGO stimulus plotted against the response to the GO stimulus in trials in which the monkey was instructed to plan a saccade to the receptive field. Data are from a 100-ms epoch starting at the onset of the probe. Solid circles are from cells in which the difference in activity was significant ($p<0.05$, t-test); hollow circles are from cells in which there was no significant difference. Across the population there was no difference in response to the two stimuli ($p>0.15$, Wilcoxon paired signed-rank test). (C) The response to the NOGO stimulus plotted against the response to the GO stimulus in trials in which the monkey was instructed to plan a saccade to the receptive field. Data are from a 150-ms epoch starting 100 ms after the onset of the probe. Across the population there was a significant difference in responses to the GO and NOGO stimuli ($p<0.001$). (D) The response to the NOGO stimulus plotted against the response to the GO stimulus in trials in which the monkey was instructed to plan a saccade away from the receptive field. Data are from a 150-ms epoch starting 100 ms after the onset of the probe. Across the population there was a significant difference in responses to the GO and NOGO stimuli ($p<0.001$). (E) The response to the complete ring plotted against the response to the GO stimulus in trials in which the monkey was instructed to plan and execute a saccade to the receptive field. Data are from a 150-ms epoch starting 100 ms after the onset of the probe. Across the population there was no difference in the responses to the GO and ring stimuli ($p>0.6$). (F) The response to the complete ring plotted against the response to the NOGO stimulus in trials in which the monkey was instructed to plan and then cancel a saccade to the receptive field. Data are from a 150-ms epoch starting 100 ms after the onset of the probe. Across the population there was a significant difference in the responses to the NOGO and circle stimuli ($p<0.001$). Adapted from Bisley and Goldberg (2003).

attentional decision must be made by looking at all of LIP, and determining the most active area in a winner-take-all manner. This implies that LIP does not make the decision, but rather like a salience map, it determines the behavioral priority of the various parts of the visual field.

The concept of LIP as a salience map provides insight into the debate about the function of LIP in the organization of behavior. Thus, studies indicate that the lateral intraparietal area (LIP) of the monkey has activity correlating with saccadic intention (Gnadt and Andersen, 1988; Barash et al., 1991; Colby et al., 1996; Snyder et al., 1997), visual attention (Gottlieb et al., 1998; Bisley and Goldberg, 2003), expected value (Sugrue et al., 2004), perceptual (Shadlen and Newsome, 2001) or economic (Platt and Glimcher, 1999) decision making, perceived motion (Assad and Maunsell, 1995; Williams et al., 2003), elapsed time (Leon and Shadlen, 2003) and stimulus shape (Sereno and Maunsell, 1998). LIP also has activity that seems not to correlate with attention (Platt and Glimcher, 1997; Snyder et al., 1997). We have shown here several examples of the dissociation of activity in LIP from a saccade plan: neurons in LIP describe distractors that are behaviorally never the targets for saccades, and they respond more to stimuli canceling a saccade to their receptive fields than they do to stimuli confirming the saccade.

If, however, we consider that LIP describes a salience map of the visual field in an agnostic manner, without specifying how that salience map is used, the seemingly contradictory views described above can be reconciled. The actual function of the salience map depends on the area to which it projects. LIP has a strong projection to two oculomotor areas: the frontal eye field (Andersen et al., 1990) and the superior colliculus (Ferraina et al., 2002). If a saccade is appropriate, the peak of the salience map in LIP can provide targeting information. If a saccade is inappropriate, for example, during the delay period of a memory-guided delayed saccade task, the oculomotor system can ignore LIP. LIP also projects to inferior temporal cortex, the ventral visual stream areas involved in pattern recognition (Baizer et al., 1991). Neurons here have large, bilateral receptive fields including the fovea, and could not be useful for targeting saccades; but attention is critical for visual perception, and neurons in inferior temporal cortex show attentional modulation (Moran and Desimone, 1985). The ventral stream can use the exactly same salience map for attention that the oculomotor system uses for targeting — the salience map in LIP. The answer to the attention versus intention debate as to the function of LIP is 'both!'

These experiments raise an important caveat: introspectively attention is both divisible and graded. Thus, human subjects can attend to multiple objects at roughly the same time (Pylyshyn and Storm, 1988; Sears and Pylyshyn, 2000) and a hallmark of parietal damage is the inability to do so (Balint, 1995). Attention can also be graded; the more likely an object is to appear at a given location the better the performance it evokes (Ciaramitaro et al., 2001). However, such distributed attentional processes only operate over relatively long periods of time. Over the time period in which we describe an attentional advantage, in one video frame, there is evidence that attention may be indivisible and ungraded (Joseph and Optican, 1996). However, the activity in LIP is graded, and so over a longer period of time the multiple peaks of the salience map in LIP may contribute to the divisibility and gradation of activity that is present in psychological studies.

References

Andersen, R.A., Asanuma, C., Essick, G. and Siegel, R.M. (1990) Corticocortical connections of anatomically and physiologically defined subdivisions within the inferior parietal lobule. J. Comp. Neurol., 296: 65–113.

Assad, J.A. and Maunsell, J.H.R. (1995) Neuronal correlates of inferred motion in primate posterior parietal cortex. Nature, 373: 518–521.

Baizer, J.S., Ungerleider, L.G. and Desimone, R. (1991) Organization of visual inputs to the inferior temporal and posterior parietal cortex in macaques. J. Neurosci., 11: 168–190.

Balint, R. (1995) Psychic paralysis of gaze, optic ataxia, and spatial disorder of attention (Translated from Monatsch. Psych. Neurol., 25: 51–81, 1909). Cogn. Neuropsychol., 12: 265–281.

Barash, S., Bracewell, R.M., Fogassi, L., Gnadt, J.W. and Andersen, R.A. (1991) Saccade-related activity in the lateral

intraparietal area. I. Temporal properties; comparison with area 7a. J. Neurophysiol., 66: 1095–1108.
Bashinski, H.S. and Bacharach, V.R. (1980) Enhancement of perceptual sensitivity as the result of selectively attending to spatial locations. Perception Psychophys., 28: 241–248.
Bisley, J.W. and Goldberg, M.E. (2003) Neuronal activity in the lateral intraparietal area and spatial attention. Science, 299: 81–86.
Bowman, E.M., Brown, V.J., Kertzman, C., Schwarz, U. and Robinson, D.L. (1993) Covert orienting of attention in macaques. I. Effects of behavioral context. J. Neurophysiol., 70: 431–443.
Bushnell, M.C., Goldberg, M.E. and Robinson, D.L. (1981) Behavioral enhancement of visual responses in monkey cerebral cortex. I. Modulation in posterior parietal cortex related to selective visual attention. J. Neurophysiol., 46: 755–772.
Ciaramitaro, V.M., Cameron, E.L. and Glimcher, P.W. (2001) Stimulus probability directs spatial attention: an enhancement of sensitivity in humans and monkeys. Vision Res., 41: 57–75.
Colby, C.L., Duhamel, J.-R. and Goldberg, M.E. (1996) Visual, presaccadic and cognitive activation of single neurons in monkey lateral intraparietal area. J. Neurophysiol., 76: 2841–2852.
Cook, E.P. and Maunsell, J.H. (2002) Attentional modulation of behavioral performance and neuronal responses in middle temporal and ventral intraparietal areas of macaque monkey. J. Neurosci., 22: 1994–2004.
Critchley, M. (1949) The phenomenon of tectile inattention with special reference to parietal lesions. Brain, 72: 538–561.
Critchley, M. (1953) The Parietal Lobes. Edward Arnold, London.
Deubel, H. and Schneider, W.X. (1996) Saccade target selection and object recognition: evidence for a common attentional mechanism. Vision Res., 36: 1827–1837.
Duhamel, J.-R., Colby, C.L. and Goldberg, M.E. (1992) The updating of the representation of visual space in parietal cortex by intended eye movements. Science, 255: 90–92.
Ferraina, S., Pare, M. and Wurtz, R.H. (2002) Comparison of cortico-cortical and cortico-collicular signals for the generation of saccadic eye movements. J. Neurophysiol., 87: 845–858.
Gnadt, J.W. and Andersen, R.A. (1988) Memory related motor planning activity in posterior parietal cortex of macaque. Exp. Brain Res., 70: 216–220.
Goldberg, M.E. and Wurtz, R.H. (1972) Activity of superior colliculus in behaving monkey. II. Effect of attention on neuronal responses. J. Neurophysiol., 35: 560–574.
Gottlieb, J.P., Kusunoki, M. and Goldberg, M.E. (1998) The representation of visual salience in monkey parietal cortex. Nature, 391: 481–484.
Heilman, K.M. and Valenstein, E. (1972) Frontal lobe neglect in man. Neurology, 22: 60–664.
James, W. (1890) The Principles of Psychology. Holt, New York.
Joseph, J.S. and Optican, L.M. (1996) Involuntary attentional shifts due to orientation differences. Perception Psychophys., 58: 651–665.
Kowler, E., Anderson, E., Dosher, B. and Blaser, E. (1995) The role of attention in the programming of saccades. Vision Res., 35: 1897–1916.
Leon, M.I. and Shadlen, M.N. (2003) Representation of time by neurons in the posterior parietal cortex of the macaque. Neuron, 38: 317–327.
Moran, J. and Desimone, R. (1985) Selective attention gates visual processing in the extrastriate cortex. Science, 229: 782–784.
Pare, M. and Wurtz, R.H. (1997) Monkey posterior parietal cortex neurons antidromically activated from superior colliculus. J. Neurophysiol., 78: 3493–3497.
Platt, M.L. and Glimcher, P.W. (1997) Responses of intraparietal neurons to saccadic targets and visual distractors. J. Neurophysiol., 78: 1574–1589.
Platt, M.L. and Glimcher, P.W. (1999) Neural correlates of decision variables in parietal cortex. Nature, 400: 233–238.
Posner, M.I. (1980) Orienting of attention. Q. J. Exp. Psychol., 32: 3–25.
Powell, K.D. and Goldberg, M.E. (2000) Response of neurons in the lateral intraparietal area to a distractor flashed during the delay period of a memory-guided saccade. J. Neurophysiol., 84: 301–310.
Pylyshyn, Z.W. and Storm, R.W. (1988) Tracking multiple independent targets: evidence for a parallel tracking mechanism. Spatial Vis., 3: 179–197.
Rizzolatti, G., Riggio, L., Dascola, I. and Umilta, C. (1987) Reorienting attention across the horizontal and vertical meridians: evidence in favor of a premotor theory of attention. Neuropsychologia, 25: 31–40.
Robinson, D.L., Goldberg, M.E. and Stanton, G.B. (1978) Parietal association cortex in the primate: Sensory mechanisms and behavioral modulations. J. Neurophysiol., 41: 910–932.
Sears, C.R. and Pylyshyn, Z.W. (2000) Multiple object tracking and attentional processing. Can. J. Exp. Psychol., 54: 1–14.
Sereno, A.B. and Maunsell, J.H. (1998) Shape selectivity in primate lateral intraparietal cortex. Nature, 395: 500–503.
Shadlen, M.N. and Newsome, W.T. (2001) Neural basis of a perceptual decision in the parietal cortex (area LIP) of the rhesus monkey. J. Neurophysiol., 86: 1916–1936.
Snyder, L.H., Batista, A.P. and Andersen, R.A. (1997) Coding of intention in the posterior parietal cortex. Nature, 386: 167–170.
Sugrue, L.P., Corrado, G.S. and Newsome, W.T. (2004) Matching behavior and the representation of value in the parietal cortex. Science, 304: 1782–1787.
Williams, Z.M., Elfar, J.C., Eskandar, E.N., Toth, L.J. and Assad, J.A. (2003) Parietal activity and the perceived direction of ambiguous apparent motion. Nat. Neurosci., 6: 616–623.
Wurtz, R.H. (1969) Visual receptive fields of striate cortex neurons in awake monkeys. J. Neurophysiol., 32: 727–742.
Yantis, S. and Jonides, J. (1984) Abrupt visual onsets and selective attention: evidence from visual search. J. Exp. Psychol. Hum. Perception Perform., 10: 601–621.
Yantis, S. and Jonides, J. (1996) Attentional capture by abrupt onsets: new perceptual objects or visual masking? J. Exp. Psychol. Hum. Perception Perform., 22: 1505–1513.

Martinez-Conde, Macknik, Martinez, Alonso & Tse (Eds.)
Progress in Brain Research, Vol. 155
ISSN 0079-6123

CHAPTER 11

Visual masking approaches to visual awareness

Stephen L. Macknik*

Departments of Neurosurgery and Neurobiology, Barrow Neurological Institute, 350 W Thomas Road, Phoenix, AZ 85013, USA

Abstract: In visual masking, visible targets are rendered invisible by modifying the context in which they are presented, but not by modifying the targets themselves. Here I summarize a decade of experimentation using visual masking illusions in which my colleagues and I have begun to establish the minimal set of conditions necessary to maintain the awareness of the visibility of simple unattended stimuli. We have established that spatiotemporal edges must be present for targets to be visible. These spatiotemporal edges must be encoded by transient bursts of spikes in the early visual system. If these bursts are inhibited, visibility fails. Target-correlated activity must rise within the visual hierarchy at least to the level of V3, and be processed within the occipital lobe, to achieve visibility. The specific circuits that maintain visibility are not yet known, but we have deduced that lateral inhibition plays a critical role in sculpting our perception of visibility, both by causing interactions between stimuli positioned across space, and also by shaping the responses to stimuli across time. Further, the studies have served to narrow the number of possible theories to explain visibility and visual masking. Finally, we have discovered that lateral inhibition builds iteratively in strength throughout the visual hierarchy, for both monoptic and dichoptic stimuli. Since binocular information is not integrated until inputs from the two eyes reach the primary visual cortex, it follows that the early visual areas contain differential levels of monoptic and dichoptic lateral inhibitions. We exploited this fact to discover that excitatory integration of binocular inputs occurs at an earlier level than interocular suppression. These findings are potentially fundamental to our understanding of all forms of binocular vision and to determining the role of binocular rivalry in visual awareness.

Keywords: consciousness; visibility; invisibility; human; primate; standing wave; electrophysiology; psychophysics; fMRI; optical imaging; awareness

Introduction

The most basic function of the visual system is to create the experience of visibility. By "visibility," I do not mean the entire process of vision, but whether a stimulus is visible, or not. In this sense, visibility is the beginning of visual perception, not the conclusion. Without visibility, the stimulus has no significance or meaning: we perceive stimulus attributes such as color, motion, and depth, only if the stimulus is also visible. This does not imply that visibility circuits must come before other types of circuits. For instance, it may not be necessary that a stimulus be visible before other types of processing take place. As far as we currently know, it may be true that the processing of visibility occurs subsequently to the processing of other visual attributes, or it may all happen at once in parallel, or there may be a disunity of processing times for different attributes (as suggested by Zeki, 2003), or there could be a mixture of all of these possibilities. Whatever the sequence of processing is, we cannot be aware of the attributes of the stimulus unless

*Corresponding author. Tel.: +1-602-406-8091;
Fax: +1-602-406-4172; E-mail: macknik@neuralcorrelate.com

DOI: 10.1016/S0079-6123(06)55011-3

they are visible. Therefore, while an unconscious visual reflex such as the accommodation of the eye may be a type of visual experience, stimuli that invoke visual reflexes are not necessarily visible: we are not aware of them despite the fact that they produce an appropriate and automatic response.

Let us assume that visual awareness correlates to brain activity within specialized neural circuits, and that not all brain circuits maintain awareness. It follows that the neural activity that leads to reflexive or involuntary motor action may not correlate with awareness because it does not reside within awareness-causing neural circuits.

By defining awareness by its correlation to standard neuronal physiological circuits (albeit of a specialized type), I confess to having sidestepped the entire issue of whether consciousness may be caused by alternative neuronal events (i.e., quantum mechanical processing within neuronal microtubules; Jibu et al., 1994), or by nonphysical processes. This may seem like a dirty trick to the supporters of nonmainstream theories of consciousness, and I apologize if I offend any sensibilities. I admit that it is not particularly scientific of me to ignore nontraditional theories, especially when some of them are potentially falsifiable and therefore serious possibilities (such as the relevance of quantum mechanical processes in microtubules). However, the experiments necessary to falsify these theories are beyond my reach, and so I will concentrate on theories that rely on traditional neurophysiological processing.

I propose that a stimulus that has become just-noticeable, or just-visible, has by definition already activated awareness-producing circuits. As I will discuss later in the chapter, our results indicate that the retina cannot maintain awareness, which excludes the possibility that awareness is a low-level process, or a process that takes place at all levels of the brain. Thus awareness is neither a final process in vision nor a very low-level process, but instead it is a mid-level process. A less than just-noticeable stimulus may activate neurons within early visual areas that do not maintain awareness, and therefore remain invisible. Awareness may be required to initiate certain higher-level visual or cognitive processes, but that issue is beyond the scope of this chapter, which addresses solely the transition between visibility and invisibility. Moreover, varying the level of salience (such as by varying level of brightness above and beyond the threshold for visibility) may equate to varying the level of visibility and awareness. However, here I have chosen to simplify the problem by focusing solely on the difference in brain activity that underlies the perception of visible vs. invisible stimuli.

My colleagues and I have set out to determine the minimal set of conditions necessary for a physical or imagined stimulus to become consciously visible. "Consciousness," as used here, solely refers to those sensory experiences of which we are aware, and it does not address other possible definitions of consciousness, such as self-consciousness. This chapter is, in part, a progress report on our attempts to discover the components of the minimal set of conditions necessary to make us aware of a visible stimulus. I will also discuss the theoretical implications of the results and future research directions, and will present some new physiological data.

Let us begin with the assumption that there is a "minimal set of conditions" necessary to achieve visibility, in the form of a specific type (or types) of neural activity within a subset of brain circuits. This minimal set of conditions will not be met if the correct circuits have the wrong type of activity (too much activity, too little activity, sustained activity when transient activity is required, etc.). Moreover, if the correct type of activity occurs, but solely within circuits that do not maintain awareness, visibility will also fail. Finding the conditions in which visibility fails is critical to the research described here: although we do not yet know what the minimal set of conditions is, we can nevertheless systematically modify potentially important conditions to see if they result in stimulus invisibility. If so, the modified condition is potentially part of the minimal set.

To establish the minimal set of conditions for visibility, we need to answer at least four questions:

(1) What stimulus parameters are important to visibility?
(2) What types of neural activity best maintain visibility (transient vs. sustained firing, rate codes, bursts of spikes, etc. — i.e., the neural code)?

(3) What brain areas must be active to maintain visual awareness?
(4) What specific neural circuits within the relevant brain areas maintain visibility?

This chapter will address our discoveries to date in answer to these four questions. While we do not have complete answers for any of the questions above, we have made some progress. The discussion section of this chapter will place these discoveries into the context of the field as a whole.

The primary tool used in my laboratory to determine the minimal set of conditions for visual awareness is a type of illusion called "visual masking." Visual masking illusions come in different flavors, but in all of them a visible target (a visual stimulus such as a rectangle) or some specific aspect of a visible target (for instance the semantic content of a word displayed visually) is rendered invisible by changing the context in which the target is presented, without actually modifying the physical properties of the target itself. That is, the target becomes less visible due solely to its spatial and/or temporal context. Visual masking illusions allow us to examine the brain's response to the same physical target under varying levels of visibility. All we need to do is measure the perceptual and physiological effects of the target when it is visible vs. invisible and we will determine many, if not all, of the minimal set of conditions that cause visibility. See Fig. 1 for a description of a type of backward visual masking called metacontrast masking.

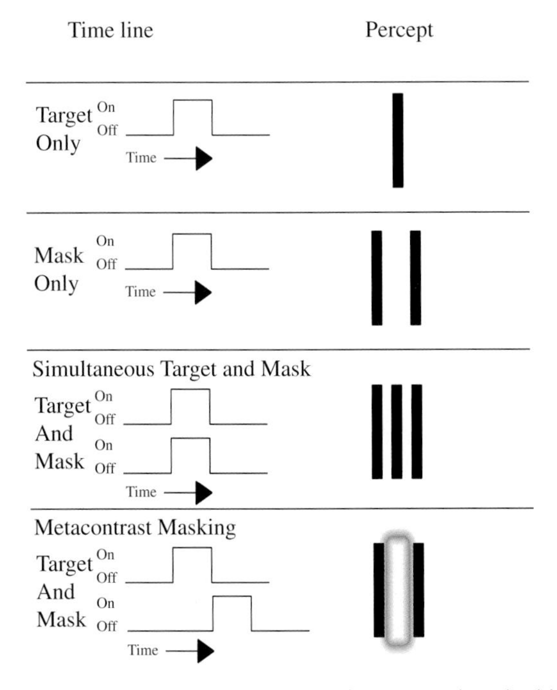

Fig. 1. Description of the perception of a target and mask with respect to temporal arrangement.

What stimulus parameters are important to visibility?

Because an imagined stimulus can be a visible stimulus, this question may seem beside the point. Why would physical stimulus parameters be important if one does not need a physical stimulus to achieve visibility? The answer may be that the same neural activity evoked by visible physical stimuli is evoked by imagined visible stimuli. Just as a physical stimulus must have certain properties to activate neural tissue in an appropriate way to achieve visibility (i.e., supra-threshold luminance contrast), so must an imagined stimulus cause equivalent neural activity to achieve visibility. Therefore, one should be able to study either imagined or physical stimuli and find equivalent answers about the neural underpinnings of visibility. The obvious advantage of studying physical rather than imagined stimuli is that they can be objectively and precisely controlled in both humans and animals (unlike a subject's imagination) in an experimental setting.

Early studies of visual masking

Early psychophysical studies of visual masking focused on the perception of spatial contrast illusions, in which simultaneously presented stimuli affected each other's brightness across space (Mueller, 1842; Hering, 1964; Mach, 1965; Helmholtz, 1985a, b, c; Chevreul, 1987). Exner accidentally discovered visual masking illusions occurring between stimuli across both time and space (Exner, 1868). His finding was serendipitous (i.e., he thought it was bad luck) since his research was meant to be focused on the time course of awareness (as reviewed by Breitmeyer, 1984). Exner was more interested in how long the brain needed to be stimulated in order to produce perception than in the quality of perception itself. Exner's contemporaries had previously assumed that two identical bars of light, when presented in different parts of space, and at different times, should appear identical. Exner found, to the contrary, that when two bars were presented with different timing, one bar sometimes seemed more visible than the other (Exner, 1868); he discovered visual masking. Stigler, a turn-of-the-20th-century German psychologist, went further to characterize some of the critical parameters by which one stimulus could actively suppress another across space and time (Stigler, 1908, 1913, 1926). These and many later studies of visibility and brightness formed the basis for our current understanding of brightness perception. I sum up below six key features of visibility and brightness perception.

Six key features of visibility and brightness perception

(1) The perceived brightness of a target is a function of its physical intensity.

(A) The just-noticeable difference in brightness (ΔI) between a target and its background is a function of the physical luminosity of the stimuli (I). This principle, $\Delta I/I = k$, first formally formulated by Fechner (1966) was first discovered by Bouguer in 1760 (Bouguer, 1961) and later rediscovered by Weber (1834, 1846) and has become known as Weber's law. By formulating Weber's law, Fechner introduced the first formally derived psychophysical principle.

(B) By assuming that just-noticeable differences were equal across sensory modalities (in units of perceptual intensity), Fechner went further to reformulate Weber's law (Uttal, 1969, 1973; Somjen, 1975) into a general statement (the Weber–Fechner's law) concerning all sensory measurements: he thus defined the magnitude of psychophysical sensations (Ψ). The equation defining psychophysical magnitude is $\Psi = k \log I$, where Ψ is defined as perceptual intensity and I defined as the physical intensity of the stimulus; k a modality or task specific constant, but it is not necessarily unique to any given modality or task. The main difference between the Weber–Fechner's law and Weber's law is that, while Weber's law is concerned with the detection of a stimulus at threshold, the Weber–Fechner's law describes the magnitude of sensation for every stimulus strength.

(C) The Weber–Fechner's law was redrafted into Stevens' power law (Stevens, 1961, 1962, 1970), which no longer assumes, as Fechner did, that the perceptual magnitude of one just-noticeable difference threshold is the same as any other. That is, depending on one's state, a given detection threshold may vary in magnitude, even though the perceived magnitude of the stimulus remains constant. In the revamped equation, the psychophysical magnitude (Ψ) is defined as proportional to the power of the physical intensity of the stimulus (I): $\Psi = k\,(I)^n$.

(D) Similar to Steven's power law, the Naka–Rushton relation linked physical intensity and intensity of the neuronal response (Naka and Rushton, 1966; Enroth-Cugell and Shapley, 1972, 1973a, b; Shapley et al., 1972; Shapley, 1974; Enroth-Cugell et al., 1975; Shapley and Enroth-Cugell, 1984; Hood and Birch, 1990; Wali and Leguire, 1991, 1992; Birch and Anderson, 1992; Roecker et al., 1992; Chung et al., 1993; Dagnelie and Massof, 1993; Kortum and Geisler, 1995). The Naka–Rushton relation was derived in order to describe the response of neural systems that saturate at high intensities (Shapley and Enroth-Cugell, 1984): $R = [I/(I+I_s)]\,R_{max}$ where R is the neuron's response (in units of change in membrane potential from total dark adaptation), R_{max} the maximal possible response of the neuron, I the illumination of the stimulus, and I_s the illumination required for the neuron to reach half its R_{max}.

(E) While these laws describe the relationship between brightness perception and the physical intensity of stimuli for most of the intensity scale, they often fail to describe the magnitude of perception at the extremes of physical intensity (Uttal, 1969, 1973; Somjen, 1975; Brown and Deffenbacher, 1979). That is, these psychophysical laws describe the relationship between physical intensity and perceptual intensity for most, but not all, visual environments.

(2) A stimulus can be made be more intense by either increasing its intensity or duration.

(A) Bloch's law (Bloch, 1885), also called the time-intensity reciprocity law, states that a short-duration visual target of high physical intensity (I) can appear to be as bright as a longer duration target of lower intensity (so long as the duration of both targets, t, is shorter than a critical duration, τ, where $\leqslant$refers to the time-integrated contrast energy): for $t \leqslant \tau$, $c = I\ t$ (Bloch, 1885; McDougall, 1904a, b; Dawson and Harrison, 1973; Kong and Wasserman, 1978; Gorea and Tyler, 1986; Duysens et al., 1991; Van Hateren, 1993).

(B) The critical duration that temporally limits Bloch's law, τ, becomes shorter as the overall intensity of the stimuli increases (Sperling and Joliffe, 1965; Roufs, 1972).

(C) The Broca–Sulzer effect (Broca and Sulzer, 1902) states that as the duration of a flashed target increases, the perceived brightness of the target first increases, but then decreases. Many people in the last century have discussed either Bloch's Law or the Broca–Sulzer effect, but none of them, to my knowledge, have explicitly discussed the discrepancy between these two claims. I can only suppose, without having studied these effects myself, that Bloch's Law describes the rising phase of the Broca–Sulzer effect and that Broca and Sulzer were unaware of Bloch's law when they published their findings.

(D) The Crawford (Crawford, 1947) visual adaptation effect illustrates the relationship between the duration of a mask and its effect on the detection threshold of a target.

(3) The perceived brightness of a target may be affected by other visual stimuli presented at different locations and/or times. This effect is exemplified by Mach bands, simultaneous contrast illusions, and visual masking.

(4) The magnitude of visual masking depends on the relative contrast energies of target and mask.

(A) Visual masking is stronger when the mask is of higher physical contrast than the target (Stigler, 1926; Kolers, 1962; Weisstein, 1971, 1972).

(B) Visual masking is stronger when the mask is of longer duration than the target (Breitmeyer, 1978).

(C) Visual masking depends on the stimulus onset asynchrony (SOA) of the target and the mask (the difference in time between the onset of the target and the onset of the mask). This is referred to as the SOA Law (Stigler, 1910; Werner, 1935; Kahneman, 1967; Turvey, 1973; Breitmeyer and Ganz, 1976; Bischof and Di Lollo, 1995).

(D) In contradiction with the SOA Law, several groups have found that optimal visual masking depends on the inter-stimulus interval (ISI) between the target and the mask (the difference in time between the end of the target and the beginning of the mask);

(Bischof and Di Lollo, 1995; Francis et al., 2004) or the stimulus termination asynchrony (STA) between the target and the mask (the difference in time between the end of the target and the end of the mask) (Macknik and Livingstone, 1998).

(5) Visual masking can be mediated exclusively through cortical processes.
   (A) Visual masking can be experienced dichoptically. That is, a mask presented solely to one eye can inhibit a target presented solely to the other eye (Stigler, 1926; Kolers and Rosner, 1960; Schiller, 1968; Weisstein, 1971; McFadden and Gummerman, 1973; Macknik and Martinez-Conde 2004a; Tse, et al., 2005).
   (B) Metacontrast masking can be experienced transcallosally. That is, a mask presented solely to one hemi-retina can inhibit a the target presented solely to the other bilaterally symmetric hemi-retina (Stigler, 1926; Schiller, 1968; McFadden and Gummerman, 1973).

(6) Visual masking is stronger with masks that are parallel or co-curvilinear to the target (Werner, 1935, 1940).

These six general findings establish that: (1) the perceived brightness of a stimulus depends on its intensity, its duration, and its temporal, spatial, and contrast relationships to other stimuli, and (2) cortical processes could be at least partly responsible for these interactions.

Studies of visibility at the spatial edge

In a lateral inhibitory network such as the retina, the lateral geniculate nucleus (LGN), or area V1, the spatial edges of stimuli excite neurons strongly, whereas the interiors of stimuli evoke relatively little response (Hubel and Wiesel, 1959; Ratliff and Hartline, 1959; Battersby and Wagman, 1962; De Weerd et al., 1995; Livingstone et al., 1996; Paradiso and Hahn, 1996). Owing to the very nature of lateral inhibition, both excitatory and inhibitory neural signals are greatest at the spatial edges of the stimulus (Ratliff et al., 1959). One visual correlate of this effect is the Mach band (Mach, 1865, 1965).

Human psychophysical studies of brightness perception and visual masking have shown that inhibition is strongest at the edge of the mask, rather than within its interior (Crawford, 1940; Rushton and Westheimer, 1962; Westheimer, 1965, 1967, 1970). This effect is known as the Westheimer function. My colleagues and I have replicated and extended these results using a

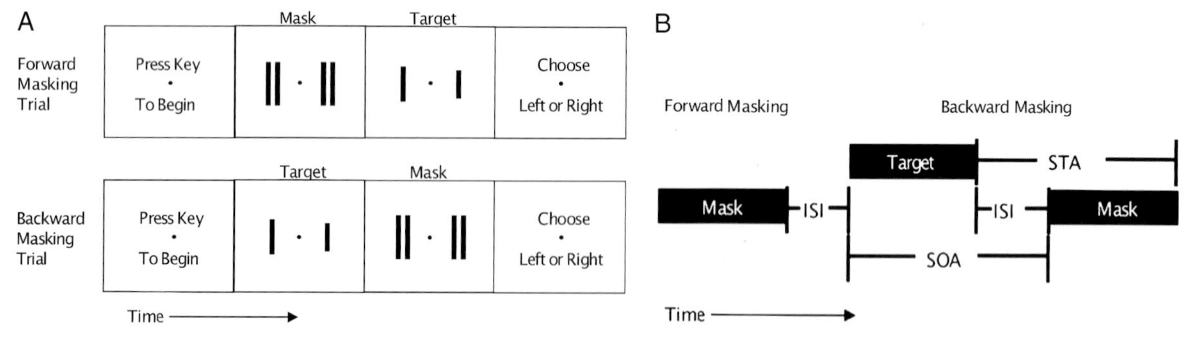

Fig. 2. (A) The sequence of events during the course of a visual masking psychophysics trial. The trial started with a delay of 500–1500 ms. In backward masking conditions, the target came before the mask. In forward masking conditions, masks came before targets. After termination of the second stimulus (mask or target), there was another 500-ms delay, after which the subject indicated which side had the longer target. (B) A schematic view of the various timing parameters used in these experiments. SOA = stimulus onset asynchrony, the interval between the onset of target and of mask; STA = stimulus termination asynchrony, the interval between termination of target and of mask; ISI = inter-stimulus interval, the interval between the termination of the target and the onset of the mask (backward masking) or between the termination of the mask and the onset of the target (forward masking). Reprinted from Macknik and Livingstone (1998), with kind permission from the Nature Publishing Group.

psychophysical visual masking experiment (in which the target and the mask were not present on the screen at the same time, unlike in Westheimer's experiments). We based the visibility assay on a length-discrimination task, in which the subject was required to determine the longer of two target bars (of 30 ms in duration) when presented at various SOAs with a visual mask of 50 ms in duration and various sizes. See Fig. 2 for a description of forward and backward masking conditions. See Fig. 3 for a description of the experimental results of spatial masking. Notice in Fig. 3 that the mask overlapped spatially the target in every condition. Masking strength was therefore a function of the distance from the mask's edge to the target's edge (see inset of Fig. 3). The results indicate that inhibition from the mask increases as the mask's edge approaches the target. The neural signal from the mask that generates the inhibition is therefore strongest at the mask's edge than within its interior.

Studies of visibility at the temporal edge

Our psychophysical studies suggest that the temporal edges of targets — the points in time at which they turn on and off — are more important to their visibility than their midlife periods (Macknik and Livingstone, 1998). Fig. 4A–C shows the performance of 25 human subjects in a backward masking task, plotted as a function of SOA, ISI, and STA. Fig. 4A shows performance as a function of SOA. The point of maximum backward masking (drop lines) does not occur at a constant

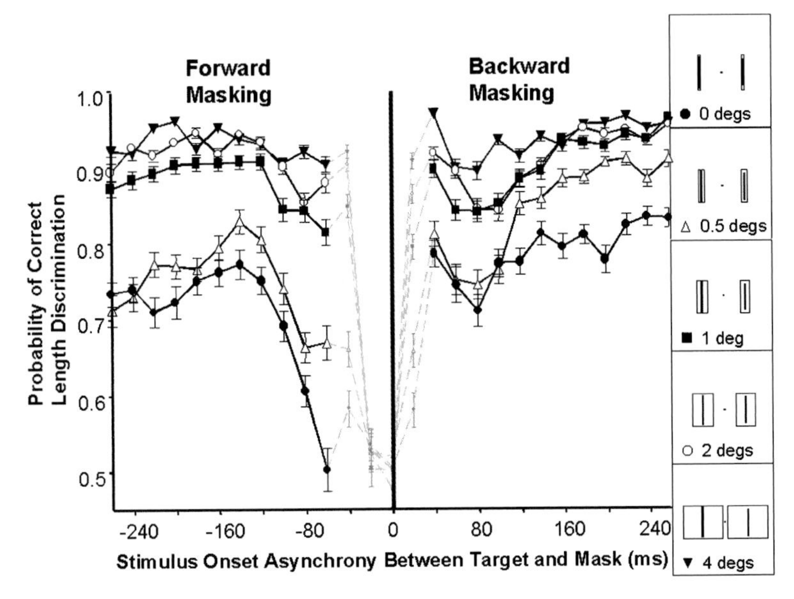

Fig. 3. Human psychophysical length-discrimination measurements of visual masking effects from 23 human subjects using overlapping opaque masks of varied size (the mask's edge distance from the target's edge was 0°, 0.5°, 1°, 2°, or 4° as indicated in the insert on the right). The subject's task was to fixate on the central black dot and choose the longer target (right or left). Targets were black bars presented for 30 ms in duration and masks were also black and presented for 50 ms: the subject's task was to fixate on the central black dot and choose the longer target (right or left). Targets turned on at time 0 ms, and masks were presented at various onset asynchronies so that they came on before, simultaneous to, or after the target in 20-ms steps. Stimulus onset asynchronies (SOAs) to the left of zero indicate forward masking conditions, and SOAs that are greater than zero indicate backward masking. Miniature gray markers with dotted connecting lines represent conditions during which the target and the mask overlapped in time and so the target was partially or completely hidden by the mask. The targets were 0.5° wide and had varied heights (5.5°, 5.0°, or 4.5°) and were placed 3° from the 0.2° wide circular fixation dot in the center of the screen. The mask was a bar 6° tall with varied widths, spatially overlapped and centered over each target. There were 540 various types of trials (2 possible choices × 2 differently sized target sets to foil local cue discrimination strategies × 5 various overlapping mask sizes × 27 stimulus onset asynchronies). Each condition was presented in random order five times to each subject, over a period of 2 days, for a total of 62,100 trials (summed over all 23 subjects). Reprinted from Macknik et al. (2000a), with kind permission from the National Academy of Sciences of the United States of America.

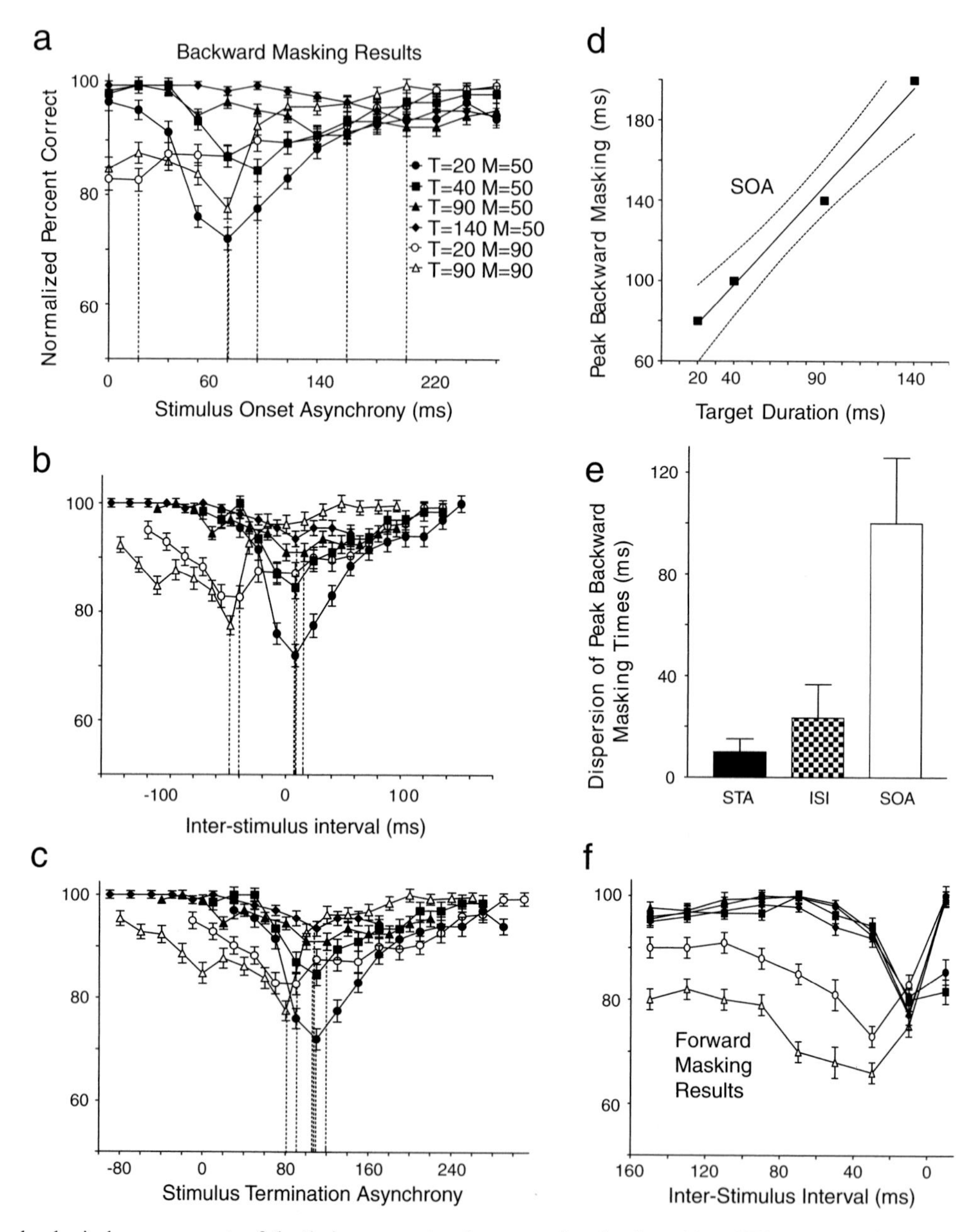

Fig. 4. Psychophysical measurements of the timing parameters important for visual masking. "T" represents the duration (in ms) of the target, and "M" represents the duration of the mask. Results represent average for 25 subjects. (A) Results from backward masking conditions plotted on a stimulus onset asynchrony (SOA) scale. Note that the points of peak masking (the x-intercepts of the drop lines) are widely dispersed. (B) Results from panel A replotted here as a function of inter-stimulus interval (ISI). The points of peak masking tend to cluster in two places, correlated with mask duration (open symbols vs. closed symbols). (C) Results from panel A replotted here on a stimulus termination asynchrony (STA) scale. The points of maximal masking are no longer dispersed, and instead cluster around an STA of about 100 ms $\pm$ 20 ms. (D) Linear regression (with 95% confidence intervals) of peak backward masking times in terms of SOA when the mask was 50 ms in duration. (E) The amount of dispersion of peak backward masking times for data tested on a scale of stimulus termination asynchrony (STA), inter-stimulus interval (ISI), and stimulus onset asynchrony (SOA). Notice that the peak backward masking times are least dispersed on an STA scale and so STA is the best predictor of backward masking. (F) Results from forward masking conditions; the optimal predictor of peak masking was the inter-stimulus interval between the termination of the mask and the onset of the target. Reprinted from Macknik and Livingstone (1998), with kind permission from the Nature Publishing Group.

SOA, but rather varies for different mask and target durations. Since SOA is determined as a function of target onset, this result suggested that backward masking was not correlated with target onset. Fig. 4D shows that optimum backward masking tends to occur at later times for longer target durations, suggesting that the termination of the target correlates with the timing of maximum backward masking (slope of linear regression $= 0.98 \pm 0.06$, $p < 0.01$). The data were therefore replotted as a function of the ISI (Fig. 4B) and the STA (Fig. 4C); since these parameters are defined as a function of target termination, a correlation in either of these parameter spaces would suggest that backward masking is tied somehow to the target's termination. We found that all backward masking conditions were correlated with an STA of about 100 ms. This suggested that the termination of the mask, 100 ms after the target disappeared, had a crucial impact on the visibility of the target, thus causing backward masking. I will discuss the effect of the mask further below; the point to take home from this result is that some (presumed physiological) event, occurring 100 ms after the target turns off, is important for target visibility.

Previous perceptual studies of backward masking had concluded, incorrectly, that there was a point of minimum visibility at some particular delay between the onsets of target and mask (the "SOA Law") (Weisstein and Haber, 1965; Smith and Schiller, 1966; Bridgeman, 1971; Matin, 1975; Breitmeyer and Ganz, 1976; Francis, 1995). The source of this confusion was that these studies did not vary the duration of the target and the mask, which would presumably have made evident that SOAs were not stable over varying target durations. A few earlier psychophysics studies *did* vary either target or mask duration, but never both (Alpern, 1953; Sperling, 1965; Breitmeyer, 1978). They reported that mask or target duration modulated the apparent *brightness* of targets, but did not discuss the effect of target or mask duration on the *timing* of peak visual masking. Nevertheless, the figures published from these earlier studies show a trend consistent with our findings (that the SOA of peak masking strength varies as a function of target duration).

We also examined the psychophysics of forward masking and found that irrespective of target and mask duration, visual masking was strongest when the mask terminated just before the beginning of the target (Fig. 4F). This suggested that some (presumed physiological) event, associated with the onset of the target, was inhibited during forward masking.

Together, our forward and backward masking results suggest that the *temporal edges* of the target (its onset and termination) are most important to its visibility. However, the factor at the target's termination that was critical to its visibility did not occur until about 100 ms after the target was turned off. I have examined the neural correlates of this effect in the next section of this chapter.

Since the signals responsible for target visibility appear concentrated at its temporal edges, it follows that masks may be most inhibitory at their temporal edges too. Previous studies had suggested that the onset and termination of the mask are more inhibitory to visual targets than the mask's midlife time points. Crawford (1947) measured psychophysically the detection threshold of spots of light flashed for 5 ms using method of adjustment (the subject could vary the spot's luminance), before, during, and after the background flashed to various luminances for 500 ms. For the purposes of this discussion, the spot can be considered the target and the background flash, the mask.

The results indicated that the subjects increased the brightness of the target (in order to detect them) at time points near the beginning and end of the mask, rather than during the mask's midlife. However, Crawford did not truly address the issue of whether the termination of the mask was more suppressive than its midlife, since he did not vary the duration of the mask. Thus, he could not know if the target inhibition related to the termination of the mask might be the delayed result of the onset of the mask. Margaret Livingstone and I addressed this issue more directly by varying the duration of the mask (Macknik and Livingstone, 1998) (Fig. 4). For the two short-duration masks (50 and 90 ms), we found that most of the inhibition was conveyed at time points near the masks' termination time. At first sight, this result seemed in conflict with previous studies suggesting that inhibition from the mask was strongest at *both* its onset and termination. To address this issue, Susana Martinez-Conde, Michael Haglund, and I conducted a psychophysical experiment to test the strength of visual masking before, during, and after the presentation of nonoverlapping masks, using a wide range of durations (100, 300, and 500 ms) (see Fig. 5). The results indicated that masks are most powerful in their ability to mask targets at their temporal edges (their onsets and terminations).

In summary, both the spatial and temporal edges of a target are critical to its visibility. Thus, in partial answer to Question #1 above (What are the stimulus parameters that are important to visibility?), spatiotemporal edges appear to be important to visibility. In addition, the spatiotemporal edges of the mask drive the strongest suppression of the target.

What types of neural activity best maintain visibility?

Studies of excitatory and inhibitory neural responses at the spatial edge

We optically imaged the neural correlates of excitatory and inhibitory signals from spatial edges in monkey area V1 (Fig. 6). Our results showed that neural signals from the edges are stronger than the signals from the interior, in agreement with previous physiological studies in the early visual system (Hubel and Wiesel, 1959; Ratliff and Hartline, 1959; Battersby and Wagman, 1962; De Weerd et al., 1995; Livingstone et al., 1996; Paradiso and Hahn, 1996).

We next wondered whether physiological *inhibition* might also be strongest at the spatial edges

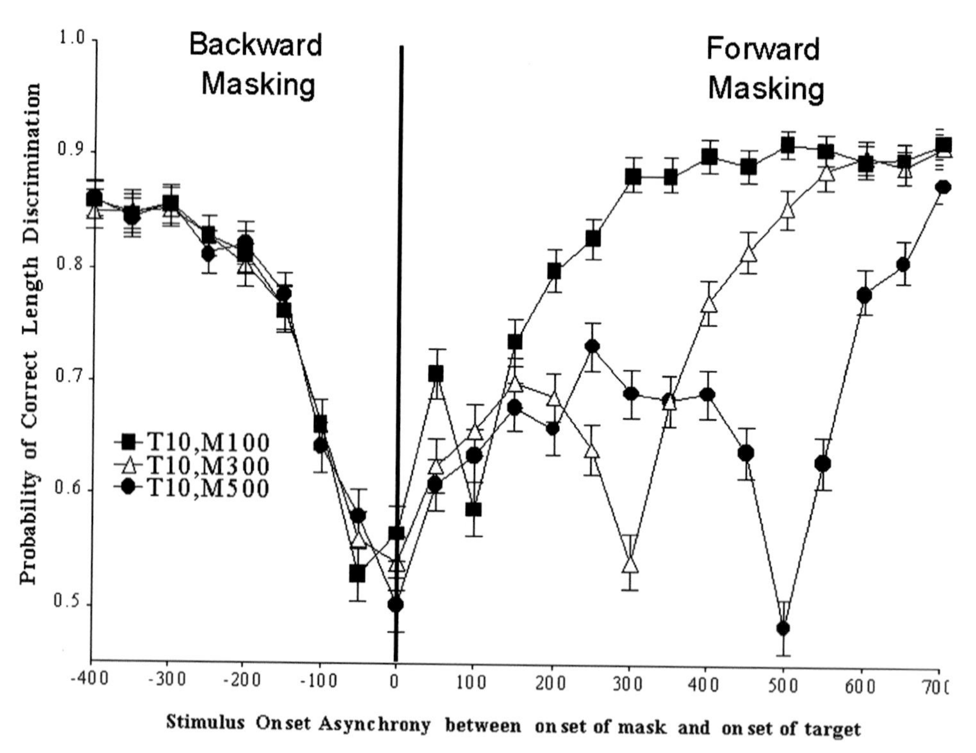

Fig. 5. Human psychophysical length-discrimination measurements of visual masking effects from 11 human subjects using nonoverlapping masks of varied duration (100, 300, or 500 ms). SOA here represents the period of time between the onset of the mask and the onset of the target (and so it has the opposite meaning than in Figs. 3 and 4). Masks (two 6° tall bars with a width of 0.5° flanking each side of each target) appear at time 0, and targets can appear earlier (backward masking), simultaneously, or later (forward masking), in 50-ms steps. Targets were black and presented for 10-ms duration, and masks were flanking black bars that abutted the target. Notice that target visibility is most greatly affected when the masks turn on and off. Reprinted from Macknik et al. (2000a), with kind permission from the National Academy of Sciences of the United States of America.

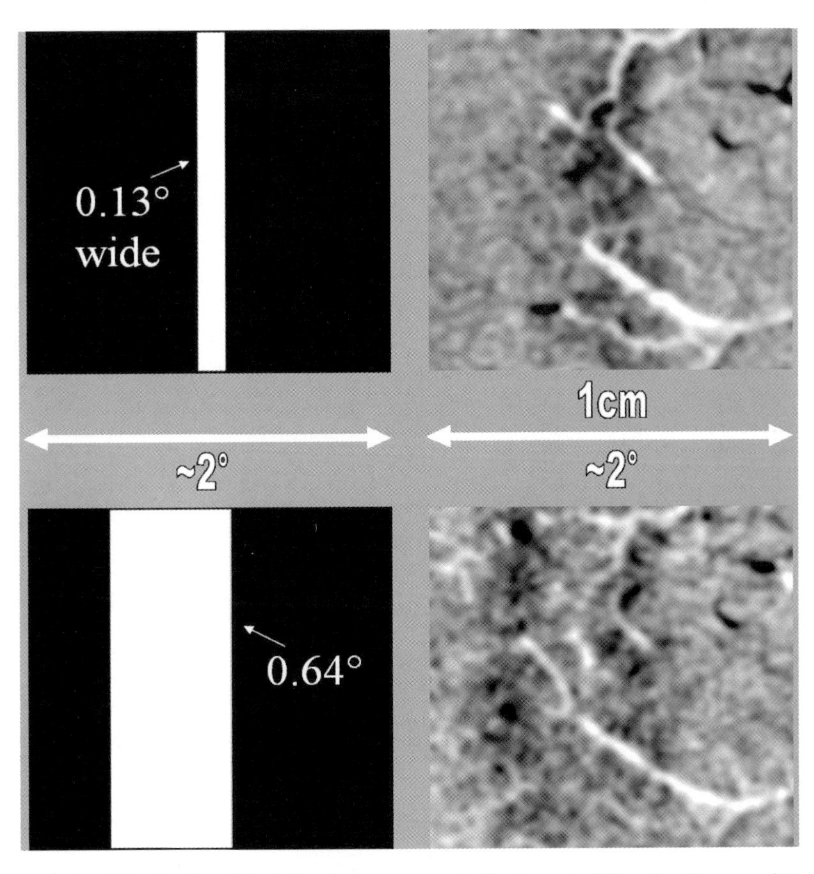

Fig. 6. The optical image of a flickering bar. The left column represents the layout of the stimuli we used to stimulate the cortical window imaged in the right column. *Top*: Image of the area V1 cortical intrinsic signal generated by a flickering bar 50 ms on and 100 ms off, 0.13° wide with an orientation of 132°. The imaged patch is 1 cm^2 square and was approximately 10°–12° below and to the left of the foveal representation, and subtended about 4° of visual angle (as measured with microelectrode penetrations at each edge of the image), at the anterior-medial border of the operculum. The vertical meridian is parallel to the lower edge of this image, the fovea is to the right. *Bottom*: Image of the intrinsic signal generated by a flickering bar 0.64° wide in the same piece of cortex (the center of the bar was also shifted here approximately 0.29° away from the fovea). Notice that the widened bar has shifted in position and split into two edges, showing that the edges have much stronger signals than the interiors of objects, at the level of area V1. Modified from Macknik et al. (2000a), with kind permission from the National Academy of Sciences of the United States of America.

(Werner, 1935; Crawford, 1940; Rushton and Westheimer, 1962; Ratliff, 1965; Westheimer, 1965; Nakayama, 1971). To answer this question, we varied the distance from the mask's edge to the target'edge. We recorded from 26 LGN neurons in the awake monkey, while presenting a visual masking illusion (the standing wave of invisibility, SWI, defined in Fig. 7) in which a flickering target (a white or black bar of 50-ms duration) is rendered invisible by a mask (two bars of 100-ms duration that flank the target to either side). The mask flickered in counterphase alternation to the target (Macknik and Livingstone, 1998; Macknik et al., 2000a) (Fig. 8).

We found that, as the distance between the target and the mask increases, the strength of the inhibition from the mask decreased. This decrease matched the perceptual decrease in visual masking found for increased mask distances (Werner, 1935). The results supported the idea that the spatial edges of the mask convey the strongest inhibitory signals to the target.

Studies of neural responses at the temporal edges

Our psychophysical studies of temporal edges suggested that there are neural events at the beginning

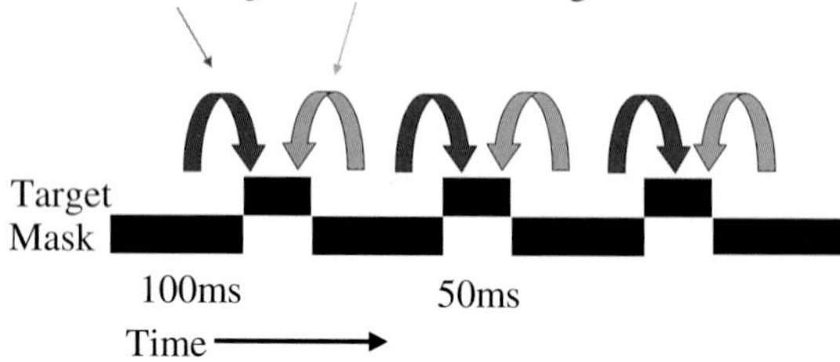

Fig. 7. The time course of events during the standing wave of invisibility illusion. A flickering target (a bar) is preceded and succeeded by two counterphase flickering masks (two bars that abut and flank the target, but do not overlap it) that are presented with optimal forward and backward timing.

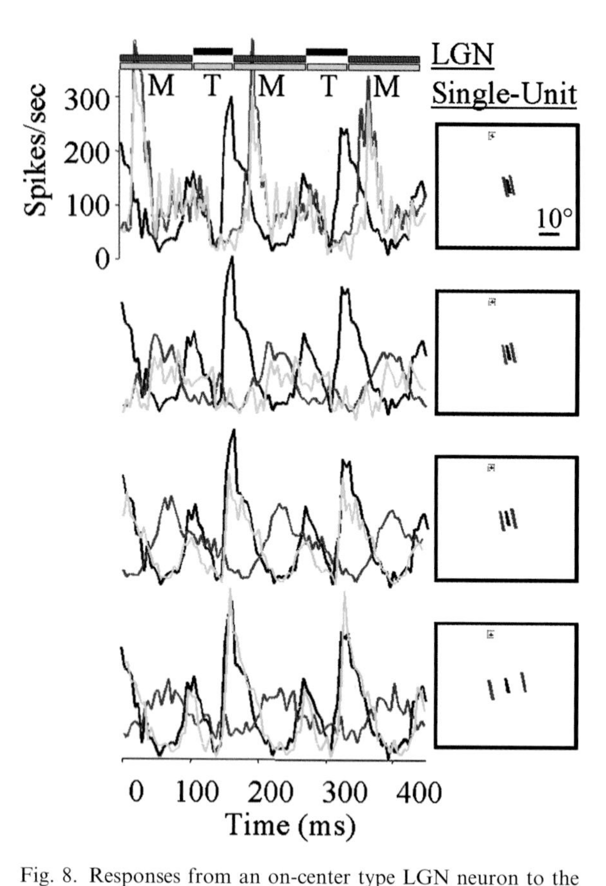

Fig. 8. Responses from an on-center type LGN neuron to the standing wave of invisibility, the timing of which is depicted in the colored bars at the top of the figure ("T" = target and "M" = mask). Inset pictures on the right represent the stimulus configuration on the display the monkey fixated on (cross near the top of each screen represents the fixation point); the mask distance from the target varied from zero in the top histogram to 8.57° of visual angle in the bottom. Stimulus size, position, and sign of stimulus varied, depending on the optimal parameters of each cell's receptive field. Each histogram represents separately the results of increasing the mask's distance (as drawn in the inset). Black traces in each row represent the target-only condition for this cell (all black traces are identical). Purple traces represent the response to the mask alone at each distance, and the blue traces in each histogram represent the firing of the cell to both the target and the mask presented cyclically (SWI illusion condition). Suppression of the target portion of the response in the SWI condition (blue lines) is the neural correlate of visual masking. As it is evident in the blue traces, increase of the mask distance decreases its inhibitory effect on the target (in correlation to perception). Notice that this neuron shows some response to the mask, although the distance between the mask and the target was 8.57° (much larger than the extent of the receptive field). This long-range effect occurred in 16 cells (62%). For all but 1 of these 16 cells the cell type was on-center, indicating that the long-range effect was probably due to light scattering within the eye. Reprinted from Macknik et al. (2000a), with kind permission from the National Academy of Sciences of the United States of America.

of stimuli and after the termination of stimuli that both convey the visibility of the target and the suppressive effects from the mask. Our physiological recordings from V1 of anesthetized monkeys showed that the neural correlates of these events are the transient onset-response to the onset of stimuli, and the transient after-discharge to the termination of stimuli (Fig. 9). Here I discuss the role that these neural events play in visibility and visual masking.

Previous studies had suggested that forward masking is physiologically best described as a function of ISI (Schiller, 1968; Judge et al., 1980). This was confirmed by our psychophysical study (Fig. 4F). Our physiological studies of forward and backward masking, moreover, showed that the neural correlate of forward masking is the inhibition of the onset-response, and the neural correlate of backward masking is the inhibition of the after-discharge (Fig. 9). These results were consistent with observations from previous physiological studies (Schiller, 1968; Bridgeman, 1975; Bridgeman, 1980; Judge et al., 1980), although these studies had not drawn conclusions about the role of the after-discharge. We proposed that the neural correlates of the temporal edges of targets were transient bursts of spikes of the onset-response and after-discharge: the suppression of these transient responses correlated to invisibility during visual masking.

It may seem somewhat counterintuitive that the after-discharge contributes to target visibility, since real-world objects do not usually turn off and should therefore not generate after-discharges. But

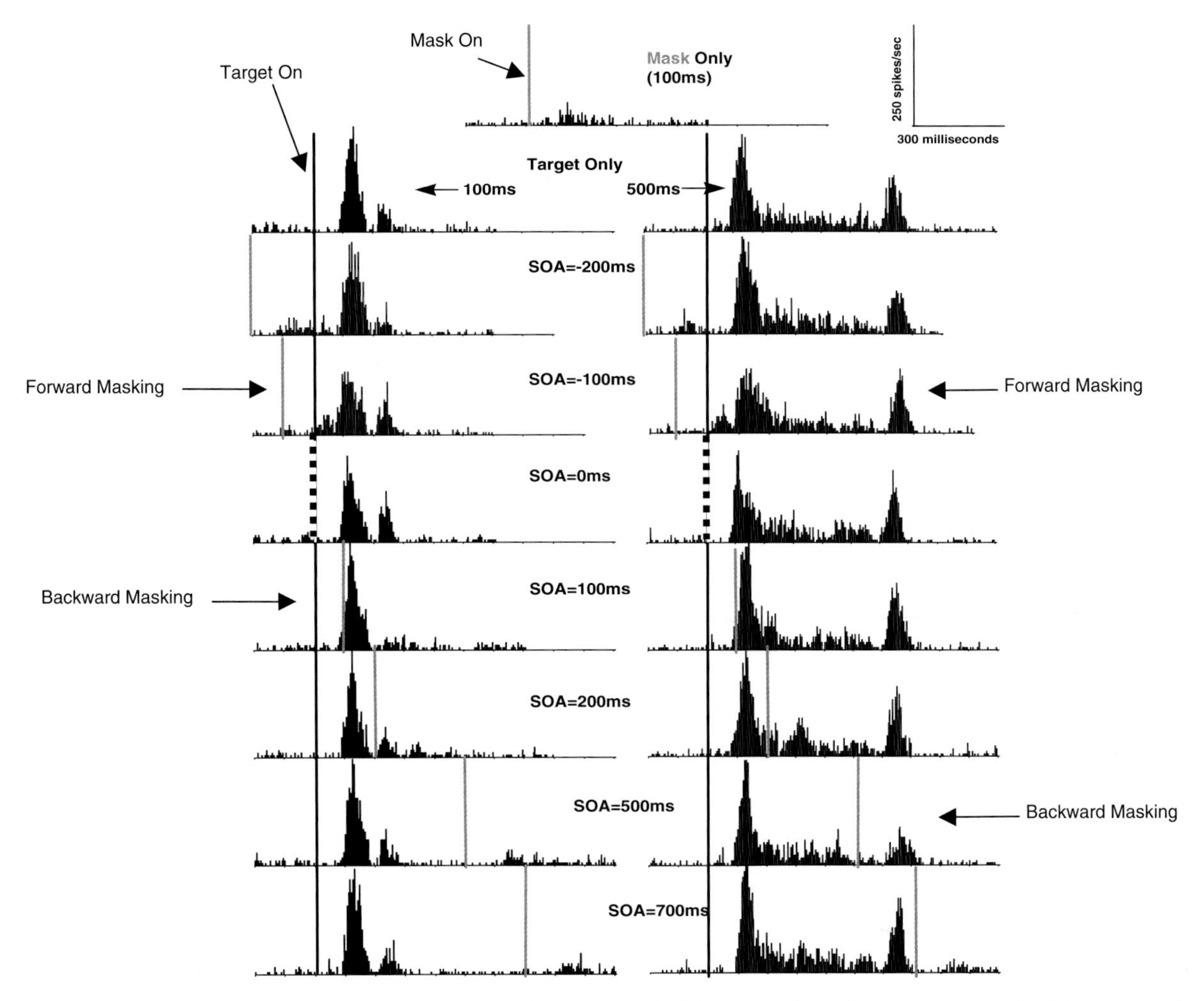

Fig. 9. Multiunit recording from upper layers of area V1 in an anesthetized rhesus monkey. The aggregate receptive field was foveal, 0.1° square, and well oriented. In contrast to the recordings from alert animals, the mask is largely outside the receptive field. The vertical colored bars (gray for mask, red for target) indicate the actual onset time of the stimulus. The transparent fields (gray for mask, black for targets) represent the time when the stimulus is expected to evoke a response (determined in the top two rows, in which the targets and mask were presented by themselves). Notice that under conditions that best correlate with human forward masking (ISI = 0 ms, here corresponding to SOA = -100 ms) the main effect of the mask is to inhibit the transient onset-response to the target. Similarly, in the condition that produces maximum backward masking in humans (STA = 100 ms; here corresponding to SOA = 100 ms for the 100 ms stimulus on the left, SOA = 500 ms for the 500 ms stimulus on the right), the after-discharge is specifically inhibited. Each histogram is an average of 50 trials with a bin width of 5 ms. Modified from Macknik and Livingstone (1998), with permission from the Nature Publishing Group.

the eyes, when open, are rarely stationary (Yarbus, 1967), so real-world stimuli *do* turn on and off several times per second from the viewpoint of each visual receptive field (Martinez-Conde et al., 2000, 2002). If images are artificially stabilized on the retina, they fade quickly (Day, 1915; Riggs and Ratliff, 1952; Coppola and Purves, 1996). Coppola and Purves (1996) showed that a stabilized image fades in as little as 80 ms, consistent with the idea that the initial onset-response can produce only a transient visible image. Moreover, Yarbus (1967) showed that after visual fading, stabilized images will reappear as a positive image if they are then turned off. This is consistent with our suggestion that the after-discharge also contributes to visibility. In our own studies, we found that microsaccades drive visibility during fixation (Martinez-Conde et al., 2004; Martinez-Conde et al., 2006), and that the neuronal activity that best correlates with microsaccades is transient bursts of spikes in the LGN and V1 (Martinez-Conde et al., 2000, 2002). See Susana Martinez-Conde's chapter in the previous vloume (Vol. 154) of Progress in Brain Research for a more in-depth review.

In summary, our results indicate that transient bursts of spikes are generated at the spatiotemporal edges of stimuli. The suppression of these transient bursts (but not the suppression of sustained tonic activity) leads to suppression of visibility in the early visual system. Thus, in partial answer to Question #2 above (What types of neural activity best communicate visibility?), it seems that transient bursts of spikes are important to the neural code for visibility.

What brain areas must be active to maintain visual awareness?

The search for the neural correlates of consciousness requires the localization of circuits in the brain that are sufficient to maintain awareness. To this end, brain areas have been sought within the ascending visual hierarchy that correlate, or more importantly, fail to correlate, with visual perception (Crick and Koch, 1990; Milner, 1995; He et al., 1996; Logothetis et al., 1996; Farah and Feinberg, 1997; Sheinberg and Logothetis, 1997; Tong et al., 1998; Zeki and Ffytche, 1998; Lee and Blake, 1999; Lamme et al., 2000; Polonsky et al., 2000; Thompson and Schall, 2000; Dehaene et al., 2001; Pascual-Leone and Walsh, 2001; Macknik and Martinez-Conde, 2004a; Lee et al., 2005; Moutoussis et al., 2005; Tse et al., 2005). Presumably, the circuits of the brain that are critical to the visibility of targets are circuits whose activity is suppressed during visual masking. A corollary to this is that, if we can identify circuits in which the target response is not suppressed during target masking, we can rule out that circuit as significant to maintaining visual awareness. This section discusses our research to identify parts of the brain whose activity correlates with visual masking.

One of the main reasons that most models of visual masking propose cortical circuits is that "dichoptic" visual masking exists (Kolers and Rosner, 1960; Weisstein, 1971; McFadden and Gummerman, 1973; Olson and Boynton, 1984; McKee et al., 1994, 1995; Harris and Willis, 2001; Macknik and Martinez-Conde, 2004a; Tse et al., 2005). To be clear about the jargon: "monocular" means "with respect to a single eye," and "monoptic" means either "monocular" or, "not different between the two eyes." "Binocular" means "with respect to both eyes" and "dichoptic" means "different in the two eyes." Thus, in dichoptic visual masking, the target is presented to one eye and the mask to the other eye, and the target is nevertheless suppressed. Since excitatory binocular processing within the geniculocortical pathway occurs first in the primary visual cortex (Minkowski, 1920; Le Gros Clark and Penman, 1934; Hubel, 1960), it has been assumed that dichoptic masking must originate from cortical circuits. The anatomical location in which dichoptic masking first begins is thus critical to our evaluation of most models of masking. It is also important to our understanding of neurons of the LGN and their relationship to the subcortical and cortical structures that feed back onto them. In order to establish where dichoptic masking first begins, we first compared the perception of monoptic to dichoptic visual masking in humans over a wide range of timing conditions never before tested (Macknik and Martinez-Conde, 2004a) (see Fig. 10). We found that dichoptic masking was as robust as monoptic masking, and that dichoptic masking exhibited the same timing characteristics previously discovered in humans for monoptic masking (Crawford, 1947; Macknik and Livingstone, 1998; Macknik et al., 2000a).

Next, we conducted recordings of LGN and V1 neurons in the awake monkey while presenting monoptic and dichoptic stimuli. To our knowledge, these were the first dichoptic masking experiments to be conducted physiologically. We found that monoptic masking occurs in all neurons of the early visual system, while dichoptic masking occurs solely cortically in a subset of binocular neurons (see Fig. 11). We also discovered that in the first binocular neurons in the visual system, excitatory responses to monocular targets are inhibited strongly only by inhibitory masks presented to the same eye, whereas interocular

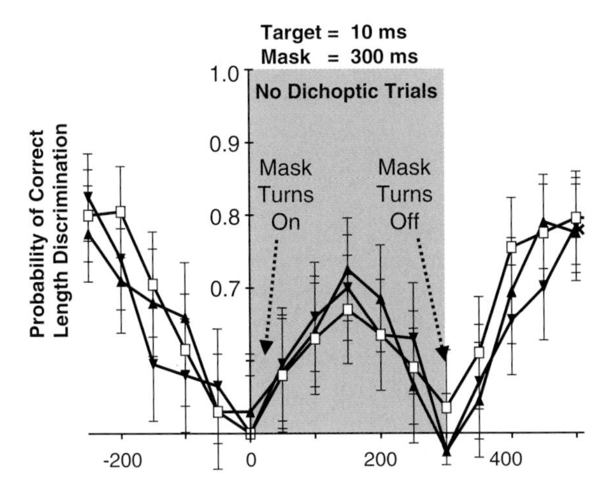

Fig. 10. Psychophysical examination of dichoptic vs. monoptic visual masking in humans. Human psychophysical measurements of visual masking when 10-ms duration target and 300-ms duration mask were presented to both eyes together (monoptic masking) and to the two eyes separately (dichoptic masking). The probability of discriminating correctly the length of two targets is diminished, in the average responses from seven subjects, when they are presented near the times of mask onset and termination. This is true regardless of whether the target and the mask are presented to both eyes (open squares), or if the target was presented to one eye only and the mask was presented to the other (target = left, mask = right: closed upright triangles; target = right, mask = left: closed upside-down triangles). Open circles signify when the target was displayed with both shutters closed, showing that the stimuli were not visible through the shutters. When the mask and the target were presented simultaneously, both eyes' shutters were necessarily open (dichoptic presentations using shutters are impossible when both stimuli are presented at the same time), and so between 0 and 250 ms all four conditions were equivalent. Dichoptic masking is nevertheless evident when the target was presented before the mask was presented (−250 to −50 ms on the abscissa), as well as when the target was presented after the mask had been terminated (300–500 ms on the abscissa). Reprinted from Macknik and Martinez-Conde (2004a), with kind permission from the MIT Press.

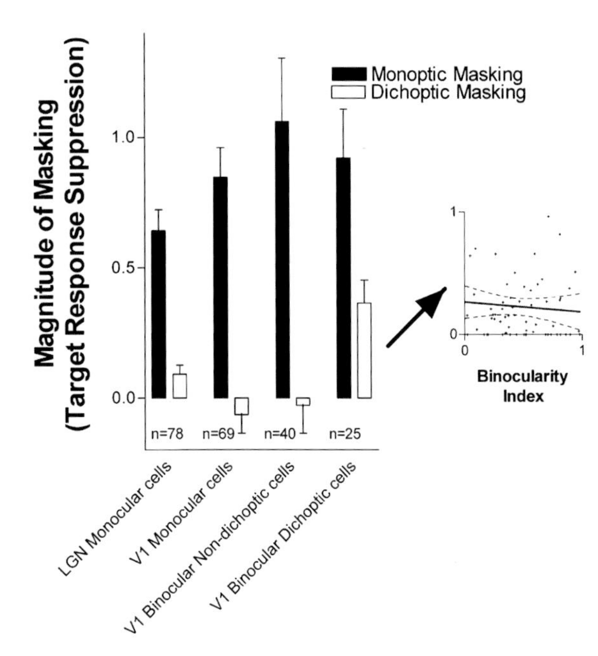

Fig. 11. Summary statistics of monoptic vs. dichoptic masking responses in the LGN and area V1. Monoptic (black bars) and dichoptic (white bars) masking magnitude as a function of cell type: LGN, V1 monocular, V1 binocular (nondichoptic masking responsive), and V1 binocular (dichoptic masking responsive) cells. Inset shows the linear regression of the dichoptic masking magnitude seen in V1 binocular neurons (dichoptic masking responsive) as a function of their degree of binocularity (all neurons displayed here were significantly binocular as measured by their relative response to monocularly presented target stimuli to the two eyes sequentially): BI of 0 would indicate that the cells were monocular, while a BI of 1 means both eyes were equally dominant. Reprinted from Macknik and Martinez-Conde (2004a), with kind permission from the MIT Press.

inhibition is surprisingly weak. Therefore, the circuits responsible for monoptic and dichoptic masking must exist in at least two places independently, one in monocular circuits and another in binocular circuits. Furthermore, the earlier circuits do not exhibit masking as a function of feedback from the later circuits, as proposed by Enns (2002). If they did, then the feedback connections would convey strong dichoptic masking from the later circuits, and the earlier circuits would inherit this trait with the feedback (Fig. 12). Since these earlier levels exhibit no dichoptic masking, we can conclude that visual masking in monoptic levels is not due to feedback from dichoptic levels.

No extant theories of visual masking propose that monoptic and dichoptic masking are generated by two different circuits: one that lies in binocular cells and another that lies solely (and specifically) within the monocular cells of V1 (and not the LGN or retina). Therefore, we believe that our results support the most parsimonious conclusion that the circuit underlying visual masking is simple lateral inhibition, which is fundamental to all known circuits of the visual system.

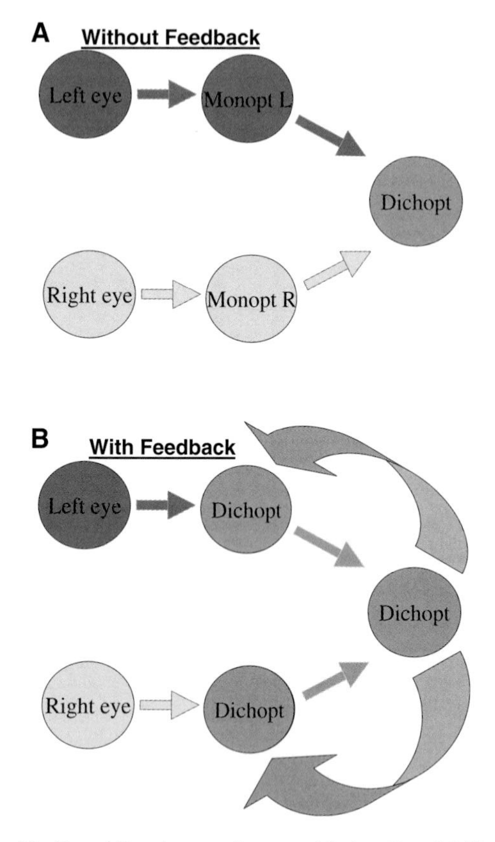

Fig. 12. Overriding issues when considering the viability of feedback mechanisms. (A) A general model of early visual system binocular integration without invoking feedback mechanisms. (B) If significant feedback exists between the initial dichoptic levels of processing and earlier levels, it means that, by definition, the earlier levels should therefore behave in the same way as the dichoptic levels (i.e., they will become dichoptic by virtue of the feedback).

The monoptic masking results presented in Fig. 11 show that the strength of monoptic masking increases, in an iterative fashion, with each successive stage of processing in the visual system. Correspondingly, Hubel and Wiesel (1961) found that inhibitory surrounds were stronger in the LGN than in the retina. We proposed that lateral inhibition mechanisms gather strength iteratively in successive stages of the visual hierarchy, as a general principle. The result that dichoptic inhibition is weak in area V1 may reflect such a general principle, given that V1 binocular neurons represent the first stage where dichoptic inhibition could exist in the ascending visual system. If this idea is correct, downstream binocular neurons in the visual hierarchy should show iteratively stronger dichoptic masking suppression effects. Whether these effects represent the discovery of a general principle of inhibitory iterative processing in the visual system, or not, we believe that dichoptic masking effects must become stronger downstream of V1, to account for the fact that the overall psychophysical magnitude of dichoptic visual masking is equivalent to that of monoptic masking (as shown in Fig. 11).

To search for the neural correlates of masking higher up in the visual hierarchy, we turned to whole brain imaging (functional magnetic resonance imaging, fMRI) techniques. Masking illusions are known to evoke reliable blood-oxygen level dependent (BOLD) signals that correlate with perception within the human visual cortex (Dehaene et al., 2001; Haynes et al., 2005; Haynes and Rees, 2005). Since the psychophysical strengths of monoptic and dichoptic masking are equivalent (Schiller, 1965; Macknik and Martinez-Conde, 2004a), we should be able to find the point in the ascending visual hierarchy in which monoptic and dichoptic masking activities are both extant, and thus determine the first point in the visual hierarchy at which awareness of visibility could potentially be maintained. Previous to this level, target responses will not be well inhibited during dichoptic masking: if these prior areas were sufficient to maintain visual awareness, the target would be perceptually visible during dichoptic masking conditions.

We mapped the retinotopic visual areas with fMRI in human subjects and measured BOLD signal in response to monoptic and dichoptic visual masking within each subject's individually mapped retinotopic areas (Fig. 13). Our results show that dichoptic masking does not correlate with visual awareness in area V1, but begins only downstream of area V2, within areas V3, V3A/B, V4 and later (Fig. 14). These results agree with previous electrophysiological results in monkeys using both visual masking and binocular rivalry stimuli (Logothetis et al., 1996; Sheinberg and Logothetis, 1997; Macknik and Martinez-Conde, 2004a), as well as with one fMRI study of binocular rivalry in humans (Moutoussis et al., 2005). Furthermore, we found that the iterative increase in lateral inhibition, which we previously

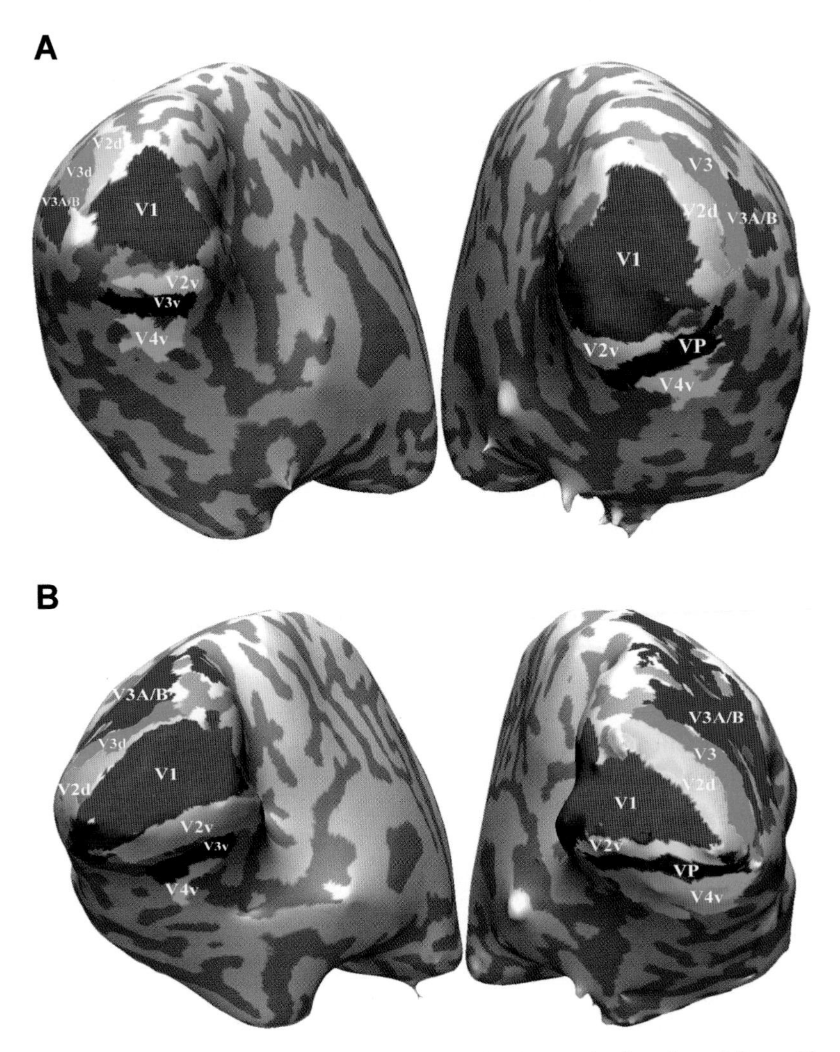

Fig. 13. (A and B) Examples of retinotopy mapping from two subjects. Visual areas that have been delineated by retinotopic mapping analysis are indicated in different colors. Reprinted from Tse et al. (2005), with kind permission from the National Academy of Sciences of the United States of America.

discovered in the subcortical visual system and in area V1 for monoptic masking, continued in the extrastriate visual areas for dichoptic masking (Fig. 14C). This fact plays an important role in localizing the circuits responsible for visibility and perception. For instance, if the brain areas that maintained visual awareness exhibited only weak suppression (i.e., areas that are early in the visual hierarchy such as the LGN and area V1), then target masking would be incomplete and targets would not be rendered invisible during masking. Since dichoptic visual masking is as strong as monoptic visual masking perceptually and target-derived neural activity is only weakly suppressed by dichoptic masks prior to area V3, it follows that the circuits responsible for visibility lie downstream of area V2, or else targets would not appear to be suppressed during dichoptic masking.

Having determined the lower boundary in the visual hierarchy for the perception of visual masking of simple targets, we set out to determine if there was also an upper boundary. To do this, we isolated the parts of the brain that showed both an increase in BOLD signal when nonillusory visible targets were displayed and a decrease in BOLD signal when the same targets were rendered less visible by visual masking. Surprisingly, only areas within the occipital lobe showed differential activation between visible and invisible targets (Fig. 15).

In addressing Question ♯3 (What brain areas must activate to achieve awareness of visibility?), our combined results suggest that visual areas beyond V2, within the occipital lobe, are responsible for maintaining our awareness of simple targets (Fig. 16). This conclusion may not hold true for complex visual stimuli or attended stimuli (see Discussion).

Our results show that masking in the early visual system is not caused by feedback from higher cortical areas that also cause dichoptic masking. It follows that the circuit that causes masking must be ubiquitous enough and simple enough that it exists at many or possibly all levels of the visual system. As lateral inhibition is the basis for all the receptive field structure that we know, it must therefore exist at all levels of the visual system that have receptive fields. Thus, lateral inhibition is a candidate circuit. This idea is strengthened by our findings that lateral inhibition is strengthened iteratively at each progressive level of the early visual system.

What specific neural circuits within the relevant brain areas maintain visibility?

We previously showed that the parts of the target that are most important to conveying its visibility are its *spatial edges*. We also found that, temporally, the parts of the target's lifetime most important to its visibility are its onset and termination, rather than its midlife. That is, the target's *temporal edges* seemed to convey the strongest signal concerning the target's visibility. Similarly, the parts of the mask that were most important to its ability to suppress the perception of the target were the mask's *spatiotemporal edges*. We moreover found that the neural correlates of the spatiotemporal edges of stimuli (both targets and masks) were transient bursts of spikes that occurred after the stimulus turned on and off, within neurons with receptive fields positioned at the spatial edges of the stimulus on the retinotopic map. Brain areas downstream of area V2, but nevertheless lying within the occipital lobe, were sufficient to maintain our awareness of visibility of simple unattended targets. Finally, we showed that visual masking and visual awareness do not rely on feedback mechanisms from higher cortical areas.

On the basis of these findings, we proposed that the peculiar timing conditions associated with visual masking may be explained through a simple lateral inhibitory network, in which the transient responses to the mask's spatiotemporal edges inhibit the transient responses to the target's spatiotemporal edges. Because the target and the mask do not overlap each other spatially, the circuit underlying masking must be called "lateral inhibition," as defined by Hartline and Ratliff (Hartline, 1949; Ratliff, 1961; Ratliff et al., 1974). The circuit has three properties:

(1) Excitatory input and output.
   (A) Monosynaptic connections between the retina and the LGN, and between the LGN and cortex, are excitatory (Cleland et al., 1971; Levick et al., 1972; Reid and Alonso, 1995).
(2) Self-inhibition.

Fig. 14. Retinotopic analysis of monoptic vs. dichoptic masking. (A) The logic underlying the analysis of masking magnitude for hypothetical retinotopic areas. The Mask Only response is bigger than the Target Only response because masks subtend a larger retinotopic angle than targets, and are moreover presented twice in each cycle for 100 ms each flash, whereas the target is single-flashed for only 50 ms. If the target response adds to the mask response in the SWI condition (no-masking percept), then the SWI response will be bigger than the Mask Only response, whereas if the target does not add (masking percept), then the SWI response will be equal or smaller (as the mask itself may also be reciprocally inhibited by the target) than the Mask Only response. (B) Monoptic and dichoptic masking magnitude (% BOLD difference of Mask Only/SWI conditions) as a function of occipital retinotopic brain area, following the analysis described in (A). Negative values indicate increased activation to the SWI condition (no masking), whereas values $\geqslant 0$ indicate decreased or unchanged SWI activation (masking). (C) Dichoptic masking magnitude (% BOLD difference of Mask Only/SWI conditions) as a function of occipital retinotopic brain area within the dorsal and ventral processing streams. The strength of dichoptic masking builds up as a function of level in the visual hierarchy for both the dorsal ($R^2 = 0.90$) and ventral ($R^2 = 0.72$) processing streams. Reprinted from Tse et al. (2005), with kind permission from the National Academy of Sciences of the United States of America.

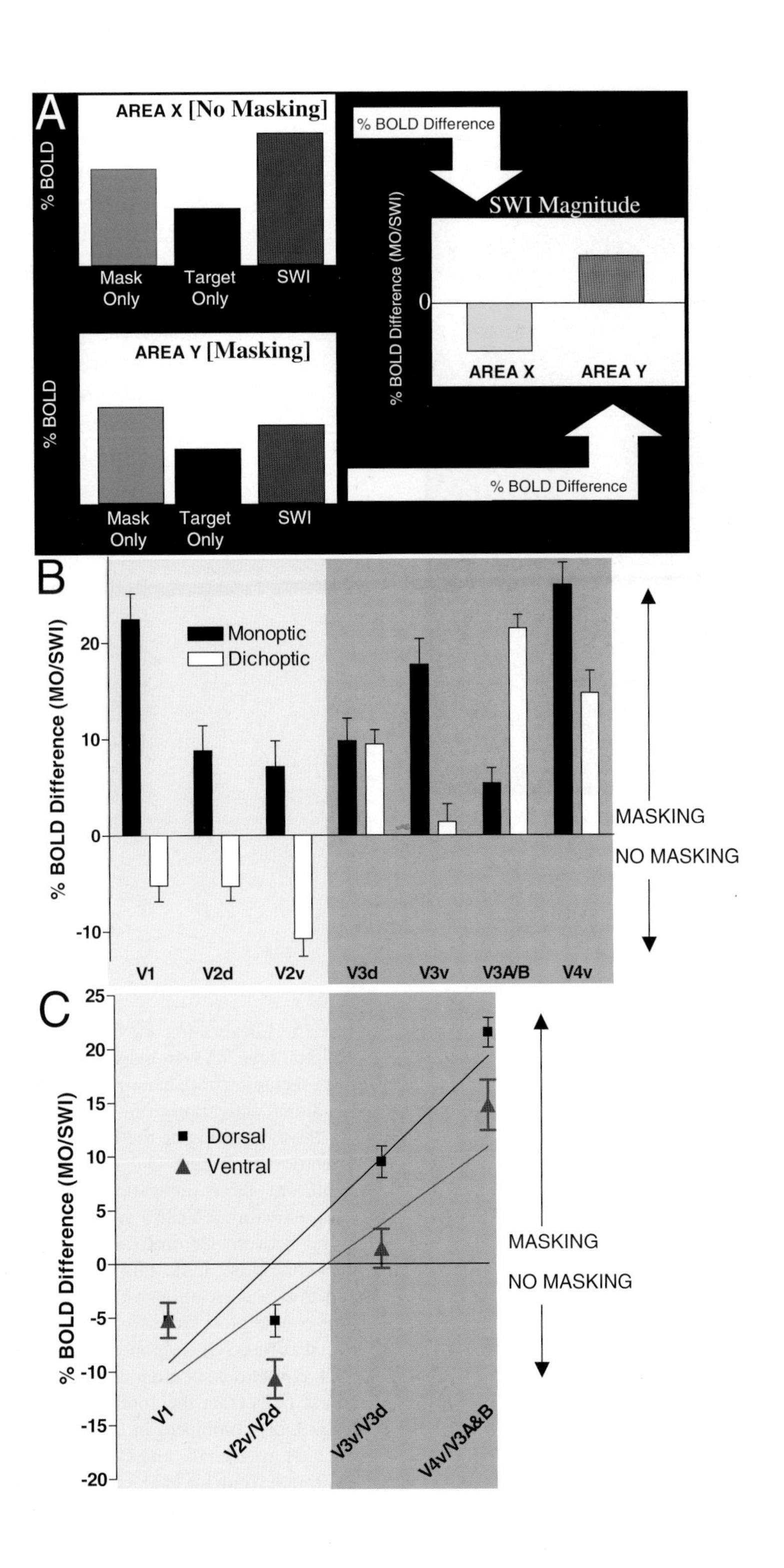
A
AREA X [No Masking]
% BOLD
Mask Only
Target Only
SWI
% BOLD Difference
SWI Magnitude
% BOLD Difference (MO/SWI)
0
AREA X
AREA Y
AREA Y [Masking]
% BOLD
Mask Only
Target Only
SWI
% BOLD Difference
B
% BOLD Difference (MO/SWI)
Monoptic
Dichoptic
MASKING
NO MASKING
V1
V2d
V2v
V3d
V3v
V3A/B
V4v
C
% BOLD Difference (MO/SWI)
Dorsal
Ventral
MASKING
NO MASKING
V1
V2v/V2d
V3v/V3d
V4v/V3A&B

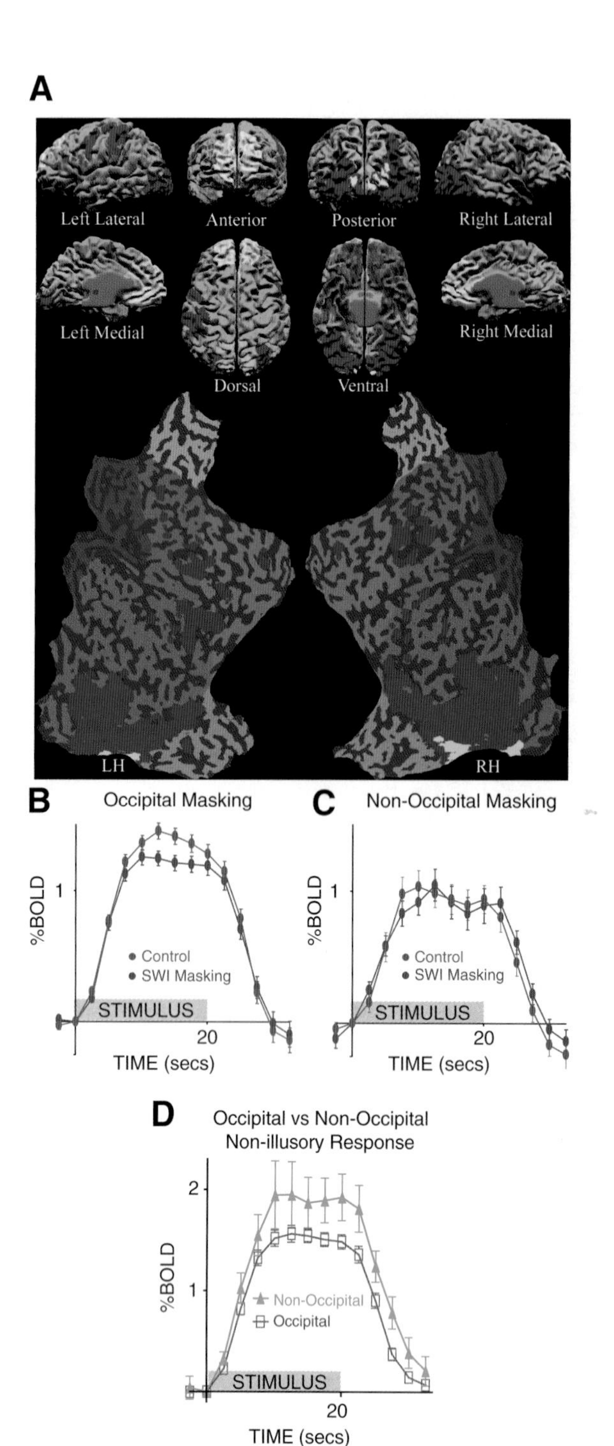

(3) Lateral inhibition as a function of excitation (thus inhibition follows excitation in time).
  (A) Spatiotemporal responses of neurons in the LGN and cortex show that excitation precedes inhibition (Ferster, 1986; Golomb et al., 1994). Fig. 17 shows a descriptive model of how a basic lateral inhibitory circuit that has these properties might account for the spatiotemporal properties found in masking.
    (a) Several previous groups have suggested that masking might be explained by lateral inhibition (Bridgeman, 1971; Lawwill, 1973; Anbar and Anbar, 1982; Francis, 1997, 1998). However, various aspects of these models do not match the timing parameters of visual masking we discovered, such as the importance of the target's after-discharge to its visibility (Macknik and Livingstone, 1998; Macknik et al., 2000a).
      (i) For instance, Breitmeyer (1984) noted that Bridgeman's model predicts that when the target and the mask are presented with nonoptimal stimulus onset

Fig. 15. Localization of visibility-correlated responses to the occipital lobe. (A) An individual brain model from all perspectives, including both hemispheres flat-mapped, overlaid with the functional activation from 17 subjects. The green-shaded areas are those portions of the brain that did not show significant activation to Target Only stimuli. The blue voxels exhibited significant target activation (Target Only activation > Mask Only activation). Yellow voxels indicate a significant difference found between Control (target-visible) and SWI (target-invisible) conditions, indicating potentially effective visual masking, and thus a correlation with perceived visibility. (B) Response time-course plots from Control vs. SWI conditions in the occipital cortex. (C) Response time-course plots from Control vs. SWI conditions in nonoccipital cortex. (D) Response time-course plots from the nonillusory conditions (Target Only and Mask Only combined) in occipital vs. nonoccipital cortex. Error bars in (B), (C), and (D) represent SEM between subjects. Reprinted from Tse et al. (2005), with kind permission from the National Academy of Sciences of the United States of America.

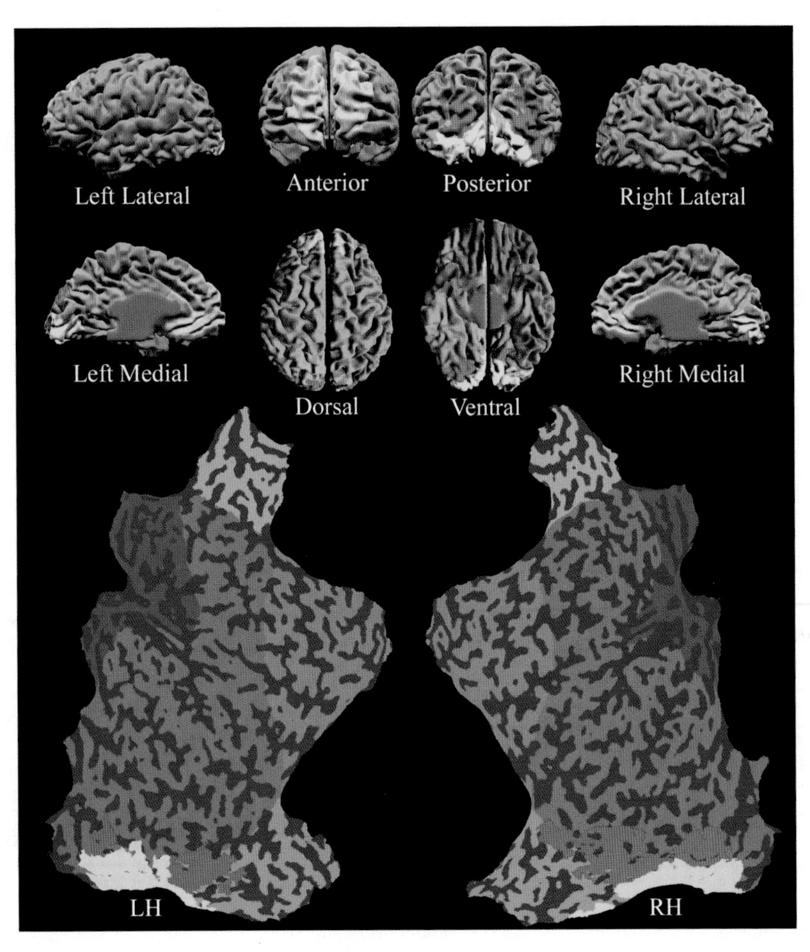

Fig. 16. Layout of retinotopic areas that potentially maintain awareness of simple targets. An individual brain model from all perspectives, including both hemispheres flat-mapped, overlaid with the functional activation from one typical subject. The yellow-shaded areas are those portions of the brain that did not show significant dichoptic masking (as in Fig. 14B and C) and thus are ruled out for maintaining visual awareness of simple targets. The pink-colored voxels represent the cortical areas that exhibited significant dichoptic masking and thus are potential candidates for maintaining awareness of simple targets. Reprinted from Tse et al. (2005), with kind permission from the National Academy of Sciences of the United States of America.

asynchronies (to produce masking), the apparent contrast of the target should beat (called "ringing" by Breitmeyer, 1984) in time (like two tuning forks of slightly different pitches) because of poor phase alignment of target and mask resonance. If this is true, then it follows that one should be able to tune the optimal SOA by noting the beat frequency of the target. At nonoptimal stimulus onset asynchronies, however, the percept of the target is quite steady (Breitmeyer, 1984).

Our model is the first to specifically propose an explanation of how the spatiotemporal edges of stimuli may interact to cause various novel

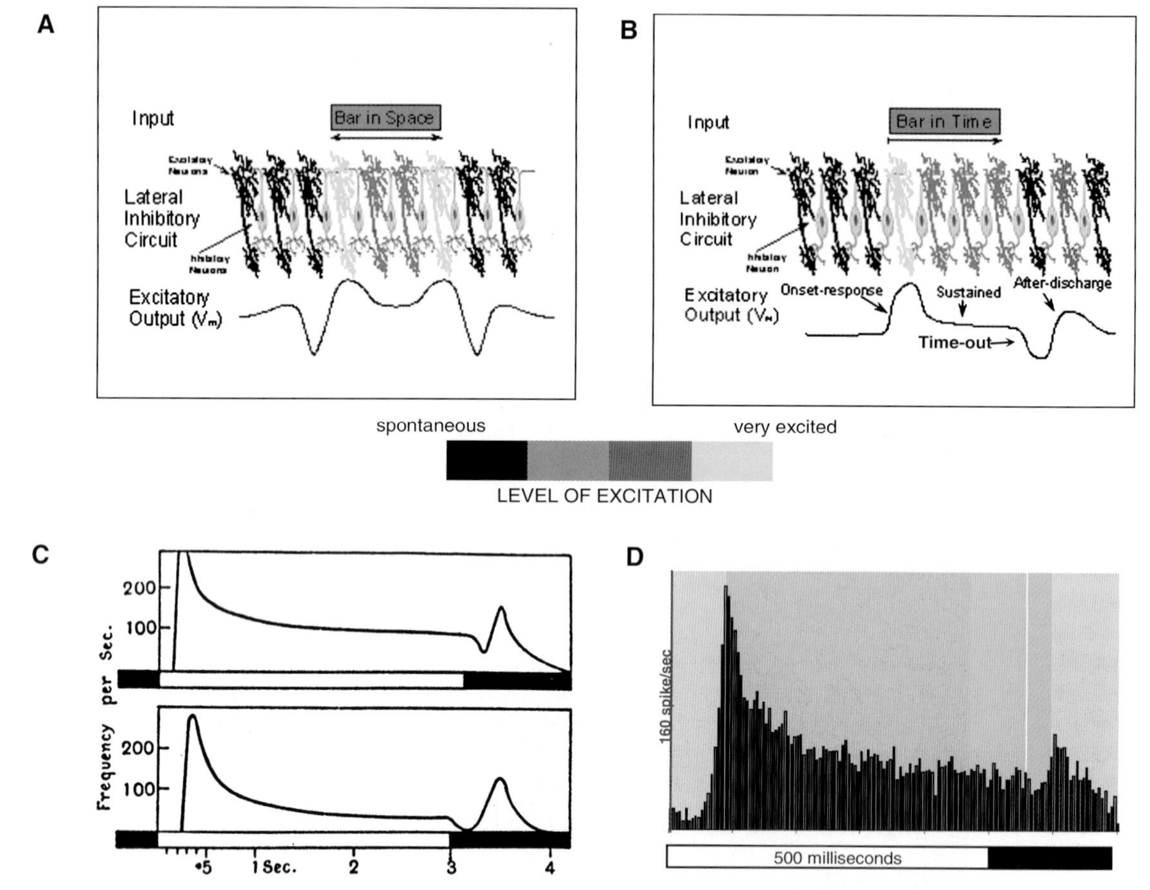

Fig. 17. (A) A representation of the spatial lateral inhibition model originally proposed by Hartline and Ratliff (Ratliff, 1961; Ratliff et al., 1974). The four excitatory neurons (highlighted in yellow and gray) in the center of the upper row receive excitatory input from a visual stimulus. This excitation is transmitted laterally in the form of inhibition, resulting in edge enhancement of the stimulus: the neuronal underpinnings of the Mach band illusion (Mach, 1965). (B) One excitatory and one inhibitory neuron taken from the spatial model in (A), now followed through an arbitrary period of time. Several response phases are predicted, including the onset-response, and the transient after-discharge (Adrian and Matthews, 1927). (C) Reprinted from Adrian and Matthews (1927); their Figs. 4 and 5). The *top* figure is the peri-stimulus time histogram of neuronal firing rate from the eel optic nerve when the retina was stimulated by a disk, 36 mm in diameter and 830 candles/m^2 in intensity. Duration of the stimulus is denoted with the white bar below the recording. The *bottom* figure is similar, except that the disk was 12.8 mm in diameter. Notice that the histograms retain their characteristic shape across different physical intensities, and that each response to the target is subsequently followed by a prominent after-discharge. (D) The average response, from 28 neurons in separate recording sites in area V1 of an anesthetized cynomolgus monkey when visually stimulated for 500 ms with an optimally oriented bar (some recording sites may not have been orientation selective, in which case orientation was arbitrarily chosen). The white bar on the bottom of the histogram represents the time in which the target was on. The onset-response period (pink) lasts for between 45 and 100 ms. The sustained period (orange) varies in duration systematically with target duration and does not appear until the target's transient onset-response peaks. The combined duration of the transient and sustained periods matches the duration of the visual stimulus. The time-out period (green) and after-discharge (blue) directly follow the cessation of the target and each lasted for about 50 ms. The magnitude of the after-discharge seems to grow in size, as the target duration increases. Reprinted from Macknik and Martinez-Conde (2004b), with kind permission from Elsevier.

illusions of invisibility (below) (Macknik and Martinez-Conde, 2004b). Quantifiable computational models of a lateral inhibitory network from the Herzog's laboratory have also shown that lateral inhibition circuits can potentially predict novel Gestalt-level masking behavior (Herzog and Fahle, 2002; Herzog et al., 2003a, b). The Herzog's computational model, based on a Wilson–Cowan algorithm, may behave so as to match our results (not yet tested).

Predictions of the lateral inhibitory model

The SWI illusion

The SWI illusion was the first perceptual prediction of the model. This illusion combined forward and backward masking together in a cyclic version of visual masking, thus suppressing all of the transient responses associated with each flicker of the target. Without the mask, the target is a highly salient flickering bar, but with the mask present, the target becomes virtually invisible (Macknik and Livingstone, 1998). The Enns and McGraw groups quantified the psychophysics of the SWI illusion (Enns, 2002; McKeefry et al., 2005). To the best of our knowledge, this is the first illusion to have been predicted from electrophysiological data, rather than the other way around (Fig. 18).

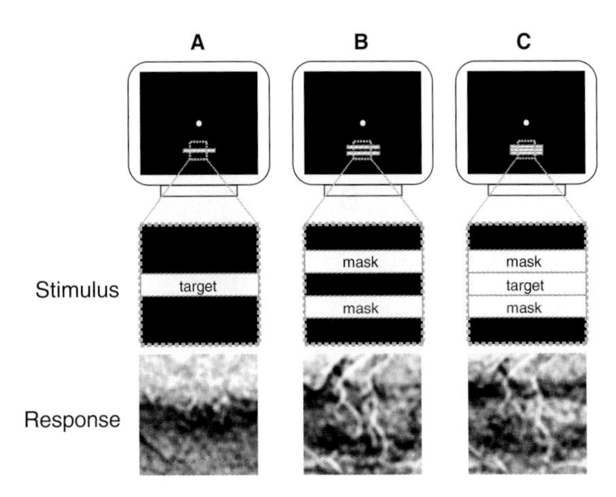

Fig. 18. Optical imaging of the SWI illusion. (A) A visual target with a width of 0.12°, and the correlated optical image. (B) Two masking stimuli presented without a target, and the correlated response. (C) The SWI stimulus (note: the target and the masks were never presented on the display at the same time) and the correlated response. Notice that the image of the target in (C) is now missing, compared to the response seen in (A), despite that both targets were displayed in an identical manner. Modified from Macknik and Haglund (1999), with kind permission from the National Academy of Sciences of the United States of America.

The temporal fusion illusion

Temporal fusion is a second illusion predicted from the model. In this illusion, two short (i.e., 50 ms) duration targets fuse perceptually into one single, long-duration target due to the presentation of a spatially and temporally abutting mask (100 ms long) during the ISI (i.e., inter-target interval) (Fig. 19). Without the intervening mask, two 50-ms targets, separated by an interval of 100 ms, are easily perceived as two sequential flashes. However, when the nonoverlapping mask (nonoverlapping either spatially or temporally) is presented during the inter-target interval, the onset-response from the mask suppresses (through backward masking) the after-discharge from the first flash of the target, while the after-discharge from the mask suppresses (through forward masking) the onset-response to the second flash of the target. Therefore, no transient responses remain to indicate that the first flash turned off, and the second flash turned on, rendering an illusory percept of a single, long-duration target. This illusion was originally suggested as a prediction of our model by Prof. David McCormick of Yale Medical School, and we reported this illusion at the SFN meeting in 2000 (Macknik et al., 2000b). The temporal fusion illusion verifies the lateral inhibitory model's predictions, but it has not yet been quantified psychophysically or physiologically.

Flicker fusion

A third consequence of the model is the well-known flicker fusion illusion. Flicker fusion (sometimes called "persistence of vision") is essential to everyday vision in the modern world (especially when viewing motion pictures, computer displays, TVs, working under artificial lighting, etc.). In 1824, Peter Mark Roget (who also wrote the famous Thesaurus) first presented the concept of "persistence of vision" to the Royal Society of London, as the ability of the retina to retain an image of an object for 1/20–1/5 s after its removal

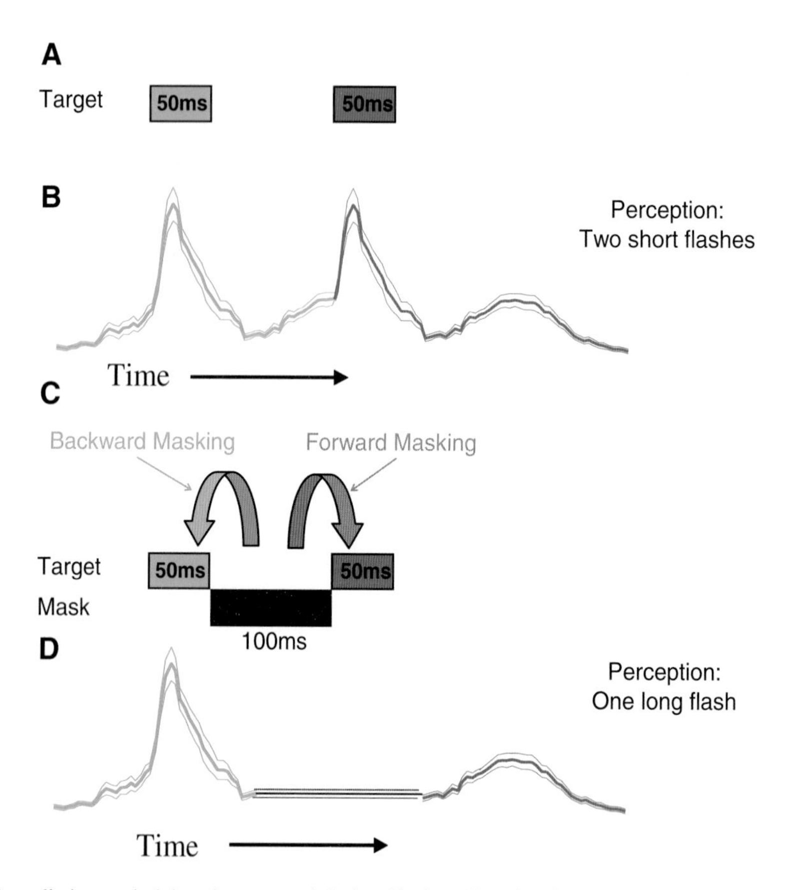

Fig. 19. Physiological prediction underlying the temporal fusion illusion. (A) Timeline depicting two sequential flashes of the same target. (B) The predicted neural responses to the stimuli in (A). The perception is of two flashes. (C) Timeline of the temporal fusion illusion. (D) The predicted neural responses to the stimuli in (C). The after-discharge to the first target and the onset-response to the second target are obliterated by the presentation of a 100-m-long mask in between both targets. The perception is of one long flash.

from our field of vision (Roget, 1825). A second principle, the "phi phenomenon" or stroboscopic effect (the basis for the famous Gestalt School of Psychology), is closely related to flicker fusion. It was first studied by Max Wertheimer (the founder of Gestalt Psychology) and Hugo Munsterberg between 1912 and 1916, who found that subjects can perceptually bridge the temporal gap in between two consecutive displays, allowing them to perceive a series of static images in a continuous movement (Munsterberg, 1916; Wertheimer, 1925).

Several excellent studies of the physiology of flicker fusion have been published (De Valois et al., 1974; Merigan, 1980; Merigan et al., 1985; Merigan and Maunsell, 1990; Gur and Snodderly, 1997). However, these studies have focused on understanding or exploiting the difference between the critical flicker fusion frequencies in the parvocellular vs. magnocellular retino-geniculo-cortical pathways (in order to understand the function of those pathways to perception), rather than on the physiology underlying the circuit that fundamentally causes flicker fusion (whatever the pathway).

It follows from our model that for two brief-duration targets presented in close succession, the after-discharge from the first target may interfere (i.e., abolish or diminish) with the onset-response from the second target. Thus the transient responses from a series of flashes would be suppressed, thereby diminishing the salience of the flashes and rendering them less visible. The flashes would therefore be more difficult to differentiate and appear to be fused.

We studied the effect of inhibition at the termination of the stimulus by flashing a target twice, with varied intervals between the first and the second flash (see Fig. 20). We found that with short ISIs, both the after-discharge of the first flash and the onset-response of the second flash were inhibited. This suggested that when flicker fusion occurs perceptually, it may be due to the lack of robust firing to the flickering stimulus (which, in a sense, forwardly and backwardly masks itself). The duration of the inhibitory effect after the first flash coincided in time and had the same duration as the time-out period of the first flash. Fig. 19 shows that suppression of the onset-response from the second flash only occurs with ISIs of 30 ms or shorter (equivalent to 33 Hz periodic). If we separate the flashes by more than 30 ms, the after-discharge to the first flash and the onset-response to the second flash begin to recover (i.e., equivalent to <3 Hz flicker). These intervals roughly coincide with the minimum flicker fusion threshold in humans for 100% contrast stimuli in the fovea (as we used in anesthetized monkeys in these studies) (Fukuda, 1979; Lennie et al., 1993; Di Lollo and Bischof, 1995; Gorea et al., 2000).

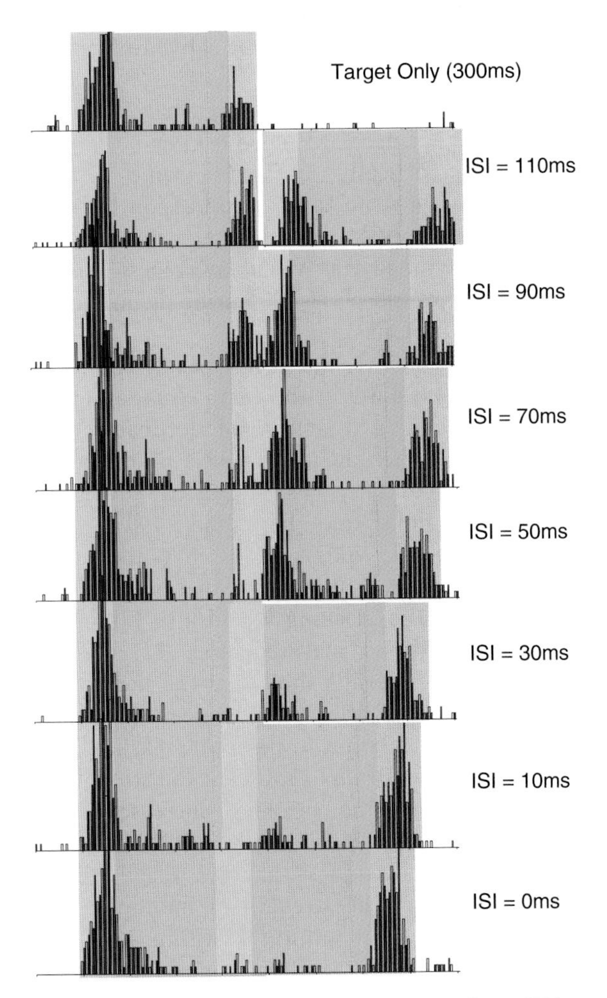

Fig. 20. Multiunit recording from upper layers of area V1 in a rhesus monkey. The receptive field was foveal, 0.1° square, and well oriented. The transparently shaded parts of the histograms represent the various phases of the cells' response to the second flash. The different phases of response are color coded as in Fig. 17D. Notice that the second-flash response is rescued as the inter-stimulus interval (ISI), or the interval between the end of the first-pulse and the beginning of the second-pulse, is increased beyond 30 ms.

In answer to Question ♯4 (What is the structure of the circuits that process visibility?), our experiments suggest that lateral inhibitory networks play a role in suppressing the circuits that maintain visibility. While the nature of the circuits that directly maintain awareness is yet unknown, the knowledge that those circuits are suppressed by lateral inhibition may help us to identify them in the future.

Discussion

This chapter describes our results to date in answer to the following four questions:

(1) What stimulus parameters are important to visibility?

   (A) Our results show that spatiotemporal edges are critical to target visibility. When we suppressed spatiotemporal edges, targets became less visible or invisible. Recent research suggests that spatiotemporal corners may be even more important than edges (Troncoso et al., 2005).

   (B) Masks produce their strongest suppressive effect at their spatiotemporal edges as well, suggesting that inhibition produced by a mask is a function of the same excitatory activity that makes stimuli visible.

(2) What types of neural activity best maintain visibility (transient vs. sustained firing, rate codes, bursts of spikes, etc — i.e., the neural code)?
   (A) Our results show that transient bursts of spikes within the early visual system are critical to maintaining visibility. These transient bursts are driven by the temporal edges (onset and termination) of targets within receptive fields that lie at the spatial edges of the targets. Other research in the field suggests that transient bursts also correlate with the visibility evoked by microsaccades during fixation (Martinez-Conde et al., 2000, 2002, 2004, 2006). Forward masking is the inhibition of the onset-response of the target, and backward masking is the inhibition of the after-discharge of the target. The after-discharge is not a function of feedback from higher levels in the visual system.
   (B) Hartline defined lateral inhibition as the neural process in which the response to one stimulus is suppressed by a second stimulus that has been presented laterally (on the retinotopic map). Therefore, by definition, most of the visual masking described here is a type of lateral inhibition. We found that the lateral inhibition from the mask is created at the spatial edges of the mask and not in its interior. Mask suppression is, moreover, strongest at the mask's onset and termination. Therefore, we conclude that the mask's suppressive effect is a function of the excitatory responses generated by the mask's spatiotemporal edges. We and others have found that classical forward masking (in which the mask's entire lifetime occurs before the target is displayed) is a function of the mask's termination suppressing the target's onset-response. We also showed that classical backward masking (in which the mask is displayed only after the target is extinguished) is essentially forward masking of the target's after-discharge. Other non-classical timing effects in masking, such as common-onset masking (Bischof and Di Lollo, 1995), temporal fusion, and flicker fusion, can be explained by the inhibitory interaction of the mask's and target's onset-responses and after-discharges.

(3) What brain areas must be active to maintain visual awareness?
   (A) Our results suggest that monoptic forms of visual masking begin subcortically, but these same neurons show no target suppression from dichoptic masks. If these neurons maintained our visual awareness, targets would be visible during dichoptic visual masking.
   (B) Our results show that monoptic masking in the early visual system is not a function of feedback from higher dichoptic levels of the visual hierarchy, as suggested previously (Enns and Di Lollo, 1997, 2000; Enns, 2002). If this hypothesis was correct, all neurons that exhibit monoptic masking would also exhibit dichoptic masking (Fig. 12). Instead our results show that the earliest levels of the visual system exhibit no correlation to dichoptic masking.
   (C) Our results show that the strength of monoptic masking increases iteratively as a function of the stage of processing within the visual hierarchy. The strength of monoptic suppression saturates within area V1. Similarly, dichoptic masking first becomes possible within V1 because it is the first binocular processing level within the ascending visual system. However, the dichoptic suppression is very weak within area V1, and we show that it rises in strength iteratively at higher levels within the visual hierarchy. This iterative buildup of

inhibition has critical implications for the formation of binocular vision of all types, including stereopsis, binocular rivalry, and dichoptic masking.

(D) In agreement with the studies by Schiller's group, we found that monoptic and dichoptic masking are equivalent in strength psychophysically. Our results further indicate that monoptic and dichoptic suppression of targets during masking only become equivalent in strength within brain areas downstream of area V2. We cannot rule out V3 as maintaining our awareness. However, because the iterative buildup of the strength of inhibition continues to increase in V4, it may be that V3 is not critical to awareness (or else we might expect that dichoptic masking targets would be more visible than in monoptic masking). Because we have not yet identified visual areas downstream of area V4 in our experiments, it remains possible that the strength of dichoptic masking continues to rise in visual areas downstream of V4. Our results do not support the hypothesis that awareness of visibility during masking is a function of coupling between V1 and the fusiform gyrus (Haynes et al., 2005).

(E) Our data show that circuits within the occipital lobe are sufficient to maintain awareness, because when we modulated the visibility of targets during masking stimulation, only occipital circuits modulated their activity as well. We showed that this finding was a verifiable localization of target-correlated awareness responses and not a null result (for instance, it may have been possible that nonoccipital responses existed, but that we missed them). We proved this by discovering that blood flow responses to nonillusory stimuli were weaker in the occipital lobe than in the target-responsive voxels outside the occipital lobe. Thus, if anything, we *underestimated* the strength of our result: masking within the occipital lobe is probably even stronger than we thought, as compared to nonoccipital areas. My laboratory will conduct future single-unit recordings to test this prediction.

(F) In combination, these results suggest that there is a lower bound within the visual hierarchy for brain areas that can maintain visual awareness (somewhere downstream of V2). Moreover, we discovered that there was also an upper bound within the visual hierarchy for brain areas that may maintain our visual awareness of simple unattended targets (somewhere within the occipital lobes). We have thus narrowed the possible areas of the brain for localizing the circuits responsible for awareness of simple unattended targets to a region between V2 and the edge of the occipital lobe.

(4) What specific neural circuits within the relevant brain areas maintain visibility?

  (A) Several perceptual predictions can be made if lateral inhibition is responsible for visual masking.

    (a) We should see a clear falloff of masking as we move the masks progressively farther from the targets. We tested this prediction by recording from neurons during the presentation of the SWI illusion with varying mask distances (Fig. 8). We found that the strength of target suppression fell off with distance.

    (b) Lateral inhibition from the mask should be produced by its spatial edges and not by its interior, since that is where excitatory signals are strongest (Fig. 6). We tested this perceptually in humans by quantifying the strength of masking during the presentation of spatially overlapping masks that grew progressively wider. As the masks grew in size, their edges increased their distance from the target (Fig. 3). The results indicated that as the overlapping masks grew wider, the strength of lateral inhibition of the targets fell off. That is, a smaller mask that just barely covers the target is the mask that most completely inhibits the target. In this condition, the mask's edges are closest to the target, so we concluded that it is the edges of the mask that produced the inhibitory effect in visual masking.

(c) Lateral inhibition from the mask is produced by its temporal edges and not its midlife period, since that is where excitatory signals are also concentrated (i.e., in the onset-response and after-discharge; Figs. 9 and 17). We tested this idea perceptually in humans, by quantifying the strength of masking during the presentation of temporally overlapping masks of increasingly longer durations. As the masks grew in duration, their temporal edges moved in time. The results indicated that irrespective of the duration of the mask, the target was always inhibited most strongly by the mask's onset and termination (Fig. 5). Moreover, as the duration of the mask increased, the strength of the suppression from the mask's termination also increased.

(d) To test the idea that lateral inhibition might play a role in forming the various phases of excitatory response, we stimulated cells with targets of varied durations (Figs. 9 and 17) (see also Macknik and Martinez-Conde, 2004b). Short-duration targets (100 ms or less) had a transient period, a time-out period, and an after-discharge longer durations targets ($>$100 ms) also exhibited a transient period, but then in addition exhibited a period of sustained firing, before the timeout period (when the target was extinguished). Very long targets differed from shorter durations only in that the sustained firing period was increased in length, and the after-discharges grew in magnitude. We supposed that the initial transient response would correspond to excitation from the stimulus, and the subsequent sustained phase of response was due to self-inhibitory delayed feedback through the lateral inhibitory circuit (as described in Fig. 17). The timeout period might then be explained by the continuation of delayed self-inhibition subsequent to the termination of the stimulus. The after-discharge would then correspond to the disinhibitory rebound of the cell after the timeout period.

(e) All of these phases of response can be observed extracellularly (and were measured intracellularly by Ferster, although he did not discuss them directly; Ferster, 1992), except for the timeout period. While Ferster's work clearly shows a period of membrane hyperpolarization after the target terminated, we wanted to establish whether or not this period of inhibition reduced the cell's ability to fire to a new stimulus. So we performed visual paired-pulse experiments in which we presented a target twice (with no change in position) with varied ISIs (Fig. 20). We found that as the ISI increased, the magnitude of the transient period of the second pulse also increased, and it eventually returned to normal when the ISI exceeded 100 ms. We could thus indirectly determine the timeout period and verify that, during this interval, the cell was relatively refractory to visual stimuli.

Repercussions for other models of masking

Models reliant on the SOA law

Our results show that models of visual masking on the basis of defunct "SOA Law," such as complicated delay-line circuits in "dual-channel" models of visual masking (Breitmeyer and Ogmen, 2000) and other models (Kahneman, 1968; Weisstein, 1968; Matin et al., 1972; Growney et al., 1975; Matin, 1975; Weisstein et al., 1975; Breitmeyer and Ganz, 1976; Enns and Di Lollo, 1997; Francis, 1997; Purushothaman et al., 2000), do not follow the timing parameters found in masking, such as the visual masking strength dependence on the STA of the target and the mask. Our results also show that visual masking effects are extant subcortically and are not caused by feedback from the cortex (Macknik et al., 2000a; Macknik and Martinez-Conde, 2004a; Tse et al., 2005), so models of visual masking that require cortical circuits or feedback mechanisms can be ruled out (Thompson and Schall, 1999; Breitmeyer and Ogmen, 2000; Lamme et al., 2000; Di Lollo et al., 2002; Enns, 2002; Haynes et al., 2005).

The object substitution model

Moreover, even seemingly high-level masking effects, such as object substitution (Enns and Di Lollo, 1997), can potentially be explained by lateral inhibition. Object substitution is an effect in which a target object is suppressed by a mask of similar shape, even though the mask does not abut the target spatially (as it is necessary in other types of masking discussed here). This effect can be explained with a feedforward lateral inhibitory model by considering that, in areas of the visual system that process object shape, the space of the visual area is no longer retinotopic, but is instead mapped for objects (Fig. 21). Lateral inhibitory networks in such an area would be expected to present a similar behavior as in lower retinotopic areas, but this activity would have effects across object-shape space, rather than retinotopic space. This hypothesis may explain why, in object substitution masking, the mask must be similar in shape to the target (which would make them near each other in object space), and why the target and the mask need not be near each other retinotopically. These ideas further suggest that visual masking can be explained with a simple feedforward (rather than feedback) lateral inhibition circuit. There is no extant evidence (yet) to invalidate the lateral inhibition based "spatiotemporal edges" theory we propose, and no other theory to date can explain the fact that visual masking

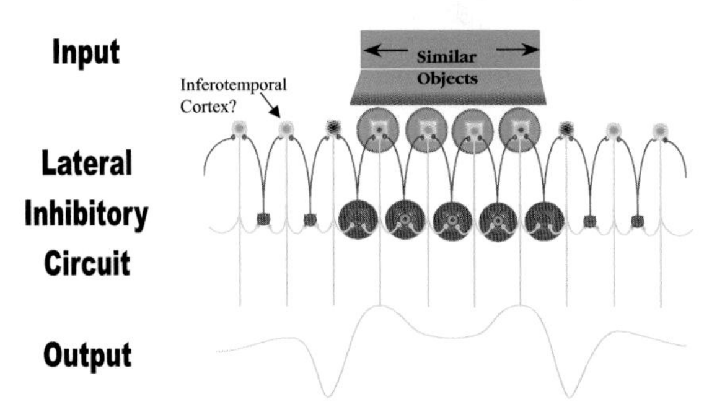

Fig. 21. A representation of the lateral inhibition model interactions within object space. The four excited neurons (highlighted in green) in the center of the upper row receive excitatory input from a visual stimulus (for instance an object or group of objects having similar shapes). This excitation is transmitted laterally in the form of inhibition, resulting in "edge enhancement" across object space that is equivalent to the retinotopic edge enhancement in earlier levels of the visual pathway (Fig. 17A). These effects could lead to object-oriented visual masking effects. Therefore, lateral inhibition may be the most parsimonious explanation of object substitution masking.

occurs subcortically and cortically (in separate circuits as I will discuss in the next section), that it occurs as a function of STA for the stimulus set used in our experiments, and that it has differential response mechanisms for monoptic and dichoptic forms of masking. Our lateral inhibition model of masking has served as the basis for explaining both forward and backward masking, and it moreover has the ability to predict novel illusions such as the SWI (Fig. 7) and temporal fusion (Fig. 19).

Confusion theories of masking

Theories in which masking occurs due to perceptual confusion have the premise that items that are similar to each other in temporal, spatial, or surface attributes are difficult to tell apart. That seems reasonable enough, except that masking is very powerful when the mask and the target differ in contrast and duration, as I discussed in the introduction of this chapter as Key Feature #4 (Boynton, 1961; Boynton and Ikeda, 1965; Erikson, 1966). Nevertheless, while the basic premise of this hypothesis might seem plausible, the more advanced, and most recently proposed, theory of confusion suggests that the reason that a target bar, when flanked by two masking bars, appears to be invisible is that the timing of the target and the mask matches the timing necessary to activate motion systems in the visual pathway, but that the visual system cannot see motion in two directions at once (Kahneman, 1967, 1968; Kahneman et al., 1967; Kahneman and Wolman, 1970). That is, the theory suggests that the visual system actively suppresses the target when it is an element of "impossible motion." Stoper and Banffy have refuted this theory by showing that the timing required to perceive apparent motion differs from the timing required to perceive backward masking (Stoper and Banffy, 1977).

Our results indicate that simple lateral inhibitory circuits, the very bases for receptive field structure, can explain the mysterious timing patterns of visual masking. We have shown that activity correlated to target visibility consists of transient bursts of impulses in response to the onset and termination of the target (its temporal edges) within neurons that have receptive fields located at the spatial edges of the target (Macknik and Livingstone, 1998; Macknik and Haglund, 1999; Macknik et al., 2000a; Macknik and Martinez-Conde, 2004a, b). Spatiotemporal edges of masks also cause onset-responses and after-discharges. Visual masking occurs when the spatiotemporal responses to the mask suppress, through a lateral inhibitory connection, the spatiotemporal responses to the target; the target then has less activity with which to encode its perceptual signal and it is thus degraded, becoming less visible.

Competing lateral inhibition models

Previous studies have proposed lateral inhibitory circuits as the basis of masking (Weisstein, 1968; Bridgeman, 1971; Weisstein et al., 1975; Francis, 1997). However, these models were designed to follow the SOA LAW and did not show responses that correlated with more recently discovered human psychophysical parameters showing the importance of spatiotemporal edges in visibility and invisibility (as in Fig. 4) (Macknik and Livingstone, 1998; Macknik et al., 2000a; Macknik and Martinez-Conde, 2004a, b).

Bridgeman model is based on the idea that visual stimuli drive oscillations (Bridgeman calls this process resonance) in cortical areas, due to the interactions of a distributed laterally inhibitory network (Bridgeman, 1971). The premise of the model is that the resonating properties of the cortical lateral inhibitory network may be crucial to encoding visibility. If one were to inhibit the predicted resonation from the target, one would inhibit the perception of the target. The model predicted that, with a backward masking stimulus, the resonance of the target cell would be impinged upon by the mask cell, if the mask's resonating inhibition was at the correct phase (SOA). Breitmeyer (1984) noted, however, that this model also predicted that the target and the mask, if presented with nonoptimal stimulus onset asynchronies for masking, should cause perceptual beating (called "ringing" by Breitmeyer, 1984). That is, Bridgeman's model predicted that nonsimultaneous stimuli (at the nonoptimal SOA) would beat visually like the sound of two tuning forks of slightly different pitches. If this was true, one should be able to tune the optimal SOA by noting the beat frequency of the target. However, at nonoptimal stimulus onset asynchronies, the percept of the target is quite steady (Breitmeyer, 1984), indicating that Bridgeman's model is not correct.

The Herzog group has recently developed a more plausible competing lateral inhibition theory of masking, as an explanation of a new family of visual masking illusions called "Shine-Through." In Shine-Through, targets are suppressed (or not) as a function of timing and the number of bars in a mask grating (Herzog et al., 2001a, b; Herzog and Koch, 2001; Herzog and Fahle, 2002). The same group has developed a computational lateral inhibitory model of visual masking that predicts the importance of the spatiotemporal edges (Herzog et al., 2003a, b, c).

Other physiological findings with visual masking
Thompson and Schall recorded from single units in the frontal lobes of the awake monkey and concluded that visual masking effects cannot be processed in the early visual system, but are instead processed at level of the frontal eye fields (FEF) (Thompson and Schall, 1999, 2000). They suggested that the neural correlate of visual masking is the "merging" of target and mask responses, rather than the inhibition of target responses. However, as their target was almost 300 times dimmer than their mask, the target and mask responses could have merged because of the differential latencies one would expect from a dim and a bright stimulus (Albrecht and Hamilton, 1982; Gawne et al., 1996). Given this latency confound, combined with the SOAs used (SOAs that were approximately equivalent to the latency difference that would be expected by a 300X luminance difference), the authors could not have differentiated whether the target was inhibited by the mask, or whether the mask's larger response occluded the latency-delayed contemporaneous dim-target response. In our experiments, our stimuli were specifically designed to avoid this confound since the target and the mask were of equal contrast. Furthermore, when the authors used either very long or short SOAs, in which the target and mask responses could be differentiated in time, they found that it was the *mask's* response that was suppressed rather than the *target's* response; this is the opposite of what one expects in visual masking. Finally, the monkey's task was to detect a blue target against a field of white distracter masks, and so it is possible that differential attentional effects on the target and mask responses would suppress the mask but not the target. These types of attentional effects are known to exists in the FEF and other parts of the brain when the primate is trained to direct its attention to particular colored stimuli (i.e., the target) and ignore others (i.e., the mask) (Reynolds et al., 1994, 1999, 2000; Bichot and Schall, 1999; Reynolds and Desimone, 1999). Thus it is possible that their results are further confounded by the effects of selective attention, rather than being a function of visual masking.

The role of cortical and intrathalamic feedback
Cortical and intrathalamic feedback is thought to play a concerted role in gating and modifying activity in the LGN. LGN activity is moreover modulated by level of arousal (Livingstone and Hubel, 1981; McCormick, 1992; Sanchez-Vives and McCormick, 1997). Our results have no bearing on the idea that thalamocortical circuits are gated through feedback to gate ascending sensory information, such as to modulate arousal or to cause other effects unrelated to visibility.

Implications for our understanding of binocular vision
We found that binocular cells in area V1 could receive strong excitatory input from both eyes (thus defining them as binocular), yet an excitatory response from a monocularly presented target was rarely suppressed (and weakly, if so) by the action of a mask presented to the other eye, even though the target responses were robustly suppressed by monoptic masks in these same binocular neurons. This suggested that interocular excitatory inputs to area V1 binocular cells are strong whereas interocular inhibitory inputs to the same cells are weak.

The monoptic masking results presented here (Fig. 11) show that the strength of monoptic masking increases iteratively with each successive stage of processing in the visual system. In fact, the sequential buildup of lateral inhibition at each step in the visual hierarchy that bears its own lateral inhibition circuit seems to be a mathematical certainty, unless some heretofore undiscovered process that somehow undoes the effects of previous lateral inhibition circuits exists in the brain.

The result that we present here (Fig. 11), showing that dichoptic inhibition is weak in area V1, may be a reflection of such a general principle because these area V1 binocular neurons represent the first stage where dichoptic inhibition could exist in the ascending visual system. If this is true, we would expect that subsequent binocular neurons in the visual hierarchy should show iteratively stronger dichoptic masking suppression effects. These results are supported by psychophysical findings suggesting that stereo matching precedes dichoptic masking (Petrov and Mckee, 2006).

Implications for binocular rivalry as a tool to study visual awareness

Binocular rivalry has been used as a tool to assess the neuronal correlates of visibility, but has generated controversy because of conflicting results. Some fMRI studies on humans report that BOLD activity in V1 correlates with visual awareness of binocular rivalry percepts (Polonsky et al., 2000; Lee et al., 2005). In contrast, other human fMRI studies (Moutoussis et al., 2005), and single-unit recording studies in primates (Leopold and Logothetis, 1996; Sheinberg and Logothetis, 1997), suggest that activity in area V1 does not correlate with visual awareness. One possible reason for this discrepancy is that none of the binocular rivalry studies have determined that the visual areas tested contained the interocular suppression circuits necessary to mediate binocular rivalry. That is, when binocular rivalry is not found in a given area, there is no extant method to control for the possibility that the circuits being studied simply do not have the components necessary to cause interocular suppression, and thus binocular rivalry. A more common conclusion is that the area cannot maintain visual awareness. However, if interocular suppression circuits do not exist in that area, then this conclusion would be as inappropriate as concluding that V1 does not maintain awareness because it does not differentiate between houses and faces. Even when an indicator of neural activity in a given area is found to correlate with perceptual state during binocular rivalry, no existing study has determined whether the interocular suppression observed in that area is weak or strong. Until a control stimulus can be developed to indicate the strength of interocular suppression in any given area, unambiguous interpretation of the neural correlates of perceptual state during binocular rivalry will not be possible, no matter how high in the visual hierarchy. Thus, using binocular rivalry stimuli alone, it is currently not possible to localize awareness circuits within the visual hierarchy.

In our fMRI study, we measured the BOLD activity underlying visual awareness during monoptic and dichoptic masking in humans. With masking, one can compare the perceptual and neural strengths of both monoptic and dichoptic target suppression. This comparison supplies a means to quantify the strength of target suppression, and also to determine whether a weak suppression of the target may be due to the weakness of interocu lar suppression in the circuit tested, or to the lack of masking components within the circuit.

Implications for occipital vs. non-occipital models of visual awareness

We found differing ratios of monoptic vs. dichoptic masking in occipital areas downstream of V2 (Fig. 14B), which may indicate that these areas do not maintain awareness of simple unattended targets, because dichoptic and monoptic masking are equivalent in strength psychophysically (Schiller, 1965; Macknik and Martinez-Conde, 2004a). It is thus possible that the activation strength of monoptic vs. dichoptic masking only fully equilibrates downstream of the retinotopic areas studied here, albeit somewhere within the occipital lobe. Nevertheless, the current data indicate that areas V1 and V2 fail to correlate with the perception of dichoptic masking, even though they contain circuits capable of processing visual masking (i.e., monoptically). These areas can, therefore, be ruled out as sufficient to maintain visual awareness, whether by themselves or in concert with earlier subcortical areas. Our results thus suggest that visual masking of simple targets can be explained through occipital models, rather than with models of masking that require feedback or nonoccipital circuits (Enns and Di Lollo, 1997; Thompson and Schall, 1999, 2000; Enns, 2002).

Care should be taken not to generalize these results to claims about the neural correlates of awareness of objects more complex than the simple targets used here. For instance, it is possible that complex visual stimuli such as faces and hands (known to be physiologically processed outside the occipital lobe; Desimone and Gross, 1979; Desimone, 1991) require activity in visual areas downstream of the occipital lobe to maintain awareness. Similarly, circuits that maintain the awareness of other types of visual processes, such as motion perception, may lie outside the occipital cortex (Williams et al., 2003). For example, Dehaene et al. (2001) found neural correlates of word-priming masking outside of the occipital lobe. It is also possible that attended stimuli, including simple targets, may invoke extraoccipital activation in a task-dependent manner (Naccache et al., 2002).

Attentional feedback and the role of area V1 in human visual awareness

Haynes and Rees (Haynes et al., 2005; Haynes and Rees, 2005) have recently published two articles showing the BOLD correlates of monoptic masking in the cortex. Our study is in agreement with one of these studies, which concluded that human area V1 alone cannot maintain awareness of stimulus orientation (Haynes and Rees, 2005). However, our results do not agree with their second study (Haynes et al., 2005), which suggested that target visibility is derived by the coupling of area V1 BOLD activity with fusiform gyrus BOLD activity. The reason for this discrepancy may lie in a potential confound of this second study: the V1 activation found by Haynes and Rees may not indicate target visibility, but rather top-down attentional feedback from the fusiform gyrus. Moreover, the correlation matrix analysis conducted in this study used only the data from the attended conditions, thus confounding attention and awareness. This possibility is suggested by their use of a behavioral task that demanded active attention to the target during scanning, which is known to cause increased BOLD activity in human V1 (Brefczynski and DeYoe, 1999). Several other fMRI studies concluding that V1 maintains awareness may also be explained by uncontrolled top-down attentional feedback (Polonsky et al., 2000; Ress et al., 2000; Lee et al., 2005). Our task, on the contrary, demanded attention to the fixation point only, precluding effects of attention to the target. Lee et al.(2005) found that peripheral binocularly rivalrous traveling waves could be tracked in V1 using fMRI and concluded that V1 must thus contribute to the maintenance of awareness. However, they also found that when the subject was attending to a different task at the fixation point, the BOLD signal from the traveling waves decreased. This suggests that the traveling wave BOLD signal seen by the authors was due to attentional feedback, and it was thus equivocal concerning a role for V1 in awareness.

Implications for theories of consciousness

The present results may lend support to some modular theories of consciousness that are sustainable with intermediate level processes (Jackendoff, 1987; Zeki and Bartels, 1999; Crick and Koch, 2000; Zeki, 2001). Our data fail to support global theories of awareness or theories that require frontal lobe activation (Damasio, 1989; Edelman, 1990; Freeman, 1991; Llinas et al., 1998; Crick and Koch, 2000; Dehaene et al., 2001; Kanwisher, 2001; Rees, 2001; Baars, 2005).

Abbreviations

BOLD	blood-oxygen level-dependent signal
FEF	frontal eye fields
fMRI	functional magnetic resonance imaging
ISI	inter-stimulus interval
LGN	lateral geniculate nucleus of the thalamus
LOC	lateral occipital cortex
SOA	stimulus onset asynchrony
STA	stimulus termination asynchrony
SWI	standing wave of invisibility
V1 (V2, V3 …)	visual cortex numbered by its level in the visual hierarchy

Acknowledgments

I gratefully thank all of my collaborators for their contributions to these studies. I am especially grateful to Susana Martinez-Conde for her helpful advice on the manuscript. These studies were funded by the NIH, Dartmouth College, and the Barrow Neurological Institute.

References

Adrian, E.D. and Matthews, R. (1927) The action of light on the eye. Part I. The discharge of impulses in the optic nerve and its relation to the electric changes in the retina. J. Physiol., 63: 378–414.

Albrecht, D.G. and Hamilton, D.B. (1982) Striate cortex of monkey and cat: contrast response function. J. Neurophysiol., 48: 217–237.

Alpern, M. (1953) Metacontrast. J. Opt. Soc. Am., 43: 648–657.

Anbar, S. and Anbar, D. (1982) Visual masking: a unified approach. Perception, 11: 427–439.

Baars, B.J. (2005) Global workspace theory of consciousness: toward a cognitive neuroscience of human experience. Prog. Brain Res., 150: 45–53.

Battersby, W.S. and Wagman, I.H. (1962) Neural limitations of visual excitability. IV. Spatial determinants of retrochaismal interaction. Am. J. Physiol., 203: 359–365.

Bichot, N.P. and Schall, J.D. (1999) Saccade target selection in macaque during feature and conjunction visual search. Vis. Neurosci., 16: 81–89.

Birch, D.G. and Anderson, J.L. (1992) Standardized full-field electroretinography. Normal values and their variation with age. Arch. Ophthalmol., 110: 1571–1576.

Bischof, W.F. and Di Lollo, V. (1995) Motion and metacontrast with simultaneous onset of stimuli. J. Opt. Soc. Am. A-Opt. Image Sci., 12: 1623–1636.

Bloch, A.M. (1885) Experience sur la vision. Comp. Rendus Seances Soc. Biol., Paris, 37: 493–495.

Bouguer, M.P. (1961) Traite d'Optique sur la gradation de la lumiere. University of Toronto Press, Toronto, Canada.

Boynton, R.M. (1961) Some temporal factors in vision. In: Rosenblith, W.A. (Ed.), Sensory Communicaton. MIT Press, Cambridge, pp. 739–757.

Boynton, R.M. and Ikeda, M. (1965) Negative flashes, positive flashes, and flicker examined by increment threshold technique. J. Opt. Soc. Am., 55: 560–566.

Brefczynski, J.A. and DeYoe, E.A. (1999) A physiological correlate of the 'spotlight' of visual attention. Nat. Neurosci., 2: 370–374.

Breitmeyer, B. (1984) Visual Masking: An integrative approach. Clarendon Press, Oxford.

Breitmeyer, B.G. (1978) Metacontrast masking as a function of mask energy. Bull. Psychonomic Soc., 12: 50–52.

Breitmeyer, B.G. and Ganz, L. (1976) Implications of sustained and transient channels for theories of visual pattern masking, saccadic suppression, and information processing. Psychol. Rev., 83: 1–36.

Breitmeyer, B.G. and Ogmen, H. (2000) Recent models and findings in visual backward masking: a comparison, review, and update. Perception Psychophys., 62: 1572–1595.

Bridgeman, B. (1971) Metacontrast and lateral inhibition. Psychol. Rev., 78: 528–539.

Bridgeman, B. (1975) Correlates of metacontrast in single cells of the cat visual system. Vis. Res., 15: 91–99.

Bridgeman, B. (1980) Temporal response characteristics of cells in monkey striate cortex measured with metacontrast masking and brightness discrimination. Brain Res., 196: 347–364.

Broca, A. and Sulzer, D. (1902) La Sensation Lumineuse en Fonction Du Temps. J. Physiol. (Paris), 4: 632–640.

Brown, E.L. and Deffenbacher, K. (1979) Perception and the Senses. Oxford University Press, New York.

Chevreul, M.E. (1987) The Principles of Harmony and Contrast of Colors: and Their Applications to the Arts. Schiffer West Chester, Pennsylvania.

Chung, N.H., Kim, S.H. and Kwak, M.S. (1993) The electroretinogram sensitivity in patients with diabetes. Korean J. Ophthalmol., 7: 43–47.

Cleland, B.G., Dubin, M.W. and Levick, W.R. (1971) Simultaneous recording of input and output of lateral geniculate neurones. Nat. — New Biol., 231: 191–192.

Coppola, D. and Purves, D. (1996) The extraordinarily rapid disappearance of entopic images. Proc. Natl. Acad. Sci. USA, 93: 8001–8004.

Crawford, B.H. (1940) The effect of field size and pattern on the change of visual sensitivity with time. Proc. R. Soc., Lnd., 129B: 94–106.

Crawford, B.H. (1947) Visual adaptation in relation to brief conditioning stimuli. Proc. Royal Soc. Lond. Ser. B, 134B: 283–302.

Crick, F. and Koch, C. (1990) Some reflections on visual awareness. Cold Spring Harb. Symp. Quant. Biol., 55: 953–962.

Crick, F. and Koch, C. (2000) The unconscious homunculus. In: Metzinger, T. (Ed.), The Neuronal Correlates of Consciousness. MIT Press, Cambridge, MA, pp. 103–110.

Dagnelie, G. and Massof, R.W. (1993) Foveal cone involvement in retinitis pigmentosa progression assessed through flash-on-flash parameters. Investig. Ophthalmol. Vis. Sci., 34: 231–242.

Damasio, A.R. (1989) Time-locked multiregional retroactivation: a systems-level proposal for the neural substrates of recall and recognition. Cognition, 33: 25–62.

Dawson, W.W. and Harrison, J.M. (1973) Bloch's law for brief flashes of large angular subtense. Perceptual Motor Skills, 36: 1055–1061.

Day, E.C. (1915) Photoelectric currents in the eye of the fish. Am. J. Physiol., 38: 369–398.

Dehaene, S., Naccache, L., Cohen, L., Bihan, D.L., Mangin, J.F., Poline, J.B. and Riviere, D. (2001) Cerebral mechanisms of word masking and unconscious repetition priming. Nat. Neurosci., 4: 752–758.

Desimone, R. (1991) Face-selective cells in the temporal cortex of monkeys. J. Cog. Neurosci., 3: 1–8.

Desimone, R. and Gross, C.G. (1979) Visual areas in the temporal cortex of the macaque. Brain Res., 178: 363–380.

De Valois, R.L., Morgan, H.C., Polson, M.C., Mead, W.R. and Hull, E.M. (1974) Psychophysical studies of monkey vision. I. Macaque luminosity and color vision tests. Vis. Res., 14: 53–67.

De Weerd, P., Gattass, R., Desimone, R. and Ungerleider, L.G. (1995) Responses of cells in monkey visual cortex during perceptual filling-in of an artificial scotoma. Nature, 377: 731–734.

Di Lollo, V. and Bischof, W.F. (1995) Inverse-intensity effect in duration of visible persistence. Psychol. Bull., 118: 223–237.

Di Lollo, V., Enns, J.T. and Rensink, R.A. (2002) Object substitution without reentry? J. Exp. Psychol. Gen., 131: 594–596.

Duysens, J., Gulyas, B. and Maes, H. (1991) Temporal integration in cat visual cortex: a test of Bloch's law. Vis. Res., 31: 1517–1528.

Edelman, G.M. (1990) The Remembered Present: A Biological Theory of Consciousness. Basic Books, New York, NY.

Enns, J.T. (2002) Visual binding in the standing wave illusion. Psychon. Bull. Rev., 9: 489–496.

Enns, J.T. and Di Lollo, V. (1997) Object substitution — a new form of masking in unattended visual locations. Psychol. Sci., 8: 135–139.

Enns, J.T. and Di Lollo, V. (2000) What's new in visual masking? Trends Cogn. Sci., 4: 345–352.

Enroth-Cugell, C., Lennie, P. and Shapley, R.M. (1975) Surround contribution to light adaptation in cat retinal ganglion cells. J. Physiol. (Lond.), 247: 579–588.

Enroth-Cugell, C. and Shapley, R.M. (1972) Cat retinal ganglion cells: correlation between size of receptive field centre and level of field adaptation. J. Physiol. (Lond.), 225: 58P–59P.

Enroth-Cugell, C. and Shapley, R.M. (1973a) Adaptation and dynamics of cat retinal ganglion cells. J. Physiol. (Lond.), 233: 271–309.

Enroth-Cugell, C. and Shapley, R.M. (1973b) Flux, not retinal illumination, is what cat retinal ganglion cells really care about. J. Physiol. (Lond.), 233: 311–326.

Erikson, C. (1966) Temporal luminance summation effects in forward and backward masking. Perception Psychophys., 1: 87–92.

Exner, S. (1868) Uber die zu einer Gesichrswahrnehmung nothige Zeit. In: Wiener Sitzungbericht der mathematisch — naturwissenschaftlichen Classe der kaiser-lichen Akademie der Wissenschaften, pp. 601–632.

Farah, M.J. and Feinberg, T.E. (1997) Consciousness of perception after brain damage. Semin. Neurol., 17: 145–152.

Fechner, G.T. (1966) Elements of Psychophysics. Holt, Rinehartd, & Winston, New York.

Ferster, D. (1986) Orientation selectivity of synaptic potentials in neurons of cat primary visual cortex. J. Neurosci., 6: 1284–1301.

Ferster, D. (1992) The synaptic inputs to simple cells of the cat visual cortex. Prog. Brain Res., 90: 423–441.

Francis, G. (1995) Neural network dynamics of inhibition: metacontrast masking. In: Gasser, M. (Ed.), Online Proceedings of the 1996 Midwest Artificial Intelligence and Cognitive Science Conference.

Francis, G. (1997) Cortical dynamics of lateral inhibition: metacontrast masking. Psychol. Rev., 104: 572–594.

Francis, G. (1998) Neural network dynamics of cortical inhibition — metacontrast masking. Inf. Sci., 107: 287–296.

Francis, G., Rothmayer, M. and Hermens, F. (2004) Analysis and test of laws for backward (metacontrast) masking. Spatial Vis., 17: 163–185.

Freeman, W.J. (1991) The physiology of perception. Sci. Am., 264: 78–85.

Fukuda, T. (1979) Relation between flicker fusion threshold and retinal positions. Percept. Mot. Skills, 49: 3–17.

Gawne, T.J., Kjaer, T.W., Hertz, J.A. and Richmond, B.J. (1996) Adjacent visual cortical complex cells share about 20-percent of their stimulus-related information. Cereb. Cortex, 6: 482–489.

Golomb, D., Kleinfeld, D., Reid, R.C., Shapley, R.M. and Shraiman, B.I. (1994) On temporal codes and the spatiotemporal response of neurons in the lateral geniculate nucleus. J. Neurophysiol., 72: 2990–3003.

Gorea, A. and Tyler, C.W. (1986) New look at Bloch's law for contrast. J. Opt. Soc. Am. A-Opt. Image Sci., 3: 52–61.

Gorea, A., Wardak, C. and Lorenzi, C. (2000) Visual sensitivity to temporal modulations of temporal noise. Vis. Res., 40: 3817–3822.

Growney, R., Weisstein, N. and Cox, S.I. (1975) Letter: measurement of metacontrast. J. Opt. Soc. Am., 65: 1379–1381.

Gur, M. and Snodderly, D.M. (1997) A dissociation between brain activity and perception: chromatically opponent cortical neurons signal chromatic flicker that is not perceived. Vis. Res., 37: 377–382.

Harris, J.M. and Willis, A. (2001) A binocular site for contrast-modulated masking. Vis. Res., 41: 873–881.

Hartline, H.K. (1949) Inhibition of activity of visual receptors by illuminating nearby retinal areas in the Limulus eye. Fed. Proc., 8: 69.

Haynes, J.D., Driver, J. and Rees, G. (2005) Visibility reflects dynamic changes of effective connectivity between V1 and fusiform cortex. Neuron, 46: 811–821.

Haynes, J.D. and Rees, G. (2005) Predicting the orientation of invisible stimuli from activity in human primary visual cortex. Nat. Neurosci., 8: 686–691.

He, S., Cavanagh, P. and Intriligator, J. (1996) Attentional resolution and the locus of visual awareness. Nature, 383: 334–337.

Helmholtz, H. (1985a) Helmholtz's Treatise on Physiological Optics. Gryphon Editions, Ltd., Birmingham, AL.

Helmholtz, H. (1985b) Helmholtz's Treatise on Physiological Optics. Gryphon Editions, Ltd., Birmingham, AL.

Helmholtz, H. (1985c) Helmholtz's Treatise on Physiological Optics. Gryphon Editions, Ltd., Birmingham, AL.

Hering, E. (1964) Outlines of a Theory of the Light Sense. Harvard University Press, Cambridge, MA.

Herzog, M.H., Ernst, U.A., Etzold, A. and Eurich, C.W. (2003a) Local interactions in neural networks explain global effects in Gestalt processing and masking. Neural Comput, 15: 2091–2113.

Herzog, M.H. and Fahle, M. (2002) Effects of grouping in contextual modulation. Nature, 415: 433–436.

Herzog, M.H., Fahle, M. and Koch, C. (2001a) Spatial aspects of object formation revealed by a new illusion, shine-through. Vis. Res., 41: 2325–2335.

Herzog, M.H., Harms, M., Ernst, U.A., Eurich, C.W., Mahmud, S.H. and Fahle, M. (2003b) Extending the shine-through effect to classical masking paradigms. Vis. Res., 43: 2659–2667.

Herzog, M.H. and Koch, C. (2001) Seeing properties of an invisible object: feature inheritance and shine-through. Proc. Natl. Acad. Sci. USA, 98: 4271–4275.

Herzog, M.H., Koch, C. and Fahle, M. (2001b) Shine-through: temporal aspects. Vis. Res., 41: 2337–2346.

Herzog, M.H., Schmonsees, U. and Fahle, M. (2003c) Timing of contextual modulation in the shine-through effect. Vis. Res., 43: 2039–2051.

Hood, D.C. and Birch, D.G. (1990) The A-wave of the human electroretinogram and rod receptor function. Investig. Ophthalmol. Visual Sci., 31: 2070–2081.

Hubel, D.H. (1960) Single unit activity in lateral geniculate body and optic tract of unrestrained cats. J. Physiol., 150: 91–104.

Hubel, D.H. and Wiesel, T.N. (1959) Receptive fields of single neurones in the cat's striate cortex. J. Physiol. (Lond.), 148: 574–591.

Hubel, D.H. and Wiesel, T.N. (1961) Integrative action in the cat's lateral geniculate body. J. Physiol., 155: 385–398.

Jackendoff, R. (1987) On beyond zebra: the relation of linguistic and visual information. Cognition, 26: 89–114.

Jibu, M., Hagan, S., Hameroff, S.R., Pribram, K.H. and Yasue, K. (1994) Quantum optical coherence in cytoskeletal microtubules: implications for brain function. Biosystems, 32: 195–209.

Judge, S.J., Wurtz, R.H. and Richmond, B.J. (1980) Vision during saccadic eye movements. I. Visual interactions in striate cortex. J. Neurophysiol., 43: 1133–1155.

Kahneman, D. (1967) An onset-onset law for one case of apparent motion and metacontrast. Perception Psychophys., 2: 577–584.

Kahneman, D. (1968) Method, findings, and theory in studies of visual masking. Psychol. Bull., 70: 404–425.

Kahneman, D., Norman, J. and Kubovy, M. (1967) Critical duration for the resolution of form: centrally or peripherally determined? J. Exp. Psychol., 73: 323–327.

Kahneman, D. and Wolman, R.E. (1970) Stroboscopic motion: effects of duration and interval. Perception Psychophys., 8: 161–164.

Kanwisher, N. (2001) Neural events and perceptual awareness. Cognition, 79: 89–113.

Kolers, P. (1962) Intensity and contour effects in visual masking. Vis. Res., 2: 277–294.

Kolers, P. and Rosner, B.S. (1960) On visual masking (metacontrast): dichoptic observations. Am. J. Psychol., 73: 2–21.

Kong, K.L. and Wasserman, G.S. (1978) Temporal summation in the receptor potential of the Limulus lateral eye: comparison between retinula and eccentric cells. Sens. Processes, 2: 9–20.

Kortum, P.T. and Geisler, W.S. (1995) Adaptation mechanisms in spatial vision — II. Flash thresholds and background adaptation. Vis. Res., 35: 1595–1609.

Lamme, V.A., Super, H., Landman, R., Roelfsema, P.R. and Spekreijse, H. (2000) The role of primary visual cortex (V1) in visual awareness. Vis. Res., 40: 1507–1521.

Lawwill, T. (1973) Lateral inhibition in amblyopia: VER and metacontrast. Doc. Ophthalmol., 34: 243–258.

Le Gros Clark, W.E. and Penman, G.G. (1934) The projection of the retina in the lateral geniculate body. Proc. R. Soc., Ser. B, 114: 291–313.

Lee, S.H. and Blake, R. (1999) Rival ideas about binocular rivalry. Vis. Res., 39: 1447–1454.

Lee, S.H., Blake, R. and Heeger, D.J. (2005) Traveling waves of activity in primary visual cortex during binocular rivalry. Nat. Neurosci., 8: 22–23.

Lennie, P., Pokorny, J. and Smith, V.C. (1993) Luminance. J. Opt. Soc. Am. A, 10: 1283–1293.

Leopold, D.A. and Logothetis, N.K. (1996) Activity changes in early visual cortex reflect monkeys' percepts during binocular rivalry [see comments]. Nature, 379: 549–553.

Levick, W.R., Cleland, B.G. and Dubin, M.W. (1972) Lateral geniculate neurons of cat: retinal inputs and physiology. Invest. Ophthalmol., 11: 302–311.

Livingstone, M.S., Freeman, D.C. and Hubel, D.H. (1996) Visual responses in V1 of freely viewing monkeys. Cold Spring Harb. Symp. Quant. Biol., LXI:, 27–37.

Llinas, R., Ribary, U., Contreras, D. and Pedroarena, C. (1998) The neuronal basis for consciousness. Philos. Trans. R. Soc. Lond. B Biol. Sci., 353: 1841–1849.

Logothetis, N.K., Leopold, D.A. and Sheinberg, D.L. (1996) What is rivalling during binocular rivalry. Nature, 380: 621–624.

Mach, E. (1865) Uber die Wirkung der raumlichen Vertheilung des Lichtreizes auf die Nethaut. Sitzungsberichte der mathematisch-naturwissenschaftlichen Classe der kaiserlichen Akademic der Wissenschaftlen, 52: 303–322.

Mach, E. (1965) On the effect of the spatial distribution of the light stimulus on the retina. MACH BANDS: Quantitative Studies on Neural Networks in the Retina. Holden-Day, San Francisco, pp. 253–271.

Macknik, S.L. and Haglund, M.M. (1999) Optical images of visible and invisible percepts in the primary visual cortex of primates. Proc. Natl. Acad. Sci. USA, 96: 15208–15210.

Macknik, S.L. and Livingstone, M.S. (1998) Neuronal correlates of visibility and invisibility in the primate visual system. Nat. Neurosci., 1: 144–149.

Macknik, S.L. and Martinez-Conde, S. (2004a) Dichoptic visual masking reveals that early binocular neurons exhibit weak interocular suppression: implications for binocular vision and visual awareness. J. Cogn. Neurosci., 16: 1–11.

Macknik, S.L. and Martinez-Conde, S. (2004b) The spatial and temporal effects of lateral inhibitory networks and their relevance to the visibility of spatiotemporal edges. Neurocomputing, 58C–60C: 775–782.

Macknik, S.L., Martinez-Conde, S. and Haglund, M.M. (2000a) The role of spatiotemporal edges in visibility and visual masking. Proc. Natl. Acad. Sci. USA, 97: 7556–7560.

Macknik, S.L., Martinez-Conde, S., and Haglund, M.M. (2000b) The role of spatiotemporal edges in visibility in the striate cortex. In: Society for Neuroscience 30th Annual Meeting, New Orleans, LA.

Martinez-Conde, S., Macknik, S., Troncoso, X. and Dyar, T. (2006) Microsaccades counteract visual fading during fixation. Neuron, 49: 297–305.

Martinez-Conde, S., Macknik, S.L. and Hubel, D.H. (2000) Microsaccadic eye movements and firing of single cells in the striate cortex of macaque monkeys. Nat. Neurosci., 3: 251–258.

Martinez-Conde, S., Macknik, S.L. and Hubel, D.H. (2002) The function of bursts of spikes during visual fixation in the awake primate lateral geniculate nucleus and primary visual cortex. Proc. Natl. Acad. Sci. USA, 99: 13920–13925.

Martinez-Conde, S., Macknik, S.L. and Hubel, D.H. (2004) Moving our eyes to keep the world visible: the role of fixational eye movements in visual perception. Nat. Rev. Neurosci., 5: 229–240.

Matin, E. (1975) The two-transient (masking) paradigm. Psychol. Rev., 82: 451–461.

Matin, E., Clymer, A.B. and Matin, L. (1972) Metacontrast and saccadic suppression. Science, 178: 179–182.

McDougall, W. (1904a) The sensations excited by a single momentary stimulation of the eye. Br. J. Psychol., 1: 78–113.

McDougall, W. (1904b) The variation of the intensity of visual sensations with the durations of the stimulus. Br. J. Psychol., 1: 151–189.

McFadden, D. and Gummerman, K. (1973) Monoptic and dichoptic metacontrast across the vertical meridian. Vis. Res., 13: 185–196.

McKee, S.P., Bravo, M.J., Smallman, H.S. and Legge, G.E. (1995) The 'uniqueness constraint' and binocular masking. Perception, 24: 49–65.

McKee, S.P., Bravo, M.J., Taylor, D.G. and Legge, G.E. (1994) Stereo matching precedes dichoptic masking. Vis. Res., 34: 1047–1060.

McKeefry, D.J., Abdelaal, S., Barrett, B.T. and McGraw, P.V. (2005) Chromatic masking revealed by the standing wave of invisibility illusion. Perception, 34: 913–920.

Merigan, W.H. (1980) Temporal modulation sensitivity of macaque monkeys. Vis. Res., 20: 953–959.

Merigan, W.H., Barkdoll, E., Maurissen, J.P., Eskin, T.A. and Lapham, L.W. (1985) Acrylamide effects on the macaque visual system. I. Psychophysics and electrophysiology. Invest. Ophthalmol. Vis. Sci., 26: 309–316.

Merigan, W.H. and Maunsell, J.H. (1990) Macaque vision after magnocellular lateral geniculate lesions. Vis. Neurosci., 5: 347–352.

Milner, A.D. (1995) Cerebral correlates of visual awareness. Neuropsychologia, 33: 1117–1130.

Minkowski, M. (1920) Uber den Verlauf, die Endigung und die zentrale Repraentation von gekreuzten und ungekreutzten Sehnervenfasern bei einigen Saugetieren und beim Menschen. Schweiz. Arch. Neurol. Psychol., 6: 201.

Moutoussis, K., Keliris, G., Kourtzi, Z. and Logothetis, N. (2005) A binocular rivalry study of motion perception in the human brain. Vis. Res., 45: 2231–2243.

Mueller, J. (1842) Elements of Physiology. Taylor & Walton, London.

Munsterberg, H. (1916) The Photoplay. A Psychological Study. D. Appelton and Company, New York and London.

Naccache, L., Blandin, E. and Dehaene, S. (2002) Unconscious masked priming depends on temporal attention. Psychol. Sci., 13: 416–424.

Naka, K.-I. and Rushton, W.A.H. (1966) S-potentials from luminosity units in the retina of fish (Cyprinidae). J. Physiol., 185: 587–599.

Nakayama, K. (1971) Local adaptation in cat LGN cells: evidence for a surround antagonism. Vis. Res., 11: 501–509.

Olson, C.X. and Boynton, R.M. (1984) Dichoptic metacontrast masking reveals a central basis for monoptic chromatic induction. Percept. Psychophys., 35: 295–300.

Paradiso, M.A. and Hahn, S. (1996) Filling-in percepts produced by luminance modulation. Vis. Res., 36: 2657–2663.

Pascual-Leone, A. and Walsh, V. (2001) Fast backprojections from the motion to the primary visual area necessary for visual awareness. Science, 292: 510–512.

Petrov, Y. and Mckee, S.P. (2006) The effect of spatial configuration on surround suppression of contrast sensitivity. J. Vis., 6: 224–238.

Polonsky, A., Blake, R., Braun, J. and Heeger, D.J. (2000) Neuronal activity in human primary visual cortex correlates with perception during binocular rivalry [In Process Citation]. Nat. Neurosci., 3: 1153–1159.

Purushothaman, G., Ogmen, H. and Bedell, H.E. (2000) Gamma-range oscillations in backward-masking functions and their putative neural correlates. Psychol. Rev., 107: 556–577.

Ratliff, F. (1961) Inhibitory interaction and the detection and enhancement of contours. In: Rosenblith, W.A. (Ed.), Sensory Communication. MIT Press, Cambridge, MA, pp. 183–203.

Ratliff, F. (1965) MACH BANDS: Quantitative Studies on Neural Networks in the Retina. Holden-Day, Inc., San Francisco.

Ratliff, F. and Hartline, H.K. (1959) The response of Limulus optic nerve fibers to patterns of illumination on the receptor mosaic. J. Gen. Physiol., 42: 1241–1255.

Ratliff, F., Knight Jr., B.W., Dodge Jr., F.A. and Hartline, H.K. (1974) Fourier analysis of dynamics of excitation and inhibition in the eye of Limulus: amplitude, phase and distance. Vis. Res., 14: 1155–1168.

Ratliff, F., Miller, W.H. and Hartline, H.K. (1959) Neural interaction in the eye and the integration of receptor activity. Ann. NY Acad. Sci., 74: 210–222.

Rees, G. (2001) Seeing is not perceiving. Nat. Neurosci., 4: 678–680.

Reid, R.C. and Alonso, J.M. (1995) Specificity of monosynaptic connections from thalamus to visual cortex. Nature, 378: 281–284.

Ress, D., Backus, B.T. and Heeger, D.J. (2000) Activity in primary visual cortex predicts performance in a visual detection task. Nat. Neurosci., 3: 940–945.

Reynolds, J., Chelazzi, L., Luck, S. and Desimone, R. (1994) Sensory interactions and effects of selective spatial attention in macaque area V2. Soc. Neurosci. Abstr. Soc. Neurosci. Abstr., 20: 1054.

Reynolds, J.H., Chelazzi, L. and Desimone, R. (1999) Competitive mechanisms subserve attention in macaque areas V2 and V4. J. Neurosci., 19: 1736–1753.

Reynolds, J.H., and Desimone, R. (1999) The role of neural mechanisms of attention in solving the binding problem. Neuron, 24: 19–29, 111–125.

Reynolds, J.H., Pasternak, T. and Desimone, R. (2000) Attention increases sensitivity of V4 neurons [see comments]. Neuron, 26: 703–714.

Riggs, L.A. and Ratliff, F. (1952) The effects of counteracting the normal movements of the eye. J. Opt. Soc. Am., 42: 872–873.

Roecker, E.B., Pulos, E., Bresnick, G.H. and Severns, M. (1992) Characterization of the electroretinographic scotopic B-wave amplitude in diabetic and normal subjects. Investig. Ophthalmol. Visual Sci., 33: 1575–1583.

Roget, P.M. (1825) Explanation of an optical deception in the appearance of the spokes of a wheel seen through vertical apertures. Philos. Trans. R. Soc. Lond., 115: 131–140.

Roufs, J.A. (1972) Dynamic properties of vision. I. Experimental relationships between flicker and flash thresholds. Vis. Res., 12: 261–278.

Rushton, W.A.H. and Westheimer, G. (1962) The effect upon the rod threshold of bleaching neighboring rods. J. Physiol., Lond., 164: 318–329.

Schiller, P.H. (1965) Monoptic and dichoptic visual masking by patterns and flashes. J. Exp. Psychol., 69: 193–199.

Schiller, P.H. (1968) Single unit analysis of backward visual masking and metacontrast in the cat lateral geniculate nucleus. Vis. Res., 8: 855–866.

Shapley, R. (1974) Gaussian bars and rectangular bars: the influence of width and gradient on visibility. Vis. Res., 14: 1457–1462.

Shapley, R., Enroth-Cugell, C., Bonds, A.B. and Kirby, A. (1972) Gain control in the retina and retinal dynamics. Nature, 236: 352–353.

Shapley, R.M., and Enroth-Cugell, C. (1984) Visual adaptation and retinal gain controls. Progress in Retinal Research. Pergamon Press, New York, pp. 263–343.

Sheinberg, D.L. and Logothetis, N.K. (1997) The role of temporal cortical areas in perceptual organization. Proc. Natl. Acad. Sci. USA, 94: 3408–3413.

Smith, M.C. and Schiller, P.H. (1966) Forward and backward masking: a comparison. Can. J. Psychol., 20: 191–197.

Somjen, G. (1975) Sensory Coding in the Mammalian Nervous System. Plenum Press, New York.

Sperling, G. (1965) Temporal and spatial visual masking. I. Masking by impulse flashes. J. Opt. Soc. Am., 55: 541–559.

Sperling, G. and Joliffe, C.L. (1965) Intensity-time relationship at threshold for spectral stimuli in human vision. J. Opt. Soc. Am., 55: 191–199.

Stevens, S.S. (1961) To honor Fechner and repeal his law. Science, 133: 80–86.

Stevens, S.S. (1962) The surprising simplicity of sensory metrics. Am. Psychol., 17: 29–32.

Stevens, S.S. (1970) Neural events and the psychophysical law. Science, 170: 1043–1050.

Stigler, R. (1908) Uber die Unterschiedsschwelle im aufsteigenden Teile einer Lichtempfindung. Pfl. arch. gesamte Physiol., 123: 163–223.

Stigler, R. (1910) Chronophotische Studien uber den Umgebungskontrast. Pfl. arch. gesamte Physiol., 135: 365–435.

Stigler, R. (1913) Metacontrast (Demonstration). In: IX Congress International de Physiologie. Arch. Int. Physiologie: Groningen, p. 78.

Stigler, R. (1926) Die Untersuchung des zeitlichen Verlaufes der optischen Erregung mittels des Metakontrates. In: Aberhslden, E. (Ed.), Handbuch der Biologischen Arbeitsmethoden. Urban and Schwarzenberg, Berlin, pp. 949–968.

Stoper, A.E. and Banffy, S. (1977) Relation of split apparent motion to metacontrast. J. Exp. Psychol. Hum. Percept. Perform., 3: 258–277.

Thompson, K.G. and Schall, J.D. (1999) The detection of visual signals by macaque frontal eye field during masking. Nat. Neurosci., 2: 283–288.

Thompson, K.G. and Schall, J.D. (2000) Antecedents and correlates of visual detection and awareness in macaque prefrontal cortex. Vis. Res., 40: 1523–1538.

Tong, F., Nakayama, K., Vaughan, J.T. and Kanwisher, N. (1998) Binocular rivalry and visual awareness in human extrastriate cortex. Neuron, 21: 753–759.

Troncoso, X., Macknik, S.L., Tse, P.U., and Martinez-Conde, S. (2005) Sharp junctions generate greater BOLD activation than shallow junctions in the human occipital cortex. In: 28th European Conference of Visual Perception. Perception, Pion: A Coruña, Spain.

Tse, P.U., Martinez-Conde, S., Schlegel, A.A. and Macknik, S.L. (2005) Visibility, visual awareness, and visual masking of simple unattended targets are confined to areas in the occipital cortex beyond human V1/V2. Proc. Natl. Acad. Sci. USA, 102: 17178–17183.

Turvey, M.T. (1973) On peripheral and central processes in vision: inferences from an information-processing analysis of masking with patterned stimuli. Psychol. Rev., 80: 1–52.

Uttal, W.R. (1969) Emerging principles of sensory coding. Perspect. Biol. Med., 12: 244–368.

Uttal, W.R. (1973) The Psychobiology of Sensory Coding. Harper & Row, New York.

Van Hateren, J.H. (1993) Spatiotemporal contrast sensitivity of early vision. Vis. Res., 33: 257–267.

Wali, N. and Leguire, L.E. (1991) Dark-adapted luminance-response functions with skin and corneal electrodes. Doc. Ophthalmol., 76: 367–375.

Wali, N. and Leguire, L.E. (1992) Fundus pigmentation and the dark-adapted electroretinogram. Doc. Ophthalmol., 80: 1–11.

Weber, E.H. (1834) De pulsu, resorptione, auditu et tactu: annotationes anatomicae et physiologiae. Leipzig: Koehler.

Weber, E.H. (1846) Der Tastsinn und das Gemeingefuhl. In: Wagner, R. (Ed.), Handworterbuch der physiologie, III, ii: 451–588.

Weisstein, N. (1968) A Rashevsky-Landahl neural net: simulation of metacontrast. Psychol. Rev., 75: 494–521.

Weisstein, N. (1971) W-shaped and U-shaped functions obtained for monoptic and dichoptic disk-disk masking. Perception Psychophys., 9: 275–278.

Weisstein, N. (1972) Metacontrast. In: Jameson, D. and Hurvich, L.M. (Eds.), Handbook of Sensory Physiology: Visual Psychophysics. Springer, New York, pp. 275–278.

Weisstein, N. and Haber, R.N. (1965) A U-shaped backward masking function. Psychon. Sci., 2: 75–76.

Weisstein, N., Ozog, G. and Szoc, R. (1975) A comparison and elaboration of two models of metacontrast. Psychol. Rev., 82: 325–343.

Werner, H. (1935) Studies on contour: I. Qualitative analysis. Am. J. Psychol., 47: 40–64.

Werner, H. (1940) Studies on contour: strobostereoscopic phenomena. Am. J. Psychol., 53: 418–422.

Wertheimer, M. (1925) Drei Abhandlungen zur Gestalttheorie. Philosophische Akademie Erlangen.

Westheimer, G. (1965) Spatial interaction in the human retina during scotopic vision. J. Physiol. Lond., 181: 881–894.

Westheimer, G. (1967) Spatial interaction in human cone vision. J. Physiol., Lond., 190: 139–154.

Westheimer, G. (1970) Rod-cone independence for sensitizing interaction in the human retina. J. Physiol., Lond., 206: 109–116.

Williams, Z.M., Elfar, J.C., Eskandar, E.N., Toth, L.J. and Assad, J.A. (2003) Parietal activity and the perceived direction of ambiguous apparent motion. Nat. Neurosci., 6: 616–623.

Yarbus, A.L. (1967) Eye Movements and Vision. Plenum Press, New York.

Zeki, S. (2001) Localization and globalization in conscious vision. Annu. Rev. Neurosci., 24: 57–86.

Zeki, S. (2003) The disunity of consciousness. Trends Cogn. Sci., 7: 214–218.

Zeki, S. and Bartels, A. (1999) Toward a theory of visual consciousness. Conscious. Cogn., 8: 225–259.

Zeki, S. and Ffytche, D.H. (1998) The Riddoch-Syndrome — insights into the neurobiology of conscious vision. Brain, 121: 25–45.

Martinez-Conde, Macknik, Martinez, Alonso & Tse (Eds.)
Progress in Brain Research, Vol. 155
ISSN 0079-6123

CHAPTER 12

Blindsight, conscious vision, and the role of primary visual cortex

Petra Stoerig*

Institute of Experimental Psychology II, Heinrich-Heine-University Düsseldorf, 40225 Düsseldorf, Germany

Abstract: What is the role the primary visual cortex (V1) in vision? Is it necessary for conscious sight, as indicated by the cortical blindness that results from V1 destruction? Is it even necessary for blindsight, the nonreflexive visual functions that can be evoked with stimuli presented to cortically blind fields? In the context of this controversial issue, I present evidence indicating that not only is blindsight possible, but that conscious vision may, to a varying degree, return to formerly blind fields with time and practice even in cases where functional neuroimaging reveals no V1 activation.

Keywords: primary visual cortex (V1); visual neuropsychology; blindness; blindsight; training; visual field recovery

Stunning controversies

The role that the occipital lobes, and specifically the primary visual cortex, play for vision has been debated for over a century. As early as 1878, Munk (1881) reported that a unilateral ablation of the monkey's occipital convexity resulted in a hemianopic, and bilateral extirpation in a complete cortical blindness. In contrast, Ferrier (1886) was convinced that bilateral removal of the greater portion of the occipital lobes caused no appreciable impairment of vision. Although methodological progress both in the precision of ablation in animals and in assessing the extent of destruction in man have led to a general acceptance of Munk's view — complete unilateral destruction or denervation of the primary visual cortex (V1) causes cortical blindness in the contralateral hemifield — the debate still continues. One of its aspects regards the permanence of the blindness. According to the classical doctrine held by the vast majority of clinicians, neurologists as well as ophthalmologists, the blindness will stay for good unless spontaneous recovery occurs within months; if it does, the damage is reversible. The contrary view traces its origins to Riddoch's (1917) early report on appreciation of movement in fields of cortical blindness, and holds that even conscious vision is possible without V1 (Zeki and ffytche, 1998). Remarkably, this divide extends to the implicit visual functions that remain in fields of cortical blindness, and has come to be known as blindsight (Sanders et al., 1974; Weiskrantz et al., 1974). According to the view advocated by Campion et al. (1983) and Fendrich et al. (1992), these nonreflexive but implicit visual functions depend on islands of surviving tissue in the primary visual cortex. This assumption is opposed by those who hold that these functions must depend on the extrageniculo-striate cortical projections rather than residual V1 (see Stoerig and Cowey, 1997, for review). Thus, V1 is at the same time presumed to be: (1) required for blindsight as well as for conscious vision and (2) required for neither conscious vision nor blindsight.

*Corresponding author. Tel.: +49-211-811-2265; Fax: +49-211-811-4522; E-mail: petra.stoerig@uni-duesseldorf.de

DOI: 10.1016/S0079-6123(06)55012-5

"Vision is a more complex function than most people realize" (Livingstone and Hubel, 1988)

In the retina, the distribution of light falling into the eye is first translated into patterns of neuronal discharges by the retinal ganglion cells. These differ morphologically as well as functionally, with major distinctions regarding spatial and temporal resolution, sensitivity to light, and specificity to wavelengths. They project the retino-recipient nuclei that lie in the hypothalamus, the thalamus, and the midbrain. The lion's share of direct retinofugal projections goes to the dorsal lateral geniculate nucleus (dLGN) of the thalamus where the axons of different ganglion cell classes contact neurons in different layers. The vast majority of dLGN projection neurons send their output to the primary visual cortex situated in the occipital lobe, about as far from the eyes as possible. In this "cortical retina," topographical relationships are preserved. In addition, inputs that originate in different types of retinal ganglion cells remain segregated not only by targeting different subdivisions of the main input layer 4, but also by virtue of their target neurons projecting onwards to functionally distinct subregions both within V1 itself and in the second visual cortical area V2. Like further stations in the feed-forward stream of information, V2 receives input not only through V1 but also from subcortical visual nuclei including the dLGN (Yukie and Iwai, 1981). The next tier of visual cortical areas includes V3, V4, and V5/MT, which preferentially process different types of visual information, indicating that they receive dominant feed-forward input originally generated by different retinal ganglion cell classes. Although functional specialization of visual cortical areas is widely accepted nowadays, its extent is still somewhat controversial on the basis of physiological evidence (Zeki, 1978; van Essen, 1985; Livingstone and Hubel, 1988; Gross, 1992; Cowey, 1994); even the processing of visual motion, for which the area V5/MT appears unequivocally specialized, has just been reported to invoke area V4 if motion adaptation takes place (Tolias et al., 2005).

As a rule of thumb, neurons have increasingly large and complex receptive field properties further upstream their visual area. Neurons in visual cortical areas extending into the temporal lobes respond preferentially to images of houses or places, to faces and gestures, and even to individuals (Perrett et al., 1982; Kanwisher et al., 1997; Kanwisher, 2001; Jellema and Perrett, 2003); they generalize over viewpoints and other stimulus particulars. In contrast, neurons in visual areas of the parietal lobes increasingly take account of where some stimulus is, or moves toward, in relation to not only to the retinal locus of stimulation but also to the subject's ever-changing bodily coordinates. These observations have given rise to the concept of two visual streams (Ungerleider and Mishkin, 1982): a dorsal, occipitoparietal one involved in visuomotor preparation; and a ventral, occipitotemporal one involved in stimulus identification. Areas in both streams receive input (1) from earlier visual areas, (2) via several routes from subcortical nuclei, and (3) from yet higher cortical areas to which they also project. In addition, they have extensive lateral connections that allow cross-talk between dorsal and ventral pathways (Fig. 1).

Lessons from lesions

Although the anatomical and physiological studies of the visual system have profoundly advanced our

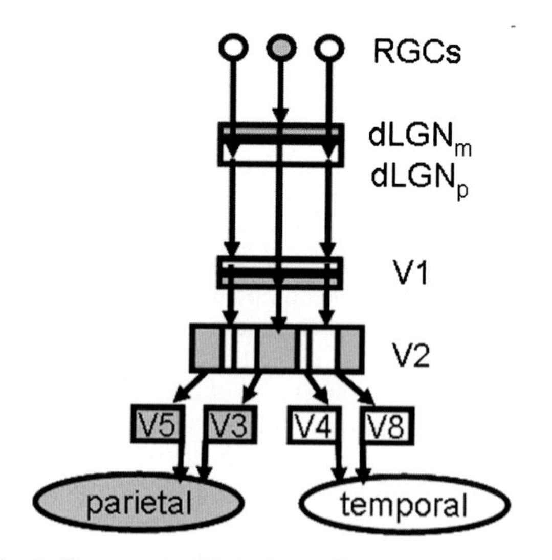

Fig. 1. The very simplified schema illustrates the functional segregation that originates in the retina and permeates the entire visual system all the way into parietal, temporal, and frontal cortices.

understanding of the visual system's functional architecture, they cannot by themselves reveal which pathways and cells contribute to conscious and which contribute to unconscious or implicit visual functions. Rather, it is through the study of the functional consequences of lesions to different parts of the system that we have learned that the primary visual system plays a much more prominent part in conscious vision than the retino-extra-geniculo-striate cortical projections. Careful neuropsychological observations of human patients with circumscribed lesion, alongside behavioral studies on animals with experimentally induced lesions, show that different levels of unconscious or blind and conscious vision remain following destruction of different parts of the system.

The neuroendocrine responses represent the lowest level of vision. They have even been demonstrated in patients who were totally blind as a result of pathologies which affect the retinas or the optic nerves. Although these patients retained no appreciation of light, their plasma melatonin levels varied in a light-dependent circadian manner. Czeisler et al. (1995) also showed that wearing a blindfold abolished this modulation, demonstrating that it is indeed through the eyes that the effect is enabled.

The next level of visual functions is that of reflexive responses. While Czeisler et al.'s (1995) patients retained no pupillary light reflexes, these and other visual reflexes may remain in patients with more centrally located lesions of the visual pathways. Such reflexive responses can also be elicited in totally blind patients; indeed, a blink reflex has even been reported in brain death (Keane, 1979).

The third level of blind visual function has been observed in patients whose lesions destroy or denervate the primary visual cortex. Most commonly, such lesions are caused by vascular incidents, but traumatic and neoplastic insults are also common causes of the ensuing cortical blindness. In the much more common cases where the lesion is confined to one hemisphere, such destruction causes a field of homonymous cortical blindness in the contralateral hemifield. Unilateral V1 destruction can spare the optokinetic nystagmus (Pasik et al., 1959; ter Braak and van Vliet, 1963) as well as a variety of pupillary reflexes that in addition to the pupillary light reflex indicate responses to chromatic and spatial information (Barbur, 2004). In addition, psychosensory effects such as the stimulus probability reflex, which elicits larger pupil dilation to rare targets (Beatty, 1986), have recently been demonstrated in response to stimuli presented in the cortically blind field (C. Loose and P. Stoerig, in preparation; Fig. 2). Moreover, patients can voluntarily initiate nonreflexive visual responses to blind field targets. They include saccadic and manual localization of stimuli presented

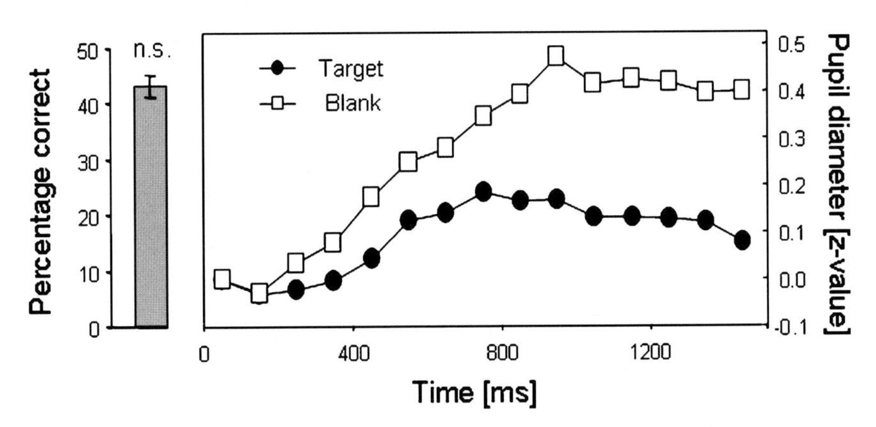

Fig. 2. Although patient KE showed no evidence of detecting the 5° gray disk at a negative log contrast of 0.61 to the white background (left), his pupil responded with a late dilation to the presentation of rare (20%) blank trials (right). As no stimulus was presented, the dilation reflects the low probability of blank stimuli, and is of psychosensory origin rather than stimulus driven. Note that pupil traces are based on "Target" responses only, so as not to confound stimulus and response probabilities.

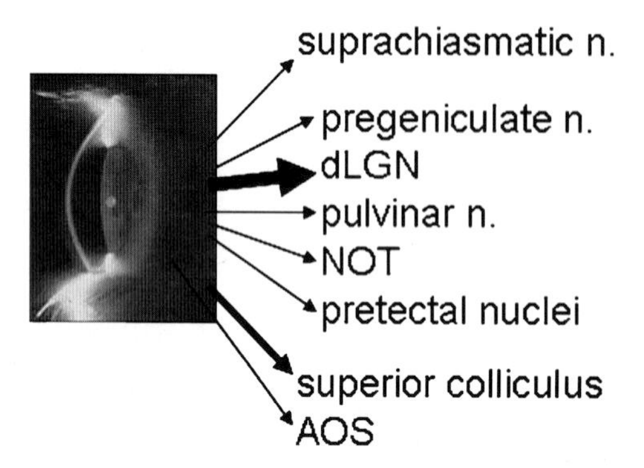

Fig. 3. The retinofugal pathways target more than 10 nuclei in different brain regions. These nuclei transmit the visual information to others, some of which also receive a direct retinal input, either via these or directly to visual cortical areas.

in the blind field (Pöppel et al., 1973; Weiskrantz et al., 1974), as well as discrimination of absence vs. presence of a stimulus (Stoerig et al., 1985), of absence or presence of stimulus motion (Perenin, 1991), and of stimuli of different orientations, flux, motion direction, and wavelength (Fig. 3; see Weiskrantz, 1986, 1990; Stoerig and Cowey, 1997, for reviews). As the patients report no perception of the stimuli, demonstration of these functions requires methods that do not rely on a considered report, but circumvent the experienced blindness. The most commonly used forced choice approach requires the patients to guess whether a stimulus was presented, where it was presented, or which one of a two possible stimuli was briefly presented in the blind field. Alternatively, processing of a stimulus in the blind field is inferred if it influences responses to targets presented in the normal hemifield. If, for example, a patient responds faster to a seen target in the normal field when simultaneously and unbeknown to him an additional stimulus is presented in the blind field, the reaction time difference indicates that the blind field stimulus has been effectively processed (Marzi et al., 1986). Capturing the dissociation between perception and performance, this phenomenon has been termed "blindsight" (Weiskrantz et al., 1974). Covering nonreflexive visual functions, it represents the highest level of visual function in the absence of a conscious stimulus representation (Stoerig, 1999).

Nonreflexive responses to unseen information have also been demonstrated in patients who, due to circumscribed lesions of extrastriate cortical areas, have lost conscious color (Meadows, 1974), form (Benson and Greenberg, 1969; Farah, 1990), or motion perception (Zihl et al., 1983). Patients with such selective visual deficits can implicitly respond to the visual feature(s) that are no longer consciously represented (e.g., Heywood et al., 1991; Milner and Goodale, 1996); indeed, the implicit processing can allow the conscious detection of stimulus properties, as when an achromatopsic patient detects the border between two abutting fields of different colors although he cannot see the colors themselves. In blindsight, where the conscious representation of all stimulus attributes are lost, such perceptual effects have been invoked by presenting the stimuli so that they straddle the border between the blind and the remaining normal visual field (see Pöppel, 1986; Marcel, 1998, for examples).

Neuronal pathways to blind vision

Anatomy and physiology

The light-dependent plasma melatonin modulation very likely depends on a sparse retinal projection to the nucleus suprachiasmaticus of the hypothalamus. It is involved in entraining the circadian rhythms to the day-and-night cycle. Of the retinorecipient nuclei, this one lies closest to the eyes, above the optic chiasm, and receives its retinal input from a small population of retinal ganglion cells that has been described in the rat (Moore et al., 1995). Whether the melanopsin-containing light-sensitive ganglion cells that have recently been described in rats and monkeys (see Foster, 2005, for summary) also project to this nucleus is presently unknown. Although functional segregation of optic fibers is more obvious in the tract than in the optic nerve (Reese and Cowey, 1988), the axons of the cells that project to the suprachiasmatic nucleus could travel more dorsally in the tract en route to their target nucleus, and thereby escape destruction in some of the pathologies that cause a peripheral blindness.

The visual reflexes are mediated by retinal projections to a variety of subcortical nuclei. The pupil light reflex involves the pretectal nuclei, which forward their output to the Edinger–Westphal nucleus receives the output of the pretectal nuclei (Magoun and Ranson, 1935; Beatty, 1986). The optokinetic nystagmus involves the retinal projections to the nucleus of the optic tract (Hoffmann, 1989), and processing of optic flow involves the three nuclei that form the accessory optic system (Simpson, 1984). Cells in the superior colliculus respond well to moving borders (Marrocco and Li, 1977), and play an important part in programming saccadic eye-movements (Mohler and Wurtz, 1976; Waitzman et al., 1991). Furthermore, collicular microstimulation improves performance in a spatially selective manner even during fixation (Müller et al., 2005), and collicular deactivation causes visual hemineglect (Sprague and Meikle, 1965; FitzMaurice et al., 2003). The nucleus is involved in crossmodal integration (Meredith and Stein, 1985), and even participates in perceptual decision-making (Horwitz et al., 2004), indicating that it contributes not only to reflexive orienting responses, but to nonreflexive functions as well.

If fibers traveling to the extra-geniculate nuclei leave the optic tract before the lesion, or escape its destructive effects in some other fashion, their target nuclei may still mediate visual responses, albeit often impaired, in patients who retain no conscious vision.

The nonreflexive visually guided responses that can be elicited in fields of cortical blindness probably make use of all pathways that survive the effects of a striate cortical lesion. This is indicated by a long series of experiments on monkeys who, in addition to occipital resection, were subjected not only to increasingly larger cortical lesions, but also to selective destruction of subcortical nuclei (see Pasik and Pasik, 1982, for review). The results show that manual and saccadic localization depend on the midbrain, and are severely disturbed when the superior colliculus is damaged in addition to V1 (Mohler and Wurtz, 1977; Feinberg et al., 1978). In contrast, rough luminance discrimination remained possible even with large cortical lesions combined with subcortical ones unless the lateral pretectum was destroyed (Pasik and Pasik, 1973). In addition to the extra-geniculate retinal pathways, the projection neurons that survive the retrograde degeneration of the dLGN that follows ablation of V1 (Van Buren, 1963; Mihailovic et al., 1971) may contribute to blindsight. Although the degeneration transneuronally affects the retinal ganglion cell of which roughly 50% die in this much slower process (Van Buren, 1963; Cowey, 1974; Cowey et al., 1989; Weller and Kaas, 1989), the survivors continue to project both to the extra-geniculate nuclei, including the pulvinar and the pregeniculate nucleus which is exempt from retrograde degeneration (Dineen et al., 1982), and to the degenerated dLGN (Kisvárday et al., 1991).

The retino-recipient nuclei that have been physiologically investigated after striate cortical ablation showed responsivity to stimuli presented in the cortically blind field (see Payne et al., 1996, for review). Furthermore, all of them project directly (like the dLGN and the pulvinar) or indirectly (like the superior colliculus) to extrastriate visual cortical areas. Different extrastriate visual cortical areas differ in the extent to which they continue to respond to information from the blind hemifield. Dorsal stream areas appear to retain more responsivity (see Bullier et al., 1994, for review). Nevertheless, despite evidence showing that area V5/MT displays the most robust responses after both cooling (Girard et al., 1992) and ablation of V1 (Rodman et al., 1989), even area MT's relative independence from V1 input is not reported consistently (Collins et al., 2005). What direction selectivity remains seems to depend on input from the superior colliculus and was abolished when it was also lesioned (Rodman et al., 1990). Whereas neither V2 nor V4 retained more than a very small number of neurons responding to the contralateral hemifield when V1 was ablated or cooled (see Bullier et al., 1994, for review), visual responses were still evoked from neurons in the polymodal cortex of the superior temporal cortex (Bruce et al., 1986).

Functional neuroimaging

Functional neuroimaging studies of human patients confirm that extrastriate visual cortex in the lesioned hemisphere continues to respond to stimulation of the blind hemifield. Although the

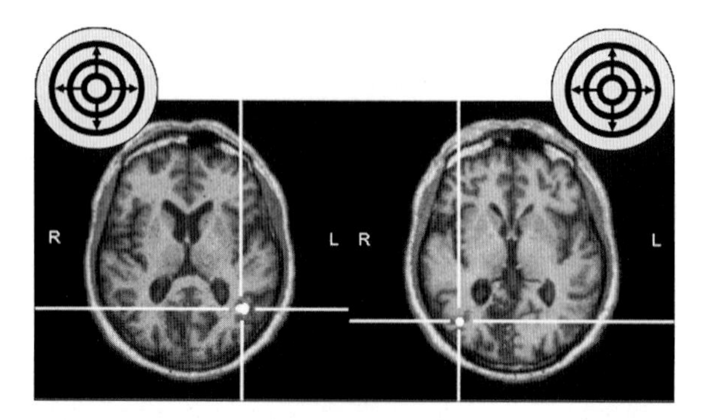

Fig. 4. Functional magnetic resonance images of patient HK obtained 3 months after a vascular lesion destroyed his right striate cortex. A circular stimulus of expanding bright and dark rings activated the human motion complex hMT+ along with early visual cortex in the normal hemisphere when presented to the normal right hemifield. When presented to the cortically blind left hemifield, the same stimulus evoked no detectable response in early visual cortex V1/V2; nevertheless, the ipsilesional motion complex was activated.

majority of published studies were performed on patients whose lesions had occurred years before the imaging took place (e.g., Goebel et al., 2001), our unpublished results also show activation of the ipsilesional human motion complex hMT+ relatively early postlesion. An example taken 3 months after the insult is shown in Fig. 4; no concomitant activation in areas V1/V2 was found. However, we have not detected hMT+ activation in all the patients we have studied early after the insult, so that the patients' neuroimaging results are in good agreement with the physiological data from monkeys. These too demonstrate the most robust responses in areas MT/V5 and its satellites, i.e., corresponding to the motion complex, but do not report it in all instances.

Blindsight has also been attributed to small islands of primary visual cortex that are supposed to survive the lesion. This explanation is based on the data of a patient who demonstrated significant evidence of detection and discrimination only if the stimuli were presented at a particular position in his blind field, but not when they were presented at a number of alternative positions (Fendrich et al., 1992). However, other patients do not show such a pattern of blindsight confined to spatially isolated islands (e.g., Stoerig and Pöppel, 1986; Stoerig, 1993; Kentridge et al., 1997). In addition, neuroimaging failed to reveal evidence for activation within the lesioned primary visual cortex in several instances (Barbur et al., 1993; Stoerig et al., 1998; Goebel et al., 2001), visually induced activation in occipitopolar cortex was not shown to correspond topographically to the "blindsight position" in Fendrich et al.'s (1993) patient, and finally, residual V1 cortex cannot explain the evidence in monkeys or patients in whom this cortical area was completely resected. Like any tissue that escapes the lesion and its degenerative consequences, small islands of V1 tissue, if they survived within the damaged region, could be recruited to serve the residual visual functions; however, the bulk of evidence on both species indicates that such a contribution is not prerequisite to blindsight.

Together, the body of anatomical, physiological, and neuroimaging data indicates that blindsight is not just mediated subcortically, but that several extrastriate visual cortical areas retain or regain visual responsivity. As this applies to both monkeys and humans, it disproves the contention that blindsight is blind because it does not invoke any cortical processing. This idea had been put forward in the context of evolutionary corticalization to explain why patients became blind following destruction of primary visual cortex, while other mammals including monkeys still displayed visually guided behavior (Klüver, 1942; Weiskrantz, 1963; Weiskrantz and Cowey, 1963, 1967). It was only by using forced choice procedures as applied in animal research that human patients were shown to also possess such capacities (Pöppel et al., 1973). To complete the cycle, by offering our hemianopic monkeys the option to respond "no target" to blind field stimuli they localized almost perfectly when not having this option, we could show that they consistently responded "no target" on target trials (Cowey and Stoerig, 1995). Like human patients, the monkeys thus behaved as if targets they could localize were not targets at all, suggesting they too had blindsight.

Functional significance of unconscious vision

The light-dependent modulation of melatonin secretion that has been demonstrated, even in patients in whom no other visual functions were found, entrains our circadian rhythms to the day-and-night cycle. Disturbances of sleep patterns result from their failure (Sack et al., 1991; Siebler et al., 1998), indicating that they influence function in an indirect but important way.

The visual reflexes protect the eyes from over-exposure to light, as the pupil light and the blink reflex, that is also elicited by fast approaching objects mediate orienting responses, and alert the organism to the sudden appearance of potentially relevant stimuli. Whereas the relatively immediate loop linking the sensory input to the reflexive motor responses is indicative of their basic relevance, the functional significance of the nonreflexive responses that are characteristic of blindsight may appear less obvious. If blindsight was exclusively a laboratory phenomenon, demonstrable in experiments which force the patients to guess the stimulus or its location, it would contribute to our understanding of the functional neuroanatomy of the visuomotor systems, but not help the patients in their daily life. Although this issue has not been extensively studied, marked improvements in the visually guided behavior have been documented in a monkey who underwent bilateral ablation of primary visual cortex. This monkey, Helen, was extensively studied (Humphrey, 1970, 1974) and showed no evidence of spatial vision for the first 19 postoperative months. Training her, first with moving, then with stationary objects, continuously improved her ability to look and reach for them, but still there was no evidence for depth vision for as long as she remained largely confined to her cage. Five years after the onset of blindness, Helen was moved to a new lab where at first no testing room was available. Humphrey then took her on a leash into the open field and woods. He describes her development as follows:

> To begin with, [...] these walks were fairly hazardous. She continually bumped into obstacles, she collided with my legs, and she several times fell into a pond. But then, day by day, there was an extraordinary change in her behaviour. On the one hand she began to systematically anticipate and skirt round obstacles in her path, while on the other she began actually to approach the trees in the field, turning towards them as we passed by, walking up and reaching out to grasp their trunks (Humphrey, 1974, p. 244).

In fact, according to video documentation, Helen eventually appeared quite like a normal monkey. Although pathology revealed a small sparing of V1 that corresponded to a peripheral part of the upper right quadrant of the visual field, she did not appear to use this remnant to locate objects, but rather looked at them directly before reaching out.

Blindsight and plasticity

Blindsight without feedback

It is obvious from the wealth of data published on blindsight that what is tested, and how it is tested, makes a big difference as to whether or not positive results are revealed. For example, the incidence of statistically significant movement detection and discrimination varies between 0 and 100% of patients tested. Perenin (1991), who tested patients with hemianopia as well as complete cortical blindness, used large fields of black dots. These would either move or remain stationary, or move from left to right or right to left, respectively. The patients solved the motion detection as well as the direction discrimination task with ~90% correct performance. At the other end of the spectrum, Barton and Sharpe (1997) tested 10 patients with cortically blind fields, using a random dot kinematogram (RDK), and found that not one was able to discriminate the direction of the moving dots even at 100% coherence. A more recent study in which three patients were tested with different types of moving patterns that included both a single bar and an RDK confirmed that all were able to detect motion in all instances, but could distinguish its direction only for the single bar (Azzopardi and Cowey, 2001). GY, a patient with long-term experience with tests of his residual visual functions, participated in this experiment, and showed this same pattern of results despite his repeatedly reported excellent ability to distinguish the direction of a single moving target (Barbur et al., 1980; Weiskrantz et al., 1995).

To specifically address the issue of blindsight learning, we used a different type of moving stimulus. A red-and-blue spiral, 5° across and with a mean luminance of 8 cd/m^2, was presented for 500 ms per trial on a white 10 cd/m^2 background. It would, or would not, rotate around its own axis in a motion detection task, and would rotate clockwise or counterclockwise in a motion discrimination task. The rotation depended on local elements and avoided global stimulus translation. As we were interested in whether patients would improve with practice, we presented both tasks for 8–12 consecutive series of 100 trials each, and analyzed performance per series as well as overall. 11 patients with fields of cortical blindness participated; one was the already mentioned GY. Upon each presentation, patients pressed one of the two response keys to indicate whether or not the spiral had been rotating, or whether it had rotated clockwise or counterclockwise. Both on- and offset of the stimuli was signalled with a brief sound to inform the patients when to attend and when to respond. The two stimuli had equal probability. Their position was adjusted to each individual field defect, and eye movements were monitored throughout to ensure that any evidence of discrimination would not reflect unstable fixation.

The results showed that of the 11 patients, 6 performed above chance ($P<0.05$ or better, χ^2 test) in the motion detection task when performance was collapsed over the series. In addition, in nine subjects the level of performance improved over the course of the series. To learn whether the improvement was significant for the group, rates of false responses to stationary stimuli ("moving"/stationary) were subtracted from those of correct responses to moving stimuli ("moving"/moving) to provide an index of discrimination that is independent of subjects' response bias. The resultant difference values for the first and the last series were used in a paired comparison test that yielded a significant advantage in favor of performance in the last ($Z = -2.045$; $P_{1\text{—tailed}} = 0.021$, Wilcoxon). Fig. 2A shows percentage correct values for three of the patients in the first and last series. Note that patient GY performed best of all subjects.

As can be seen in Fig. 5B, the corresponding values were lower for the motion direction task. This finding agrees with the published results of Perenin (1991) and Azzopardi and Cowey (2001), and confirms that in the blind field direction discrimination is more difficult than detection of motion. Nevertheless, six patients performed significantly in this task. Rotation direction discrimination, albeit clearly difficult under the present conditions, is thus possible in cortically blind fields. Whether the discrimination reflects the use of rotation rather than RDKs, whether the spiral's color contrast made a difference, or whether the size of the individual elements that move together

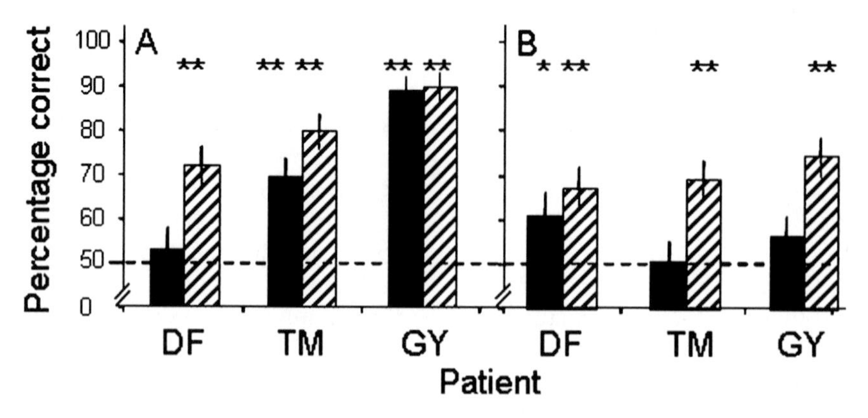

Fig. 5. Results of rotation detection (A) and rotation direction discrimination (B) in three of the patients. For each, performance in the first (black bars) and last (striped bars) of the 100 trial series is shown. Note that all patients performed better in the rotation detection than in the rotation direction discrimination, and the improvement was significant for the group only in the former task. Asterisks indicate whether performance was significant (*$p<0.05$; **$p<0.005$).

plays a role cannot be decided at present. Group analysis of the rotation direction discrimination revealed no statistically meaningful difference between first and last series; nevertheless, two of the three patients shown in Fig. 5B, TM and GY, showed evidence for perceptual learning in the form of a significant positive correlation between the discrimination index ("clockwise"/clockwise − "clockwise"/counterclockwise) and the series (TM: $\rho = 0.738$, $P_{1\text{-tailed}} = 0.018$; GY: $\rho = 0.745$, $P_{1\text{-tailed}} = 0.011$).

To learn which of the many factors that are likely to underlie the differences in blindsight performance contributed most to the interindividual variability in the two rotation tasks, we performed a hierarchical regression analysis. Factors entered to explain the performance included the patients' age, their age at lesion, the size of the lesion as estimated on the basis of the T1-weighted MR images, the rough size of the field defect, the length of time during which each had regularly or irregularly participated in blindsight testing, and the age at which they had begun to do so. The results showed that of these factors, the overall length of blindsight experience, which varied from 0.2 to 20 years, correlated highest with performance in the motion detection task ($p = 0.002$); this factor alone explained 57% of the variance ($R^2 = .568$; $p = 0.005$). For the motion direction discrimination, none of the same factors explained a significant part of the variance.

In addition to confirming that some tasks are more difficult to solve than others, the results of the spiral experiment support the notion that blindsight is subject to learning. This is in agreement with results of Bridgeman and Staggs (1982) as well as Zihl and colleagues (Zihl, 1980; Zihl and von Cramon, 1980; Zihl and Werth, 1984). Both groups tested localization in the cortically blind fields of their patients, and found that performance improved with practice when pointing (Bridgeman and Staggs, 1982) or saccadic responses were required (Zihl, 1980; Zihl and von Cramon, 1980; Zihl and Werth, 1984). It is interesting in the context of perceptual learning that the practice effect transferred to stimuli of lower contrast in the patient tested by Bridgeman and Staggs (1982), and even to a different function in the three patients of Zihl and von Cramon (1980). The latter patients were first required to blink whenever a 116 in., 100 ms stimulus was presented to the blind field; blank stimuli were used for control comparison. 480–600 trials were given, and all patients' blink responses to targets, but not to blank control stimuli, increased. Then the same stimulus was presented at seven different positions from 10° to 40° off fixation on the horizontal meridian, and the patients were asked to initiate a saccade to where they guessed the stimulus had appeared. Localization accuracy was compared to results of a similar series conducted before the blink response experiment, and was found considerably improved in two of the patients. In addition to task- and stimulus-specific perceptual learning, transfer of practice effects has thus been demonstrated.

Blindsight with feedback

It is noteworthy that these effects have been reported although the patients did not receive feedback in any of these studies. This is a general feature of testing blindsight in humans, but not in monkeys with primary visual cortex ablation. Monkeys, unlike humans, are usually rewarded for responding correctly, as part of the procedure required to "explain" their task. As feedback is likely to facilitate learning, we have used it in the following localization task.

The patient is seated in front of a hemi-cylindrical training perimeter. It is studded with red light emitting diode (LED) buttons which are spaced ~7° apart laterally. Fixating the central one that is lit throughout the series, the subject presses a start key. This causes a second LED to light up; simultaneously, a sound is emitted from a central loudspeaker integrated into the setup; this sound informs the patient that a LED is on but does not provide a cue to its position. The patient is told beforehand which meridian(s) is active, and has to find the lit LED as quickly as he can with either hand, but without moving his eyes. As targets appear on either side of fixation, half of them will be visible to a patient with a complete hemianopia. The sound continues until the proper LED has been pressed, which extinguishes both the light and the sound, thereby informing the patient that he has hit the correct LED. The computer registers all LEDs pressed per trial as well as the search time from stimulus onset to the hit.

In November, 2004, patient BT presented with a complete hemianopia to the left that resulted from a large lesion (an arteriovenous malformation had been embolized and extirpated in 1999) that destroyed the largest part of his right occipital lobe. He was tested with the LEDs on the horizontal meridian, which are schematically presented in Fig. 6. Median search times for an early test conducted in February 2005 are compared to the corresponding data from a similar series collected 4 months later; BT had in the meantime been trained in both target detection and localization. The results shown in Fig. 7 are based on 15 responses per LED, and reveal a reduction in search times for the data from the blind field ($Z = -2.173$, $P_{1\text{—tailed}} = 0.0014$; Wilcoxon).

The improvements observed over time in the few studies devoted to blindsight learning may explain why positive results are somewhat more common in cortically blind monkeys. Whereas patients are often tested on a very limited number of presentations, monkeys may receive hundreds or thousands of trials to learn whether they will reach the criterion or not. Clearly, all the factors that influence the results in studies of human patients also affect the monkeys'. The task is important, and so is the extent of the lesion that is often not confined to primary visual cortex. Lesion size affects the extent of transneuronal retrograde retinal ganglion cell degeneration (Cowey et al., 1999), as well as visually guided behaviors, as has been systematically studied in monkeys (see the section on "Neuronal pathways to blind vision"). In addition, the age at lesion, usually lower for the monkeys, contributes. Considerably more extensive rewiring has been demonstrated in monkeys following early striate cortical damage "which results in neuronal compensations, and in the sparing of certain classes of visually guided behaviors that do not occur following equivalent damage sustained in adulthood" (Payne et al., 1996, p. 742). Indeed, the age at lesion, and not the age at the beginning of blindsight testing which is closely correlated with age at lesion in monkeys but not human patients, was the second most predictive factor of individual performance in the rotation detection task. The amount of experience with blindsight testing was the first.

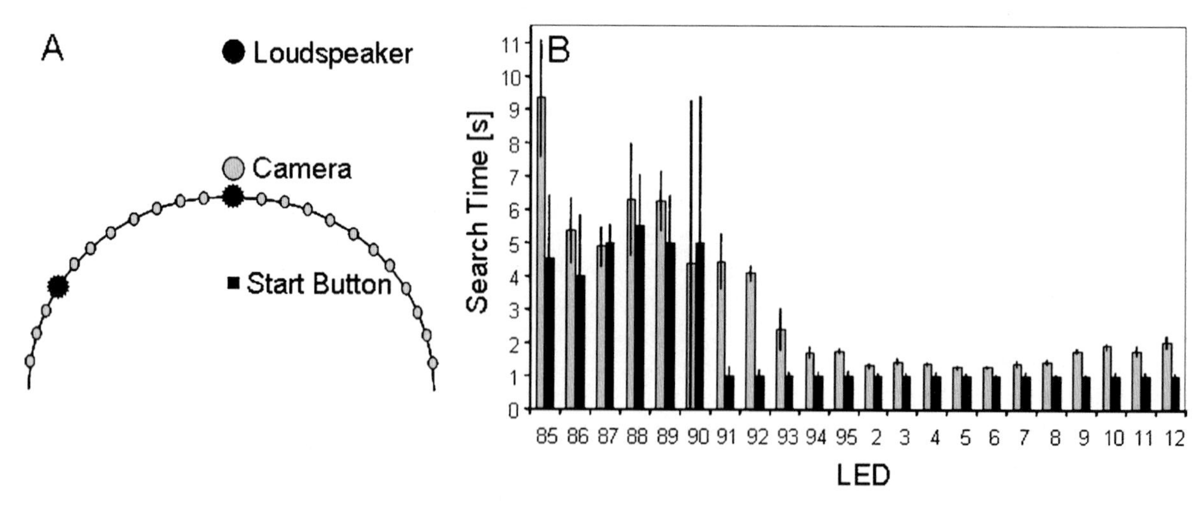

Fig. 6. Manual localization was tested in a training perimeter. The horizontal meridian that was used for patient BT is drawn on the left. The central LED served as fixation. When the patient pressed the start button, the target LED lit up, and a sound buzzed for as long as it took the patient to find and press it; this extinguished both the light and the sound. Median search times ($n = 15$ per LED, $\pm$the Semi Interquartile Range) represent data collected ~4 months apart. BT reported detecting the first LED to the left of fixation when it lit up, because of light emanating from it, but insisted that he had no information as to the location of the other targets, and found himself to be only guessing.

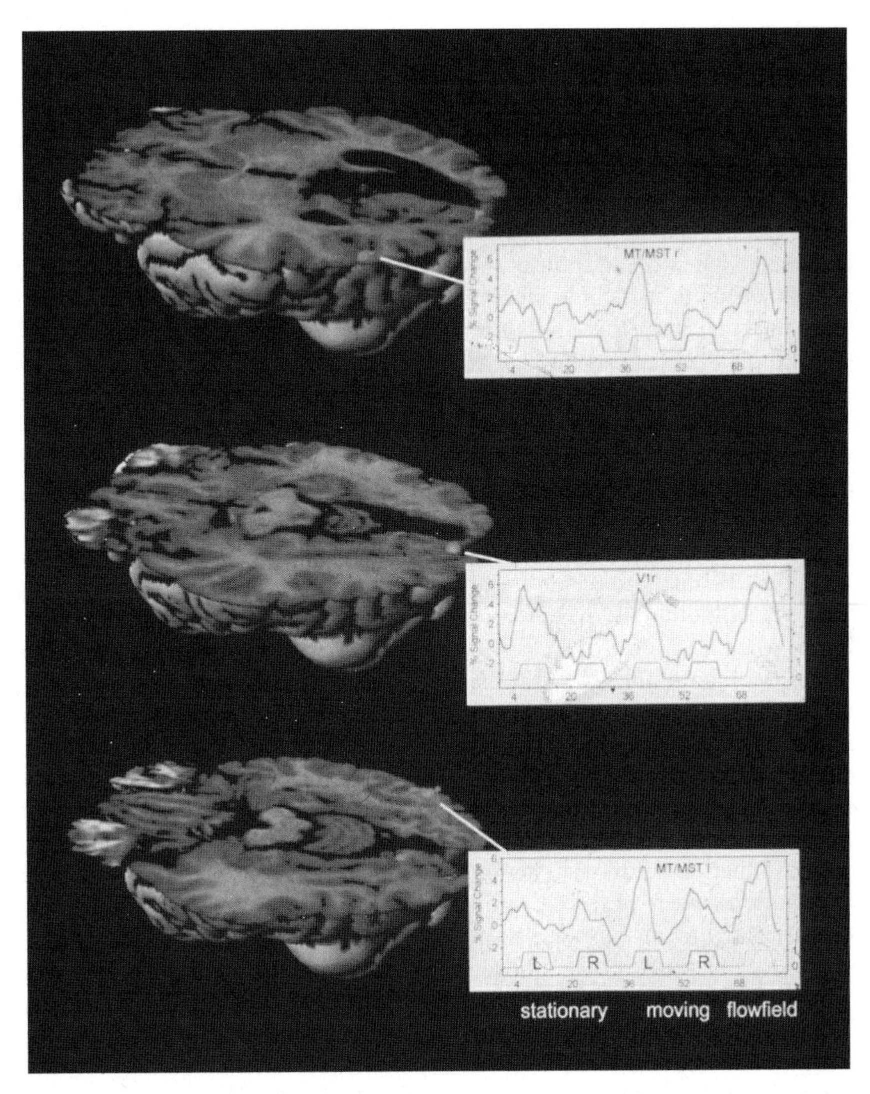

Fig. 7. Functional neuroimaging data of patient GY show strong ipsilesional activation in response to a rotating spiral presented in the hemianopic field. Unlike presentation of the same stimulus in the normal field which caused activation both in the motion complex and in V1, no V1 activation is detected anywhere within the lesioned calcarine cortex.

The blind in blindsight

The hypotheses that have been put forward regarding the blind in blindsight address the role of the primary visual cortex in conscious vision. The major ones suggest that (1) conscious vision is lost because the extrastriate visual cortex depends on the massive geniculo-striate cortical input, (2) conscious vision depends on the back-propagation of signals from the higher visual cortical areas, and (3) both output of and input to primary visual cortex are required.

According to the first hypothesis, V1 is important because the quantitatively lesser extra-geniculo-striate cortical routes to the higher visual cortex are insufficient to render vision conscious. A critical amount, or organization, of visual cortical activation would be required for that purpose, but is not reached without the massive geniculo-striate-extrastriate input. Whether the processing that V1 performs before it transfers its output to the next stages of processing is also important is not known. Neuroimaging data from human subjects performing a challenging detection task at contrast threshold showed that a large, stimulus-independent response in the early visual cortical areas V1–V3 predicted behavioral performance, which was best when this response was large (Ress et al., 2000). Due to its being stimulus independent, the authors interpret this finding as manifestation of attentional processes. Whether it is attention or some other endogenous fluctuation, its co-varying with performance indicates that the state of the early cortices determines whether weak stimuli are processed in a manner that allows them to become consciously represented. Whether it is the output signal of these cortices, or whether it is the signals that the higher extrastriate regions send back to the early areas including V1, which is in some way insufficient when V1 is destroyed, is still open; destruction of V1 compromises both the feed-forward and the feedback signals. A role for the latter is suggested by elegant physiological studies in awake behaving monkeys, where activity patterns in V1 differentiated figure and ground in difficult segmentation tasks only when the animals succeeded in the task (Supér et al., 2001). The figure–ground differentiation, when observed, only started at ~90 ms after figure onset, suggesting that back-propagated signals with their longer latencies play an important role in performance. Lamme et al. (1998) proposed that feedback connections are essentially involved in rendering vision conscious, while the feed-forward flow is required for fast behavioral responses that are not always linked to stimulus awareness. However, since we presently have no procedures that selectively disrupt either the back-projections or the feed-forward ones, it is not known whether loss of back-propagated information by itself produces blindness. Consequently, we do not know whether one type of transfer is more important than the other, or whether, as assumed in hypothesis (3), both are required for conscious vision.

More precisely, this should read veridical conscious vision because nonveridical vision, such as hallucinations (Kölmel, 1984; Lepore, 1990) and phosphenes (Cowey and Walsh, 2000), can occur in fields of cortical blindness. This demonstrates that the brain retains the capacity to produce some kinds of vivid conscious vision even when V1 is destroyed (Stoerig, 2001). Therefore it may not be the primary visual cortex that is inexpendable for conscious vision as such, but rather a sufficient amount of appropriate activity. According to this view, such activity would arise spontaneously in release excitation, or be induced by magnetic or electrical stimulation, but it would not be evoked by the retinal input that still reaches the visual cortex via extra-geniculo-striate cortical routes. This suggestion not only agrees with the evidence but also allows for the possibility of some conscious vision to return to the cortically blind field. If the plastic changes that depend on training and age at lesion strengthen the extrastriate cortical activation, shrinkage or thinning out of the blind field might ensue.

The sight in blindsight

Patient GY was already mentioned in the section "Blindsight and plasticity". He suffered his lesion when he was 8 years of age. This relatively early lesion should, together with his long-lasting and intensive experience with testing his hemianopic field, predestine him for a recovery of at least some conscious vision, if that is possible when just a small macular sparing remains of the affected hemifield. In fact, and despite his often, and somewhat unfortunately, being referred to as a blindsight subject, GY has been known to acknowledge awareness of stimuli for a long time. Indeed, Barbur et al. (1980) already mentioned that he sees "dark shadows" in response to flashed targets, and entitled a later positron emission tomography (PET) study of GY "Conscious visual perception without V1" (Barbur et al., 1993). However, GY has not only used visual terms such as "dark shadows" to describe his sensations, but also denied "seeing" the stimuli he detected, discriminated, and acknowledged to be aware of. It is on the basis of studying this subject that Weiskrantz (1998) suggested a distinction of blindsight type I and II; type I is the original form of blindsight which is observed in the absence of acknowledged awareness, whereas type II allows for "nonvisual" awareness of visual stimuli. Nonvisual, abstract rather than phenomenal awareness is not experienced in normal veridical vision; therefore, if GY's hemianopic visual functions were indeed "nonvisual" in this sense, visual stimuli presented to his normal field should not evoke a similar type of awareness. Arguing that a perceptual match, if it existed, would show that his hemianopic vision was visual, Stoerig and Barth (2001) attempted to find a stimulus condition which when presented to GY's normal hemifield would evoke the same kind of sensation that more prominent stimuli evoke in his hemianopic field. Having presented a series of degraded stimuli to the good field which GY always discarded upon first sight as being "visual", we eventually hit upon a match when we presented a moving bar not defined by luminance contrast but by coherent shifts in a grainy texture (see <www.ebarth.de/demos/gy>). Comparing this stimulus in the good field to a moving luminance-defined bar in the hemianopic field produced not only a perceptual match, but also closely similar discrimination performance (Stoerig and Barth, 2001). We concluded that GY has some type of low-level phenomenal vision rather than abstract conscious access (Block, 1995) in his affected hemifield.

As none of the functional neuroimaging studies in which he participated (Barbur et al., 1993; Sahraie et al., 1997; Zeki and ffytche, 1998; Baseler et al., 1999; Goebel et al., 2001) produced any evidence for activation within his lesioned V1, GY's vision in the hemianopic field must — at least as far as this conclusion is warranted when no pathological data are available — be mediated by the extra-geniculo-striate cortical pathways. If, as argued above, the strength of the extrastriate activation was important for the conscious rendering of stimuli, GY ought to show relatively strong patterns of activation in response to targets presented to his hemianopic field. This expectation gained support when a stronger response was observed in GY's motion complex when he discriminated the direction of a single moving dot in his hemianopic field in his "aware" rather than "unaware" mode (Zeki and ffytche, 1998). In addition, the ipsilesional activation that occurred in roughly the same region when the red-and-blue spiral was presented to GY's hemianopic field was remarkably strong not only when compared to that evoked by presenting the same stimulus to the good field, but also to data from other patients (Goebel et al., 2001; see also Fig. 4). Both findings agree with the hypothesis that the strength of extrastriate cortical activation is at least one of the critical factors determining whether or not a stimulus is consciously represented.

More evidence for a route to extrastriate cortex that bypasses V1 comes from a study of a different patient who suffered a stroke at an age of 19 (Schoenfeld et al., 2002). Like GY, this patient performed above chance in motion and color change experiments, and reported perceiving moving objects in his affected hemifield. Presentation of motion and color change stimuli in the hemianopic field during functional neuroimaging yielded pronounced activation in ipsilesional extrastriate cortical regions, although even at low threshold ($p<0.1$) no activity was detected within the lesioned V1, and magnetoencephalography revealed earlier stimulus-locked responses in hMT+ than in V2.

Our own data on a patient, who suffered a stroke in the territory of the posterior cerebral artery when he was 58 years, both confirm and extend these results. HK's ipsilesional motion complex responded to moving blind field stimuli already within 3 months of his lesion when no color-evoked responses could be detected (see Fig. 4). A year and regular weekly tests later, his ipsilesional ventral cortex also responded to color. It is probably safe to assume that the blindsight training contributed to this change despite his higher age at lesion. Moreover, HK also increasingly reports receiving "signals" from his hemianopic field, and performs at >90% correct in some tasks (but not others).

Blindsight and recovery

While these three patients — GY, HK, and Schoenfeld et al.'s patient — appear to have recovered some more or less low-level vision in large regions of their hemianopic fields, shrinkage of the blind field has been observed in others. Zihl and von Cramon (1985) reviewed 55 cases of patients who had undergone systematic localization training, and reported that this led "in the majority of patients, to an enlargement of the visual field" (p. 335). Like Sabel and colleagues who trained their patients specifically along the borders between the normal and the blind field (Kasten et al., 1998a), the authors attribute the enlargement to recovery of reversibly damaged striate cortical tissue (Kasten et al., 1998b). While both groups have reported specific effects in the trained regions, our own blindsight training rather seems to produce shrinkage of the absolute portion of the defect from the periphery inwards (see Stoerig, 1998). A remarkable case is that of patient FS whose trauma-induced damage destroyed the temporal lobe and deprived the visual cortex of its afferents. This caused a hemianopia that had already become incomplete, albeit with macular splitting, when I first saw him several years after his accident. He continued to participate in experiments (see Pöppel, 1985, 1986; Stoerig, 1987; Goebel et al., 2001, for examples), and now presents with a strip of homonymous blindness extending along the horizontal meridian. This strip is still broadest in the central hemifield, but remarkably, the vision that slowly returned over the years has become normal in the recovered regions above and below (Fig. 8).

Together, these results demonstrate not only that the visual system is capable of remarkable plasticity, but also that different mechanisms are likely to mediate the different types of recovery.

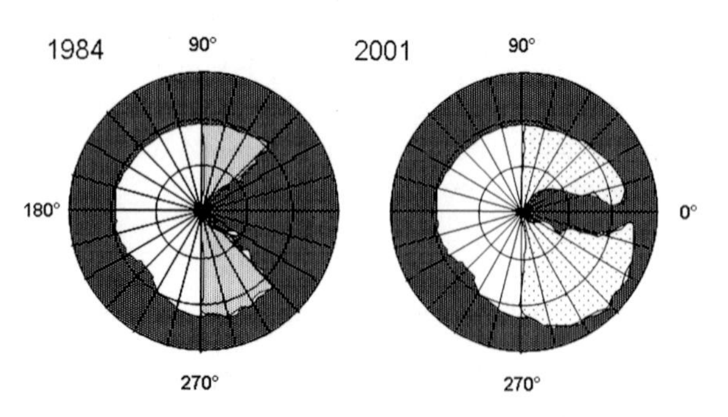

Fig. 8. Visual field plots of patient FS are based on a combination of dynamic and static perimetry using a 116 in., 320 cd/m^2 white target on a 10 cd/m^2 white background.

These range from the recovery of low-level vision within largely unchanged defective field borders to shrinkage of the blind field in stimulated regions or in extensive, predominantly peripheral portions of the defect. Striate cortex is very likely involved when it is still available, and may enable recovery of visual awareness that more closely resembles that of normal vision. However, the low-level conscious visual functions seen in Scheonfeld et al.'s patient, in GY, and in HK appear to be possible without a contribution from ipsi- or contralesional V1. Thus, V1 does not seem necessary for the recovery of some low-level vision. Probably the most remarkable case to date is that of FS, where visual stimulation of the recovered visual field produced only extrastriate cortical activation (Kleiser et al., 2001). How the brain's plasticity is engaged to invoke these changes, and how they can best be strengthened and harnessed to serve recovery of vision, requires further investigation.

Conclusion

Although the controversies regarding the role of V1 in conscious vision are bound to continue, a large body of data from monkeys and human patients demonstrates that blindsight does not depend on surviving islands of striate cortex. Furthermore, varying degrees of conscious vision can return to previously absolute fields of blindness. This not only is true within weeks or months postlesion, but can evolve in a long-term process that, like blindsight, seems to depend on challenging the system's plasticity by forcing the subject to respond to blind field targets. While striate cortex will be involved whenever the lesion allows it, the brain's plastic capacities allow functional improvements even when V1 is destroyed or denervated and the lesion occurs late in life.

Abbreviations

dLGN	dorsal lateral geniculate nucleus
V1	primary visual cortex or striate cortex
V2	second visual cortical area
V3, V4, V5	extrastriate visual cortical areas
MT (= V5)	middle temporal area

hMT+ human motion complex that includes adjacent motion-sensitive areas like MST and FST

RDK random dot kinematogram

Acknowledgments

Much of the work here presented could not have been carried out without the patients' long-term commitment. Thanks are also extended to Sandro Sardina, Manfred Mittelstaedt, and Andrea Lauber who contributed to the experiments. The work was supported by the Deutsche Forschungsgemeinschaft (SFB 194 — A15).

References

Azzopardi, P. and Cowey, A. (2001) Motion discrimination in cortically blind patients. Brain, 124: 30–46.

Barbur, J.L. (2004) Learning from the pupil. Studies of basic mechanisms and clinical application. In: Chalupa, L.M. and Werner, J.S. (Eds.), The Visual Neurosciences. Bradford Books, Cambridge, pp. 641–656.

Barbur, J.L., Ruddock, K.H. and Waterfield, V.A. (1980) Human visual responses in the absence of the geniculo-calcarine projection. Brain, 103: 905–928.

Barbur, J.L., Watson, J.D.G., Frackowiak, R.S.J. and Zeki, S. (1993) Conscious visual perception without V1. Brain, 116: 1293–1302.

Barton, J.J. and Sharpe, J.A. (1997) Motion direction discrimination in blind hemifields. Ann. Neurol., 41: 255–264.

Baseler, H.A., Morland, A.B. and Wandell, B.A. (1999) Topographic organization of human visual areas in the absence of input from primary cortex. J. Neurosci., 19: 2619–2627.

Beatty, J. (1986) The pupillary system. In: Coles, M.G.H., Donchin, E. and Porges, S.W. (Eds.), Psychophysiology. Systems, Processes, and Application. Guilford Press, New York, pp. 43–50.

Benson, D.F. and Greenberg, J.P. (1969) Visual form agnosia. Arch. Neurol., 20: 82–89.

Block, N. (1995) On a confusion of a function of consciousness. Behav. Brain Sci., 18: 227–287.

Bridgeman, B. and Staggs, D. (1982) Plasticity in human blindsight. Vision Res., 22: 1199–1203.

Bruce, C.J., Desimone, R. and Gross, C.G. (1986) Both striate cortex and superior colliculus contribute to visual properties of neurons in superior temporal polysensory area of macaque monkey. J. Neurophysiol., 55: 1057–1075.

Bullier, J., Girard, P. and Salin, P.-A. (1994) The role of area 17 in the transfer of information to extrastriate cortex. In: Peters, A. and Rockland, K.S. (Eds.) Cerebral Cortex, Vol.10. Plenum Press, New York, pp. 301–330.

Campion, J., Latto, R. and Smith, Y.M. (1983) Is blindsight due to scattered light, spared cortex and near threshold effects? Behav. Brain Sci., 6: 423–486.

Collins, C.E., Xu, X., Khaytin, K., Kaskan, P.M., Casagrande, V.A. and Kaas, J.H. (2005) Optical imaging of visually evoked responses in the middle temporal area after deactivation of primary visual cortex in adult primates. Proc. Natl. Acad. Sci. USA, 102: 5594–5599.

Cowey, A. (1974) Atrophy of retinal ganglion cells after removal of striate cortex in a rhesus monkey. Perception, 3: 257–260.

Cowey, A. (1994) Cortical visual areas and the neurobiology of higher visual processes. In: Farah, M.J. and Ratcliff, G. (Eds.), Object Representation and Recognition. Lawrence Erlbaum, Hillsdale, NJ, pp. 3–31.

Cowey, A. and Stoerig, P. (1995) Blindsight in monkeys? Nature, 373: 247–249.

Cowey, A., Stoerig, P. and Perry, V.H. (1989) Transneuronal retrograde degeneration of retinal ganglion cells after damage of striate cortex in macaque monkeys: selective loss of Pβ cells. Neuroscience, 29: 65–80.

Cowey, A., Stoerig, P. and Williams, C. (1999) Variance in transneuronal retrograde ganglion cell degeneration in monkeys after removal of striate cortex: effects of the size of the cortical lesion. Vision Res., 39: 3642–3652.

Cowey, A. and Walsh, V. (2000) Magnetically induced phosphenes in sighted, blind and blindsighted observers. Neuroreport, 11: 3269–3273.

Czeisler, C.A., Shanahan, D.L., Klerman, E.B., Martens, H., Brotman, D.J., Emens, J.S., Klein, T. and Rizzo III, J.F. (1995) Suppression of melatonin secretion in some blind patients by exposure to bright light. New Engl. J. Med., 322: 6–11.

Dineen, J., Hendrickson, A. and Keating, E.G. (1982) Alterations of retinal inputs following striate cortex removal in adult monkey. Exp. Brain Res., 47: 446–456.

Farah, M.J. (1990) Visual Agnosias. MIT Press, Cambridge, MA.

Feinberg, T.E., Pasik, T. and Pasik, P. (1978) Extrageniculostriate vision in the monkey. VI. Visually guided accurate reaching behavior. Brain Res., 152: 422–428.

Fendrich, R., Wessinger, C.M. and Gazzaniga, M.S. (1992) Residual vision in a scotoma: implications for blindsight. Science, 258: 1489–1491.

Fendrich, R., Wessinger, C.M. and Gazzaniga, M.S. (1993) Authors' response. Science, 261: 494–495.

Ferrier, D., 1886^2. The Functions of the Brain. Smith, Elder and Co., London.

FitzMaurice, M.C., Ciaramitaro, V.M., Palmer, L.A. and Rosenquist, A.C. (2003) Visual detection deficits following inactivation of the superior colliculus. Vis. Neurosci., 20: 687–701.

Foster, R.G. (2005) Bright blue times. Nature, 433: 698–699.

Girard, P., Salin, P.-A. and Bullier, J. (1992) Response selectivity of neurons in area MT of the macaque monkey during reversible inactivation of area V1. J. Neurophysiol., 67: 1–10.

Goebel, R., Muckli, L., Zanella, F.E., Singer, W. and Stoerig, P. (2001) Sustained extrastriate cortical activation without visual awareness revealed by fMRI studies of hemianopic patients. Vision Res., 41: 1459–1474.

Gross, C. (1992) Representation of visual stimuli in inferior temporal cortex. Philos. Trans. Roy. Soc. London, B, 335: 3–10.

Heywood, C.A., Cowey, A. and Newcombe, F. (1991) Chromatic discrimination in a cortically colour blind observer. Eur. J. Neurosci., 3: 802–812.

Hoffmann, K.-P. (1989) Control of the optokinetic reflex by the nucleus of the optic tract in primates. In: Allum, J.H.J. and Hullinger, M. (Eds.) Progress in Brain Research, Vol. 80. Elsevier, Amsterdam, New York, pp. 173–182.

Horwitz, G.D., Batista, A.P. and Newsome, W.T. (2004) Representation of an abstract perceptual decision in macaque superior colliculus. J. Neurophysiol., 91: 2281–2296.

Humphrey, N.K. (1970) What the frog's eye tells the monkey's brain. Brain, Behav. Evol., 3: 324–337.

Humphrey, N.K. (1974) Vision in a monkey without striate cortex: a case study. Perception, 3: 241–255.

Jellema, T. and Perrett, D.I. (2003) Cells in monkey STS responsive to articulated body motions and consequent static posture: a case of implied motion? Neuropsychologia, 41: 1728–1737.

Kanwisher, N. (2001) Faces and places: of central (and peripheral) interest. Nature Neurosci., 4: 455–456.

Kanwisher, N., McDermott, J. and Chun, M.M. (1997) The fusiform face area: a module in human extrastriate cortex specialized for face perception. J. Neurosci., 17: 4302–4311.

Kasten, E., Wüst, S., Behrens-Baumann, W. and Sabel, B.A. (1998b) Computer-based training for the treatment of partial blindness. Nat. Med., 4: 1083–1087.

Kasten, E., Wüst, S. and Sabel, B.A. (1998a) Residual vision in transition zones in patients with cerebral blindness. J. Clin. Exp. Neuropsychol., 20: 581–598.

Keane, J.R. (1979) Blinking to sudden illumination. A brain stem reflex present in neocortical death. Arch. Neurol., 36: 52–53.

Kentridge, R.W., Heywood, C.A. and Weiskrantz, L. (1997) Residual vision in multiple retinal locations within a scotoma: implications for blindsight. J. Cogn. Neurosci., 9: 191–202.

Kisvárday, Z., Cowey, A., Stoerig, P. and Somogyi, P. (1991) The retinal input into the degenerated dorsal lateral geniculate nucleus after removal of striate cortex in the monkey: Implications for residual vision. Exp. Brain Res., 86: 271–292.

Kleiser, R., Niedeggen, M., Wittsack, J., Goebel, R. and Stoerig, P. (2001) Is V1 necessary for conscious vision in areas of relative cortical blindness? Neuroimage, 13: 654–661.

Klüver, H. (1942) Functional significance of the geniculostriate system. Biol. Symp., 7: 253–299.

Kölmel, H.W. (1984) Coloured patterns in hemianopic fields. Brain, 107: 155–167.

Lamme, V.A.F., Supèr, H. and Spekreijse, H. (1998) Feedforward, horizontal, and feedback processing in the visual cortex. Curr. Opin. Neurobiol., 8: 529–535.

Lepore, F.E. (1990) Spontaneous visual phenomena with visual loss: 104 patients with lesions of retinal and neural afferent pathways. Neurology, 40: 444–447.

Livingstone, M.S. and Hubel, D. (1988) Segregation of form, color, movement, and depth: Anatomy, physiology, and perception. Science, 240: 740–749.

Magoun, H.W. and Ranson, S.W. (1935) The central path of the light reflex. Arch. Ophthalmol., 13: 791–811.

Marcel, A.J. (1998) Blindsight and shape perception: deficit of visual consciousness or of visual function? Brain, 121: 1565–1588.

Marrocco, R.T. and Li, R.H. (1977) Monkey superior colliclus: properties of single cells and their afferent inputs. J. Neurophysiol., 40: 844–860.

Marzi, C.A., Tassinari, G., Lutzemberger, L. and Aglioti, A. (1986) Spatial summation across vertical meridian in henianopics. Neuropsychologia, 24: 749–758.

Meadows, J.C. (1974) Disturbed perception of colours associated with localized cerebral lesions. Brain, 97: 615–632.

Meredith, M.A. and Stein, B.E. (1985) Interactions among converging sensory inputs in the superior colliculus. Science, 221: 389–391.

Mihailovic, L.T., Cupic, D. and Dekleva, N. (1971) Changes in the numbers of neurons and glia cells in the lateral geniculate nucleus of the monkey during retrograde cell degeneration. J. Comp. Neurol., 142: 1510–1526.

Milner, A.D. and Goodale, M.A. (1996) The Visual Brain in Action. Oxford University Press, Oxford.

Mohler, C.W. and Wurtz, R.H. (1976) Organization of monkey superior colliclus: intermediate layer cells discharging before eye movements. J. Neurophysiol., 39: 722–744.

Mohler, C.W. and Wurtz, R.H. (1977) Role of striate cortex and superior colliclus in visual guidance of saccadic eye movements in monkeys. J. Neurophysiol., 40: 74–94.

Moore, R.Y., Speh, J.C. and Card, J.P. (1995) The retinohypothalamic tract originates from a distinct subset of retinal ganglion cells. J. Comp. Neurol., 352: 351–366.

Müller, J.R., Philiastides, M.G. and Newsome, W.T. (2005) Microstimulation of the superior colliculus focuses attention without moving the eyes. Proc. Natl. Acad. Sci. USA, 102: 524–529.

Munk, H., 1881. Über die Functionen der Grosshirnrinde. Verlag von August Hirschwald, Berlin, pp. 76–95.

Pasik, T. and Pasik, P. (1973) Extrageniculostriate vision in the monkey. IV. Critical structures for light vs. no-light discrimination. Brain Res., 56: 165–182.

Pasik, P. and Pasik, T. (1982) Visual functions in monkeys after total removal of visual cerebral cortex. Contr. Sens. Physiol., 7: 147–200.

Pasik, P., Pasik, T. and Krieger, H.P. (1959) Effects of cerebral lesions upon optokinetic nystagmus in monkeys. J. Neurophysiol., 22: 297–304.

Payne, B.R., Lomber, S.G., Macneil, M.A. and Cornwell, P. (1996) Evidence for greater sight in blindsight following damage of primary visual cortex early in life. Neuropsychologia, 34: 741–774.

Perenin, M.-T. (1991) Discrimination of motion direction in perimetrically blind fields. Neuroreport, 2: 397–400.

Perrett, D.I., Rolls, E.T. and Caan, W. (1982) Visual neurons responsive to faces in the monkey temporal cortex. Exp. Brain Res., 47: 329–342.

Pöppel, E. (1985) Bridging a neuronal gap. Naturwissenschaften, 72: 599–600.

Pöppel, E. (1986) Long-range colour-generating interactions across the retina. Nature, 320: 523–525.

Pöppel, E., Held, R. and Frost, D. (1973) Residual visual function after brain wounds involving the central visual pathways in man. Nature, 243: 295–296.

Reese, B.E. and Cowey, A. (1988) Segregation of functionally distinct axons in the monkey's optic tract. Nature, 331: 350–351.

Ress, D., Backus, B.T. and Heeger, D.J. (2000) Activity in primary visual cortex predicts performance in a visual detection task. Nat. Neurosci., 3: 940–945.

Riddoch, G. (1917) Dissociations of visual perception due to occipital injuries, with especial reference to appreciation of movement. Brain, 40: 15–57.

Rodman, H.R., Gross, C.G. and Albright, T.D. (1989) Afferent basis of visual response properties in area MT of the macaque: I. Effects of striate cortex removal. J. Neurosci., 9: 2033–2050.

Rodman, H.R., Gross, C.G. and Albright, T.D. (1990) Afferent basis of visual response properties in area MT of the macaque: II Effects of superior colliculus removal. J. Neurosci., 10: 1154–1164.

Sack, R.L., Lewy, A.J., Blood, M.L., Stevenson, J. and Keith, L.D. (1991) Melatonin administration to blind people: phase advances and entrainment. J. Biol. Rhythms, 6: 249–261.

Sahraie, A., Weiskrantz, L., Barbur, J.L., Simmons, A., Williams, S. and Brammer, M.J. (1997) Pattern of neuronal activity associated with conscious and unconscious processing of visual signals. Proc. Natl. Acad. Sci. USA, 94: 9406–9411.

Sanders, M.D., Warrington, E.K., Marshall, J. and Weiskrantz, L. (1974) 'Blindsight': vision in a field defect. Lancet, 1: 707–708.

Schoenfeld, M.A., Noesselt, T., Poggel, D., Tempelmann, C., Hopf, J.-M., Woldorff, M.G., Heinze, H.-J. and Hillyard, S.A. (2002) Analysis of pathways mediating preserved vision after striate cortex lesions. Ann. Neurol., 52: 814–824.

Siebler, M., Steinmetz, H. and Freund, H.-J. (1998) Therapeutic entrainment of circadian rhythm disorder by melatonin in a non-blind patient. J. Neurol., 245: 327–328.

Simpson, J.I. (1984) The accessory optic system. Annu. Rev. Neurosci., 7: 13–41.

Sprague, J.M. and Meikle, T.H. (1965) The role of the superior colliculus in visually guided behaviour. Exp. Neurol., 11: 115–146.

Stoerig, P. (1987) Chromaticity and achromaticity. Evidence for a functional differentiation in visual field defects. Brain, 110: 869–886.

Stoerig, P. (1993) Sources of blindsight. Science, 261: 493–494.

Stoerig, P. (1998) Blindsight. In: Huber, A. and Koempf, D. (Eds.), Klinische Neuroophthalmologie. Georg Thieme Verlag, Stuttgart, New York, pp. 375–377.

Stoerig, P. (1999) Blindsight. In: Wilson, R. and Keil, F. (Eds.), The MIT-Encyclopedia of the Cognitive Sciences. MIT-Press, Cambridge/MA, pp. 88–90.

Stoerig, P. (2001) The neuroanatomy of phenomenal vision. Ann. New York Acad. Sci., 929: 176–194.

Stoerig, P. and Barth, E. (2001) Low-level phenomenal vision despite unilateral destruction of primary visual cortex. Conscious Cogn., 10: 574–587.

Stoerig, P. and Cowey, A. (1997) Blindsight in man and monkey. Brain, 120: 535–559.

Stoerig, P. and Pöppel, E. (1986) Eccentricity-dependent residual target detection in visual field defects. Exp. Brain Res., 64: 469–475.

Stoerig, P., Hübner, M. and Pöppel, E. (1985) Signal detection analysis of residual target detection in a visual field defect due to a post-geniculate lesion. Neuropsychologia, 23: 589–599.

Stoerig, P., Kleinschmidt, A. and Frahm, J. (1998) No visual responses in denervated V1: high-resolution functional magnetic resonance imaging of a blindsight patient. Neuroreport, 9: 21–25.

Supér, H., Spekreijse, H. and Lamme, V.A. (2001) Two distinct modes of sensory processing observed in monkey primary visual cortex (V1). Nat. Neurosci., 4: 304–310.

Ter Braak, J.W. and van Vliet, A.G.M. (1963) Subcortical nystagmus in the monkey. Psychiat. Neurol. Neurochir., 66: 277–283.

Tolias, A.S., Keliris, G.A., Smirnakis, S.M. and Logothetis, N.K. (2005) Neurons in macaque area V4 acquire directional tuning after adaptation to motion stimuli. Nat. Neurosci., 8: 591–593.

Ungerleider, L.G. and Mishkin, M. (1982) Two cortical visual systems. In: Ingle, D.J., Goodale, M.A. and Mansfield, R.J.W. (Eds.), Analysis of Visual Behavior. MIT-Press, Cambridge, MA, pp. 549–586.

Van Buren, J.M. (1963) Trans-synaptic retrograde degeneration in the visual system of primates. J. Neurol. Neurosurg. Psychiat., 26: 402–409.

van Essen, D.C. (1985) Functional organization of primate visual cortex. In: Peters, A. and Jones, E.G. (Eds.) Cerebral Cortex, Vol. 3. Plenum Press, New York, pp. 259–329.

Waitzman, D.M., Ma, T.P., Optican, L.M. and Wurtz, R.H. (1991) Superior colliculus neurons mediate the dynamic characteristics of saccades. J. Neurophysiol., 66: 1716–1737.

Weller, R.E. and Kaas, J.H. (1989) Parameters affecting the loss of ganglion cells of the retina following ablation of striate cortex in primates. Vis. Neurosci., 3: 327–342.

Weiskrantz, L. (1963) Contour discrimination in a young monkey with striate cortex ablation. Neuropsychologia, 1: 145–164.

Weiskrantz, L. (1986) Blindsight: A Case Study and Implications. Oxford University Press, Oxford.

Weiskrantz, L. (1990) Outlooks for blindsight: explicit methods for implicit processes [Review]. Proc. Roy. Soc. Lond. B Biol. Sci., 239: 247–278.

Weiskrantz, L. (1998) Consciousness and commentaries. In: Hameroff, S., Kaszniak, A. and Scott, A. (Eds.), Towards a Science of Consciousness II—The Second Tucson Discussion and Debates. MIT Press, Cambridge, MA, pp. 371–377.

Weiskrantz, L., Barbur, J.L. and Sahraie, A. (1995) Parameters affecting conscious versus unconscious visual discrimination with damage to the visual cortex (V1). Proc. Natl. Acad. Sci. USA, 92: 6122–6126.

Weiskrantz, L. and Cowey, A. (1963) Striate cortex lesions and visual acuity of the rhesus monkey. J. Comp. Physiol. Psychol., 56: 225–231.

Weiskrantz, L. and Cowey, A. (1967) Comparison of the effects of striate cortex and retinal lesions on visual acuity in the monkey. Science, 155: 104–106.

Weiskrantz, L., Warrington, E.K., Sanders, M.D. and Marshall, J. (1974) Visual capacity in the hemianopic field following restricted occipital ablation. Brain, 97: 709–728.

Yukie, M. and Iwai, E. (1981) Direct projection from the dorsal lateral geniculate nucleus to the prestriate cortex in macaque monkeys. J. Comp. Neurol., 201: 81–97.

Zeki, S. (1978) Functional specialisation in the visual cortex of the rhesus monkey. Nature, 274: 423.

Zeki, S. and ffytche, D.H. (1998) The Riddoch syndrome: insights into the neurobiology of conscious vision. Brain, 121: 25–45.

Zihl, J. (1980) "Blindsight": Improvement of visually guided eye movements by systematic practice in patients with cerebral lesions. Neuropsychologia, 18: 71–77.

Zihl, J. and von Cramon, D. (1980) Registration of light stimuli in the cortically blind hemifield and its effect on localization. Behav. Brain Res., 1: 287–298.

Zihl, J. and von Cramon, D. (1985) Visual field recovery from scotoma in patients with postgeniculate damage. A review of 55 cases. Brain, 108: 335–365.

Zihl, J., von Cramon, D. and Mai, N. (1983) Selective disturbance of movement vision after bilateral brain damage. Brain, 106: 313–340.

Zihl, J. and Werth, R. (1984) Contributions to the study of "blindsight". II. The role of specific practice for saccadic localization in patients with postgeniculate visual field defects. Neuropsychologia, 22: 13–22.

Martinez-Conde, Macknik, Martinez, Alonso & Tse (Eds.)
Progress in Brain Research, Vol. 155
ISSN 0079-6123

CHAPTER 13

Bilateral frontal leucotomy does not alter perceptual alternation during binocular rivalry

Fernando Valle-Inclán* and Emma Gallego

Department of Psychology, University of La Coruña, Campus de Elviña, La Coruña 15071, Spain

Abstract: When discrepant stimuli are presented to each eye and fusion is impossible, perception spontaneously oscillates between the two patterns (binocular rivalry). Functional MRI (fMRI) research identified a frontoparietal network in the right hemisphere associated with perceptual transitions, and it has been proposed that this network is at the origin of the perceptual alternations. Neuroimaging results, however, do not imply causality and lesion studies are needed. Here, we studied one patient who had most of the prefrontal cortex disconnected from the rest of the brain after a bilateral frontal leucotomy. His performance in two binocular rivalry tasks was indistinguishable from that of the controls. The results indicate that prefrontal cortex is unnecessary for perceptual alternations during binocular rivalry.

Keywords: binocular rivalry; prefrontal cortex; leucotomy; consciousness; perceptual alternation

Introduction

When the monocular images cannot be fused, perception alternates between the two or more possible interpretations. This phenomenon (binocular rivalry) posits two basic questions, both related to the neural machinery of consciousness. One question is at what level within the visual pathways is rivalry resolved. Single unit recordings (reviewed in Logothetis, 1998) indicate that the proportion of neurons following the percept, instead of the sensory stimulation, increases from V1 (20% of recorded neurons correlated their firing rate with the perceptual report) to areas in the temporal cortex (80% of the recorded neurons). Early fMRI studies on binocular rivalry (Lumer et al., 1998; Tong et al., 1998; Lumer and Rees, 1999) also found neural activity related to consciousness in extraestriate cortex, and not in V1, in agreement with single unit recordings. More recent results, however, indicate that blood oxygen level dependent (BOLD) oscillations in V1 (Polonsky et al. 2000; Tong and Engel, 2001) and also in LGN (Haynes et al., 2005; Wunderlich et al., 2005) are correlated with consciousness (see also the study by Kleinschmidt et al., 1998, with ambiguous figures). The reasons for the discrepancy between single unit and fMRI results might lie in the origins of the BOLD signal, more related to local field potentials than to spike firing (Logothetis et al., 2001; Logothetis and Wandell, 2004). In agreement with this interpretation, Gail et al. (2004) found that local field potential activity in V1 was correlated with consciousness, but firing rate was not. Taking all the studies together, it can be concluded that the neural correlates of consciousness are distributed along the ventral pathway, including V1 and quite possibly LGN, and overlap with the anatomy of the perceptual machinery.

The second question raised by binocular rivalry concerns the control of the perceptual alternation. It is well known that the alternation rate can be influenced by low-level factors (luminosity,

*Corresponding author. Tel.: +34-981-167000; Fax: +34-981-167153; E-mail: fval@udc.es

DOI: 10.1016/S0079-6123(06)55013-7

contrast, size, and motion), suggesting that perceptual alternance is controlled early in the visual pathway. Recent experiments also demonstrated that rivalry only appears when local (not global) elements are discordant (Carlson and He, 2004), and that the alternation rate depends on local neural adaptation (Blake et al., 2003; Chen and He, 2004). The lack of attentional or volitional effects (Meng and Tong, 2004) also supports that rivalry alternation is controlled at early stages of visual processing.

Neuroimaging studies, however, suggest a top-down control of perceptual changes. Lumer et al. (1998) found that activity in frontal, parietal, and extrastriate cortex in the right hemisphere was larger during rivalry than during the replay condition (a movie reproducing the perceptual transitions previously indicated by the observer in the binocular rivalry condition). Since the phenomenal changes were assumed to be similar in the two conditions, the authors proposed that the frontoparietal network was involved in the generation of the perceptual oscillation. In partial agreement with these results, Kleinschmidt et al. (1998) found bilateral activation in ventral prefrontal, frontal eye-fields, and parietal areas associated with perceptual reversal of ambiguous figures. Lumer and Rees (1999) compared rivalry with a stable-viewing condition and also found frontoparietal activation, presumably related to perceptual switching. In summary, neuroimaging results indicate that association areas generally involved in visual attention, become active during image transitions, and it has been proposed that the source of the perceptual oscillation lies in this frontoparietal network. Causality, however, cannot be claimed without lesion studies, and the few published experiments do not allow such strong claim. Bonneh et al. (2004) found that right hemisphere lesions slowed the alternation rate only when neglect was evident. O'Shea and Corballis (2003, 2005) showed that rivalry can be produced in left and right hemispheres in a split-brain patient, which contradicts a special role for the right hemisphere. Here we present the first neuropsychological study, to our knowledge, about the involvement of prefrontal cortex on binocular rivalry.

Methods

The experiment was approved by the ethics committee of the Institut Pere Mata. The patient was a man, 70 years old, who was diagnosed of schizophrenia and suffered a bilateral leucotomy (Friedman's procedure) in 1956. A clinical MRI scanning (see Fig. 1) showed an enormous subcortical lesion that apparently isolated most of the prefrontal cortex from the rest of the brain. The dorsolateral prefrontal cortex was also damaged. The Wisconsin Card Sorting Test (WCST) evidenced a marked perseveration (93%), confirming that frontal functions were severely compromised. Despite the extensive lesion, and the years in the psychiatric institution, the patient was collaborative.

The stimuli were drawn in green and red shades and presented on a laptop screen viewed from about 70 cm. The patient wore red and green filters that were switched between the eyes at the beginning of each observation period. He was instructed to say aloud what he perceived and the experimenter (EG) held down one of two keys depending on the patient's perception. Timing and stimulus presentation was controlled using Psychophysics Toolbox (Brainard, 1997; Pelli, 1997).

In the first experiment, the stimuli were two orthogonal gabor gratings (5° in diameter, spatial frequency 3 cpd). At the end of each 1 min observation period, one of the gratings was removed during 10 s. The purpose of this manipulation was to test the reliability of the patient's perceptual reports. There were two sessions each comprising ten 1 min observation periods.

In the second experiment, the stimuli were a house in red shades and a face in green shades (Tong et al., 1998) presented in the center of the screen. The approximate size was 5° × 5°. There were four experimental sessions on different days. Each session comprised ten 1 min observation periods.

Results

In Experiment 1, the patient always correctly reported the orientation of the grating presented at the end of each observation period, confirming the

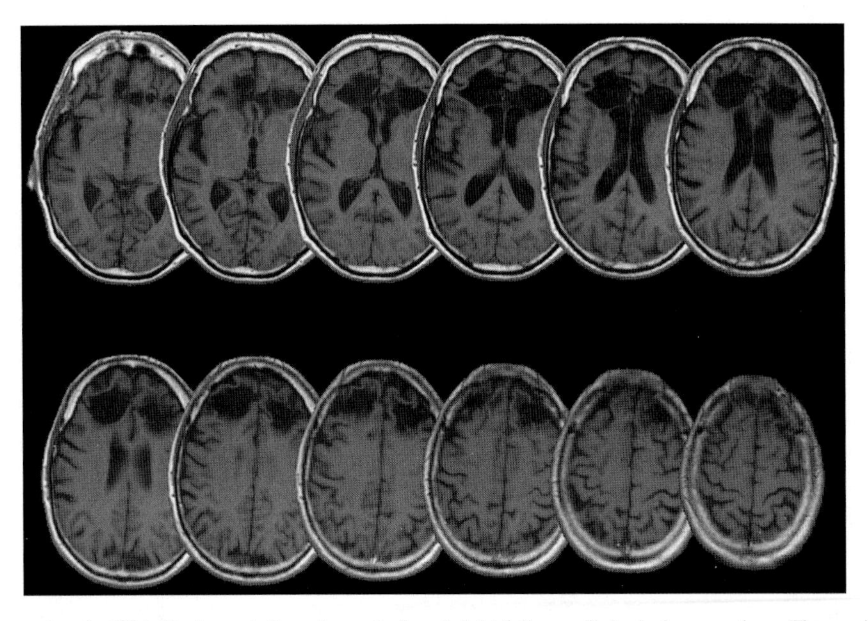

Fig. 1. MRI of the patient in 2004. Horizontal slices, 5 mm. Left and right follow radiological conventions. The surgical operation was performed in 1956.

accuracy of his perceptual reports. The mean duration of dominance phases (3.16 s), the percentage of time in dominance (28.12%), and the distribution of dominance phase durations (Fig. 2A), all of them closely corresponded with what could be expected for normal observers.

The results of Experiment 2 are summarized in Fig. 2B. Mean dominance duration was 3.97 s, the proportion of time in dominance was 82.29%, and the distribution of dominance phases were indistinguishable from control subjects (the two authors) depicted in Fig. 2C.

Discussion

Most long-range connections between prefrontal cortex and the rest of the brain were severed by the surgical procedure, and despite this extensive disconnection, binocular rivalry showed the typical perceptual alternation. This finding strongly suggests that prefrontal cortex does not play a causal role in the perceptual switching, in contradiction with previous interpretations of fMRI results implicating a frontoparietal network (Kleinschmidt et al., 1998; Lumer et al., 1998; Lumer and Rees, 1999). Although the lesion did not affect the right inferior frontal gyrus (the area identified in some of the previous fMRI studies), the subcortical connections in that area were damaged and consequently, the frontoparietal network should have been disrupted. It could also be argued that this patient's brain had undergone a massive reorganization (he was operated in 1956) and the functions performed by the inferior frontal gyrus were assumed by other area(s). This interpretation, however, does not go well with the results on the WCST, that clearly shows an important deficit in frontal-lobe functioning (i.e., those functions tested by the WCST were not assumed by other areas).

It seems fair to assume that there is some contradiction between the present results and previous fMRI studies that adds up to an increasing number of contradictory findings in fMRI and lesion studies (Petersen et al., 1988; Rorden and Karnath, 2004; Fellows and Farah, 2005). A general explanation for these discrepancies would be that some fMRI activations might be epiphenomenal, or nonessential, for performing the task under study (see Rorden and Karnath, 2004, for an extended discussion).

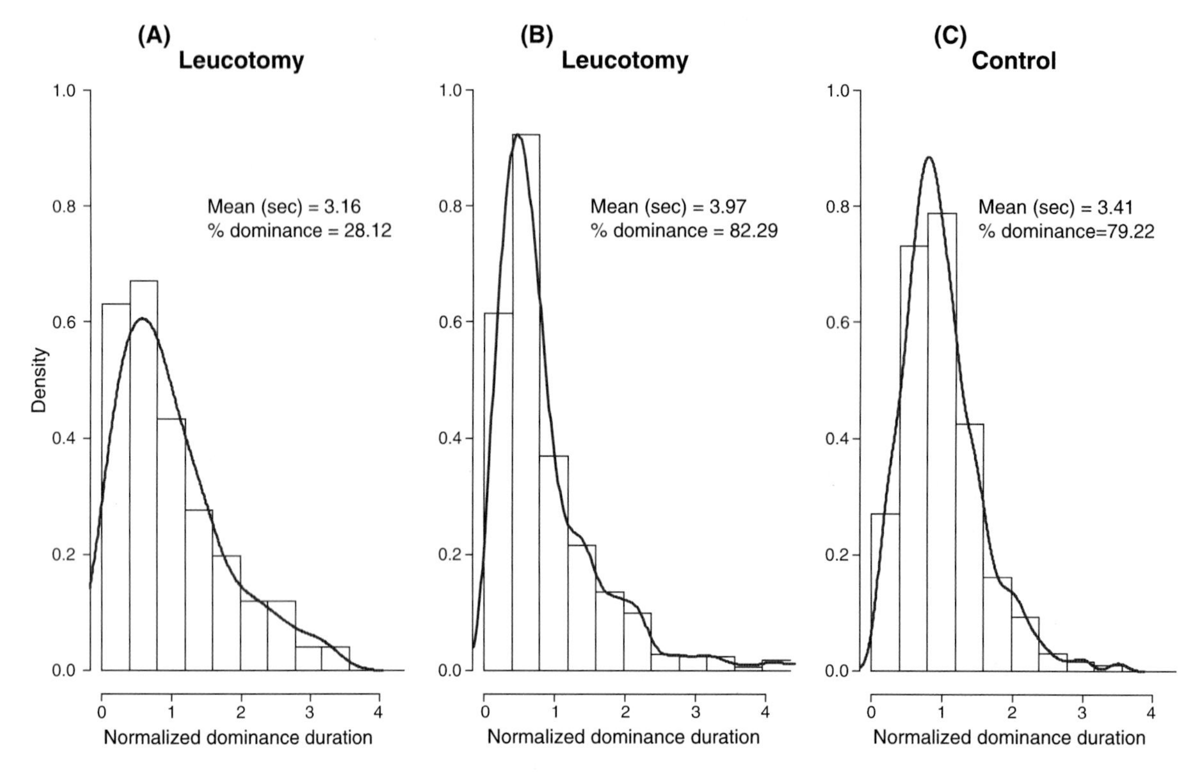

Fig. 2. Results. (A) Left panel: Histogram of the normalized phase durations (each phase duration divided by the mean duration) in the first experiment. (B) Middle panel: Histogram of the normalized phase durations in the second experiment. (C) Results for the control subjects (the two authors) with the stimuli used in the second experiment.

A different explanation would be that there was some confounding variable in the fMRI experiments. For example, when rivalry was compared to a stable perception condition (Lumer and Rees, 1999), perceptual changes were present only in the rivalry condition, and change detection was accompanied by frontoparietal activation (Beck et al., 2001). Therefore, the frontoparietal activation identified in Lumer and Rees (1999) might not be related to rivalry, but to detection of changes in the environment. When rivalry condition is compared to a movie mimicking subject's perception during rivalry (replay condition), as in Lumer et al. (1998), there are two other possible confoundings. The replay condition is run after the binocular rivalry task, thus if practice reduced activation in frontal lobes (Maccotta and Buckner, 2004; Tomasi et al., 2004; Kelly and Garavan, 2005) frontal activation might be larger during rivalry than during the replay condition. Still another possible source of confusion would be that the two conditions were different in difficulty, being perceptual changes during rivalry much more difficult to track than during the replay condition (O'Shea and Corballis, 2005b), and task difficulty is known to correlate with frontoparietal activations (Giesbrecht et al., 2003).

Our results, together with those of O'Shea and Corballis (2003, 2005a) and Bonneh et al. (2004) cast serious doubts on the special role of the right hemisphere during binocular rivalry, and the role of prefrontal areas. This conclusion may not apply to other multistable phenomena, such as ambiguous figures, that are much more prone to attention (Meng and Tong, 2004) and are affected by prefrontal lesions (Ricci and Blundo, 1990; Meenan and Miller, 1994).

Acknowledgements

The research was financed by the Spanish Ministry of Education (BS02001-0415/PSCE). We thank the Institut Pere Mata (Ctra. Institut Pere Mata, Reus, Tarragona 43206, Spain) for the facilities provided for this study.

References

Beck, D.M., Rees, G., Frith, C.D. and Lavie, N. (2001) Neural correlates of change detection and change blindness. Nat. Neurosci., 4: 645–650.

Blake, R., Sobel, K.V. and Gilroy, L. (2003) Visual motion retards alternations between conflicting perceptual interpretations. Neuron, 39: 869–878.

Bonneh, Y.S., Pavlovskaya, M., Ring, H. and Soroker, N. (2004) Abnormal binocular rivalry in unilateral neglect: evidence for a non-spatial mechanism of extinction. Neuroreport, 15: 473–477.

Brainard, D.H. (1997) The psychophysics toolbox. Spatial Vis., 10: 433–436.

Carlson, T.A. and He, S. (2004) Competing global representations fail to initiate binocular rivalry. Neuron, 43: 907–914.

Chen, X. and He, S. (2004) Local factors determine the stabilization of monocular ambiguous and binocular rivalry stimuli. Curr. Biol., 14: 1013–1017.

Fellows, L.K. and Farah, M.J. (2005) Is anterior cingulate cortex necessary for cognitive control? Brain, 128: 788–796.

Gail, A., Brinksmeyer, H.J. and Eckhorn, R. (2004) Perception-related modulations of local field potential power and coherence in primary visual cortex of awake monkey during binocular rivalry. Cereb. Cortex, 14: 300–313.

Giesbrecht, B., Woldorff, M.G., Song, A.W. and Mangun, G.R. (2003) Neural mechanism of top-down control during spatial and feature attention. NeuroImage, 19: 496–512.

Haynes, J.D., Deichman, R. and Rees, G. (2005) Eye-specific effects of binocular rivalry in the human lateral geniculate nucleus. Nature, 438: 496–499.

Kelly, A.M. and Garavan, H. (2005) Human functional neuroimaging of brain changes associated with practice. Cereb. Cortex, doi: 10.1093/cercor/bhi005.

Kleinschmidt, A., Büchel, C., Zeki, S. and Frackowiak, R.S.J. (1998) Human brain activity during spontaneously reversing perception of ambiguous figures. Philos. Trans. R. Soc. Lond. B Biol. Sci., 265: 2427–2433.

Logothetis, N.K. (1998) Single units and conscious vision. Philos. Trans. R. Soc. Lond. B Biol. Sci., 353: 1801–1818.

Logothetis, N.K., Pauls, J., Augath, M., Trinath, T. and Oeltermann, A. (2001) Neurophysiological investigation of the basis of fMRI signal. Nature, 412: 150–157.

Logothetis, N.K. and Wandell, B.A. (2004) Interpreting the BOLD signal. Annu. Rev. Physiol., 66: 735–769.

Lumer, E.D., Friston, K.J. and Rees, G. (1998) Neural correlates of perceptual rivalry in the human brain. Science, 280: 1930–1934.

Lumer, E.D. and Rees, G. (1999) Covariation of activity in visual and prefrontal cortex associated with subjective visual perception. Proc. Natl. Acad. Sci. USA, 96: 1669–1673.

Maccotta, L. and Buckner, R.L. (2004) Evidence for neural effects of repetition that directly correlate with behavioral priming. J. Cogn. Neurosci., 16: 1625–1632.

Meenan, J.P. and Miller, L.A. (1994) Perceptual flexibility after frontal or temporal lobectomy. Neuropsychologia, 32: 1145–1149.

Meng, M. and Tong, F. (2004) Can attention selectively bias bistable perception? Differences between binocular rivalry and ambiguous figures. J. Vis., 4: 539–551.

O'Shea, R.P. and Corballis, P.M. (2003) Binocular rivalry in split brain observers. J. Vis., 3: 610–615.

O'Shea, R.P. and Corballis, P.M. (2005a) Visual grouping on binocular rivalry in a split-brain observer. Vision Res., 45: 247–261.

O'Shea, R.P. and Corballis, P.M. (2005b). In: Alais, D. and Blake, R. (Eds.), Binocular rivalry in the divided brain. MIT Press, Cambridge, MA, pp. 301–316.

Pelli, D.G. (1997) The video toolbox software for visual psychophysics: transforming numbers into movies. Spatial Vis., 10: 437–442.

Petersen, S.E., Fox, P.T., Posner, M.I., Mintun, M. and Raichle, M.E. (1988) Positron emission tomographic studies of the cortical anatomy of single word processing. Nature, 331: 585–589.

Polonsky, A., Blake, R., Braun, J. and Heeger, D.J. (2000) Neuronal activity in human primary visual cortex correlates with perception during binocular rivalry. Nat. Neurosci., 3: 1153–1159.

Ricci, C. and Blundo, C. (1990) Perception of ambiguous figures after focal brain lesions. Neuropsychologia, 28: 1163–1173.

Rorden, C. and Karnath, H-O. (2004) Using human brain lesions to infer function: a relic from a past era in the fMRI age? Nat. Rev. Neurosci., 5: 813–819.

Tomasi, D., Ernst, T., Caparelli, E.C. and Chang, L. (2004) Practice-induced changes of brain function during visual attention: a parametric fMRI study at 4 Tesla. Neuroimage, 23: 1414–1421.

Tong, F. and Engel, S.A. (2001) Interocular rivalry revealed in the human cortical blind-spot representation. Nature, 411: 195–199.

Tong, F., Nakayama, K., Vaughan, J.T. and Kanwisher, N. (1998) Binocular rivalry and visual awareness in human extrastriate cortex. Neuron, 21: 753–759.

Wunderlich, K., Schneider, K.A. and Kastner, S. (2005) Neural correlates of binocular rivalry in the human lateral geniculate nucleus. Nat. Neurosci., 8: 1595–1602.

SECTION IV

Crossmodal Interactions in Visual Perception

Introduction

Our global sensory experience is not generally compartmentalized to rigid perceptual categories corresponding to the different sensory domains. However, traditional research in perception has not often crossed sensory modality boundaries. The last couple of decades have changed this landscape, and crossmodal research has now become one of the richest fields of neuroscientific enquiry.

This final section explores how visual perception is influenced by the other senses, and how our brain integrates visual information with information from the other sensory modalities. The five chapters herein attack the problem of crossmodal perception from multiple fronts, ranging from intracellular recordings in visually deprived animals to fMRI studies in human synesthetes.

David Burr and David Alais discuss how visual and auditory signals are combined, how they are aligned in time, and how attentional resources are allocated to these two modalities.

Noam Sagiv and Jamie Ward review several types of crossmodal interactions in synesthesia and discuss how they may relate to crossmodal interactions in normal perception.

Salvador Soto-Faraco and colleagues find that crossmodal dynamic capture (whereby the perceived direction of motion in the auditory modality is influenced by visual motion) is a relatively automatic process, robust to various top-down factors.

Maria V. Sanchez-Vives and colleagues present new physiological and anatomical data concerning audio-visual interactions in the primary visual cortex of the visually deprived cat.

Kristin Porter and Jennifer Groh discuss their recent findings that sound location in the primate inferior colliculus is encoded by a rate code, which is a different format from the place code used to represent the location of visual stimuli.

Susana Martinez-Conde

Martinez-Conde, Macknik, Martinez, Alonso & Tse (Eds.)
Progress in Brain Research, Vol. 155
ISSN 0079-6123

CHAPTER 14

Combining visual and auditory information

David Burr[1,2,*] and David Alais[3]

[1]*Dipartimento di Psicologia, Università degli Studi di Firenze, Via S. Nicolò 89, Firenze, Italy*
[2]*Istituto di Neuroscience del CNR, Via Moruzzi 1, Pisa 56100, Italy*
[3]*Department of Physiology and Institute for Biomedical Research, School of Medical Science, University of Sydney, Sydney, NSW 2006, Australia*

Abstract: Robust perception requires that information from by our five different senses be combined at some central level to produce a single unified percept of the world. Recent theory and evidence from many laboratories suggests that the combination does not occur in a rigid, hardwired fashion, but follows flexible situation-dependent rules that allow information to be combined with maximal efficiency. In this review we discuss recent evidence from our laboratories investigating how information from auditory and visual modalities is combined. The results support the notion of Bayesian combination. We also examine temporal alignment of auditory and visual signals, and show that perceived simultaneity does not depend solely on neural latencies, but involves active processes that compensate, for example, for the physical delay introduced by the relatively slow speed of sound. Finally, we go on to show that although visual and auditory information is combined to maximize efficiency, attentional resources for the two modalities are largely independent.

Keywords: crossmodal integration; vision; audition; ventriloquist effect; flash-lag effect; attention

As Ernst and Bülthoff (2004) point out in their excellent review, the key to robust perception is the efficient combination and integration of multiple sources of sensory information. How the brain achieves this integration — both within and between sensory modalities — to form coherent perceptions of the external environment is one of the more challenging questions of sensory and cognitive neuroscience. Neurophysiologically, sensory interactions have become well documented over several decades. More recently, perceptual research combined with solid modeling is beginning to complement the neurophysiology. This chapter summarizes some recent psychophysical work on audiovisual interactions from our laboratories.

*Corresponding author. Tel.: +39-050-3153175; Fax: +39-050-315-3210; E-mail: dave@in.cnr.it

Pitting sight against sound: the ventriloquist effect

Ventriloquism is the ancient art of making one's voice appear to come from elsewhere, exploited by the Greek and Roman oracles, and possibly earlier (Connor, 2000). We regularly experience the effect when watching television and movies, where the voices seem to emanate from the actors' lips rather than from the actual sound source. The original explanations for ventriloquism (dating back to the post-Newtonian scientific efforts of early 18th century) assumed that it was based on the physical properties of sound, that performers somehow *projected* sound waves in a way to appear to emanate from their puppets, using special techniques (Connor, 2000). Only relatively recently has the alternative been considered, that ventriloquism is a sensory illusion created by our neural systems. These explanations assume that vision

DOI: 10.1016/S0079-6123(06)55014-9

predominates over sound, and somehow *captures* it (Pick et al., 1969; Warren et al., 1981; Mateeff et al., 1985; Caclin et al., 2002).

More recently, another approach has been suggested for combination of information. Several authors (Clarke and Yuille, 1990; Ghahramani et al., 1997; Jacobs, 1999; Ernst and Banks, 2002; Battaglia et al., 2003) have suggested and shown that multimodal information may be combined in an optimal way by summing the independent stimulus estimates from each modality according to an appropriate weighting scheme. The weights are given by the inverse of the variance (σ^2) of the underlying noise distribution (which can be assessed separately from the width of the psychometric function). For auditory and visual combination this can be expressed as

$$\hat{S} = w_A \hat{S}_A + w_V \hat{S}_V \quad (1)$$

where $\hat{S}$ is the optimal estimate, $\hat{S}_A$ and $\hat{S}_V$ are the independent estimates for audition. w_A and w_V are the weights by which the unimodal estimates are scaled, and are inversely proportional to the auditory and visual variances σ_A^2 and σ_V^2

$$w_A = 1/\sigma_A^2, \quad w_V = 1/\sigma_V^2, \quad (2)$$

Normalizing the sums of the weights to unity

$$k = 1/\sigma_A^2 + 1/\sigma_V^2 \quad (3)$$

This model is "optimal" in that it combines the unimodal information to produce a multimodal stimulus estimate with the lowest possible variance (i.e., with the greatest reliability see Clarke and Yuille, 1990).

We (Alais and Burr, 2004b) tested the predictions of Eq. (1) directly by asking observers to localize in space brief light "blobs" or sound "clicks," presented first separately (unimodally) and then together (bimodally). The purpose of the unimodal presentation was to measure the precision of these judgments under various conditions to provide estimates of variances σ_A^2 and σ_V^2. Fig. 1A shows typical results for four different stimuli: visual blobs of various degrees of blur and auditory tones. The data are fitted by cumulative Gaussian curves from which one can extract two parameters: the best estimate of perceived position $\hat{S}$ (often also referred to as the "point of subjective

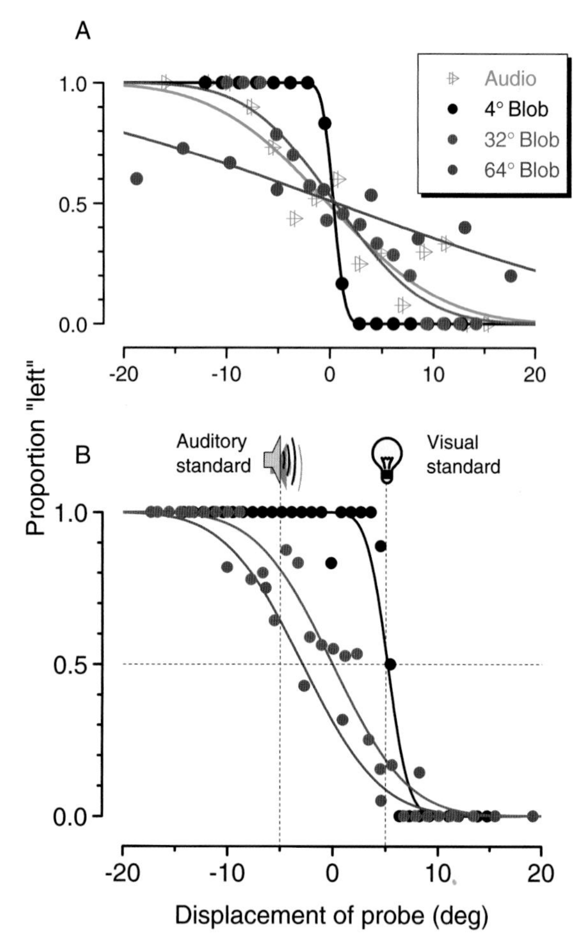

Fig. 1. (A) Unimodal psychometric functions for localization of an auditory stimulus (green), and visual Gaussian blobs of variable size. Localization for fine blobs is very good (as indicated by the steep psychometric functions), but is far poorer for very blurred blobs. Auditory localization is in between, similar to visual localization with 32° blobs. The curves are best-fitting cumulative Gaussian functions. (Reproduced with permission from Alais and Burr, 2004b.) (B) Bimodal psychometric functions for dual auditory and visual presentations. In the "conflict" presentation, the visual stimulus was displaced rightward by 5° and the auditory stimulus leftward by the same amount (as indicated by vertical lines). The 4° stimulus (black symbols) tend to follow the visual standard, the 64° stimulus (blue symbols) the auditory standard and the 32° stimulus (red symbols) falls in between. The curves are not best fits to the data, but predictions from the Bayesian model described in Eqs. (1)–(4). Modified from Alais and Burr (2004b, p. 258), Copyright, with permission from Elsevier.

equality" or PSE), given by the point where the curves crosses 50%, and the threshold for making the judgment, given by the width or standard deviation (σ). $\hat{S}$ was near zero for all conditions, implying that the observer, on average, saw the stimuli where it was actually displayed (at zero). However, the steepness of the curves varied considerably from condition to condition. They were steepest (small estimate of σ) for the small (4°) visual stimuli, becoming much broader for the blurred stimuli. The steepness of the auditory curves was in between, similar to the visual curve at 32°.

In the bimodal condition two different types of presentation were made on each trial, a *conflict* presentation, where the visual stimulus was displaced $+\Delta°$ and the auditory stimulus $-\Delta°$ from center, and a *probe* presentation, where the visual and auditory stimuli covaried around a mean position. Subjects were asked to judge which stimulus appeared more "rightward." Example results are shown in Fig. 1B, for $\Delta = 5°$ (meaning that the visual stimulus was displaced 5° rightward and the auditory stimulus 5° leftward, as indicated by the vertical dashed lines of Fig. 1B). The effect of the conflict clearly depended on the size of the visual blob stimuli. For 4° blobs (black symbols), the curves are clearly shifted to the right so the mean (PSE) lines up with the position of the visual stimuli. This is the classic ventriloquist effect. However, for 64° blobs (blue symbols) the reverse holds, and the curves shift leftward toward (but not quite reaching) the auditory standard. For the intermediate blur (32°, red symbols) the results are intermediate, with the PSE of the bimodal presentation falling midway between the visual and auditory standard.

The results of all the conflicts used are summarized in Fig. 2A. For each conflict and each subject, curves similar to those shown in Fig. 1B were plotted and fitted with cumulative Gaussian distributions, and the PSE (apparent coincidence of conflict and probe) was defined as the mean (50% point) of the distribution. As the example of Fig. 1B shows, for relatively unblurred visual blobs (4° blur: filled squares), vision dominated totally, while for extremely blurred blobs (128°: filled triangles), the opposite occurred, suggesting that audition dominates. At intermediate levels of blur (32°: open circles), neither stimulus dominated completely, with the points falling between the two extremes. The continuous lines are model predictions from Eq. (1), with variances σ_A^2 and σ_V^2 estimated from unimodal presentations of the auditory and visual stimuli (from curves like Fig. 1). These predictions are remarkably close to the data, providing strong evidence that Eq. (1) is applicable in these circumstances.

An even stronger test for optimal combination is that the discrimination thresholds (square root of the variances) of the bimodal presentation increases should increase

$$\sigma_{VA}^2 = \frac{\sigma_V^2 \sigma_A^2}{\sigma_A^2 + \sigma_V^2} < \min(\sigma_V^2, \sigma_A^2) \qquad (4)$$

where σ_{VA} is the threshold of the combined presentation that can never be greater than either the visual or the auditory thresholds. When visual or auditory variances differ greatly, σ_{AV} will be given by the lower threshold. But when they are similar, σ_{AV} will be about $\sqrt{2}$ less than either σ_A or σ_V.

Fig. 2B shows average normalized thresholds for six observers in the crossmodal task with medium-blur levels (blob size 32°), where one expects the greatest crossmodal improvement. To reduce subject variability, all crossmodal thresholds were normalized to unity, and the visual and auditory thresholds averaged with the same normalization factor. Both visual and auditory thresholds are about 1.4 (i.e., $\sqrt{2}$) times higher than the crossmodal thresholds. The predicted averaged crossmodal thresholds (calculated by applying Eq. (4) to the individual data, and then averaging) are very close to the obtained data.

These results strongly suggest that the ventriloquist effect is a specific example of optimal combination of visual and auditory spatial cues, where each cue is weighted by an inverse estimate of its variability, rather than one modality capturing the other. As visual localization is usually far superior to auditory location, vision normally dominates, apparently "capturing" the sound source and giving rise to the classic ventriloquist effect. However, if the visual estimate is corrupted sufficiently by blurring the visual target over a large region of space, vision can become worse than audition, and optimal localization correctly predicts that sound will effectively capture sight. This is broadly consistent with other reports of integration of sensory

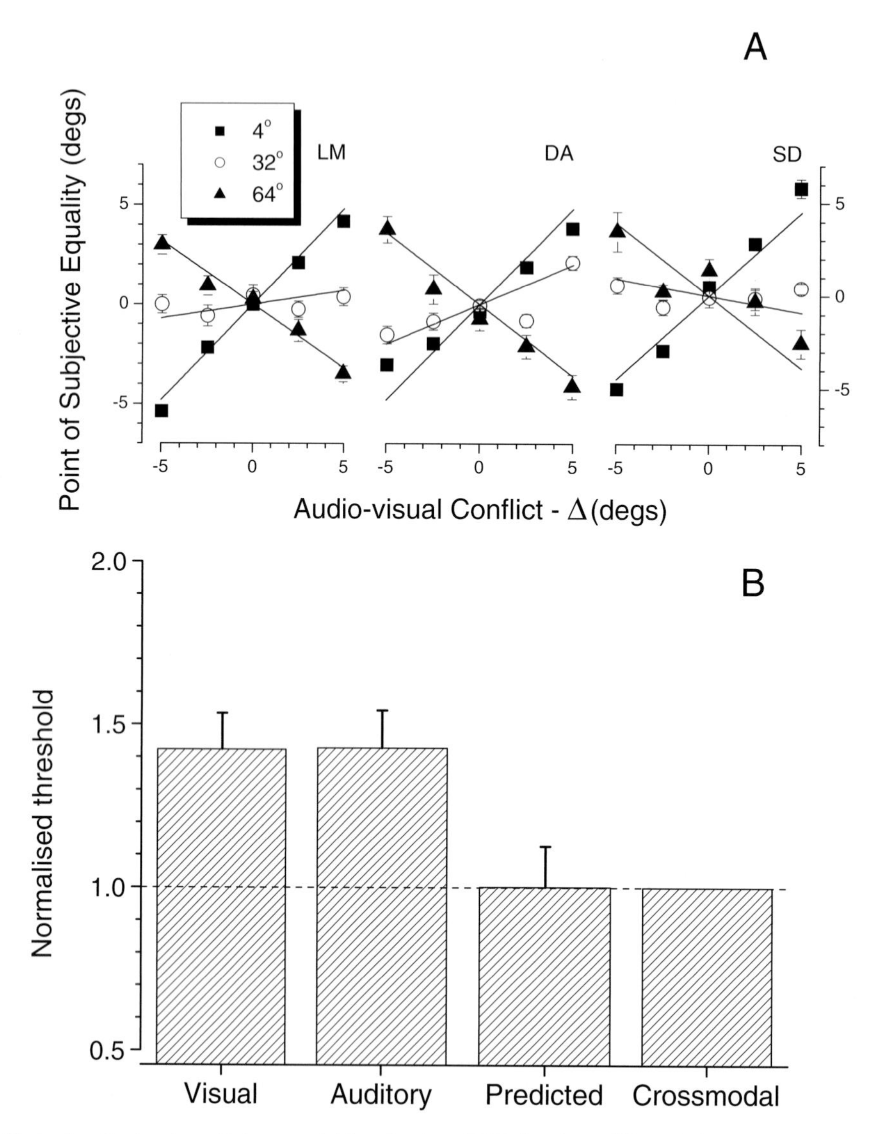

Fig. 2. (A) The effect of audiovisual conflict on spatial localization (PSE) for three different observers, and three different sizes of visual stimuli: 4° (filled squares), 32° (open circles) and 64° (filled triangles). The solid lines are the predictions of Eq. (1) using individual estimates of σ_A and σ_V for the three different sized blobs (from Fig. 1 and similar curves for the other subjects). (B) Average normalized thresholds of six subjects, for the condition where visual thresholds were similar to auditory thresholds (blob size 32°). All individual thresholds were normalized to the threshold in the crossmodal condition before averaging.
Modified from Alais and Burr (2004b, pp. 259–260), Copyright, with permission from Elsevier.

information (Clarke and Yuille, 1990; Ghahramani et al., 1997; Jacobs, 1999; Ernst and Banks, 2002; Alais and Burr, 2003). However, it differs slightly from the results of Battaglia et al. (2003) who found that vision tended to dominate more than predicted by Eq. (1): they introduced a hybrid Bayesian model to explain their effects.

Note that for auditory localization to be superior to vision, the visual targets needed to be blurred extensively, over about 60°, enough to blur most scenes beyond recognition. However, the location of the audio stimulus was defined by only one cue (interaural timing difference) and was not time varying, so auditory localization was only about one-sixth as accurate as normal hearing (Mills, 1958; Perrott and Saberi, 1990). If the effect were to generalize to natural hearing conditions, then 10° blurring would probably be sufficient. This is still a gross visual distortion, explaining why the reverse ventriloquist effect is not often noticed for spatial events. There are cases, however, when it does become relevant, not so much for blurred as for ambiguous stimuli, such as when a teacher tries to make out which child in a large class was speaking.

There is one previously reported case where sound does capture vision; this is for temporal localization where a small continuous (and peripherally viewed) light source seems to pulse when viewed together with a pulsing sound source (Shams et al., 2000; Shams et al., 2002). Furthermore, the presence of the clicks do not only make the light appear to flash, but can improve performance on visual discrimination tasks (Berger et al., 2003; Morein-Zamir et al., 2003). Although no model was offered to account for this phenomenon, it may well result from sound having far better temporal acuity than vision, resulting in the sound information being heavily weighted and appearing to capture the visual stimulus. Sounds can also modulate visual potentials in early visual areas (Shams et al., 2001), mimicking closely the modulation caused by visual stimuli, suggesting a direct interaction at an early level. Indeed preliminary evidence from our laboratories suggests that optimal, Bayesian combination of sight and sound, where the auditory temporal acuity is superior to vision, may also explain these effects (Alais and Weston, 2005; Burr et al., 2005).

An important and difficult remaining question is how the nervous system "knows" the variances associated with individual estimates. Must it "learn" these weights from experience, or could a direct estimate of variance be obtained from neural activity of a population, for example, by observing the spread of activation along a spatiotopic map? Previous studies have shown that observers can learn cue-integration strategies (Jacobs and Fine, 1999) and that the learning can be very rapid (Triesch et al., 2002). We can only guess at the neural mechanisms involved, but it is not implausible that the central nervous system encodes an estimate of measurement error along with every estimate of position, or other attribute (Ernst and Banks, 2002).

Integration of audio and visual motion

Following on from the integration of static positional cues, we asked whether auditory and visual information about motion could be effectively combined, and what are the rules of combination (Alais and Burr, 2004a). In particular, we were interested whether the combination may be "compulsory," or whether observers had access to the unimodal information (see Hillis et al., 2002). Motion seemed an interesting area to study, as a key neural area involved in the multisensory combination is the superior colliculus (Stein, 1998), particular the deep layers. The superior colliculus has strong reciprocal links, via the pulvinar, with the middle-temporal (MT) cortical area (Standage and Benevento, 1983). MT is an area specialized for processing visual movement whose activity is strongly correlated with visual motion perception (Britten et al., 1992; Britten et al., 1996). MT outputs project directly to the area ventral intraparietal (VIP) where they combine with input from auditory areas to create bimodal cells with strong motion selectivity (Colby et al., 1993; Bremmer et al., 2001; Graziano, 2001). Motion perception, therefore, seemed a good area to look for strong bimodal interactions.

In order to maximize audiovisual interactions, we first measured motion detection thresholds unimodally (in two alternative forced choice) for vision and for audition, and matched them for strength. Subjects identified which interval contained the movement, without judging the direction of motion. Visual and auditory stimulus strengths were then scaled in the individual unimodal thresholds so as to be equally effective, and presented bimodally, with coherence varying together to determine the joint threshold. In separate conditions, auditory and visual stimuli moved in the same direction (and speed), or in the opposite direction (with matched speed).

Fig. 3 plots thresholds on a two-dimensional plot, with auditory coherence on the ordinate and

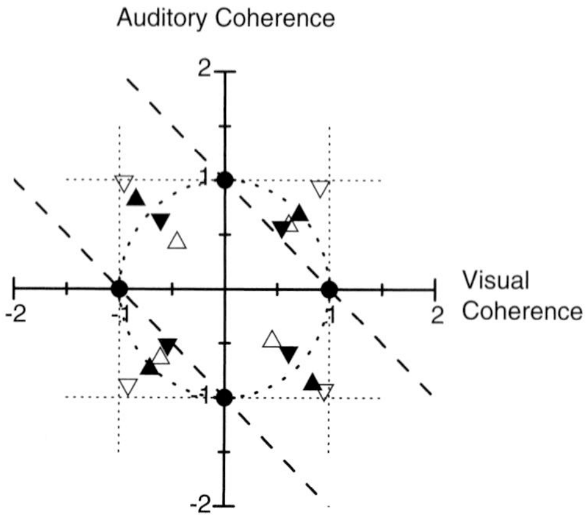

Fig. 3. Nondirectional bimodal facilitation for motion detection. The four separate subjects are indicated by different symbols on the two-dimensional plot, plotting coherence of the auditory moving stimulus against coherence of the visually moving stimulus. All thresholds are normalized so the unimodal thresholds are one. The dashed diagonal lines show the prediction for linear summation and the dashed circle for Bayesian, "statistical" summation of information. Clearly the data follow the Bayesian prediction, with no tendency whatsoever to elongate in the direction predicted by mandatory summation (–45°). Mean thresholds for same direction was 0.83, for opposite direction 0.84, with none of the observers exhibiting a significant difference.

Reproduced from Alais and Burr (2004a, p. 190), Copyright, with permission from Elsevier.

visual coherence on the abscissa. By definition, all unimodal thresholds are unity. For all observers except one (inverted triangular symbol), thresholds were lower in the bimodal than unimodal condition. However, the improvement was as good when the motion was in the opposite direction (second and fourth quadrants) as when it was in the same direction (first and third quadrants). Averaging over the four observers, mean threshold for the same-direction motion (0.83) and opposite-direction motion (0.84) were virtually identical. Clearly, the direction of the unimodal motions it was not important for bimodal motion detection.

The pattern of results is clearly not consistent with a model of linear summation of signed motion signals. The level of summation observed is too small for this (ideal prediction would be 0.5), and more importantly does not show the asymmetry toward like direction that would be expected. Perfect linear summation would follow the dashed lines oriented at –45°. Of course, this prediction is somewhat extreme, but any form of mandatory fusion should lead to an elongation of the threshold ellipse, so it is longer alone the –45° axis (where the visual and auditory directions are opposed, and should tend to annul each other). Our results give no indication whatsoever of this elongation, agreeing with Hilis et al. (2002) who demonstrated mandatory fusion within a sensory system (vision) but not between vision and touch.

The summation is, however, consistent with a statistically optimal combination of signals based on maximum likelihood estimation of Eq. (4) discussed in the previous section, and indicated in Fig. 3 by a dashed circle (Clarke and Yuille, 1990; Ghahramani et al., 1997; Jacobs, 1999; Ernst and Banks, 2002). As the auditory and visual weights were equated by equating the unimodal thresholds, the expected improvement from Eq. (4) is a factor of $1/\sqrt{2}$ $(=0.71)$, not very different from the observed 0.84. Importantly, the prediction is the same for like and opposite motion, as both carry the same amount of information, although they are perceptually very distinct.

Taken together, these results show a small nondirectional gain in bimodal movement detection for bimodal motion, consistent with statistical combination, but not with a direct summation of signed audio and visual motion signals. This held true both for coherently moving visual objects and for spatially distributed motions, in central and in peripheral vision (Alais and Burr, 2004a), agreeing with two recent studies using similar methods and stimuli (Meyer and Wuerger, 2001; Wuerger et al., 2003).

Temporal synchrony — the flash-lag effect

It has long been known that the order in which perceptual events are perceived does not always reflect the order in which they were presented. For example, Titchener (1908) showed that salient, attention-grabbing stimuli are often perceived to have occurred before less salient stimuli (the "prior entry effect"). More recently, Moutoussis and Zeki (1997) showed that different attributes of the same object can appear to change at different times: if the color and direction of motion change simultaneously, color seems to lead. But perhaps the clearest example of a systematic temporal mislocalization is the so-called "flash-lag effect," first observed by MacKay (1958) and more recently revised and extensively studied by Nijhawan (1994); for review see (Krekelberg and Lappe, 2001). If a stationary disk is briefly flashed at the exact moment when a moving disk passes it, the stationary disk seems to "lag" behind the moving disk. Many explanation of the flash-lag effect have been suggested, including spatial extrapolation (Nijhawan, 1994), attention (Baldo and Klein, 1995), differential neural latencies (Purushothaman et al., 1998), spatial averaging (Krekelberg and Lappe, 2000) and "postdiction" (Eagleman and Sejnowski, 2000).

Whatever the explanation for the effect, an interesting question is whether it is specific for visual stimuli, or whether it also occurs in other senses, and crossmodally, and whether these effects could reasonably be attributed to neural latencies. We therefore measured the flash-lag effect for auditory stimuli, both for spatial motion and for spectral motion in frequency. In both cases a strong flash-lag effect was observed (Fig. 4): the stationary stimulus seemed to lag 160–180 ms behind the moving stimulus, whether the motion was in space or in frequency. This effect is in the same direction as that observed for vision, but far stronger: visual effects under the conditions of this experiments were about 20 ms. It was also possible to measure the effect crossmodally: using a visual flash as probe to a moving sound or a sound burst as probe to a moving visual stimulus. Both these conditions produced large and reliable flash-lag effects, roughly midway between the purely visual and purely auditory effects.

These results show that the flash-lag effect is not peculiar to vision, but occurs in audition, and also crossmodally. They also provide the possibility of investigating the mechanisms producing the effects, by comparing the magnitudes under the various audio and visual conditions. If the flash-lag effect were simply due to differences in neural latencies and processing time, then the relative latencies necessary to produce the results of Fig. 4 are easily calculated. As the auditory–auditory effects were the largest, the neural response to auditory motion would have to be much faster than that to an auditory flash (by about 180 ms). As the visual–visual effects were small, the response to visual motion should be only about 20 ms faster than that to a visual flash. And as the auditory–visual and visual–auditory effects were of comparable size, the visual latencies should be between the auditory motion and flash latencies. The best estimates to give the results of Fig. 4 are shown in Fig. 5A, normalizing the visual latency estimate arbitrarily to 100 ms. Fig. 5B shows recent results measuring neural delays for visual and auditory moving and stationary stimuli with three different techniques: an integration measure, perceptual alignment and reaction times (Arrighi et al., 2005). These three measures all agree quite well with each other, in suggesting that they are measuring the same thing. However, the order of the latencies measured directly is quite different from that required for the flash-lag effect. For audition, motion latencies were systematically longer than flash latencies, whereas the reverse is required for the flash-lag effect, both in audition and crossmodally.

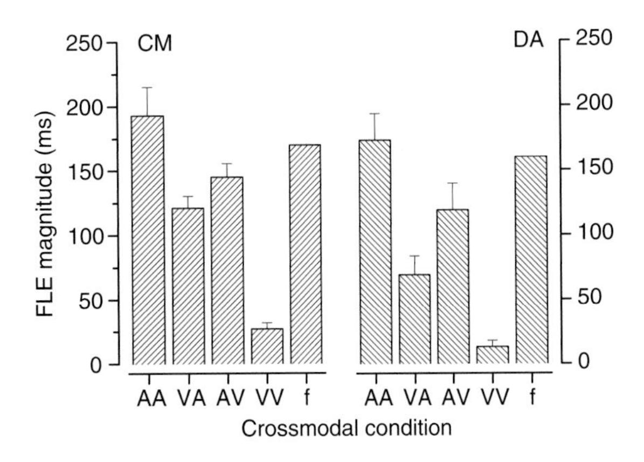

Fig. 4. Magnitude of the flash-lag effect for various auditory–visual conditions. The column indicated "*f*" refers to "motion" up and down the scales played to one ear, to which subjects had to align a tone played to the other ear. For all other bars, the first symbol refers to the modality of the moving stimulus and the second to that of the stationary "flash." Reproduced from Alais and Burr (2003, p. 60), Copyright, with permission from Elsevier.

These results reinforce previous work showing that the flash-lag effect does not result directly

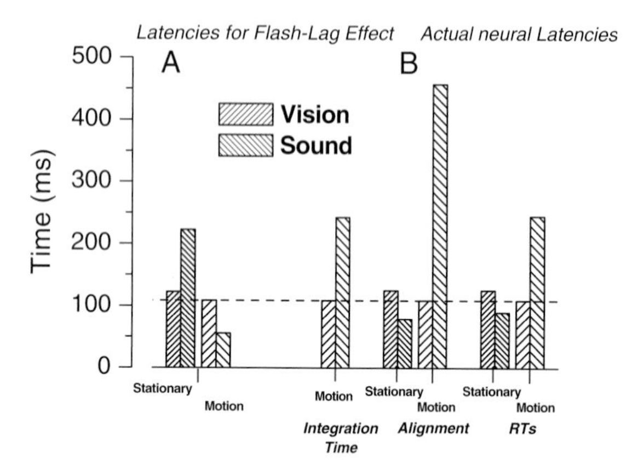

Fig. 5. (A) The relative hypothetical neural latencies necessary to account for the flash-lag data of Fig. 4, assuming simple linear accumulation of delays. Rightward hatching refers to vision, leftward to audition, sparse hatching to motion and dense hatching to stationary stimuli. Auditory motion needs to be processed the fastest, auditory "flashes" the slowest and vision in between. (B) Actual latencies measured with three different techniques: integration, perceptual alignment and reaction times (aligning all visual motion results to the reaction time data, indicated by the dashed line, so only relative latencies are shown). The results are self-consistent between the three techniques, but go in the opposite direction from those required to explain the flash-lag effect (A).
Reproduced from Arrighi et al. (2005, p. 2922), Copyright, with permission from Elsevier.

from neural latencies, but clearly reflect sensory processing strategies, possibly related to calibrating motor and sensory input (Nijhawan, 1994). It is interesting that the effects should be much larger with hearing than vision. This may be related to the fact that auditory localization of position is much less precise than visual localization (see Fig. 1). This is consistent with more recent work by Nijhawan (personal communication) showing that the flash-lag effect also occurs for touch, and is much larger when measured on the forearm (where receptive fields are large and localization imprecise) than on the finger (with small receptive fields and fine localization).

Compensating for the slow propagation speed of sound

Studies of audiovisual temporal alignment have generally found that an auditory stimulus needs to be delayed by several tens of milliseconds in order to be perceptually aligned with a visual stimulus (Hamlin, 1895; Bald et al., 1942; Bushara et al., 2001). This temporal offset is thought to reflect the slower processing times for visual stimuli. This arises because acoustic transduction between the outer and inner ears is a direct mechanical process and is extremely fast at just 1 ms or less (Corey and Hudspeth, 1979; King and Palmer, 1985), while phototransduction in the retina is a relatively slow photochemical process followed by several cascading neurochemical stages and lasts around 50 ms (Lennie, 1981; Lamb and Pugh, 1992). Thus, differential latencies between auditory and visual processing generally agree quite well with the common finding that auditory signals must lag visual signals by around 40–50 ms if they are to be perceived as temporally aligned.

Most studies of audiovisual alignment, however, are based on experiments in the near field, meaning auditory travel time is a negligible factor. Studies conducted over greater distances have produced contradictory results (Sugita and Suzuki, 2003; Kopinska and Harris, 2004; Lewald and Guski, 2004) regarding whether brain can compensate for the slow travel time of sound. We recently tested whether knowledge of the external distance of an auditory source could be used to compensate for the slow travel time of sound relative to light (Alais and Carlile, 2005). We reasoned that to compensate for auditory travel time would require a robust cue to auditory source distance, since it involves overriding the temporal difference between the signals as they arrive at the listener. We therefore used the most powerful auditory depth cue — the direct-to-reverberant energy ratio (Bronkhorst and Houtgast, 1999) — to indicate source distance.

To create a suitable sound stimulus, we recorded the impulse response function of a large concert auditorium (the Sydney Opera House) and convolved it with white noise. This stimulus sounded like a burst of white noise played in a large reverberant environment (Fig. 6). It began with a direct (i.e., anechoic) portion lasting 13 ms, followed by a long reverberant tail that dissipated over 1350 ms. To vary the apparent distance of the sound burst, we varied the amplitude of the initial part of the

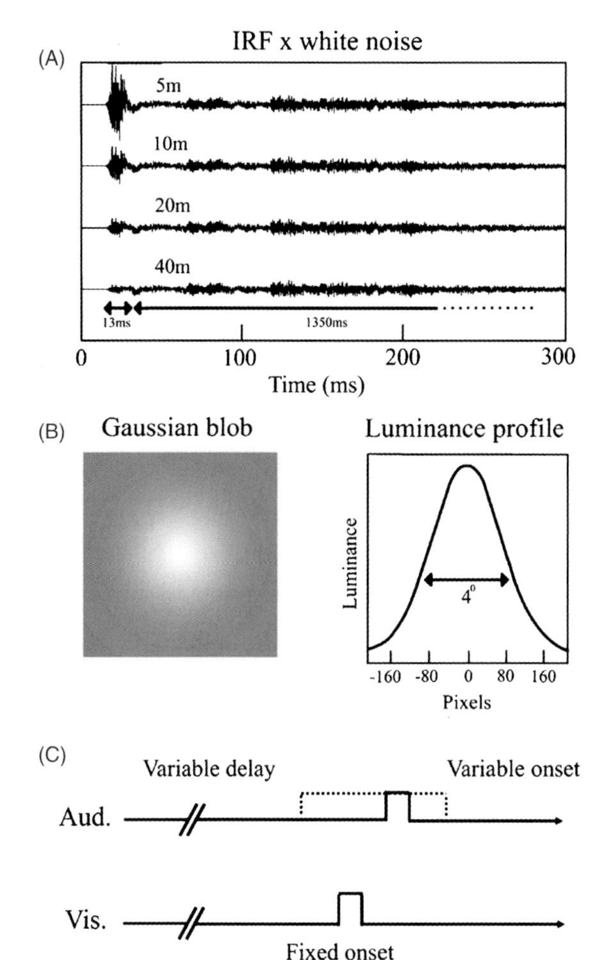

Fig. 6. The stimuli and procedures used to measure visual-acoustic synchrony. (A) The impulse response function on the top row (5 m) is the original function recorded in the Sydney Opera House convolved with white noise. The direct sound is the initial portion of high amplitude, and the long tail reverberant signal, which lasted 1.35 ms (identical for all four stimuli). Because the direct-to-reverberant energy ratio is a very strong cue to auditory source distance, attenuating the direct portion by 6 dB (halving amplitude) simulates a source distance of 10 m (see *Methods*). Further 6 dB attenuations simulated auditory distances of 20 and 40 m. (B) The visual stimulus was similar to that shown (*Left*), a circular luminance patch that was presented for 13 ms. The spatial profile of the stimulus (*Right*) was Gaussian with a full half-width of 4° of visual angle. (C) The onset of the auditory stimulus (*Upper*) was varied by an adaptive procedure to find the point of subjective alignment with the visual stimulus (*Lower*). A variable random period preceded the stimuli after the subject initiated each trial. Reproduced from Alais and Carlile (2005, p. 2245), Copyright, with permission.

stimulus, while leaving the reverberant tail fixed for all simulated depths. Since the energy ratio of the early direct portion to the later reverberant tail is a powerful cue to auditory depth, we could effectively simulate a situation in which a sound source was heard at various distances in a constant reverberant environment, in a darkened high-fidelity anechoic chamber. To measure perceived audiovisual alignment, a brief spot of was light flashed on a dark computer screen and served as a temporal reference point. The sound onset was advanced or retarded in time using an adaptive staircase method until the onset of the sound burst was perceived to be synchronous with the light flash.

The original recording in the auditorium was made 5 m from the sound source, and successive 6 dB scaling of the early direct portion simulated stimuli at 10, 20 and 40 m (see Fig. 6a). In enclosed reverberant environments, the direct-to-reverberant energy ratio is the strongest cue to auditory sound source distance because the incident level decreases by 6 dB with each doubling of distance while the level of the reverberant tail is approximately invariant (Bronkhorst and Houtgast, 1999; Zahorik, 2002; Kopinska and Harris, 2004).

The results were clear: the point of subjective alignment of auditory and visual stimuli depended on the source distance simulated in the auditory stimulus. Sound onset times had to be increasingly delayed to produce alignment with the visual stimulus as perceived acoustic distance increased (Fig. 7A). Best-fitting linear functions describe the data well, with slopes varying between observers from 2.5 to 4.2 ms/m, with the average (3.2 ms/m, shown by the dotted line of Fig. 7B) approximately consistent with the delay needed to compensate for the speed of sound (2.9 ms/m at 20°C, indicated by the dashed line). These results suggest that subjects were attempting to compensate for the travel time from the simulated source distance using a subjective estimate of the speed of sound.

Various controls were performed to show that the reverberant tail of the sound wave was essential for the subjective audiovisual alignment to shift in time (Alais and Carlile, 2005). In a further control, the observers' attention was focused on the onset burst by requiring them to make speedier responses (slow responses were rejected). Under this condition

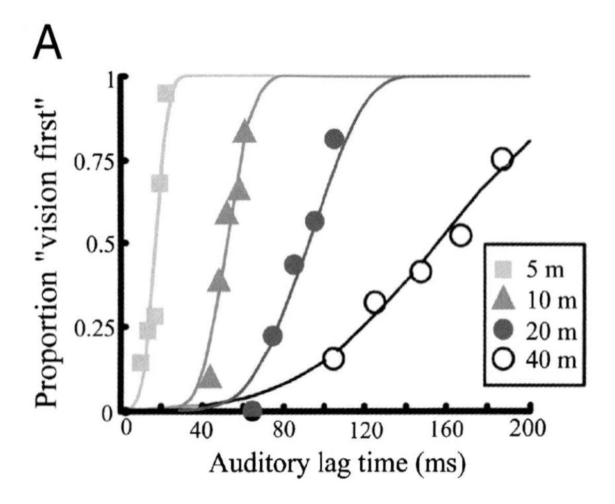

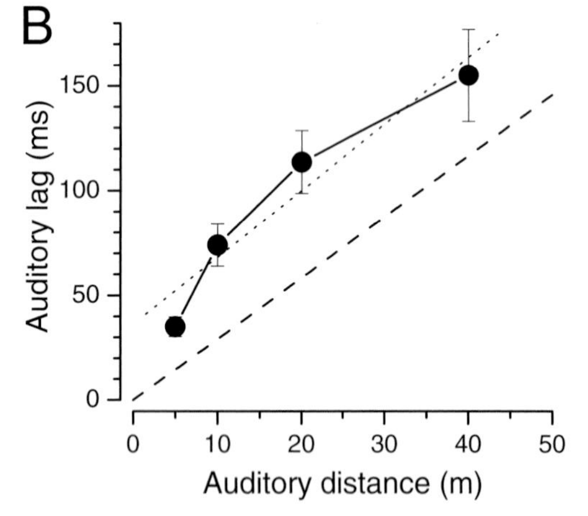

Fig. 7. (A) Psychometric functions for one observer at each of the four simulated auditory distances plotting the proportion of trials in which the visual stimulus was judged to have occurred before the auditory stimulus, as a function of the delay of the auditory stimulus. From left to right, the curves represent the 5-, 10-, 20- and 40-m conditions. The abscissa shows time measured from the onset of the visual stimulus. (B) Average points of subjective audiovisual alignment (the half-height of the psychometric functions) for four observers at each of the four auditory distances. As auditory distance simulated by the direct-to-reverberant energy ratio increased, the auditory stimulus was perceptually aligned with earlier visual events, consistent with subjects using the energy ratio in their alignment judgments. The dotted line shows the best-fitting linear regression to the data. The slope of the regression is 3.2 ms/m, consistent with the speed of sound (2.9 ms/m at 20°C, indicated by the lower dashed line).
Reproduced from Alais and Carlile (2005, p. 2245), Copyright, with permission.

(where the reverberant tail is not attended) there is no systematic variation across auditory depth, showing that use of this cue is strategic and task dependent, rather than an automatic integration.

The essential finding from these experiments is that the brain is able to compensate for the fact that, with increasing source distance, the acoustic signal arising from a real bimodal event will arrive at the perceiver's head at progressively later times than the corresponding visual signal. These studies clearly refute any simple account of audiovisual alignment based solely on neural latencies, which would predict a common auditory lag for all simulated source distances, determined by the differential neural processing latencies for vision and audition. However, we show that the point of subjective alignment became systematically delayed as simulated auditory distance increased. Thus, the data suggest an active, interpretative process capable of exploiting auditory depth cues to temporally align auditory and visual signals at the moment they occur at their external sources.

This process could be termed "external" alignment, in contrast to "internal" alignment based on time of arrival and internal latencies. Because external alignment requires the brain to ignore a considerable temporal asynchrony between two neural signals (specifically, the late arrival of the auditory signal), it is unlikely to do so unless there is a robust depth cue to guide it. The direct-to-reverberant energy ratio appears to be a powerful enough cue to permit this, provided it is relevant to do so. Without a reliable depth cue, the brain seems to default to aligning signals internally, demonstrating flexibility in determining audiovisual alignment. External alignment would require knowledge of source distance and speed of sound. The direct-to-reverberant energy ratio provides a reliable auditory distance cue, and listeners presumably derive an experience-based estimate of the speed of sound, which is validated and refined through interaction with the environment.

Crossmodal attention

With the environment providing much competing input to the sensory system, selecting relevant information for further processing by limited neural resources is important. Cells in the deep layers of the superior colliculus play an important role in exogenous attention. However, attention can also be deployed voluntarily (endogenous attention) to select certain stimuli from the array of input stimuli (Desimone and Duncan, 1995). Attentional selection improves performance on many tasks, as limited cognitive resources are allocated to the selected location or object to enhance its neural representation. This is true both for tasks that may be considered to be "high-level" and for those considered "low level" (for review see Pashler, 1998).

Evidence from neurophysiology, neuropsychology and neuroimaging suggests that attention acts at many cortical levels, including primary cortices. Neuroimaging and single-unit electrophysiology point to attentional modulation of both V1 and A1 (Woodruff et al., 1996; Grady et al., 1997; Luck et al., 1997; Brefczynski and DeYoe, 1999; Gandhi et al., 1999; Jancke et al., 1999; Kanwisher and Wojciulik, 2000; see also Corbetta and Shulman, 2002).

Some psychophysical studies also show crossmodal attentional effects. For example, shadowing a voice in one location while ignoring one in another is slightly improved by watching a video of moving lips in the shadowed location (Driver and Spence, 1994), and performance can be worsened by viewing a video of the distractor stream (Spence et al., 2000). Also, precuing observers to the location of an auditory stimulus can also increase response speed to a visual target, and vice versa (Driver and Spence, 2004). On the other hand, several studies from the older psychological and human factors literature show substantial independence between visual and auditory attention (Triesman and Davies, 1973; Wickens, 1980), and some more recent studies also point in this direction (Bonnel and Hafter, 1998; Ferlazzo et al., 2002). In addition, the "attentional blink" (the momentary reduction in attention following a perceptual decision) is modality specific, with very littler transfer between vision and audition (Duncan et al., 1997).

Overall, the evidence relating to whether attention is supramodal or whether it exists as a separate resource for each modality is equivocal. We therefore measured basic discrimination thresholds for low-level auditory and visual stimuli while dividing attention between concurrent tasks of the same or different modality. If attention is a single supramodal system, then a secondary distractor task should reduce performance equally for intramodal and extramodal distractor tasks. However, if there are separate attentional resources for vision and audition, then extramodal distractors should not impair performance on the primary task. Our results suggest that vision and audition have their own attentional resources.

We measured discrimination thresholds for visual contrast and pitch, initially on their own, then while subjects did a concurrent secondary task that was either intramodal or extramodal. The secondary (distractor) task for the visual modality was to detect whether one element in a brief central array of dots was brighter than the others, and the secondary task in audition was to detect whether a brief triad of tones formed a major or a minor chord. Stimuli for the secondary tasks had a fixed level of difficulty (1 standard deviation above threshold level, as determined in a pilot experiment).

Fig. 8 shows psychometric functions from one observer showing performance on the primary visual task (contrast discrimination, left-hand panel) and on the primary auditory task (frequency discrimination, right-hand panel). In each panel, filled circles represent performance on the primary task when measured alone, while the two other curves show performance on the primary task when measured in the dual task context. The filled squares in each panel show primary task performance measured in the presence of a concurrent intramodal distractor task. The psychometric functions in this case are shifted to the right, showing a marked increase in the contrast (or frequency) increment required to perform the primary task. For all subjects, increment thresholds were at least twofold larger for intramodal distractors, and as much as fivefold. The critical condition is shown by the open triangles. These show primary task performance measured when the distractor task was extramodal. Psychometric functions in this case are very similar to those obtained without any distractor task (filled circles) indicating that for both audition and vision, primary

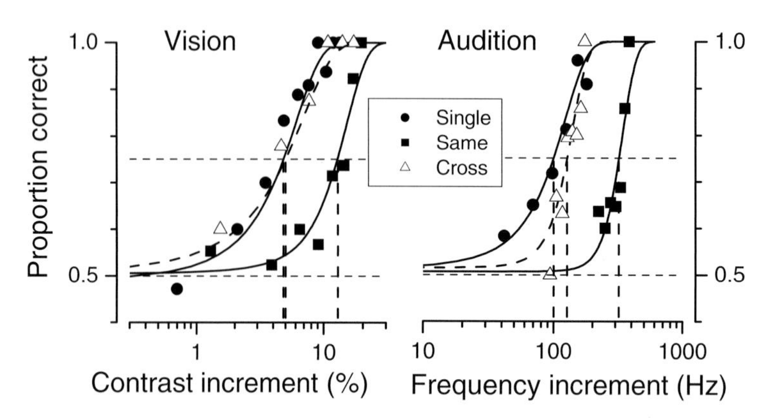

Fig. 8. Examples of psychometric functions for visual contrast and auditory frequency discriminations for one naïve observer (R.A.). The filled circles show the thresholds for the primary task alone, the filled squares when performed together with the secondary task in the same modality and the open triangles when performed with the secondary task in the other modality. Chance performance was 50% (lower dashed line). The curves are best-fitting cumulative Gaussians, from which thresholds were calculated (taken as the 75% correct point, indicated by the vertical dashed lines). The secondary task in the same modality clearly impeded performance, shifting the psychometric functions toward higher contrasts and frequencies, without greatly affecting their slope or general form. In this experiment the secondary tasks were adjusted in difficulty to produce 92% correct performance when presented alone ($d' = 2$).

task performance was largely unaffected by a competing task presented to another modality. Importantly, the psychometric functions remained orderly during the dual tasks, without decreasing slope or increasing in noise, implying a real change in the threshold limit. A marked change in slope or noisiness would have suggested that the subjects were "multiplexing" and attempting to alternate between tasks from trial to trial. This would have compromised their performance on the primary task and produced noisier data with a shallower slope.

Fig. 9 summarizes the primary thresholds in the dual-task conditions for three observers. The dual-task thresholds are shown as multiples of the primary thresholds that were measured in the single-task conditions (i.e., filled circles of Fig. 8), so that a value of 1 (dashed line) would indicate no change at all. In all cases secondary tasks that were intramodal raised primary thresholds considerably, while the extramodal secondary tasks had virtually no effect. The average increase in primary threshold produced by intramodal distractors was a factor of 2.6 for vision and a factor of 4.2 for audition, while the average threshold increase produced by extramodal distractors was just 1.1 for vision and 1.2 for audition.

The final cluster of columns in Fig. 9 shows the same data averaged over observers. The large effects of intramodal distractors are clear. Statistical tests on the two extramodal conditions (the two middle columns) showed that the mean increase in the primary auditory threshold produced by the extramodal (visual) distractor was statistically significant ($p = 0.002$); however, the mean increase in the primary visual threshold produced by the extramodal (auditory) distractor was not significantly greater than 1.0 ($p > 0.05$).

The results of these experiments clearly show that basic auditory and visual discriminations of the kind used here are not limited by a common central resource. A concurrent auditory task dramatically increased thresholds for auditory frequency discriminations, and a concurrent visual task dramatically increased thresholds for visual contrast discrimination. However, a concurrent task in a different modality had virtually no effect on primary task thresholds in vision or audition, regardless of whether the tasks were spatially superimposed or separated, and irrespective of task load.

Several previous studies have reported interactions between visual and auditory attentional resources (Driver and Spence, 1994; Spence and Driver, 1996; Spence et al., 2000; Driver and Spence, 2004). However, these studies involved directing attention to different regions of space, whereas we took care to ensure that the spatial

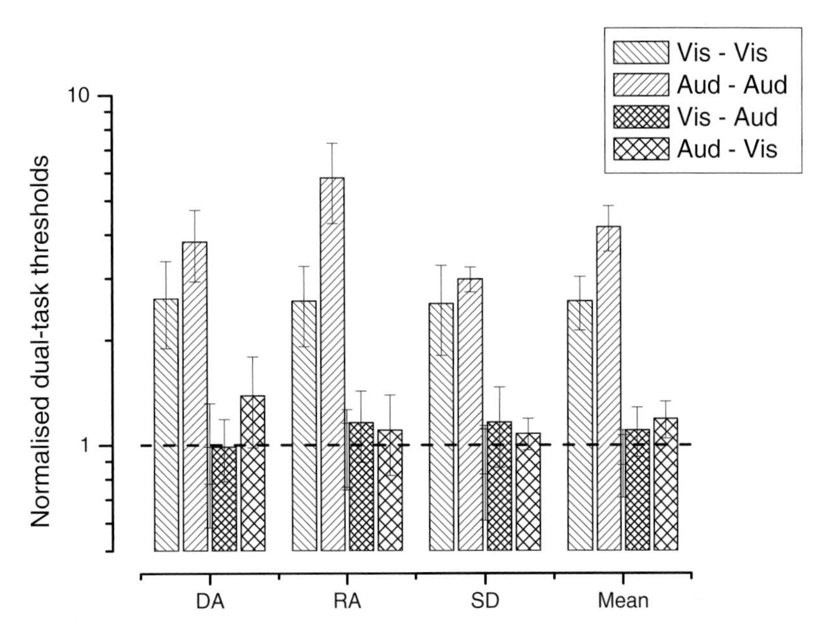

Fig. 9. Threshold performance for three observers (author D.A. and two naïve subjects) for visual and auditory discriminations, all normalized by the single-task threshold. Rightward hatch show visual thresholds for dual tasks with a visual secondary task, leftward hatch auditory thresholds with a visual auditory task. Dense crosshatching shows visual thresholds with an auditory secondary task and sparse crosshatching auditory thresholds with a visual secondary task. The only large effects are for dual tasks in the same modality. Error bars refer to standard errors, calculated by 500 iterations of bootstrapping (Efron and Tibshirani, 1993).

regions stimulated by our visual and auditory stimuli were as similar as possible, and that attention was distributed over the whole field. Furthermore, many of the reported effects were quite small, with d' improving from about 0.1 to 0.5 at most (as calculated from their reported error scores). These effects are nearly an order of magnitude less than the intramodal effects we observed. One of our crossmodal conditions showed a very small effect of attention (auditory thresholds measured with visual secondary task), although not the other. However, while statistically significant, the decrement in discriminability caused by the extramodal distractor task was only about 20%, compared to 420% for the intramodal distractor task. So while we cannot totally exclude the existence of crossmodal leakage of attentional limitations, these effects must be considered to be very much secondary compared with the magnitude of intramodal attentional effects.

Although our results are at odds with the conclusions of several recent reports indicating supramodal attentional processes, there is a growing body of evidence indicating independent attentional processes. Our conclusions are in broad agreement with some of the older psychological and human factors literature (Triesman and Davies, 1973; Wickens, 1980), and also agree with those of more recent crossmodal attentional studies using psychophysical and behavioral paradigms quite different to ours (Duncan et al., 1997; Bonnel and Hafter, 1998; Ferlazzo et al., 2002). In addition, a recent transcranial magnetic stimulation experiment that disrupted areas within parietal cortex during visual and somatosensory orienting revealed modality-specific attentional substrates (Chambers et al., 2004), rather than the region being a supramodal attention network (e.g., Macaluso et al., 2002). Other support for our findings comes from recent evidence suggesting that attention is not a unitary phenomenon, but acts at various cortical levels, including early levels of sensory processing and the primary cortical areas of V1 and A1 (Kanwisher and Wojciulik, 2000). Attentional modulation of primary cortices is particularly relevant to our study because the contrast and pitch discrimination tasks used in our experiment are probably mediated by primary cortical areas (Recanzone et al., 1993; Boynton et al., 1999; Zenger-Landolt and Heeger, 2003).

Our results are therefore quite consistent with the notion that each primary cortical area is modulated by its own attentional resources, with very little interaction across modalities. This does not exclude the possibility that attentional effects could also occur at higher levels, after visual and auditory information is combined. Depending on the nature of the task demands, the most sensible strategy might well be to employ a supramodal attentional resource for a given task. For example, speech comprehension in a noisy environment would improve if spatially colocated visual (lip movements) and auditory (speech) signals were combined via a supramodal spatial attention system.

Concluding remarks

Overall, two important general points can be taken from the experiments summarized above. The first is that the Bayesian approach provides a very useful structure for modeling crossmodal interactions. It should be stressed, however, that this approach is largely descriptive, and addresses primarily the issue of how to weight the information from different sources for cue combination. Much work remains to be done to understand how the brain obtains the reliability estimates it needs to use such a framework. Moreover, the model does not address important issues such as the disparity or "conflict" limits beyond which the perceptual system vetoes crossmodal integration.

The second important issue concerns the role of attention. Attention clearly plays an important role in crossmodal interactions, but its nature seems to be more complex than has been previously appreciated. Using low-level stimuli, we found very strong evidence for independent attentional mechanisms for audition and vision. However, important work from other groups shows evidence for supramodal attention in crossmodal tasks. These apparently conflicting results are probably best understood as different aspects of a complex and distributed attentional system that varies in its network organization form one task to another, tailoring itself to optimally perform a particular task. Accordingly, attention will appear to be low level and duplicated unimodally for audiovisual tasks such as we used that are mediated in the primary cortices, but will appear supramodal for tasks involving higher level processes or for tasks where there is uncertainty over which sense should be monitored.

The burgeoning activity in crossmodal research will no doubt shed light on these important matters of attention and Bayesian combination. The flexible nature of attentional processes needs to be understood more fully, and the preattentive aspects of crossmodal interactions need to be specified. As for the Bayesian approach, there is clearly a growing body of evidence highlighting its enormous potential as a model of multisensory combination. Important remaining questions concern the role of knowledge, expectation and attention, and how these factors can be built into a Bayesian framework, most likely by exploiting prior distributions to complement the maximum likelihood combination of ascending sensory signals.

References

Alais, D. and Burr, D. (2003) The "flash-lag" effect occurs in audition and cross-modally. Curr. Biol., 13: 59–63.

Alais, D. and Burr, D. (2004a) No direction-specific bimodal facilitation for audiovisual motion detection. Brain Res. Cogn. Brain Res., 19(2): 185–194.

Alais, D. and Burr, D. (2004b) The ventriloquist effect results from near-optimal bimodal integration. Curr. Biol., 14(3): 257–262.

Alais, D. and Carlile, S. (2005) Synchronizing to real events: subjective audiovisual alignment scales with perceived auditory depth and speed of sound. Proc. Natl. Acad. Sci. USA, 102(6): 2244–2247.

Alais, D. and Weston, E. (2006) Temporal ventriloquism: perceptual shifts in temporal position and improved audiovisual precision predicted by maximum likelihood estimation. J. Vis., 6: 171.

Arrighi, R., Alais, D. and Burr, D. (2005) Neural latencies do not explain the auditory and audio-visual flash-lag effect. Vision Res., 45: 917–2925.

Bald, L., Berrien, F.K., Price, J.B. and Sprague, R.O. (1942) Errors in perceiving the temporal order of auditory and visual stimuli. J. Appl. Psychol., 26: 382–388.

Baldo, M.V. and Klein, S.A. (1995) Extrapolation or attention shift? Nature, 378(6557): 565–566.

Battaglia, P.W., Jacobs, R.A. and Aslin, R.N. (2003) Bayesian integration of visual and auditory signals for spatial localization. J. Opt. Soc. Am. A Opt. Image. Sci. Vis., 20(7): 1391–1397.

Berger, T.D., Martelli, M. and Pelli, D.G. (2003) Flicker flutter: is an illusory event as good as the real thing? J. Vis., 3(6): 406–412.

Bonnel, A.M. and Hafter, E.R. (1998) Divided attention between simultaneous auditory and visual signals. Percept. Psychophys., 60(2): 179–190.

Boynton, G.M., Demb, J.B., Glover, G.H. and Heeger, D.J. (1999) Neuronal basis of contrast discrimination. Vision Res., 39(2): 257–269.

Brefczynski, J.A. and DeYoe, E.A. (1999) A physiological correlate of the 'spotlight' of visual attention. Nat. Neurosci., 2(4): 370–374.

Bremmer, F., Schlack, A., Shah, N.J., Zafiris, O., Kubischik, M., Hoffmann, K., Zilles, K. and Fink, G.R. (2001) Polymodal motion processing in posterior parietal and premotor cortex: a human fMRI study strongly implies equivalencies between humans and monkeys. Neuron, 29(1): 287–296.

Britten, K.H., Newsome, W.T., Shadlen, M.N., Celebrini, S. and Movshon, J.A. (1996) A relationship between behavioral choice and the visual responses of neurons in macaque MT. Vis. Neurosci., 13(1): 87–100.

Britten, K.H., Shadlen, M.N., Newsome, W.T. and Movshon, J.A. (1992) The analysis of visual motion: a comparison of neuronal and psychophysical performance. J. Neurosci., 12(12): 4745–4765.

Bronkhorst, A.W. and Houtgast, T. (1999) Auditory distance perception in rooms. Nature, 397(6719): 517–520.

Burr, D.C., Morrone, M.C. and Banks, M.S. (2006) The ventriloquist effect in time is consistent with optimal combination across senses. J. Vis., 6: 387.

Bushara, K.O., Grafman, J. and Hallett, M. (2001) Neural correlates of auditory-visual stimulus onset asynchrony detection. J. Neurosci., 21(1): 300–304.

Caclin, A., Soto-Faraco, S., Kingstone, A. and Spence, C. (2002) Tactile "capture" of audition. Percept. Psychophys., 64(4): 616–630.

Chambers, C.D., Stokes, M.G. and Mattingley, J.B. (2004) Modality-specific control of strategic spatial attention in parietal cortex. Neuron, 44(6): 925–930.

Clarke, J.J. and Yuille, A.L. (1990) Data Fusion for Sensory Information Processing. Kluwer Academic, Boston.

Colby, C.L., Duhamel, J.R. and Goldberg, M.E. (1993) Ventral intraparietal area of the macaque: anatomic location and visual response properties. J. Neurophysiol., 69(3): 902–914.

Connor, S. (2000) Dumbstruck: A Cultural History of Ventriloquism. Oxford University Press, Oxford.

Corbetta, M. and Shulman, G.L. (2002) Control of goal-directed and stimulus-driven attention in the brain. Nat. Rev. Neurosci., 3(3): 201–215.

Corey, D.P. and Hudspeth, A.J. (1979) Response latency of vertebrate hair cells. Biophys. J., 26(3): 499–506.

Desimone, R. and Duncan, J. (1995) Neural mechanisms of selective visual attention. Ann. Rev. Neurosci., 18: 193–222.

Driver, J. and Spence, C. (1994) Spatial synergies between auditory and visual attention. In: Umiltà, C. and Moscovitch, M. (Eds.) Attention and Performance: Conscious and Nonconscious Information Processing, Vol. 15. MIT Press, Cambridge, MA, pp. 311–331.

Driver, J. and Spence, C. (2004) Crossmodal spatial attention: evidence from human performance. In: Spence, C. and Driver, J. (Eds.), Crossmodal Space and Crossmodal Attention. Oxford University Press, Oxford.

Duncan, J., Martens, S. and Ward, R. (1997) Restricted attentional capacity within but not between sensory modalities. Nature, 387(6635): 808–810.

Eagleman, D.M. and Sejnowski, T.J. (2000) Motion integration and postdiction in visual awareness. Science, 287(5460): 2036–2038.

Efron, B. and Tibshirani, R.J., 1993. An Introduction to the Bootstrap. Monographs on Statistics and Applied Probability, Vol. 57. Chapman & Hall, New York.

Ernst, M.O. and Banks, M.S. (2002) Humans integrate visual and haptic information in a statistically optimal fashion. Nature, 415(6870): 429–433.

Ernst, M.O. and Bülthoff, H.H. (2004) Merging the senses into a robust percept. Trends Cogn. Sci., 8(4): 162–169.

Ferlazzo, F., Couyoumdjian, M., Padovani, T. and Belardinelli, M.O. (2002) Head-centred meridian effect on auditory spatial attention orienting. Q. J. Exp. Psychol. A, 55(3): 937–963.

Gandhi, S.P., Heeger, D.J. and Boynton, G.M. (1999) Spatial attention affects brain activity in human primary visual cortex. Proc. Natl. Acad. Sci. USA, 96(6): 3314–3319.

Ghahramani, Z., Wolpert, D.M. and Jordan, M.I. (1997) Computational models of sensorimotor integration. In: Morasso, P.G. and Sanguineti, V. (Eds.), Self-Organization, Computational Maps and Motor Control. Elsevier Science, Amsterdam, pp. 117–147.

Grady, C.L., Van Meter, J.W., Maisog, J.M., Pietrini, P., Krasuski, J. and Rauschecker, J.P. (1997) Attention-related modulation of activity in primary and secondary auditory cortex. Neuroreport, 8(11): 2511–2516.

Graziano, M.S. (2001) A system of multimodal areas in the primate brain. Neuron, 29(1): 4–6.

Hamlin, A.J. (1895) On the least observable interval between stimuli addressed to disparate senses and to different organs of the same sense. Am. J. Psychol., 6: 564–575.

Hillis, J.M., Ernst, M.O., Banks, M.S. and Landy, M.S. (2002) Combining sensory information: mandatory fusion within, but not between, senses. Science, 298(5598): 1627–1630.

Jacobs, R.A. (1999) Optimal integration of texture and motion cues to depth. Vision Res., 39(21): 3621–3629.

Jacobs, R.A. and Fine, I. (1999) Experience-dependent integration of texture and motion cues to depth. Vision Res., 39(24): 4062–4075.

Jancke, L., Mirzazade, S. and Shah, N.J. (1999) Attention modulates activity in the primary and the secondary auditory cortex: a functional magnetic resonance imaging study in human subjects. Neurosci. Lett., 266(2): 125–128.

Kanwisher, N. and Wojciulik, E. (2000) Visual attention: insights from brain imaging. Nat. Rev. Neurosci., 1(2): 91–100.

King, A.J. and Palmer, A.R. (1985) Integration of visual and auditory information in bimodal neurones in the guinea-pig superior colliculus. Exp. Brain. Res., 60(3): 492–500.

Kopinska, A. and Harris, L.R. (2004) Simultaneity constancy. Perception, 33(9): 1049–1060.

Krekelberg, B. and Lappe, M. (2000) A model of the perceived relative positions of moving objects based upon a slow averaging process. Vision Res., 40(2): 201–215.

Krekelberg, B. and Lappe, M. (2001) Neuronal latencies and the position of moving objects. Trends Neurosci., 24(6): 335–339.

Lamb, T.D. and Pugh Jr., E.N. (1992) A quantitative account of the activation steps involved in phototransduction in amphibian photoreceptors. J. Physiol., 449: 719–758.

Lennie, P. (1981) The physiological basis of variations in visual latency. Vision Res., 21(6): 815–824.

Lewald, J. and Guski, R. (2004) Auditory-visual temporal integration as a function of distance: no compensation for sound-transmission time in human perception. Neurosci. Lett., 357(2): 119–122.

Luck, S.J., Chelazzi, L., Hillyard, S.A. and Desimone, R. (1997) Neural mechanisms of spatial selective attention in areas V1, V2, and V4 of macaque visual cortex. J. Neurophysiol., 77(1): 24–42.

Macaluso, E., Frith, C.D. and Driver, J. (2002) Supramodal effects of covert spatial orienting triggered by visual or tactile events. J. Cogn. Neurosci., 14(3): 389–401.

Mackay, D.M. (1958) Perceptual stability of a stroboscopically lit visual field containing self-luminous objects. Nature, 181(4607): 507–508.

Mateeff, S., Hohnsbein, J. and Noack, T. (1985) Dynamic visual capture: apparent auditory motion induced by a moving visual target. Perception, 14(6): 721–727.

Meyer, G.F. and Wuerger, S.M. (2001) Cross-modal integration of auditory and visual motion signals. Neuroreport, 12(11): 2557–2560.

Mills, A. (1958) On the minimum audible angle. J. Acoust. Soc. Am., 30: 237–246.

Morein-Zamir, S., Soto-Faraco, S. and Kingstone, A. (2003) Auditory capture of vision: examining temporal ventriloquism. Brain Res. Cogn. Brain. Res., 17(1): 154–163.

Moutoussis, K. and Zeki, S. (1997) A direct demonstration of perceptual asynchrony in vision. Proc. R. Soc. Lond. B Biol. Sci., 264(1380): 393–399.

Nijhawan, R. (1994) Motion extrapolation in catching. Nature, 370(6487): 256–257.

Pashler, H.E. (1998) The Psychology of Attention. MIT Press, Cambridge, MA.

Perrott, D. and Saberi, K. (1990) Minimum audible angle thresholds for sources varying in both elevation and azimuth. J. Acoust. Soc. Am., 87: 1728–1731.

Pick, H.L., Warren, D.H. and Hay, J.C. (1969) Sensory conflict in judgements of spatial direction. Percept. Psychophys., 6: 203–205.

Purushothaman, G., Patel, S.S., Bedell, H.E. and Ogmen, H. (1998) Moving ahead through differential visual latency. Nature, 396(6710): 424.

Recanzone, G.H., Schreiner, C.E. and Merzenich, M.M. (1993) Plasticity in the frequency representation of primary auditory cortex following discrimination training in adult owl monkeys. J. Neurosci., 13(1): 87–103.

Shams, L., Kamitani, Y. and Shimojo, S. (2000) Illusions. What you see is what you hear. Nature, 408(6814): 788.

Shams, L., Kamitani, Y. and Shimojo, S. (2002) Visual illusion induced by sound. Brain Res. Cogn. Brain Res., 14(1): 147–152.

Shams, L., Kamitani, Y., Thompson, S. and Shimojo, S. (2001) Sound alters visual evoked potentials in humans. Neuroreport, 12(17): 3849–3852.

Spence, C. and Driver, J. (1996) Audiovisual links in endogenous covert spatial attention. J. Exp. Psychol. Hum. Percept. Perform., 22(4): 1005–1030.

Spence, C., Ranson, J. and Driver, J. (2000) Cross-modal selective attention: on the difficulty of ignoring sounds at the locus of visual attention. Percept. Psychophys., 62(2): 410–424.

Standage, G.P. and Benevento, L.A. (1983) The organization of connections between the pulvinar and visual area MT in the macaque monkey. Brain. Res., 262(2): 288–294.

Stein, B.E. (1998) Neural mechanisms for synthesizing sensory information and producing adaptive behaviors. Exp. Brain. Res., 123(1-2): 124–135.

Sugita, Y. and Suzuki, Y. (2003) Audiovisual perception: implicit estimation of sound-arrival time. Nature, 421(6926): 911.

Titchener, E.B. (1908) Lectures on the Elementary Psychology of Feeling and Attention. MacMillan, New York.

Triesch, J., Ballard, D.H. and Jacobs, R.A. (2002) Fast temporal dynamics of visual cue integration. Perception, 31(4): 421–434.

Triesman, A.M. and Davies, A. (1973) Divided attention to ear and eye. In: Kornblum, S. (Ed.) Attention and Performance, Vol. 4. Academic Press, New York.

Warren, D.H., Welch, R.B. and McCarthy, T.J. (1981) The role of visual-auditory "compellingness" in the ventriloquism effect: implications for transitivity among the spatial senses. Percept. Psychophys., 30(6): 557–564.

Wickens, C.D., 1980. The structure of attentional resources. In: Attention and Performance, Vol. VIII. Erbaum, Hillsdale, NJ.

Woodruff, P.W., Benson, R.R., Bandettini, P.A., Kwong, K.K., Howard, R.J., Talavage, T., Belliveau, J. and Rosen, B.R. (1996) Modulation of auditory and visual cortex by selective attention is modality-dependent. Neuroreport, 7(12): 1909–1913.

Wuerger, S.M., Hofbauer, M. and Meyer, G.F. (2003) The integration of auditory and visual motion signals at threshold. Percept. Psychophys., 65(8): 1188–1196.

Zahorik, P. (2002) Direct-to-reverberant energy ratio sensitivity. J. Acoust. Soc. Am., 112(5 Pt 1): 2110–2117.

Zenger-Landolt, B. and Heeger, D.J. (2003) Response suppression in v1 agrees with psychophysics of surround masking. J. Neurosci., 23(17): 6884–6893.

Martinez-Conde, Macknik, Martinez, Alonso & Tse (Eds.)
Progress in Brain Research, Vol. 155
ISSN 0079-6123

CHAPTER 15

Crossmodal interactions: lessons from synesthesia

Noam Sagiv* and Jamie Ward

Department of Psychology, University College London, 26 Bedford Way, London WC1H 0AP, UK

Abstract: Synesthesia is a condition in which stimulation in one modality also gives rise to a perceptual experience in a second modality. In two recent studies we found that the condition is more common than previously reported; up to 5% of the population may experience at least one type of synesthesia. Although the condition has been traditionally viewed as an anomaly (e.g., breakdown in modularity), it seems that at least some of the mechanisms underlying synesthesia do reflect universal crossmodal mechanisms. We review here a number of examples of crossmodal correspondences found in both synesthetes and non-synesthetes including pitch-lightness and vision-touch interaction, as well as cross-domain spatial-numeric interactions. Additionally, we discuss the common role of spatial attention in binding shape and color surface features (whether ordinary or synesthetic color). Consistently with behavioral and neuroimaging data showing that chromatic–graphemic (colored-letter) synesthesia is a genuine perceptual phenomenon implicating extrastriate cortex, we also present electrophysiological data showing modulation of visual evoked potentials by synesthetic color congruency.

Keywords: synesthesia; number forms; crossmodal perception; multisensory integration; anomalous experience

Introduction

Narrowly defined, synesthesia is a condition in which stimulation in one sensory modality evokes an additional perceptual experience in another modality. For example, sounds may evoke colors for some individuals (e.g., Ginsberg, 1923). While various sensory combinations have been described (e.g., Cytowic, 2002; Day, 2004), a common type of synesthesia in which seen letters and numbers induce color experience actually occurs intramodally in vision (e.g., Ramachandran and Hubbard, 2001a). Furthermore, it is clear that the meaning of the inducing stimulus is implicated in synesthesia at least sometimes rather than its lower level physical features (Myles et al., 2003). Moreover, certain concepts (e.g., days of the week) may induce synesthesia when synesthetes are thinking about them, hearing about them, or reading about them. Ordinal sequences (e.g., letters, numbers, and time units) often serve as inducers. In the recent literature, these have been regarded as variants of synesthesia. Nevertheless, it is still a matter of some debate how inclusive the definition of synesthesia should be. For example, there is no consensus concerning synesthesia-like cross-domain phenomena involving spatial forms or personification (e.g., Calkins, 1895). We do not use the term synesthesia to describe any crossmodal correspondences and associations, but rather only those cases in which a perceptual experience is involved, i.e., what Martino and Marks (2001) call *strong synesthesia*.

Synesthesia exists in developmental and acquired forms. The former runs in the family (e.g., Baron-Cohen et al., 1996; Ward and Simner, 2005), while the latter has been described in a

*Corresponding author. Tel.: +44-1895-265341; Fax: +44-1895-269724; E-mail: noam127@yahoo.com

DOI: 10.1016/S0079-6123(06)55015-0

variety of conditions as well as altered states of consciousness. Some examples include synesthesia following brain or nerve injury (Jacobs et al., 1981), synesthesia in late-blind individuals (Armel and Ramachandran, 1999; Steven and Blakemore, 2004). Hallucinogens such as lysergic acid diethylamide (LSD), mescaline (Hartman and Hollister, 1963), or ayahuasca (Shanon, 2002) often induce synesthetic hallucination. It is also reported by healthy individuals between sleep and wakefulness (Sagiv and Ben-Tal, in preparation) and a high proportion of meditators (Walsh, 2005).

Developmental synesthesia is thought to be characterized by a remarkable consistency of synesthetic correspondence within an individual across time (which has, in fact, been used as one diagnostic criterion). Still, synesthetes rarely agree on particular synesthetic correspondence (see, e.g., Pat Duffy and Carol Steen's colored alphabets; Duffy, 2001). However, some trends can be traced in large synesthete populations (e.g., Shanon, 1982; Day, 2004; Rich et al., 2005). Beyond disagreement on particular correspondences, substantial heterogeneity is often found in phenomenological description, for example, in the spatial extent of synesthetic percepts (e.g., Dixon et al., 2004; Sagiv, 2004). Indeed we are only beginning to appreciate individual difference among synesthetes currently grouped together under single labels (Dixon and Smilek, 2005; Hubbard et al., 2005a).

Synesthesia is involuntary and automatically evoked (in contrast to imagery at will). It is unclear whether similar underlying mechanisms give rise to developmental and acquired synesthesia. Reports of acquired synesthesia are largely anecdotal; however, it is possible that some predisposition is required. Additionally, positive symptoms are often under-reported, however, and thus prevalence of acquired synesthesia is harder to establish. Similarly, individuals with developmental synesthesia may have not realized that how they perceive the world is unusual in any way.

It has been suggested that synesthesia is unidirectional (Mills et al., 1999). Phenomenologically, this seems to be the case (e.g., pain may evoke a bright orange color; however, orange objects do not usually induce pain). However, at least in numerical cognition, evidence for bidirectional interaction between color and magnitude processing is available (Cohen-Kadosh et al., 2005; Knoch et al., 2005).

Prevalence

Estimate concerning the prevalence of (developmental) synesthesia vary widely. Early surveys, relying on subjective reports, suggested that up to one in five individuals may have some form of synesthesia (e.g., Uhlich, 1957). Estimates based on self-referred samples have been far more pessimistic. Cytowic (1997) estimated that only 1 in 25,000 might experience the condition, while Baron-Cohen et al (1996) estimated that the condition is present in *at least* 1 in 2000 individuals. Baron-Cohen and his colleagues combined for the first time subjective reports with an objective test of genuineness (testing for consistency synesthetic correspondences) thus ensuring low rate of false alarms. However, because the prevalence estimate was based on the number of people responding to a newspaper advertisement (divided by readership figures), many cases were probably missed either because they did not see the ad or saw it but chose not to respond.

The best estimate we have so far comes from two large-scale surveys we recently conducted (Simner et al., in press). In both we combined objective and subjective methods for ascertaining synesthesia, but minimized self-referral bias. One survey in London's Science Museum included 1190 unsuspecting visitors. The study was conducted in an interactive gallery in the museum, and visitor cooperation was high, although they did not know what the purpose of the experiment was. Here we looked at chromatic–graphemic synesthesia (the most common type in self-referred samples). Participants were requested to choose a color that "goes best" with letters and numbers. The process was repeated immediately, yielding a measure of consistency. Following this participants were asked direct questions concerning synesthetic experience. 1.1% of the individuals tested were identified as chromatic–graphemic synesthetes.

In the second survey of multiple types of synesthesia among 500 Scottish students who where given more information on the varieties of synesthesia, we identified 23 synesthetes (4.6%) of which about half only reported colored weekdays. The prevalence of chromatic–graphemic synesthesia in this sample (1.4%) was similar to that obtained in the museum study. In summary, the condition may not be so unusual and appears to be almost two orders of magnitude more common than previously regarded.

Indeed, some view synesthesia not as an anomalous phenomenon, but rather as one reflecting a normal mode of cognition (for a discussion, see Sagiv, 2004) that remains implicit in most individuals. Consistent with this view is the high prevalence of synesthesia in a number of altered states of consciousness suggesting that many of us have some predisposition to synesthesia. In the next section we explore the idea that synesthesia shares much in common with normal perception and discuss some common mechanism and processes.

What does synesthesia share in common with normal perception?

All contemporary accounts of synesthesia propose that there is some anomalous connectivity or cross-activation between different regions of the brain. Aside from this broad consensus,[1] there is disagreement about: the nature of the connectivity between regions (e.g., horizontal vs. feedback connections); whether or not the pathways implicated in synesthesia are present in the normal population or whether synesthetes have some privileged pathways; and differences in how anomalous connectivity could be instantiated at the neural level (e.g., literal increases in white matter connections, or neurochemical differences). At present, nothing is known about possible structural differences between the brains of synesthetes and other members of the population (at either a micro- or macroscale). As such, the processes underlying synesthesia have tended to be inferred from studies at a functional (e.g., fMRI) or cognitive level. For example, Nunn et al. (2002) reported V4/V8 activation induced by synesthetic color in synesthetes listening to spoken words (eyes closed). Behavioral evidence demonstrating that synesthesia could influence perception of the evoking stimulus (e.g., grouping, detection, and apparent motion) also strongly suggests that visual synesthetic experience taps into existing visual mechanisms (Ramachandran and Hubbard, 2001a; Smilek et al., 2001; Blake et al., 2004).

In the remainder of this chapter, we will consider the extent to which synesthesia may utilize some of the same mechanisms found in other members of the population in situations of crossmodal perception. For example, it is generally accepted that there are crossmodal mechanisms for linking together sight and touch (e.g., Spence, 2002) and sight and sound (e.g., King and Calvert, 2001) when both modalities are stimulated. But could these same mechanisms be implicated in synesthetic experiences of touch given a unimodal visual stimulus, or synesthetic experiences of vision given a unimodal auditory stimulus?

Sound–vision synesthesia

According to a large self-referred sample of synesthetes, at least 18% of them experience color induced by auditory stimuli such as music and noise (Day, 2004). Many more synesthetes experience colors from speech but not from other types of auditory stimulus. However, for these synesthetes the color tends to depend on linguistic properties of the stimulus (e.g., the graphemic composition) rather than acoustic properties (e.g., Baron-Cohen et al., 1993; Paulesu et al., 1995). For example, heard words beginning with the letter "P" may all tend to elicit the same color even when not starting by a /p/ sound (e.g., "psychology," "photo," and "potato" may have the same color). The present discussion will restrict itself to sound-vision synesthesia that depends on perceptual properties of the auditory stimulus, notably pitch.

One reason why this type of synesthesia has been regarded as of great interest is the suggestion that it may be present in all human infants. This raises the possibility that adult synesthetes have retained some pathways/processes that most other members of the population lost during development (e.g., Baron-Cohen, 1996; Maurer, 1997). In human infants, visual event-related

[1] But cf. Shanon (2003).

potentials (ERPs) to auditory stimuli decrease after 6 months but are still present at between 20 and 30 months of age (Neville, 1995). In kittens there is evidence of anatomical connections between primary auditory and visual areas that are lost as part of a pruning mechanism (Dehay et al., 1987; Innocenti et al., 1988). It may be genetically programmed given that it does not appear to depend on the presence of visual input (Innocenti et al., 1988). However, caution must be exerted in linking these findings to human synesthesia as interspecies differences are likely. More recent findings in the macaque suggest a lifelong presence of connections from primary auditory cortex to peripheral area V1, as well as from a polysensory temporal region to peripheral V1 (Falchier et al., 2002). It is also conceivable that the ERP effects in humans reflect developmental changes in the balance between feedback connections from multimodal to unimodal regions rather than direct pathways between unimodal areas (Kennedy et al., 1997).

Our own research has addressed the question of shared processes between synesthetes and the normal population by looking at ways in which auditory properties (e.g., pitch and timbre) link with visual properties (e.g., the chromaticity and luminance of colors; Ward, Huckstep, and Tsakanikos (2006)). Our reasoning was that if both synesthetes and controls perform the task in similar ways then this would point to some shared processes between them. Such a finding would be more consistent with the view that synesthesia uses some of the same processes as normal crossmodal perception than the view that synesthetes have privileged mechanisms not present in others (except possibly in the earliest phase of development). Ward et al. (2006) presented 10 synesthetes and 10 controls with 70 tones of different pitch and timbre on several occasions. Given that controls do not report actual visual experiences, their task was to choose the "best" hue to go with each tone. It was found that both groups showed an identical trend to associate low pitch with dark colors and high pitch with light colors (with luminance measured on the Munsell scale). This pattern was unaffected by differences in timbre. By contrast, differences in

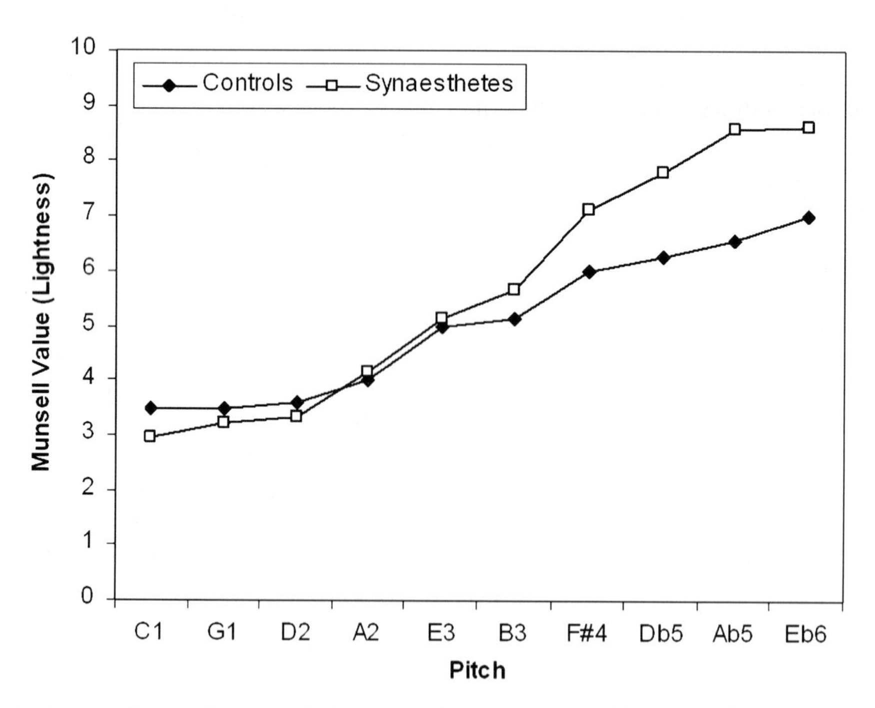

Fig. 1. Both sound-color synesthetes and nonsynesthetic controls show the same trend (no group difference: $F(1,18) = 2.51$, N.S.) to associate low pitch tones with darker colors (low Munsell values) and high pitch tones with lighter colors (higher Munsell values). Figure reprinted from Ward et al. (2006), by permission of Masson.

chroma (ranging from grayscale to high-saturation colors) were associated with differences in timbre, with pure tones judged to be less "colorful" than musical tones. The highest values of chroma were found around middle C. Again, there were no differences between synesthetes and controls in the way that properties of the auditory stimulus were mapped on to visual properties (Fig. 1).

Our findings are consistent with previous research in the nonsynesthesia literature showing a relationship between pitch and lightness (e.g., Marks, 1974, 1987; Hubbard, 1996), and with previous anecdotes from synesthetes (e.g., Marks, 1975). However, it represents the first direct comparison between these groups. Moreover, we were able to establish that in other key respects our synesthetes behaved differently. The synesthetes showed far greater internal consistency than controls (e.g., a control may choose light blue or light pink for the same tone on two occasions). Synesthetes, but not controls, appeared to generate colors automatically insofar as they showed Stroop-like interference for naming a colored patch incongruent with the color of a simultaneously presented tone (e.g., a blue color paired with a "red" tone). In total, our results point to some sharing of processes between synesthetes and the normal population but with some additional features required to explain the greater reliability and automaticity of the synesthetes.

There are a number of accounts of acquired sound-vision synesthesia. In some instances they are associated with visual field loss and the synesthetic experiences appear in the "blind" region (Jacobs et al., 1981). All these cases had lesions to the peripheral visual system (optic nerve or chiasm). One patient exhibited this form of synesthesia with a medial temporal and midbrain tumor (Vike et al., 1984). There was no visual dysfunction and the synesthesia disappeared when the tumor was removed. These forms of acquired synesthesia possibly reflect an adjustment of the relative weightings between pre-existing unimodal auditory, unimodal visual, and multimodal regions but is less likely to reflect the development of entirely new pathways. Nonsynesthetes who have been blindfolded for several days begin to experience visual hallucinations, some of which appear to be elicited by auditory stimuli (Merabet et al., 2004). In nonsynesthetes, primary auditory areas can be activated by a unimodal visual stimulus (in silent lip reading; Calvert, 1997). Conversely, in nonsynesthetes fitted with a cochlear implant due to deafness, activity in low-level visual areas (V1 and V2) is found when listening to unimodal auditory stimuli (Giraud et al., 2001). In both the studies of Calvert et al. (1997) and Giraud et al. (2001) the activity in primary auditory and visual areas was not reported to be associated with conscious auditory or visual experiences (although neither study explicitly discusses it). This may imply that the level of activity is below a threshold for conscious perceptual experience in these participants but rose above the threshold in the synesthetes. This remains speculative but plausible.

Vision–touch synesthesia

There is a wealth of evidence in the cognitive neuroscience literature pointing to a strong cross-modal interaction between vision and touch (for a review see Spence, 2002). For example, tactile acuity for stimuli applied to the arm is improved if the arm is visible, even though the mechanism producing the tactile stimulation cannot be seen (Kennett et al., 2001). Similarly, single-cell recordings of monkeys have identified bimodal cells that respond to both touch and vision (Graziano, 1999).

Armel and Ramachandran (1999) documented an acquired case of synesthesia arising following retinal damage in which tactile stimulation of the arms induced color photisms. Interestingly, the intensity of the photisms was related to the location of the arms within the "visual" field (remembering that the patient was blind) such that the synesthetic experiences were greater when the arms were in a normal field of view. This type of synesthesia has also been noted after cortical blindness (Goldenberg et al., 1995). The synesthetic images induced by tactile stimuli may have lead to a denial of blindness (anosognosia or Anton's syndrome).

Conversely, synesthesia-like tactile and kinesthetic sensations have been induced in amputated limbs using mirrors (Ramachandran and Rogers-Ramachandran, 1996), in a limb with hemi-anesthesia whilst viewing real or video-recorded tactile stimulations (Halligan et al., 1996), or in healthy limbs using adapting prisms (Mon-Williams et al., 1997). The fact that synesthesia-like sensations can be turned on or off (e.g., depending on the presence or absence of a mirror) implies that it reflects pre-existing mechanisms rather than reflecting longer term cortical changes.

We have recently documented a case of vision-touch synesthesia, C, that has a developmental rather than an acquired origin (she reports having the sensations all her life and other family members possess synesthesia) (Blakemore et al., 2005). When C sees another person being touched, she experiences tactile sensations on her own body. She does not report tactile sensations when inanimate objects are touched. This was investigated using fMRI. The study contrasted C's brain activity to that of 12 controls when viewing a human face or neck being touched relative to viewing parts of inanimate objects (e.g., a fan) being touched. For controls, a "mirror touch" circuit was activated when viewing humans touched relative to objects that included premotor regions, the primary and secondary somatosensory cortex, and the superior temporal sulcus (for related but not identical findings see Keysers et al., 2004). In C, the same circuit was activated but some regions (left premotor, primary somatosensory cortex) were activated to a much greater level than that observed in any of the individual control subjects (Fig. 2). This implies that her synesthesia arises largely from the same system of mirror touch used by others but above a threshold for conscious tactile perception. There is, however, one caveat to this, namely that C showed bilateral activation in the anterior insula that was not present in the control group. This may be linked to a self-attribution mechanism given that the region is activated when actions (e.g., Farrer and Frith, 2002) and memories (Fink et al., 1996) are attributed to oneself rather than another person.

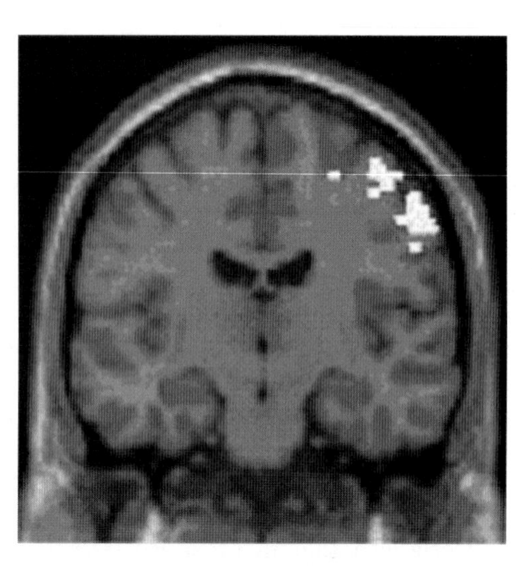

Fig. 2. Activation of the SI head area ($x = 60$, $y = -15$, $z = 48$) in nonsynesthetic controls arising from observing touch to a human face (relative to a human neck). The same region was over-activated in synesthete C, and was also active when participants were physically touched on the face. Figure reprinted from Blakemore et al., by permission of Oxford University Press.

Synesthesia, feature binding, and spatial attention

Most synesthetes describe visual synesthetic experiences as seen in their mind's eye. However, some synesthetes project these experiences externally (Dixon et al., 2004). In the case of chromatic–graphemic synesthesia, this usually means that the surface of visually presented graphemes appears colored. We therefore have a rather unusual instance of feature binding: one in which color and shape are combined together although only the latter is actually presented to the perceiver.

The binding problem — how we combine together color, shape, and other surface features into objects — has been the center of much controversy (Wolfe and Cave, 1999). According to one influential view, binding of surface features is achieved by engaging spatial attention mechanisms (Treisman and Gelade, 1980). Indeed, under conditions of divided attention, binding may fail, giving rise to illusory conjunctions (e.g., Treisman and Schmidt, 1982) — correct registration of features of two or more objects, but incorrectly combining them (e.g., a lavender X and a chartreuse O might be perceived as a chartreuse X and a lavender O). Illusory conjunctions are more common when spatial attention is disrupted after brain damage (for review, see Robertson, 1999, 2003). In one striking case, RM — a Balint's syndrome patient, exhibits illusory conjunctions even under free view conditions (e.g., Friedman-Hill et al., 1995).

In a recent study we examined whether binding of externally projected synesthetic color to the evoking shape in chromatic–graphemic synesthesia obeys the rules of normal feature binding, i.e., whether spatial attention is necessary for binding of synesthetic color and the evoking grapheme (Sagiv, Heer, and Robertson, 2006a). One obvious reason to think that synesthetic color serves as another surface feature is that "projecting" synesthetes describe it as such. Consistent with this, neuroimaging studies show that synesthetic color also activates the color area V4/V8 (Nunn et al., 2002; Sperling et al., 2006).

First, we found that unlike ordinary (wavelength-derived) color, synesthetic color does not pop-out in visual search, i.e., synesthetic color is not available preattentively. Consistent with this, Laeng et al. (2004) showed that what seemed like pop-out of synesthetic color was actually restricted to trials in which the target was within the focus of attention. Edquist et al. (2006) also reached the conclusion that the evoking grapheme must be attended before synesthetic color arises.

Second, we showed that spatial attention modulates synesthesia: we presented irrelevant digit primes, followed by a colored target. While prime location was fixed throughout the experiment, target distance from fixation alternated between blocks (such that attention was either focused around fixation or distributed across a wider region). As expected, congruency of target color and the synesthetic color of the digit primes modulated color judgment times. Importantly, this effect was larger when attention was distributed across a wide area including the digit prime, than in the case in which spatial attention was allocated to a narrow region around fixation, leaving the synesthetic inducers outside the focus of attention.

We concluded that spatial attention does indeed play a key role in binding projected synesthetic color to the evoking grapheme and hypothesized that parietal mechanisms may be implicated. Indeed, in a follow-up transcranial magnetic stimulation (TMS) study, Esterman et al. (2004) reduced the interference of synesthetic color in a color-naming task by stimulating the parieto-occipital junction. In total, binding synesthetic colors to graphemes does involve attentional mechanisms necessary for ordinary feature binding.

The time course of chromatic–graphemic synesthesia

Although synesthetic color may not be available preattentively, it does seem to arise as soon as the evoking stimulus is identified according to synesthetes' reports. Consistently with this, a wide range of studies suggest that the processes underlying synesthesia must be fairly rapid. At least, rapid enough to influence perception (Ramachandran and Hubbard, 2001a; Smilek et al., 2001; Palmeri et al., 2002; Blake et al., 2004), likely modulating activity in visual areas. While neuroimaging studies confirmed that synesthetic experience does involve activation of visual cortex (Aleman et al., 2001; Nunn et al., 2002; Hubbard et al., 2005a; Sperling et al., 2006), little is known about the time course of processing in synesthesia.

Schiltz et al. (1999) recorded ERPs while synesthetes and nonsynesthete control subjects viewed achromatic letters and numbers. They found that synesthetes had larger P300 components over frontal sites (for both target and nontarget graphemes) compared with the control group. This relatively late time course seems at odds with the dramatic effects synesthesia can have on perception of the evoking stimulus. Furthermore, Schiltz et al. failed to demonstrate the involvement of posterior cortex. Because Schiltz et al. did not use a within-subject design, we cannot rule out the possibility that the group differences reflect some co-morbid neuropsychological factors rather than synesthesia per se.

Recently we ran another ERP study of synesthesia using a within-subject design. We tested AD, a chromatic–graphemic synesthete (described further in Sagiv and Robertson, 2004). We recorded ERPs while she viewed centrally presented letters (one at a time). The letter stimuli, F and K, were either colored congruently with AD's synesthetic color (green and red, respectively) or incongruently with her synesthetic color (the color of the other letter). On 10% of the trials we presented a target letter (I, synesthetic color white). To ensure that AD's attention was focused on the presented letters, we requested her to report, at the end of each block, the number of I's presented.

Earlier electroencephalography (EEG) and magnetoencephalography (MEG) studies showed that responses to orthographic vs. nonorthographic material diverge at posterior temporal locations as early as 150–200 ms after stimulus presentation (Bentin et al., 1999; Tarkiainen et al., 1999). We examined the N1/N170 component (150–170 ms for AD) elicited by congruently and incongruently colored letters. Mean ERP amplitudes (relative to a nose reference) at this time range, in eight blocks of trials were used as random factor. These were measured at PO7 and PO8, the posterior scalp locations at which the negative component N1 evoked by letters was maximal.

AD's N1 was significantly more negative in congruent than in the incongruent condition, showing for the first time an early effect of synesthesia on evoked potentials recorded over the posterior scalp, within an individual subject. Additionally, the N1 was larger on the right (PO8) than on the left (PO7) side, a somewhat unusual pattern for a right-handed participant; however, there was no interaction between congruency and hemisphere (Fig. 3).

What are origins of this congruency effect? Note that in both conditions a synesthetic color is induced. The only difference is that we present letter–color combinations that match or mismatch

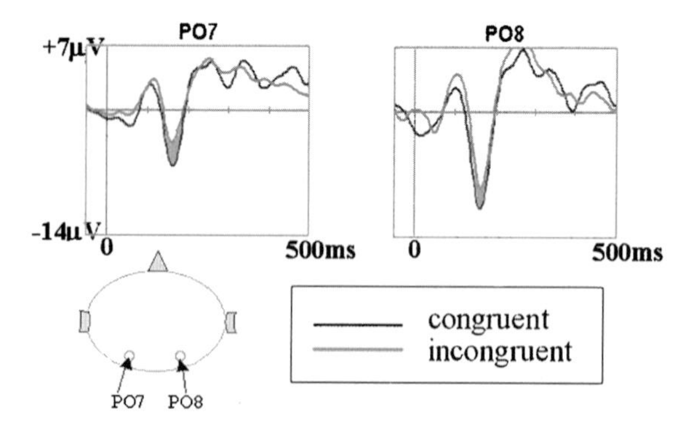

Fig. 3. Event-related potentials evoked by letter stimuli colored either congruently or incongruently with AD's synesthetic colors. The approximate location of scalp locations (PO7 and PO8) is shown on the schematic drawing on the lower left.

AD's individual synesthetic correspondences. This could have implications on both perceived stimulus contrast and stimulus categorization. The larger N1 recorded in the congruent condition is more consistent with the latter. Because AD reports seeing both her synesthetic color and the actual stimulus color at the same time, the perceived contrast in the congruent condition (e.g., red on red) is lower than that in the incongruent condition (e.g., red on green). On the other hand, the congruently colored letter may be easier to categorize. Thus, a more canonical stimulus form (at least for AD) may evoke larger N1. Alternatively, the effect may be due to attentional modulation.

While further studies will be required in order to understand the nature of this congruency effect, it is clear that it can serve as a marker for the time course of synesthesia. It may be useful as another tool for assessing individual differences among synesthetes (Hubbard et al., 2005a). Finally, the early modulation of posterior ERPs is consistent with the claim that synesthesia is a genuine perceptual phenomenon.

Number–space synesthesia

As much as 12% of the population experiences numbers as occupying a particular spatial configuration (Seron et al., 1992). These have been termed number forms (Galton, 1880a, b). Although the overall direction of these forms is often left-to-right, the precise configuration can be idiosyncratic (Fig. 4), as can their locations in space. For some, the number forms occupy peri-personal space around their body, for others it is in their "minds eye." For some, the number forms are reported to move through space according to the number attended to; for others, it is a static representation. We have recently provided some evidence for the authenticity of these subjective reports by showing that the task of deciding which of the two numbers is larger is biased according to whether the two numbers are displayed in an arrangement that is congruent or incongruent with their number form (Sagiv, Simner, Collins, butterworth, and Ward (2006b)).

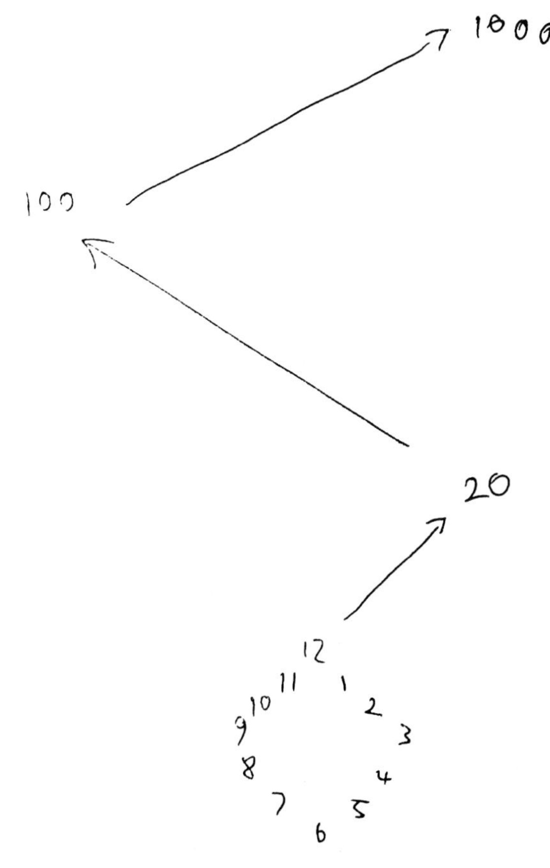

Fig. 4. An example of a more convoluted number form drawn by one of our synesthetes (colors not shown).

Although number-space synesthesia does not involve one of the traditional five sensory modalities, it shares a number of properties with other types of synesthesia. They are reported to be consistent over time, to come to mind automatically, and to have no known origin. Our sense of space is clearly a perceptual dimension, although it is not tied to any one specific sense and may be represented at multiple levels in the brain (e.g., egocentric vs. allocentric space) (e.g., Robertson, 2004). Sagiv et al., (2006b) found that number forms are far more prevalent in synesthetes who experience colors in response to numbers than in other members of the population or in other types of synesthesia. One account of this association is that the spatial attributes of numbers are applied to the associated synesthetic colors, thus leading to a heightened awareness of a number-space relationship that, in most others, remains implicit. An alternative explanation is that number-space and number-color synesthesia are caused by the same underlying mechanisms (e.g., cross-activation of brain areas, in the case of numbers forms – in the parietal lobe). Indeed, similar regions in the parietal lobes are known to mediate both aspects of numerical cognition and spatial processing (Hubbard et al., 2005b).

Evidence for a spatial (but typically implicit) mental number line in the normal population comes from the SNARC effect — the Spatial-Numerical Association of Response Codes (Dehaene et al., 1993). If participants are asked to make number judgments of parity (i.e., odd or even) about the numbers 1 to 9 then they are faster at making judgments about small numbers (<5) with their left hand and faster at making judgments about larger numbers (>5) with their right hand. Hence, participants perform as if reliant on a spatially based mental number line running from left to right. In addition, it has been shown that passive viewing of numbers can induce spatial shifts of attention (Fischer et al., 2003) and that spatial attention deficits can bias numerical judgments (Vuilleumier et al., 2004). Consciously perceived number forms also tend to run from left to right, although they sometimes twist and turn (Sagiv et al., 2006b). The extent to which this is culturally biased is not entirely clear (the SNARC effect is reduced in Persian immigrants living in Paris; Dehaene et al., 1993). Number forms also occasionally point to cultural biases (e.g., 1–12 arranged like a clock). Nevertheless, it is conceivable that an association between numbers and space is universal even if direction in space is not.

Summary and conclusions

The literature reviewed here points to a significant number of similarities between synesthetes and nonsynesthetes in the way that different perceptual dimensions are linked together. This suggests that synesthesia is based on universal mechanisms rather than being based on mechanisms found solely in synesthetes. Although the review has been rather selective, there is evidence to suggest that the same holds in many other types of synesthesia including emotion-color correspondences (Ward, 2004), grapheme-color synesthesia (Rich et al., 2005; Simner et al., 2005), and the spatial representation of calendar time (Gevers et al., 2003; Sagiv et al., in press). Of course, synesthesia is different and any account of synesthesia must explain the differences between synesthetic and normal perception as well as the similarities. At least three differences are in need of explanation: phenomenology, automaticity, and reliability. At present, it is unclear whether the fact that synesthetes have conscious perceptual experiences reflects quantitative increases in activity in critical brain regions or whether it reflects a more complex integration of several regions. Understanding these differences may provide some insights into the relationship between brain function and perceptual experience.

Crossmodal integration is obviously very useful for making inferences about objects and events in our environment. It seems, however, that this involves more than pathway convergence. In fact, there is a large body of evidence suggesting that activity in unimodal brain areas is modulated by information coming from other senses (e.g., Macaluso and Driver, 2005). One wonders whether the question should be why do so many people fail to experience synesthesia under normal conditions?

Still, lessons from synesthesia have even wider implications. Many aspects of cognition involve making some form of cross-domain correspondences. Language is one obvious example. Indeed, links between synesthesia, metaphor, creativity, and the origins of language have been suggested (e.g., Ramachandran and Hubbard, 2001b). Quite a few metaphors seem so intuitive that we may have forgotten that they link otherwise unrelated sensory modalities (e.g., "high pitch," "future lies ahead," "to be touched by your sentiments," etc.). Our ability to empathize is another example of cross-domain mapping. In this case an analogy is made between the self and others. Mirror-touch synesthesia may simply be an extreme form of this very basic capacity (yet still an interesting test case for theories of embodied cognition).

Number forms (as well as spatial descriptions of time) are found to be even more common than "standard" synesthesia involving vision, touch, taste, smell, or sound. In this case, space is used not only as a "common currency" for crossmodal interactions and binding different stimulus properties, but also as a dimension along which concepts can be mapped. How does space facilitate understanding of quantity? According to Walsh (2003), space, time, and quantity are all represented by a general magnitude processing system in the parietal lobe. We find, however, that synesthesia and spatial forms are commonly induced by ordinal sequences, including the letters of the alphabet — a category that is harder to describe in terms of magnitude. Instead we propose that use of space as an organizing principle may be understood as a useful tool for grasping abstract concepts by constructing concrete spatial representations. Indeed, this too could reflect a basic feature of human cognition, not only responsible for the SNARC effect (Dehaene et al., 1993), but also the precursor of the use of graphic representation so prevalent in human culture.

Acknowledgment

The authors are supported by The Leverhulme Trust.

References

Aleman, A., Rutten, G.J., Sitskoorn, M.M., Dautzenberg, G. and Ramsey, N.F. (2001) Activation of striate cortex in the absence of visual stimulation: an fMRI study of synesthesia. Neuroreport, 12: 2827–2830.

Armel, K.C. and Ramachandran, V.S. (1999) Acquired synesthesia in retinitis pigmentosa. Neurocase, 5: 293–296.

Baron-Cohen, S. (1996) Is there a normal phase of synaesthesia in development? Psyche, 2. See http://psyche.cs.monash.edu.au/ [This is an online-ONLY journal; no page numbers are available].

Baron-Cohen, S., Burt, L., Smith-Laittan, F., Harrison, J. and Bolton, P. (1996) Synaesthesia: prevalence and familiality. Perception, 25: 1073–1079.

Baron-Cohen, S., Harrison, J., Goldstein, L.H. and Wyke, M. (1993) Coloured speech perception: is synaesthesia what happens when modularity breaks down? Perception, 22: 419–426.

Bentin, S., Mouchetant-Rostaing, Y., Giard, M.H., Echallier, J.F. and Pernier, J. (1999) ERP manifestations of processing printed words at different psycholinguistic levels: time course and scalp distribution. J. Cogn. Neurosci., 11: 235–260.

Blake, R., Palmeri, T., Marois, R. and Kim, C.-Y. (2004) On the perceptual reality of synesthesia. In: Robertson, L.C. and Sagiv, N. (Eds.), Synesthesia: Perspectives from Cognitive Neuroscience. Oxford University Press, New York, pp. 47–73.

Blakemore, S.-J., Bristow, D., Bird, G., Frith, C. and Ward, J. (2005) Somatosensory activations during the observation of touch and a case of vision-touch synesthesia. Brain, 128: 1571–1583.

Calkins, M.W. (1895) Synaesthesia. Am. J. Psychol., 7: 90–107.

Calvert, G.A., Bullmore, E.T., Brammer, M.J., Campbell, R., Williams, S.C.R., McGuire, P.K., Woodruff, P.W.R., Iverson, S.D. and David, A.S. (1997) Activation of auditory cortex during silent lipreading. Science, 276: 593–596.

Cohen-Kadosh, R., Sagiv, N., Linden, D.E.J., Robertson, L.C., Elinger, G. and Henik, A. (2005) When blue is larger than red: Colors influence numerical cognition in synesthesia. J. Cogn. Neurosci., 17: 1766–1773.

Cytowic, R. (1997) Synaesthesia: phenomenology and neuropsychology? In: Baron-Cohen, S. and Harrison, J.E. (Eds.), Synaesthesia: Classic and Contemporary Readings. Blackwell, Oxford, pp. 17–39.

Cytowic, R.E. (2002) Synaesthesia: A Union of the Senses (2nd Edition). MIT Press, Cambridge.

Day, S. (2004) Some demographic and socio-cultural aspects of synesthesia. In: Robertson, L.C. and Sagiv, N. (Eds.), Synesthesia: Perspectives from Cognitive Neuroscience. Oxford University Press, New York, pp. 11–33.

Dehaene, S., Bossini, S. and Giraux, P. (1993) The mental representation of parity and numerical magnitude. J. Exp. Psychol. Gen., 122: 371–396.

Dehay, C., Kennedy, H. and Bullier, J. (1987) Characterization of transient cortical projections from auditory, somatosensory and motor cortices to visual areas 17, 18, and 19 in the kitten. J. Comp. Neurol., 272: 68–89.

Dixon, M.J. and Smilek, D. (2005) The importance of individual differences in grapheme-color synesthesia. Neuron, 45: 821–823.

Dixon, M.J., Smilek, D. and Merikle, P.M. (2004) Not all synaesthetes are created equal: projector vs. associator synaesthetes. Cogn. Affect. Behav. Neurosci., 4: 335–343.

Duffy, P.L. (2001). Blue Cats and Chartreuse Kittens: How Synaesthetes Color Their World. Times Books, New York, pp. 84–85.

Edquist, J., Rich, A.N., Brinkman, C. and Mattingley, J.B. (2006) Do synaesthetic colors act as unique features in visual search? Cortex, 42: 222–231.

Esterman, M., Verstynen, T., Ivry, R. and Robertson, L.C. (2004) Attenuating the synesthetic experience with rTMS. Presented at the Fourth Annual American Synesthesia Association meeting, Berkeley, CA, November 5–7.

Falchier, A., Clavagnier, S., Barone, P. and Kennedy, H. (2002) Anatomical evidence of multimodal integration in primate striate cortex. J. Neurosci., 22: 5749–5759.

Farrer, C. and Frith, C.D. (2002) Experiencing oneself vs another person as being the cause of an action: the neural correlates of the experience of agency. Neuroimage, 15: 596–603.

Fink, G.R., Markowitsch, H.J., Reinkemeier, M., Bruckbauer, T., Kessler, J. and Heiss, W.D. (1996) Cerebral representation of one's own past: neural networks involved in autobiographical memory. J. Neurosci., 16: 4275–4282.

Fischer, M.H., Castel, A.D., Dodd, M.D. and Pratt, J. (2003) Perceiving numbers causes spatial shifts of attention. Nat. Neurosci., 6: 555–556.

Friedman-Hill, S.R., Robertson, L.C. and Treisman, A. (1995) Parietal contributions to visual feature binding: evidence from a patient with bilateral lesions. Science, 269: 853–855.

Galton, F. (1880a) Visualised numerals. Nature, 21: 252–256.

Galton, F. (1880b) Visualised numerals. J. Anthropol. Inst., 10: 85–102.

Gevers, W., Reynvoet, B. and Fias, W. (2003) The mental representation of ordinal sequences is spatially organized. Cognition, 87: B87–B95.

Ginsberg, L. (1923) A case of synaesthesia. Am. J. Psychol., 34: 582–589.

Giraud, A.L., Price, C.J., Graham, J.M., Truy, E. and Frackowiak, R.S.J. (2001) Cross-modal plasticity underpins language recovery after cochlear implantation. Neuron, 30: 657–663.

Goldenberg, G., Mullbacher, W. and Nowak, A. (1995) Imagery without perception: a case study of anosognosia for cortical blindness. Neuropsychologia, 33: 1373–1382.

Graziano, M.S. (1999) Where is my arm? The relative role of vision and proprioception in the neuronal representation of limb position. Proc. Natl. Acad. Sci. USA, 96: 10418–10421.

Halligan, P.W., Hunt, M., Marshall, J.C. and Wade, D.T. (1996) When seeing is feeling: acquired synaesthesia or phantom touch? Neurocase, 2: 21–29.

Hartman, A.M. and Hollister, L.E. (1963) Effect of mescaline, lysergic acid diethylamide and psilocybin on color perception. Psychopharmacolgia, 4: 441–451.

Hubbard, T.L. (1996) Synesthesia-like mappings of lightness, pitch and melodic interval. Am. J. Psychol., 109: 219–238.

Hubbard, E.M., Arman, A.C., Ramachandran, V.S. and Boynton, G.M. (2005a) Individual differences among grapheme-color synaesthetes: brain–behavior correlations. Neuron, 45: 975–985.

Hubbard, E.M., Piazza, M., Pinel, P. and Dehaene, S. (2005b) Interactions between number and space in the parietal lobe. Nat. Rev. Neurosci., 6: 435–448.

Innocenti, G.M., Berbel, P. and Clarke, S. (1988) Development of projections from auditory to visual areas in the cat. J. Comp. Neurol., 272: 242–259.

Jacobs, L., Karpik, A., Bozian, D. and Gothgen, S. (1981) Auditory-visual synesthesia. Sound-induced photisms. Arch. Neurol., 38: 211–216.

Kennedy, H., Batardiere, A., Dehay, C. and Barone, P. (1997) Synaesthesia: implications for developmental neurobiology. In: Baron-Cohen, S. and Harrison, J.E. (Eds.), Synaesthesia: Classic and Contemporary Readings. Blackwell, Oxford, pp. 243–256.

Kennett, S., Taylor-Clarke, M. and Haggard, P. (2001) Noninformative vision improves the spatial resolution of touch in humans. Curr. Biol., 11: 1188–1191.

Keysers, C., Wicker, B., Gazzola, V., Anton, J.L., Fogassi, L. and Gallese, V. (2004) A touching sight: SII/PV activation during the observation and experience of touch. Neuron, 42: 335–346.

King, A.J. and Calvert, G.A. (2001) Multisensory integration: perceptual grouping by eye and ear. Curr. Biol., 11: R322–R325.

Knoch, D., Gianotti, L.R., Mohr, C. and Brugger, P. (2005) Synesthesia: when colors count. Cogn. Brain Res., 25: 372–374.

Laeng, B., Svartdal, F. and Oelmann, H. (2004) Does color synesthesia pose a paradox for early selection theories of attention? Psychol. Sci., 15: 277–281.

Macaluso, E. and Driver, J. (2005) Multisensory spatial interactions: a window onto functional integration in the human brain. Trends Neurosci., 28: 264–271.

Marks, L.E. (1974) On associations of light and sound: the mediation of brightness, pitch, and loudness. Am. J. Psychol., 87: 173–188.

Marks, L.E. (1975) On coloured-hearing synaesthesia: cross-modal translations of sensory dimensions. Psychol. Bull., 82: 303–331.

Marks, L.E. (1987) On cross-modal similarity: auditory–visual interactions in speeded discrimination. J. Exp. Psychol. Hum. Percept. Perform., 13: 384–394.

Martino, G. and Marks, L.E. (2001) Synesthesia: strong and weak. Curr. Dir. Psychol. Sci., 10: 61–65.

Maurer, D. (1997) Neonatal synaesthesia: implications for the processing of speech and faces. In: Baron-Cohen, S. and Harrison, J.E. (Eds.), Synaesthesia: Classic and Contemporary Readings. Blackwell, Oxford, pp. 224–242.

Merabet, L.B., Maguire, D., Warde, A., Alterescu, K., Stickgold, R. and Pascual-Leone, A. (2004) Visual hallucinations during prolonged blindfolding in sighted subjects. J. Neuroopthalmol., 24: 109–113.

Mills, C.B., Howell Boteler, E. and Oliver, G.K. (1999) Digit synaesthesia: a case study using a stroop-type test. Cogn. Neuropsychol., 16: 181–191.

Mon-Williams, M., Wann, J.P., Jenkinson, M. and Rushton, K. (1997) Synaesthesia in the normal limb. Proc. Biol. Sci., 264: 1007–1010.

Myles, K.M., Dixon, M.J., Smilek, D. and Merikle, P.M. (2003) Seeing double: the role of meaning in alphanumeric-color synaesthesia. Brain Cogn., 53: 342–345.

Neville, H.J. (1995) Developmental specificity in neurocognitive development in humans. In: Gazzaniga, M. (Ed.), The Cognitive Neurosciences. MIT Press, Cambridge.

Nunn, J.A., Gregory, L.J., Brammer, M., Williams, S.C.R., Parslow, D.M., Morgan, M.J., Morris, R.G., Bullmore, E.T., Baron-Cohen, S. and Gray, J.A. (2002) Functional magnetic resonance imaging of synesthesia: activation of V4/V8 by spoken words. Nat. Neurosci., 5: 371–375.

Palmeri, T.J., Blake, R., Marois, R., Flanery, M.A. and Whetsell Jr., W. (2002) The perceptual reality of synesthetic colors. Proc. Natl. Acad. Sci. U S A, 99: 4127–4131.

Paulesu, E., Harrison, J., Baron-Cohen, S., Watson, J.D.G., Goldstein, L., Heather, J., Frackowiak, R.S.J. and Frith, C.D. (1995) The physiology of coloured hearing: a PET activation study of color-word synaesthesia. Brain, 118: 661–676.

Ramachandran, V.S. and Hubbard, E.M. (2001a) Psychophysical investigations into the neural basis of synaesthesia. Proc. Biol. Sci., 268: 979–983.

Ramachandran, V.S. and Hubbard, E.M. (2001b) Synaesthesia — a window into perception, thought and language. J. Conscious. Stud., 8: 3–34.

Ramachandran, V.S. and Rogers-Ramachandran, D. (1996) Synaesthesia in phantom limbs induced with mirrors. Proc. Biol. Sci., 263: 377–386.

Rich, A.N., Bradshaw, J.L. and Mattingley J.B. (2005). A systematic, large-scale study of synaesthesia: implications for the role of early experience in lexical-color associations. Cognition, 98: 53–84.

Robertson, L.C. (1999) What can spatial deficits teach us about feature binding and spatial maps? Vis. Cogn., 6: 409–430.

Robertson, L.C. (2003) Binding, spatial attention and perceptual awareness. Nat. Rev. Neurosci., 4: 93–102.

Robertson, L.C. (2004) Space, Objects, Minds and Brains. Psychology Press, Hove.

Sagiv, N. (2004) Synesthesia in perspective. In: Robertson, L.C. and Sagiv, N. (Eds.), Synesthesia: Perspectives from Cognitive Neuroscience. Oxford University Press, New York, pp. 3–10.

Sagiv, N. and Ben-Tal, O. (in preparation) Nocturnal synesthesia.

Sagiv, N., Heer, J. and Robertson, L.C. (2006a) Does binding of synesthetic color to the evoking grapheme require attention? Cortex, 42: 232–242.

Sagiv, N. and Robertson, L.C. (2004) Synesthesia and the binding problem. In: Robertson, L.C. and Sagiv, N. (Eds.), Synesthesia: Perspectives from Cognitive Neuroscience. Oxford University Press, New York, pp. 90–107.

Sagiv, N., Simner, J., Collins, J., Butterworth, B. and Ward, J. (2006b) What is the relationship between synaesthesia and visuo-spatial Number Forms? Cognition, 101: 114–128.

Schiltz, K., Trocha, K., Wieringa, B.M., Emrich, H.M., Johannes, S. and Munte, T.F. (1999) Neurophysiological aspects of synesthetic experience. J. Neuropsych. Clin. Neurosci., 11: 58–65.

Seron, X., Pesenti, M., Noel, M.-P., Deloche, G. and Cornet, J.A. (1992) Images of numbers, or "when 98 is upper left and 6 sky blue". Cognition, 44: 159–196.

Shanon, B. (1982) Color associates to semantic linear orders. Psychol. Res., 44: 75–83.

Shanon, B., 2002. The Antipodes of the Mind: Charting the Phenomenology of the Ayahuasca Experience. Oxford University Press, Oxford, pp. 189–190, 337–338.

Shanon, B. (2003) Three stories concerning synesthesia. A reply to Ramachandran and Hubbard. J. Consc. Stud., 10: 69–74.

Simner, J., Mulvenna, C., Sagiv, N., Tsakanikos, E., Witherby, S.A., Fraser, C., Scott, K. and Ward, J. (in press) Synaesthesia: the prevalence of atypical cross-modal experiences. Perception.

Simner, J., Ward, J., Lanz, M., Jansari, A., Noonan, K., Glover, L. and Oakley, D.A. (2005) Non-random associations of graphemes to colours in synaesthetic and non-synaesthetic populations. Cogn. Neuropsychol., 22: 1069–1085.

Smilek, D., Dixon, M.J., Cudahy, C. and Merikle, P.M. (2001) Synaesthetic photisms influence visual perception. J. Cogn. Neurosci., 13: 930–936.

Spence, C. (2002) Multisensory attention and tactile information-processing. Behav. Brain Res., 135: 57–64.

Sperling, J.M., Prvulovic, D., Linden, D.E.J., Singer, W. and Stirn, A. (2006) Neuronal correlates of graphemic color synaesthesia: a fMRI study. Cortex, 42: 295–303.

Steven, M.S. and Blakemore, C. (2004) Visual synesthesia in the blind. Perception, 33: 855–868.

Tarkiainen, A., Helenius, P., Hansen, P.C., Cornelissen, P.L. and Salmelin, R. (1999) Dynamics of letter string perception in the human occipitotemporal cortex. Brain, 122: 2119–2132.

Treisman, A. and Gelade, G. (1980) A feature integration theory of attention. Cogn. Psychol., 12: 97–136.

Treisman, A. and Schmidt, H. (1982) Illusory conjunctions in the perception of objects. Cogn. Psychol., 14: 107–141.

Uhlich, E. (1957) Synesthesia in the two sexes. Z. Exp. Angew. Psychol., 4: 31–57.

Vike, J., Jabbari, B. and Maitland, C.G. (1984) Auditory-visual synesthesia: report of a case with intact visual pathways. Arch. Neurol., 41: 680–681.

Vuilleumier, P., Ortigue, S. and Brugger, P. (2004) The number space and neglect. Cortex, 40: 399–410.

Walsh, R. (2005) Can synesthesia be cultivated? Indications from surveys of meditators. J. Consc. Stud., 12: 5–17.

Walsh, V. (2003) A theory of magnitude: common cortical metrics of time, space and quantity. Trends Cogn. Sci., 7: 483–488.

Ward, J. (2004) Emotionally mediated synaesthesia. Cogn. Neuropsychol., 21: 761–772.

Ward, J., Huckstep, B. and Tsakanikos, E. (2006) Sound-color synaesthesia: to what extent does it use cross-modal mechanisms common to us all? Cortex, 42: 264–280.

Ward, J. and Simner, J. (2005) Is synaesthesia an X-linked dominant trait with lethality in males? Perception, 34: 611–623.

Wolfe, J.M. and Cave, K.R. (1999) The psychophysical evidence for a binding problem in human vision. Neuron, 24: 11–17.

Martinez-Conde, Macknik, Martinez, Alonso & Tse (Eds.)
Progress in Brain Research, Vol. 155
ISSN 0079-6123

CHAPTER 16

Integrating motion information across sensory modalities: the role of top-down factors

Salvador Soto-Faraco[1,*], Alan Kingstone[2] and Charles Spence[3]

[1]*ICREA and Parc Científic de Barcelona – Universitat de Barcelona, Barcelona, Spain*
[2]*Department of Psychology, University of British Columbia, 2136 West Mall, Vancouver, BC V6T 1Z4, Canada*
[3]*Department of Experimental Psychology, University of Oxford, South Parks Road, Oxford OX1 3UD, UK*

Abstract: Recent studies have highlighted the influence of multisensory integration mechanisms in the processing of motion information. One central issue in this research area concerns the extent to which the behavioral correlates of these effects can be attributed to late post-perceptual (i.e., response-related or decisional) processes rather than to perceptual mechanisms of multisensory binding. We investigated the influence of various top-down factors on the phenomenon of crossmodal dynamic capture, whereby the direction of motion in one sensory modality (audition) is strongly influenced by motion presented in another sensory modality (vision). In Experiment 1, we introduced extensive feedback in order to manipulate the motivation level of participants and the extent of their practice with the task. In Experiment 2, we reduced the variability of the irrelevant (visual) distractor stimulus by making its direction predictable beforehand. In Experiment 3, we investigated the effects of changing the stimulus–response mapping (task). None of these manipulations exerted any noticeable influence on the overall pattern of crossmodal dynamic capture that was observed. We therefore conclude that the integration of multisensory motion cues is robust to a number of top-down influences, thereby revealing that the crossmodal dynamic capture effect reflects the relatively automatic integration of multisensory motion information.

Keywords: motion; multisensory; audition; vision; perception; attention

Introduction

The brain represents the world through the combination of information available to a variety of different sources, often arising from distinct sensory modalities. The mechanisms of multisensory integration that afford this combination are thought to play a central role in human perception and attention. Indeed, mounting evidence from the field of neuroscience reveals that neural interactions between sensory systems are more pervasive than had been traditionally thought (e.g., see; Stein and Meredith, 1993; Calvert et al., 2004, for reviews) and some researchers have even gone as far as to claim that it is necessary to account for multisensory integration if one wants to achieve a comprehensive understanding of human perception (e.g., Churchland et al., 1994; Driver and Spence, 2000).

The intimate link between the senses in terms of information processing has often been studied through the consequences of *intersensory conflict* (i.e., Welch and Warren, 1986; Bertelson, 1998). Whereas real-world objects usually produce correlated information to several sensory modalities (i.e., the auditory and visual correlates of speech provide but one everyday example), in the

*Corresponding author. Tel.: +34-93-6009769;
Fax: +34-93-6009768; E-mail: Salvador.Soto@icrea.es

DOI: 10.1016/S0079-6123(06)55016-2

laboratory, one can create a situation of conflict between the cues coming from different senses in order to assess the resulting consequences for perception. One famous example of this approach is the McGurk illusion (McGurk and MacDonald, 1976), whereby the acoustic presentation of a syllable such as /ba/ dubbed onto a video recording of someone's face uttering the syllable [ga] can result in the illusory perception of hearing DA, a sound that represents a phonetic compromise between what was presented visually and what was presented auditorily. In the domain of spatial perception, the well-known ventriloquist illusion provides another illustration of intersensory conflict. In this illusion, the presentation of a beep and a flash at the same time but from disparate locations creates the impression that the sound came from somewhere near the location of the light (i.e., Howard and Templeton, 1966; Bertelson, 1998; de Gelder and Bertelson, 2003, for reviews; see Urbantschitsch, 1880, for an early report).

Multisensory integration in motion processing: the crossmodal dynamic capture task

One especially interesting case of multisensory integration in the spatial domain is associated with motion information. Recent studies have revealed that motion cues, such as the direction of motion, are subject to strong interactions when various sensory modalities convey conflicting information. For example, people will often report hearing a sound that is physically moving from left to right as if it were moving from right to left, when presented with a synchronous visual stimulus moving leftward (e.g., Zapparoli and Reatto, 1969; Mateef et al., 1985; Soto-Faraco et al., 2002, 2003, 2004a, 2004b; Soto-Faraco and Kingstone, 2004, for reviews). This crossmodal dynamic capture phenomenon has also been studied using adaptation aftereffects, whereby after the repeated presentation of a visual stimulus moving in one direction, observers can experience a static sound as if it appeared to move in the opposite direction (Kitagawa and Ichihara, 2002; see also Vroomen and de Gelder, 2003).

In recent years, one of the most frequently used paradigms in the study of the crossmodal integration of motion cues has been the crossmodal dynamic capture task (see Soto-Faraco et al., 2003; Soto-Faraco and Kingstone, 2004, for reviews). In this task, participants are typically asked to judge the direction of an auditory apparent (or real) motion stream while at the same time they attempt to ignore an irrelevant visual motion stream. The visual motion stream can be presented in the same direction as the sounds or in the opposite direction (varying on a trial-by-trial basis), and is sometimes presented at the same time as the sound or 500 ms after the onset of the sound. The typical result is that there is a strong congruency effect when sound and light streams are presented at the same time: while performance is near perfect in the same-direction condition, auditory direction judgments drop by around 50% when the direction of the visual motion stimulus is presented in the direction opposite to the sound. When the visual motion is desynchronized with the sounds (i.e., by 500 ms), no congruency effects are seen and performance is very accurate overall, thus demonstrating that auditory motion by itself provides a robust and unambiguous direction signal. These effects have been demonstrated using both apparent motion and continuous motion streams (Soto-Faraco et al., 2004b).

Levels of processing of multisensory integration in motion perception

A point of contention in the literature concerning the multisensory integration of motion perception regards the extent to which these interactions occur at early (perceptual) stages of processing, or else they reflect the results of somewhat later (post-perceptual) processes related to response biases and/or cognitive strategies (e.g., Meyer and Wuerger, 2001; Vroomen and de Gelder, 2003; Wuerger et al. 2003; Soto-Faraco et al., 2005). Several researchers have attempted to demonstrate that interactions between visual and auditory motion signals can produce adaptation after-effects, which are generally thought to arise from the fatigue of sensory receptors at early stages of information processing (Kitagawa and Ichihara, 2002; Vroomen and de Gelder, 2003). Other researchers have demonstrated audiovisual interactions for motion stimuli using paradigm designed to be resistant to any influence of response or cognitive biases through the use of various psychophysical methods (e.g., Alais and Burr, 2004; Soto-Faraco et al., 2005). However, evidence for the likely influence of post-perceptual stages of processing in multisensory motion interactions has also been demonstrated in studies using psychophysical models that separate sensitivity from bias in response criterion. In particular, some researchers have suggested that there is no effect of directional congruency at the perceptual stage of information processing (crossmodal dynamic capture), and that any congruency effects observed are to be attributed to stimulus–response compatibility biases that arise when the target sensory signal is ambiguous (e.g., Meyer and Wuerger, 2001; Wuerger et al., 2003; Alais and Burr, 2004).

As this brief review hopefully conveys, the emerging picture is a rather complex one. On the one hand, several sources of evidence converge on the conclusion that there is a perceptual basis for multisensory integration of motion direction cues (i.e., crossmodal dynamic capture experiments, and after-effects). This would mean that, at least in some cases, motion cues from different sensory modalities are bound together in an automatic and obligatory fashion, prior to any awareness concerning the presence of two independent motion cues (Soto-Faraco et al., 2005). On the other, there are clear demonstrations that, as for other crossmodal illusions (see Choe et al., 1975; Bertelson and Aschersleben, 1998; Welch, 1999; Caclin et al., 2002), the influence of cognitive and response biases can also play a role (Meyer and Wuerger, 2001; Wuerger et al., 2003; Sanabria et al., submitted). These post-perceptual influences appear to be especially strong when the target signal provides ambiguous information. Moreover, recent studies suggest that even the separation that has been traditionally made between early (perceptual) and late (post-perceptual) processes may not be as clear-cut as had been thought previously (e.g., Wohlschläger, 2000). Indeed, the influence of cognitive biases may pervade even the earliest stages of information processing, and therefore the so-called perceptual and post-perceptual processes may, in fact, interact to produce the resulting percept which observers are aware of.

Scope of the present study

The majority of previous studies have sought evidence for the perceptual basis of crossmodal integration in motion processing by attempting to demonstrate interactions between the senses in situations where cognitive biases and/or response biases are controlled for and therefore unlikely to confound the results (Kitagawa and Ichihara, 2002; Vroomen and de Gelder, 2003; Soto-Faraco et al., 2005). The few direct attempts to measure the influence of cognitive or response biases on the perception of motion (or on other stimulus features) have involved the use of somewhat ambiguous target stimuli. For example, Meyer and Wuerger (2001) presented participants with random dot kinematograms (RDKs) where a certain percentage of the dots (0–32%) moved in a given predetermined direction, whereas the rest moved randomly. They evaluated the influence of a moving sound source (i.e., white noise directionally congruent or conflicting with the dots) on participants' responses and found a strong response bias but barely any perceptual effect (see Wuerger et al., 2003, for similar conclusions). However, the interpretation of these studies may be compromised by the particular type of stimuli used, which may have favored segregation between modalities on the basis of perceptual grouping principles (see Sanabria et al., 2004, 2005).

In the present study, our approach was to find out whether we could induce a top-down modulation of the crossmodal dynamic effect by manipulating cognitive and task variables directly. In Experiment 1, we presented a typical version of the crossmodal dynamic capture task with the addition that participants received trial-by-trial feedback regarding the accuracy of their responses. In Experiment 2, we performed a comparison between a condition in which the visual motion distractors could move in either of two possible directions (as in all previous experiments using this paradigm) and a condition in which distractors moved in a predetermined direction throughout a block of trials (i.e., always toward the left or always toward the right). Finally, in Experiment 3, we tested the crossmodal dynamic capture effect using two different tasks: one involved reporting the direction in which the sounds moved and the other reporting the origin of the first sound. The prediction being that if there is any degree of voluntary or strategic control over the multisensory motion integration that is mediated by cognitive or response variables then it should be revealed in one or more of these manipulations as a modulation (in particular, a reduction) of the basic congruency effect.

Experiment 1

The goal of the first experiment was to introduce conditions that would induce participants to perform to the best of their ability in the crossmodal dynamic capture task, and give them the opportunity to improve their performance. Specifically, participants were given auditory feedback followed by a 1-s penalty delay every time they made an erroneous response, and twice as many trials as compared to previous crossmodal dynamic capture experiments (see Soto-Faraco et al., 2002, 2004b). We reasoned that the inclusion of feedback and the more extended exposure to the experimental task should have provided participants with the motivation and the opportunity to apply strategic top-down control in order to maximize their performance on their judgments of sound direction (see also Rosenthal et al., 2004, for a related argument).

Method

Participants

Twelve undergraduate students at the University of British Columbia participated in Experiment 1. None reported problems in hearing, and they all reported normal or corrected vision. They received either course credit or financial payment (CAN $4) for their participation.

Apparatus and materials

A set of three pure tones (450, 500, and 550 Hz) was generated using the software Cool Edit 96 (Syntrillium Software Corp.). Each tone was 50 ms long (with a 5 ms amplitude envelope) and was presented from each of two loudspeaker cones (placed 30 cm apart) through the PC SoundCard (ProAudio Basic 16). One foam cube (7 cm side), with one orange light-emitting diode (LED) attached, was placed in front of each loudspeaker cone (see Fig. 1). In addition, a red LED was placed between the two loudspeakers to serve as the central fixation point. The input to the LEDs (+5 v) was controlled via a custom-built relay box operating through the PC parallel port. Two footpedals placed beneath the table were used to collect responses. The experiment was programmed using the Expe6 software (Pallier et al., 1997).

Procedure

The participants sat in front of the two loudspeaker cones and held the foam cubes (the right cube in their right hand, and the left cube in their left hand) with the index fingers placed beside the LEDs during the experimental session (see Fig. 1a). This was done to maintain the participant's posture relatively constant throughout the experiment and across participants. The participants were asked to rest their toes on the footpedals and to respond by releasing either the left or right pedal to indicate leftward or rightward auditory motion, respectively. The room was completely dark during the experiment. Two 50 ms tones were presented sequentially from different loudspeakers with a 100 ms silent interstimulus interval (ISI) between them (these parameters have been shown to yield a reliable impression of motion in most people; Soto-Faraco et al., 2004a, 2004b; Strybel and Vatakis, 2005). The frequency of the two tones was held constant on any given trial, but varied randomly from trial to trial (from the three possible frequencies), in order to discourage participants from using low-level acoustic information, other than direction of motion, that might have arisen from any peculiarities associated with the room's acoustics or any minor equipment imperfections. The direction of the tones (right to left or left to right) was chosen at random on every trial. Two light flashes (50 ms each, with a 100 ms ISI) were presented either synchronously with the sounds or else lagging the onset of the sounds by 500 ms (i.e., the stimulus

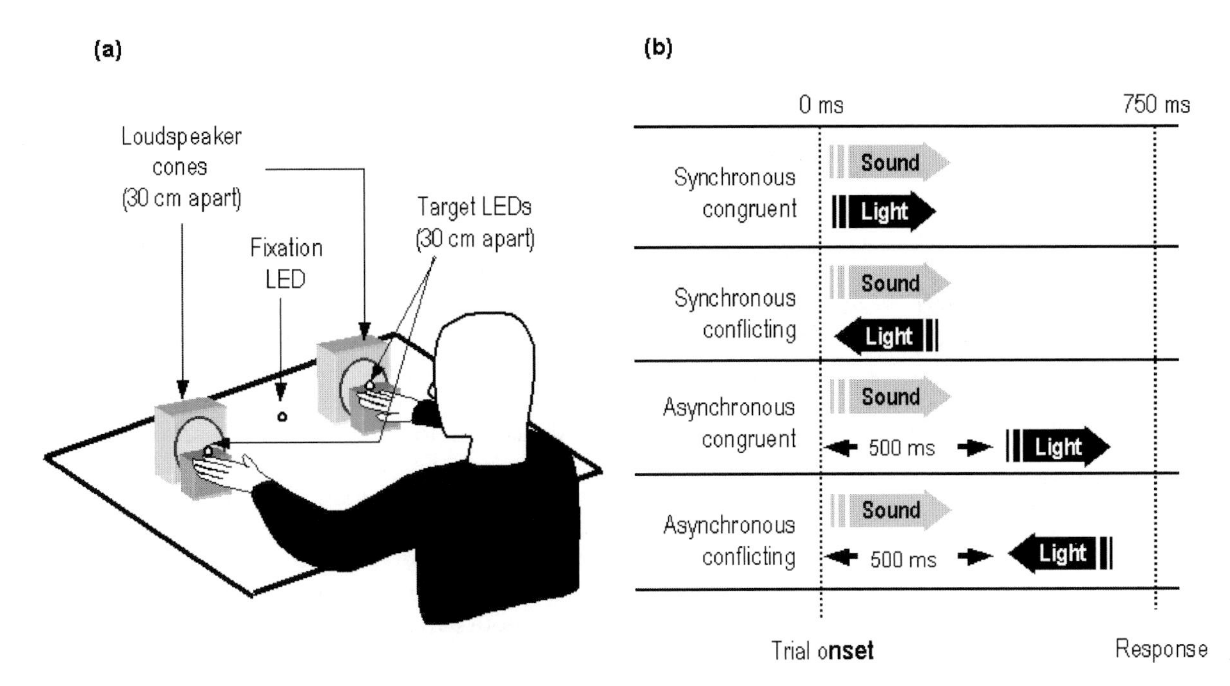

Fig. 1. Experimental procedures. (a) Shows a schematic representation of the experimental setup used in Experiments 1–3. Participants responded to the direction of the auditory apparent motion streams (from the left loudspeaker to the right, or vice versa) while attempting to ignore the visual apparent motion presented from the LEDs placed just in front of the loudspeakers. Responses were made by means of two footpedals placed on the floor (not shown). (b) A summary of the main conditions used in Experiments 1–3 (the left-right direction of the sounds and lights was counterbalanced in the experiments). All four types of displays were mixed randomly throughout the experiment (except for Experiment 2, see main text for details).

onset asynchrony, SOA, was 500 ms). The direction of the light flashes, and whether they were synchronous or asynchronous with the tones varied randomly, and with equal probability, from trial to trial (see Fig. 1b).

Participants responded by momentarily raising their foot off one of two footpedals attached to the floor, beneath the toes of either foot. Responses were unspeeded, and there was a mandatory delay of 750 ms after the onset of the first tone in order for the response to be registered and the next trial to begin. This ensured that participants complied with the instruction to make unspeeded responses and that even in the asynchronous condition participants experienced both pairs of tones and lights before making their response. The central red LED was illuminated as soon as a pedal response had been recorded, and was switched off when the footpedal was pressed again. The next trial was not started until a response had been recorded.

There were 24 trials per condition, yielding a total of 96 trials in the experiment. When the footpedal was pressed down after a response had been made, a 1500 ms interval occurred before the onset of the next trial. Every time participants made an error they heard a 500 ms low-frequency buzz and the start of the next trial was delayed by an additional 1000 ms. Prior to the start of the experiment, participants received a 10-trial training block in the auditory motion task including feedback on errors but without visual distractors. If a participant was not confident of his/her performance, the training block was repeated.

Results and discussion

An analysis of variance (ANOVA) was conducted on the accuracy data (see Fig. 2), with congruency (distractor light moving in a direction congruent with the target sound vs. distractor light moving in

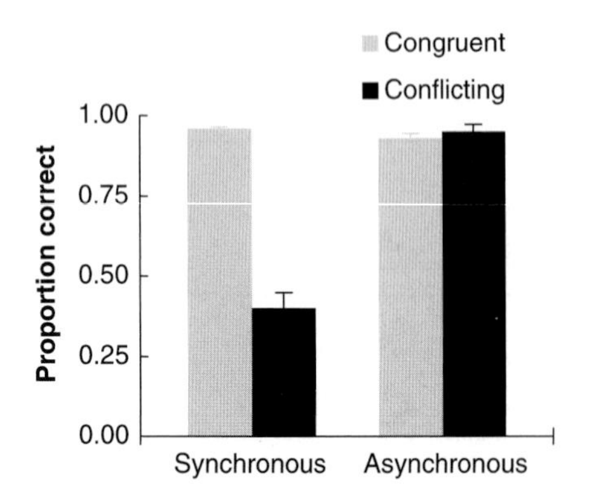

Fig. 2. Proportion of correct responses to the direction of the auditory apparent motion stream as a function of the relative direction of the distractor visual apparent motion stream (congruent vs. incongruent) and the relative onset time of the two streams (synchronous vs. asynchronous). Error bars show the standard error of the mean for each condition.

a conflicting direction) and synchrony (synchronous sound–light presentation vs. asynchronous sound–light presentation) as the within-participants factors. The ANOVA revealed significant main effects of both synchrony ($F[1, 11] = 105, p<0.001$) and congruency ($F[1, 11] = 114, p<0.001$), as well as a significant interaction between these factors ($F[1, 11] = 142, p<0.001$). The main effects and the interaction can all be accounted for by the fact that performance in the synchronous-conflicting condition (accuracy, $M = 40\%$, SD = 15) was significantly worse than that in any other condition (where performance was above 93% correct, on average, for all conditions). Therefore, these results indicate that when conflicting visual motion information is presented concurrently with auditory motion, it leads to a significant modulation in the ability of participants to determine the direction of the auditory source. This is the typical result that has been reported in a number of our previous crossmodal dynamic capture experiments. No trace of a decrease in the strength of the effect was observed here regardless of the incorporation of informative feedback. The average size of the crossmodal dynamic capture effect (measured as the accuracy in congruent trials minus the accuracy in conflicting trials, in the synchronous condition) seen in previous experiments was 47%, ranging between 35% and 61%. This average has been computed from data in the synchronous condition of 13 different experiments ($N = 156$ participants) included in comparable studies (see, Soto-Faraco et al., 2002, 2003, 2004b; Sanabria et al., 2004, 2005). The size of the effect observed in the present experiment (56%) is no smaller than what would be expected in a typical crossmodal dynamic capture experiment ($\chi^2(\mathrm{df} = 1) = 0.35, p>0.5$).

We conducted a further analysis on the data from Experiment 1, now including Practice as a factor (first vs. second half of the experiment) to examine whether experience with the error feedback led to any noticeable improvement in performance (i.e., in order to assess whether participants could somehow override the capture of auditory motion by concurrent conflicting visual motion after experiencing the displays a number of times). The pattern of results remained unchanged across both halves of the experiment, with practice having no main effect or interaction with either congruency or synchrony (all Fs < 1).

The results of Experiment 1 indicate that the crossmodal dynamic capture of audition is not weakened when the participants are placed in a situation that enhances their opportunity, and possibly their motivation given the penalty delay period suffered after every error, to perform accurately. Thus, it does not appear that strategic voluntary top-down control can be used to improve performance even when informative response feedback (consisting of an audible signal and a delay in the inter-trial interval) was provided after every error. Note that in the present study, participants had ample opportunity to use any strategy that they could think of in order to improve performance during the experimental session (save for failing to maintain central fixation or closing their eyes), yet no trace of any overall improvement or improvement with practice (i.e., in the second last half of the experiment) was detected.

Experiment 2

The results of Experiment 1 show that participants were unable to take advantage of the extra cues provided to them in order to get acquainted with the nature of the stimuli and overcome, at least to a certain extent, the effect of visual motion distractors during performance of the crossmodal dynamic capture task. The goal of Experiment 2 was to test whether reducing the variability of the visual distractors by making their direction predictable in advance would allow participants to ignore them more effectively. The crossmodal dynamic capture experiments can be considered as a form of crossmodal selective attention task, where participants have to respond according to the auditory component of the stimulus while at the same time filtering out the irrelevant visual component. Previous selective attention research has revealed that the deleterious effect of task-irrelevant distractors is stronger if they vary from trial to trial than if they are held constant throughout a block of trials (i.e., Garner, 1974; Soto-Faraco et al., 2004a; see Marks, 2004, for a review of this methodology in crossmodal research). Here, we compared the typical crossmodal dynamic capture task (where the direction of visual motion distractors was randomly chosen from trial to trial) with a situation in which the direction of visual motion distractors was maintained constant throughout a block of trials and was therefore predictable.

Method

Participants

Thirty new participants from the same population as in Experiment 1 were tested. All had normal hearing and normal or corrected-to-normal vision. They were given course credit in exchange for their participation.

Materials, apparatus, and procedure

Every aspect of the method was identical to Experiment 1 except for the following. The participants were tested in three blocks of 96 trials each (with congruency and synchrony chosen randomly and equiprobably on a trial-by-trial basis), giving rise to a total of 288 trials. In one block (mixed block), the visual distractor could move to the left or to the right unpredictably, just as in Experiment 1. In the other two blocks, the visual distractor always moved in a predetermined direction (fixed left or fixed right). The order of these three types of blocks was counterbalanced across participants according to a Latin square design. In this experiment, no error–feedback was provided, except in the training session.

Results and discussion

Analyses on the proportion scores

The results of the two fixed blocks were pooled in the analyses, after checking that there were no significant differences between the fixed-left and fixed-right blocks (Fig. 3). We submitted the proportion correct data to an ANOVA with the following within-participants factors: visual motion direction (mixed vs. fixed), congruency (congruent vs. conflicting), and synchrony (synchronous vs. asynchronous). Both the main effects of congruency and of synchrony were significant ($F(1, 29) = 66.9$, $p<0.001$, and $F(1, 29) = 75.2$, $p<0.001$, respectively), as in Experiment 1. Moreover, the typical crossmodal dynamic capture effect was also replicated with the interaction between congruency and synchrony being significant ($F(1,29) = 78.2$, $p<0.001$). There was a strong congruency effect in synchronous trials (40.1%, $t(29) = 8.8$, $p<0.001$) but not in the asynchronous trials (1.5%, $t(29) = 1.0$, $p = 0.317$). The factor of visual motion direction (mixed vs. fixed) did not reach significance ($F<1$), nor were there any significant interactions involving this factor. In particular, the three-way interaction between congruency, synchrony, and visual motion direction was far from significance[1] ($F(2,58) = 1.1$, $p = 0.297$). Additional analyses confirmed that neither the fixed-left nor the fixed-right visual motion direction conditions differed from the mixed visual motion condition when tested separately (both Fs<1 for the interaction).

[1]Power calculations (e.g., Howell, 1992) revealed that with the number of participants tested here ($N = 30$) and the variability in the present data, this analysis is sensitive ($p = 0.80$) to effect sizes of 9.5% or larger in percentage points of difference between the congruency effect in the mixed and fixed direction visual motion distractor conditions. The actual difference between congruency effects in the fixed vs. mixed conditions in the data was only 2.5%.

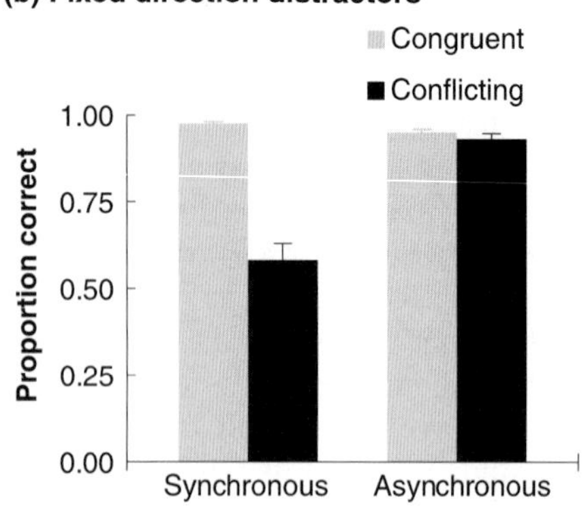

Fig. 3. Proportion correct responses to the direction of the auditory apparent motion stream as a function of the relative direction of the distractor visual apparent motion stream (congruent vs. incongruent) and the relative onset time of the two streams (synchronous vs. asynchronous). Error bars show the standard error of the mean for each condition. Panel (a) shows the results for the mixed condition (the direction of the visual distractor was chosen at random), and panel (b) shows the results for the fixed condition (the direction of the visual distractor was fixed throughout each block).

Table 1. Signal detection results for the synchronous (Synch) and asynchronous (Asynch) conditions in Experiment 2

	Sensitivity (d')		Criterion (c)	
	Synch	Asynch	Synch	Asynch
Signal-congruent	2.09	3.05	0.39	1.54
Noise-congruent	1.75	3.06	1.62	1.62

Note: These signal detection results are based only on the fixed condition. Sounds moving to the left were arbitrarily chosen as the signal, and sounds moving to the right as noise. The signal detection parameters were assessed independently for each participant, and then averaged for trials in which the visual distractor was consistent with the signal and across trials in which the visual distractor was consistent with the noise.

Signal detection analyses

The present experimental design enabled us to perform an additional analysis on the basis of signal detection theory (e.g., MacMillan and Creelman, 1991), which separates the contribution of perceptual sensitivity (d' parameter) from response criterion (c parameter) (see Table 1). Because the experiment included two blocks in which the distractors were presented in a fixed direction, we were able to arbitrarily label "left moving" target sounds as signal, and "right moving" target sounds as noise, and then compare the sensitivity and the response criterion across the two types of distractors (one which was always congruent with the signal, the other which was always congruent with the noise). First, we submitted the d' data to an ANOVA with the factors of distractor direction (signal-consistent or noise-consistent) and synchrony (synchronous and asynchronous). The main effect of distractor direction was significant ($F(1,29) = 91.8$, $p<0.001$), while the main effect of synchrony was only marginally significant ($F(1,29) = 3.2$, $p = 0.082$). Critically, the interaction between distractor direction and synchrony reached significance ($F(1,29) = 4.3$, $p<0.05$). This interaction was caused by a significant effect of distractor direction on d' in the synchronous trials ($d' = 2.09$ in the signal-congruent trials vs. $d' = 1.75$ in the noise consistent trials, $t(29) = 2.15$, $p<0.05$), combined with a null effect of distractor direction in the asynchronous trials (3.05 vs. 3.06, respectively, $t<1$). The ANOVA on the criterion (c) parameter included the same factors. Here, both main terms ($F(1,29) = 50.0$, $p<0.001$, for distractor direction; $F(1,29) = 42.4$, $p<0.001$, for synchrony) as well as their interaction ($F(1,29) = 52.6$, $p<0.001$) were significant. The interaction was explained by a significant effect of distractor direction on the c parameter in the synchronous trials ($c = 0.39$ in the signal-congruent trials vs. $c = 1.62$ in the noise consistent trials, $t(29) = 7.25$, $p < 0.001$) combined with a null effect of distractor direction in the asynchronous trials (1.54 vs. 1.62, respectively, $t(29) = 1.16$, $p = 0.254$).

Two relevant conclusions emerge from the results of Experiment 2: first, the proportion correct scores showed that the crossmodal dynamic capture effect was not modulated by advance knowledge concerning the direction in which the visual distractor moved. That is, contrary to attentional effects that are usually explained as a result of response-conflict (perhaps the *flanker task* being the most well-known example, e.g., Eriksen and Eriksen, 1974; Chan et al., 2005), in this case variability in the distractor dimension (visual motion direction) did not have any significant influence on the net congruency effect observed. This again suggests that the effect of crossmodal conflict in motion cues leaves little room for top-down control strategies to modulate the final perceptual outcome (see Marks, 2004).

The other major conclusion to emerge from the results of Experiment 2 comes from the signal detection analyses. Here, we were able to calculate an unbiased indicator of the participants' actual sensitivity in discriminating the direction of the target motion (auditory) in two situations: both when the direction of the visual distractor was consistent with the signal, and when it was consistent with the noise. For synchronous trials, there was a clear decrement in the participants' ability to discriminate the direction of auditory motion, with respect to the asynchronous (baseline) condition. This indicates that visual motion influenced perceptual sensitivity to the direction of auditory motion. Moreover, in blocks where the visual stimulus always moved in the opposite direction to the to-be-detected signal, it was even more difficult to detect the target sound, as compared to blocks where it constantly moved in a direction congruent with the signal. Our results also revealed that there was a significant change in response criterion between the signal-congruent and noise-congruent blocks: As has been seen in previous studies, the criterion was more conservative in blocks where the distractor moved in the opposite direction to the target sound. None of these directional congruency effects (either in the sensitivity or criterion measure) were seen for asynchronous trials.

Experiment 3

Thus far, our manipulations have been directed toward trying to modulate crossmodal dynamic capture effects by inducing a change in the voluntary strategies of the participants (Experiment 1) or in the distracting nature of the irrelevant stimuli (Experiment 2). In our final experiment, we varied the task itself by asking the participants to respond to the location of the first sound, rather than to the direction of movement of the auditory stream (though note that, in principle, given the constraints of the experimental design either source of information can be used to infer the answer to the direction task). However, by introducing this manipulation we hoped to promote a strategy based on selectively attending to the first of the two component events in the apparent motion stream. If participants are able to single out just the first event, they should perform better in this task. This prediction was based on previous findings that performance improves considerably when participants are presented with only one sound (rather than two) combined with one light flash (rather than two) (see Soto-Faraco et al., 2004b). In this experiment, participants were explicitly asked to concentrate on just the first event in the stream in order to establish the origin of the moving stream (and not the direction). If participants were able to select the information on the first event of the apparent motion stream, then they should be able to improve their performance considerably.

Methods

Participants

Twelve new participants from the same population as in Experiments 1 and 2 were selected for this experiment. None reported auditory or visual deficits (other than those corrected by wearing lenses). They received course credit for their participation.

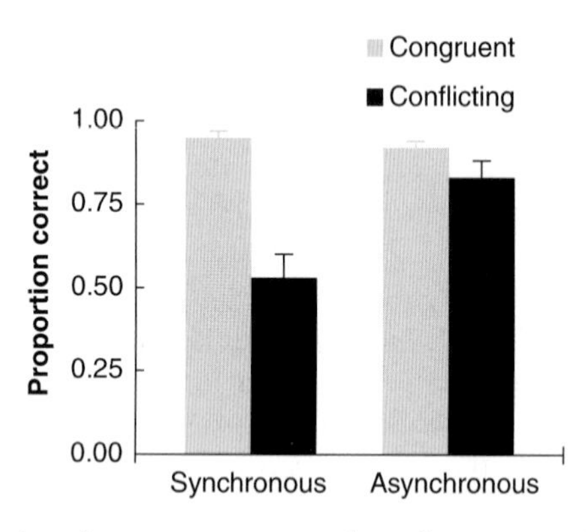

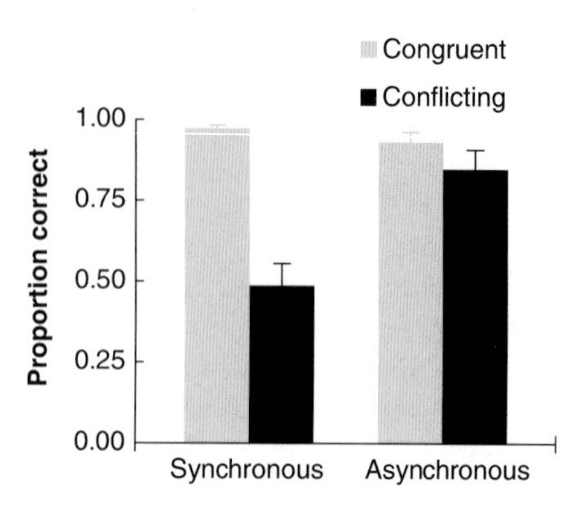

Fig. 4. Proportion of correct responses to the auditory target stimulus as a function of the relative direction of the distractor visual apparent motion stream (congruent vs. incongruent) and the relative time of onset the two streams (synchronous vs. asynchronous). Error bars show the standard error of the mean for each condition. Panel (a) shows the results when the participants had to report the direction of the auditory apparent motion stream, and panel (b) when the participants reported location of the first sound.

Materials, apparatus and procedure

All aspects of the method were the same as in Experiment 1, except for the following. Participants responded to a total of 96 trials, divided into two blocks of 48 trials, where no feedback was provided. The two blocks were identical, with the only change from one block to the other being in terms of the instructions given to the participants; "respond to the direction of the auditory motion stream" (as in Experiment 1), or "respond to the location of the first sound." In the latter type of block, the participants were encouraged to concentrate on the first event of the apparent motion stream. The order of the blocks was counterbalanced across participants. Note that in both types of block, and just as in Experiments 1 and 2, participants had to wait until 750 ms from the onset of the trial in order to respond (see Fig. 4).

Results and discussion

The data were submitted to an ANOVA with three within-participant factors: task (report origin vs. report direction), congruency, and synchrony. As in Experiments 1 and 2, both congruency and synchrony were significant ($F(1,11) = 53.9$, $p<0.001$; and $F(1,11) = 57.6$, $p<0.001$, respectively), as was their interaction ($F(1,11) = 25.8$, $p<0.001$). The interaction was caused by a large congruency effect (45%, $t(11) = 7.5$, $p<0.001$) in the synchronous condition, whereas the asynchronous condition only showed a small, albeit still significant, effect of congruency (9%, $t(11) = 2.2$, $p = 0.05$). Neither the main effect of task ($F<1$), nor any of the interactions involving this factor was significant (all two-way interactions $F<1$, and $F(1,11) = 1.66$, $p = 0.224$, for the three-way interaction).

Despite the fact that participants in the "respond to the location of the first sound" block presumably concentrated on the initial event (sound) of the apparent motion stream, the size of the crossmodal congruency effect was comparable to that obtained in the baseline task of "respond to the direction of the sound" (i.e., the typical task in previous studies of the crossmodal dynamic capture effect). This result is perhaps somewhat surprising given the previous finding that, when presenting only one isolated sound paired with a spatially congruent or incongruent light flash, the size of the congruency effect decreased dramatically (Soto-Faraco et al., 2004b). The results of Experiment 3 therefore suggest that participants could not effectively focus their attention on the first of the two events in the apparent motion stream, in order to improve their accuracy in the task. This is yet another line of evidence suggesting that the effects of visual motion on the perceived direction of an auditory apparent motion stream are difficult to overcome even when the task lends itself to a potentially successful strategy.[2]

General discussion

The goal of the present study was to address whether the effects of multisensory integration of motion (indexed by performance in the crossmodal dynamic capture task) could be modulated by top-down processes. We used three different experimental manipulations that are known to promote top-down (attentional) strategies. In Experiment 1, we encouraged participants to perform as accurately as they could by providing them with feedback on trials where they made an erroneous response. However, the size of the crossmodal dynamic capture effect remained as strong as would be expected, based on the averaged data from previous experiments, even after prolonged practice (i.e., in the last block of trials). This result converges with a recent study reported by Rosenthal et al. (2004) where the authors found no effects on the amount of sound-induced flash illusion (see Shams et al., 2000) as a function of whether feedback information was provided or not. Parallel to our own argument, the authors of that study concluded that the sound-induced flash illusion "reflects auditory-visual interactions at an early perceptual processing level, and is too dominant to be modified by higher level influences of feedback" (from their abstract). In Experiment 2, we reduced the variability of the irrelevant distractor by making its direction completely predictable beforehand. However, contrary to the results of numerous other previous selective attention paradigms (e.g., see Marks, 2004), participants in our study were unable to take advantage of this regularity. Finally, in Experiment 3, we used a task-related manipulation by asking participants to respond to the location of the first sound, so that they could attempt to focus on the first event in the stream and filter out any influence of the subsequently presented stimuli. Again, no sign of a decrement in the size of the crossmodal capture effect was seen, as compared to the standard sound motion discrimination task. In itself, the result of Experiment 3 is somewhat paradoxical because a stimulus that, on its own, should have been perceived correctly most of the time (location of the first event in the apparent motion stream) was strongly influenced by stimuli that arrived later in time. This adds to a number of other phenomena demonstrating the occurrence of within- and crossmodal integration effects in time, such as the *saltation* illusion or cutaneous "rabbit" (e.g., Geldard and Sherrick, 1972) and the temporal ventriloquism (e.g., Morein-Zamir et al., 2002).

The results of all three experiments reported here converge on the conclusion that the binding of auditory and visual motion cues is resistant to a variety of voluntary top-down factors. This is not to say that there is no possible influence of shifts in response criterion, as the results of the signal detection measures in Experiment 2 attest. Indeed, potential biases at the level of response selection (i.e., induced by stimulus–response compatibility effects) are clearly possible in the present paradigm, as well as in other related studies. This is always the case when the task-irrelevant distractor bears some congruency relationship with the responses available to the participant in the task (see Bertelson, 1999; de Gelder and Bertelson, 2003, for a review). These effects have been clearly pointed out in studies using threshold target stimuli and suprathreshold distractor motion stimuli (i.e., Meyer and Wuerger, 2001). However, what is clear from the results reported here is that this type of post-perceptual factors cannot, by themselves alone, explain the key effects usually seen in crossmodal dynamic capture experiments. This conclusion is in

[2]Despite the fact that the responses in this task were unspeeded, we thought it worthwhile to perform an analysis of the reaction time (RT) data for this experiment, in order to exclude the possibility that participants may have responded faster in the "respond to sound origin" task, thereby introducing a speed difference with respect to the "respond to sound motion direction" task. The average RT was 1073 ms. There was no main effect of task nor any other significant interaction in this analysis (all $p > 0.3$) except for the congruency factor, which resulted marginally significant ($F(1,11) = 4.8, p = 0.05$). This marginally significant effect was caused by responses to incongruent trials being somewhat slower than responses to congruent trials, overall (1148 vs. 999 ms, respectively), as one would expect.

line with what has been reported previously in several other studies using the complementary approach, that is, measuring audiovisual motion integration with psychophysical paradigms designed to be resistant to a variety of post-perceptual influences (e.g., Kitagawa and Ichihara, 2002; Vroomen and de Gelder, 2003; Soto-Faraco et al., 2005). Invariably in these studies, where the chance for post-perceptual influences was reduced, a strong crossmodal influence in the perception of motion direction still remained.

As discussed in the Introduction, there are several results suggesting that crossmodal directional congruency is not a relevant factor for motion perception (Meyer and Wuerger, 2001; Alais and Burr, 2004; see also Allen and Kolers, 1981, for similar conclusions). How can the results of these studies be reconciled with the present results and the results of previous investigations suggesting strong directional congruency effects (e.g., Kitagawa and Ichihara, 2002; Soto-Faraco et al., 2002, 2004a, 2004b, 2005; Vroomen and de Gelder, 2003)? We believe that there may be various potential reasons for this discrepancy. First, several of the studies that have failed to find congruency effects at a perceptual level have used visual motion as the target stimulus (e.g., Meyer and Wuerger, 2001; Alais and Burr, 2004, Experiment 2). It has been shown that the influence of auditory motion on visual motion processing can be considerably weaker than the reverse situation (i.e., auditory target and visual distractor), even for stimuli that have been equalized in terms of their quality (or strength) of apparent motion (see Soto-Faraco et al., 2004b). Second, in some, if not all of these discrepant studies, there were powerful cues for perceptual segregation between the information presented in the two modalities. For example, in certain older experiments (i.e., Allen and Kolers, 1981) the visual and auditory streams were presented from different spatial locations. In Meyer and Wuerger's (2001) more recent study, as well as in Alais and Burr's (2004; Experiment 1), the visual stimulus consisted of RDKs. Various studies have shown that perceptual grouping within sensory modalities precedes multisensory grouping, and that its strength can modulate crossmodal effects precisely in this type of task (Sanabria et al., 2004, 2005). Finally, another potentially important factor in some of the experiments is the task that was used. In the investigation by Alais and Burr (2004; see also Allen and Kolers, 1981), the participant's task was to detect the presence of motion (not to judge the direction of motion); therefore, it is unclear whether influences on the perception of motion direction could be measured easily, if at all. In summary, at this stage it is not possible to pinpoint the relative contribution of each of these potentially confounding factors on the failure to observe directional effects in motion processing across modalities in some experiments. However, it remains clear that under the appropriate conditions, the processing of motion is a strongly multisensory phenomenon. Moreover, it appears that these multimodal processes are highly robust and relatively insensitive to the influence of a variety of top-down processes.

Abbreviations

ANOVA	analysis of variance
ISI	interstimulus interval
LED	light-emitting diode
RDK	random dot kinematogram
SOA	stimulus onset asynchrony

Acknowledgments

This research was funded by a grant from the Ministerio de Educación y Ciencia to SS-F (Spain, TIN2004-04363-C03-02), by grants from the Michael Smith Foundation for Health Research, the National Science and Engineering Research Council of Canada and the Human Frontier Science Program to AK, and a grant from the Oxford McDonnell Centre for Cognitive Neuroscience to CS and SS-F.

References

Alais, D. and Burr, D. (2004) No direction-specific bimodal facilitation for audiovisual motion detection. Cogn. Brain Res., 19: 185–194.

Allen, P.G. and Kolers, P.A. (1981) Sensory specificity of apparent motion. J. Exp. Psychol. Hum. Percept. Perform., 7: 1318–1326.

Bertelson, P. (1998) Starting from the ventriloquist: the perception of multimodal events. In: Sabourin M., and Fergus C., et al., (Eds.), Advances in Psychological Science, Vol. 2, Biological and Cognitive Aspects, Psychology Press, Hove, England, pp. 419–439.

Calvert, G.A., Spence, C., and Stein, B.E. (Eds.), (2004) The Handbook of Multisensory Processes, MIT Press, Cambridge, MA.

Chan, J.S., Merrifield, K. and Spence, C. (2005) Auditory spatial attention assessed in a flanker interference task. Acta Acust., 91: 554–563.

Churchland, P.S., Ramachandran, V.S. and Sejnowski, T.J. (1994) A critique of pure vision. In: Koch, C. and Davis, J.L. (Eds.), Large-Scale Neuronal Theories of the Brain. MIT Press, Cambridge, MA, pp. 23–60.

de Gelder, B. and Bertelson, P. (2003) Multisensory integration, perception, and ecological validity. Trends Cogn. Sci., 7: 460–467.

Driver, J. and Spence, C. (2000) Multisensory perception: beyond modularity and convergence. Curr. Biol., 10: R731–R735.

Eriksen, B.A. and Eriksen, C.W. (1974) Effects of noise letters upon the identification of a target letter in a nonsearch task. Percept. Psychophys., 16: 143–149.

Garner, W.R. (1974) The Processing of Information and Structure. Erlbaum, Hillsdale, NJ.

Geldard, F. and Sherrick, C. (1972) The cutaneous 'rabbit': a perceptual illusion. Science, 178: 178–179.

Howard, I.P. and Templeton, W.B. (1966) Human Spatial Orientation. Wiley, New York.

Howell, D.C. (1992) Statistical Methods for Psychology (3rd ed). Duxbury Press, Belmont, CA.

Kitagawa, N. and Ichihara, S. (2002) Hearing visual motion in depth. Nature, 416: 172–174.

Macmillan, N.A. and Creelman, C.D. (1991) Detection Theory: A User's Guide. Cambridge University Press, Cambridge.

Marks, L.E. (2004) Cross-modal interactions in speeded classification. In: Calvert, G.A., Spence, C. and Stein, B.E. (Eds.), The Handbook of Multisensory Processes. MIT Press, Cambridge, MA, pp. 85–106.

Mateef, S., Hohnsbein, J. and Noack, T. (1985) Dynamic visual capture: apparent auditory motion induced by a moving visual target. Perception, 14: 721–727.

McGurk, H. and MacDonald, J. (1976) Hearing lips and seeing voices. Nature, 265: 746–748.

Meyer, G.F. and Wuerger, M. (2001) Cross-modal integration of auditory and visual motion signals. Neuroreport, 12: 2557–2560.

Morein-Zamir, S., Soto-Faraco, S. and Kingstone, A. (2002) Auditory capture of vision: examining temporal ventriloquism. Cogn. Brain Res., 17: 154–163.

Pallier, C., Dupoux, E. and Jeannin, X. (1997) EXPE: an expandable programming language for on-line psychological experiments. Behav. Res. Methods Instrum. Comput., 29: 322–327.

Rosenthal, O., Shimojo, S. and Shams, L. (2004, June) Sound-induced flash illusion is resistant to feedback training. Paper presented at 5th Annual Meeting of the International Multisensory Research Forum, Barcelona, Spain. Available at www.science.mcmaster.ca/˜IMRF/2004_submit/viewabstract.php?id = 101

Sanabria, D., Soto-Faraco, S. and Spence, C. (2004) Exploring the role of visual perceptual grouping on the audiovisual integration of motion. Neuroreport, 15: 2745–2749.

Sanabria, D., Spence, C. and Soto-Faraco, S. (in press) Perceptual and decisional contributions to audiovisual interactions in the perception of apparent motion: a signal detection study. Cognition.

Sanabria, D., Soto-Faraco, S., Chan, J. and Spence, C. (2005) Intramodal perceptual grouping modulates multisensory integration: evidence from the crossmodal dynamic capture task. Neurosci. Lett., 377: 59–64.

Shams, L., Kamitani, Y. and Shimojo, S. (2000) What you see is what you hear. Nature, 408: 788.

Soto-Faraco, S. and Kingstone, A. (2004) Multisensory integration of dynamic information. In: Calvert, G.A., Spence, C. and Stein, B.E. (Eds.), The Handbook of Multisensory Processes. MIT Press, Cambridge, MA, pp. 49–68.

Soto-Faraco, S., Kingstone, A. and Spence, C. (2003) Multisensory contributions to motion perception. Neuropsychologia, 41: 1847–1862.

Soto-Faraco, S., Lyons, J., Gazzaniga, M.S., Spence, C. and Kingstone, A. (2002) The ventriloquist in motion: illusory capture of dynamic information across sensory modalities. Cogn. Brain Res., 14: 139–146.

Soto-Faraco, S., Navarra, J. and Alsius, A. (2004a) Assessing automaticity in audiovisual speech integration: evidence from the speeded classification task. Cognition, 92: B13–B23.

Soto-Faraco, S., Spence, C. and Kingstone, A. (2004b) Cross-modal dynamic capture: congruency effects in the perception of motion across sensory modalities. J. Exp. Psychol. Hum. Percept. Perform., 30: 330–345.

Soto-Faraco, S., Spence, C. and Kingstone, A. (2005) Automatic capture of auditory apparent motion. Acta Psychol., 118: 71–92.

Stein, B.E. and Meredith, M.A. (1993) The Merging of the Senses. MIT Press, Cambridge, MA.

Strybel, T.Z. and Vatakis, A. (2005) A comparison of auditory and visual apparent motion presented individually and with crossmodal moving distractors. Perception, 33: 1033–1048.

Urbantschitsch, V. (1880) Ueber des Einfluss einer Sinneserregung auf die übrigen Sinnesempfingungen. [On the influence of one sensory percept on the other sensory percepts]. Arch. geschichte Physiol., 42: 154–182.

Vroomen, J. and de Gelder, B. (2003) Visual motion influences the contingent auditory motion aftereffect. Psychol. Sci., 14: 357–361.

Welch, R.B. and Warren, D.H. (1986) Intersensory interactions. In: Boff K.R., Kaufman L., and Thomas J.P. (Eds.), Handbook of Perception and Human Performance. Vol. 1. Sensory Processes and Perception, Wiley, New York, pp. 25.1–25.36.

Welch, R.B. (1999) Meaning, attention, and the "unity assumption" in the intersensory bias of spatial and temporal perceptions. In: Ascherlseben, G., Bachmann, T. and Musseler, J. (Eds.), Cognitive Contributions to the Perception of Spatial and Temporal Events. Elsevier Science B. V, Amsterdam, pp. 371–387.

Wohlschläger, A. (2000) Visual motion priming by invisible actions. Vision Res., 40: 925–930.

Wuerger, S.M., Hofbauer, M. and Meyer, G.F. (2003) The integration of auditory and visual motion signals at threshold. Percept. Psychophys., 65: 1188–1196.

Zapparoli, G.C. and Reatto, L.L. (1969) The apparent movement between visual and acoustic stimulus and the problem of intermodal relations. Acta Psychol., 29: 256–267.

Martinez-Conde, Macknik, Martinez, Alonso & Tse (Eds.)
Progress in Brain Research, Vol. 155
ISSN 0079-6123

CHAPTER 17

Crossmodal audio–visual interactions in the primary visual cortex of the visually deprived cat: a physiological and anatomical study

M.V. Sanchez-Vives[1,*], L.G. Nowak[2], V.F. Descalzo[1], J.V. Garcia-Velasco[1], R. Gallego[1] and P. Berbel[1]

[1]*Instituto de Neurociencias de Alicante, Universidad Miguel Hernández—CSIC, Apartado 18, 03550 San Juan de Alicante, Spain*
[2]*Centre de recherche "Cerveau et Cognition". CNRS-Université Paul Sabatier 133, route de Narbonne, 31062 Toulouse Cedex, France*

Abstract: Blind individuals often demonstrate enhanced non-visual perceptual abilities. Neuroimaging and transcranial magnetic stimulation experiments have suggested that computations carried out in the occipital cortex may underlie these enhanced somatosensory or auditory performances. Thus, cortical areas that are dedicated to the analysis of the visual scene may, in the blind, acquire the capacity to participate in other sensory processing. However, the neural substrate that underlies this transfer of function is not fully characterized. Here we studied the synaptic and anatomical basis of this phenomenon in cats that were visually deprived by dark rearing, either early visually deprived after birth (EVD), or late visually deprived after the end of the critical period (LVD); data were compared with those obtained in normally reared cats (controls). The presence of synaptic and spike responses to auditory stimulation was examined by means of intracellular recordings in area 17 and the border between areas 17 and 18. While none of the cells recorded in control and LVD cats showed responses to sound, 14% of the cells recorded in EVD cats showed both subthreshold synaptic responses and suprathreshold spike responses to auditory stimuli. Synaptic responses were of small amplitude, but well time-locked to the stimuli and had an average latency of 30 ± 12 ms. In an attempt to identify the origin of the inputs carrying auditory information to the visual cortex, wheat germ agglutinin-horseradish peroxidase (WGA-HRP) was injected in the visual cortex and retrograde labeling examined in the cortex and thalamus. No significant retrograde labeling was found in auditory cortical areas. However, the proportion of neurons projecting from supragranular layers of the posteromedial and posterolateral parts of the lateral suprasylvian region to V1 was higher than that in control cats. Retrograde labeling in the lateral geniculate nucleus showed no difference in the total number of neurons between control and visually deprived cats, but there was a higher proportion of labeling in C-laminae in deprived cats. Labeled cells were not found in the medial geniculate nucleus, a thalamic relay for auditory information, in either control or visually deprived cats. Finally, immunohistochemistry of the visual cortex of deprived cats revealed a striking decrease in pavalbumin- and calretinin-positive neurons, the functional implications of which we discuss.

Keywords: plasticity; crossmodal; multisensory; visually deprived; critical period

---

*Corresponding author. Phone: (+34)965-919368;
E-mail: mavi.sanchez@umh.es

DOI: 10.1016/S0079-6123(06)55017-4

Introduction

Blind individuals often demonstrate enhanced non-visual perceptual abilities. Such enhanced perception has been widely documented for auditory spatial discrimination tasks (Ashmead et al., 1998; Lessard et al., 1998; Roder et al., 1999) as well as for tactile discrimination tasks (Merabet et al., 2004; Roder et al., 2004). Several lines of evidence support the suggestion that these improved performance could be due, at least in part, to the recruitment of the visual cortex for somatosensory or auditory processing following a process of cross-modal plasticity. First, there is an increased electrical activation of the visual cortex during pitch discrimination (Kujala et al., 1995) and during spatial discrimination for peripheral sound sources (Roder et al., 1999) in the early blind, which is not observed in sighted subjects. Second, functional imaging studies of people who were blind from an early age revealed that their primary visual cortex could be activated by Braille reading and other tactile discrimination tasks (Sadato et al., 1996). Third, the activation of the visual cortex appears to be functionally relevant, since the blockade of occipital processing by means of transcranial magnetic stimulation interferes with tactile performance in early blind but not in sighted individuals (Cohen et al., 1997; Gougoux et al., 2004; Merabet et al., 2004). An additional support for a role of the occipital cortex has recently been provided by a functional brain imaging study in which the activation of occipital cortex in early blind individuals was found to be positively correlated with performance in a monaural auditory localization task (Gougoux et al., 2005). Although some authors have found evidence of auditory-visual cross-modal plasticity in late-onset blindness (Kujala et al., 1997), it appears that the performance improvement in non-visual tasks is larger in subjects with early onset blindness (e.g. Gougoux et al., 2004). This observation is in agreement with the fact that the cerebral cortex is more prone to plasticity during the early years.

Human studies therefore show that visual cortex of blind people undergoes some modifications that allow its recruitment in the performance of non-visual tasks. Understanding how the visual cortex is taken over by other sensory modalities requires examining the neuronal basis of cross-modal plasticity, which can only be explored in animal models. Since the seminal studies of Wiesel and Hubel (1965), it is known that visual deprivation early in life had a large impact on the structure and the physiology of the visual system (Hubel et al., 1977; Sherman and Spear, 1982; Fregnac and Imbert, 1984). In addition to alterations in the visual system development and functioning, early visual deprivation also results in the appearance of response to non-visual modalities in areas that were exclusively or mainly visual in normal animals. For example, Hyvarinen et al. (1981) found somatosensory responses in 19% of the cells recorded in area 19 of monkeys deprived of sight by bilateral eyelid suture shortly after birth. The same study also mentions weak somatosensory responses in area 17. Toldi et al. (1988, 1994) also showed that, in rat monocularly enucleated at birth, somatosensory evoked responses invaded the visual cortex, including the anterior part of area 17. Rauschecker and Korte (1993) found that, in cats that had been visually deprived for several years, most of the cells located in the caudal part of the anterior ectosylvian cortex, which is normally dominated by the visual modality, had acquired auditory responses. Furthermore, these neurons showed better spatial tuning for sound source localization than those recorded in adjacent auditory territories in normal cats (Korte and Rauschecker, 1993). Recently, Yaka et al. (1999) found that the number of cells responding to auditory stimulation was larger in two extrastriate areas (areas ALLS and AMLS) in cats that were visually deprived shortly after birth, either by eyelid suturing or by enucleation. Auditory responses were also obtained in 6% of the cells recorded in area 17 of the enucleated cats, but not in normal and eyelid-sutured cats (Yaka et al., 1999, 2000). Crossmodal plasticity also has been documented in the visual cortex of neonatally enucleated hamsters (Izraeli et al., 2002), where 63% of the cells were found to respond to auditory stimuli.

In this study we examined, at the physiological and anatomical levels, how the auditory modality becomes able to generate neuronal activity in the visual cortex of visually deprived cats. Cats were deprived of vision by rearing them in the dark for several months to 1 year. Deprivation started either immediately after birth (early visually deprived cats, EVD group) or at a mature age (late visually deprived cats, LVD group). The physiological study consisted in recording intracellularly from neurons in area 17 and in the 17/18 border region (Fig. 1). Our aim was to detect synaptic, subthreshold responses to sound stimulation, which could have remained undetected if recorded extracellularly. To identify possible new inputs to the visual cortex of deprived cats, retrograde tracers were injected in the equivalent areas in the hemisphere contralateral to the one from which recordings were obtained. The number of retrogradely labeled neurons were quantified in various cortical and thalamic areas. Finally, immunohistochemical staining of inhibitory neurons allowed us to examine the consequences of visual deprivation on inhibition in cortex, changes that explain a number of functional observations that are found in visually deprived animals.

Methods

Thirteen cats of either sex were included in this study, including physiological and diverse anatomical studies (see below). Five of these cats were born and reared in a normal environment (control cats) and eight were visually deprived. Among these eight visually deprived cats, three were deprived as adults (aged 4 months to 3 years) by placing them in a dark room for 3–7 months until the day of the experiment. We will refer to them as the LVD group, since deprivation took place after the end of the critical period. The remaining five cats were born and raised in a dark environment until the time of recording (aged 6 months to 1 year) and they correspond to the EVD group.

All but one of the anatomical studies (immunohistochemistry and retrograde labeling) were carried out in the animals in which the electrophysiological studies were done.

Visual deprivation

Control cats were reared in a 12 h light/dark cycle. Visually deprived cats were kept in a pitch-black dark room. We chose this method rather than enucleation for its reversibility, and rather than lid-suture, since that method does not guarantee total absence of visual stimulation (Spear et al., 1978). Animal caretakers walked into the cats' room from a darkened hallway, and were provided with a helmet carrying a video camera with infrared lighting (GL 380, SHARP®; λ peak = 950 nm) that allowed them to maintain the room clean, check the state of the animals and replace food and water. Another video camera sensitive to infrared, located inside the animal room, allowed us to check the animals' state and behavior on a monitor from outside the room. In order to create a sound-rich environment, four different sounds (pure tones between 2 and 20 kHz; 65 dB) coming from four speakers in the different walls of the room were used for alternate stimulation for 2–6 h/day.

The project was approved by the local ethical committee. Local rules comply with the EU guidelines on protection of vertebrates used for experimentation (Strasbourg 3/18/1986, Spanish law BOE 256; 10/25/1990).

Animal preparation for in vivo recording

Intracellular recordings "*in vivo*" from the visual cortex of cats were obtained following the methodology that we have previously described (Sanchez-Vives et al., 2000b), except for the anesthetic used during the recordings. In short, adult cats were anesthetized with ketamine (12–15 mg/kg, i.m.) and xylazine (1 mg/kg, i.m.) and then mounted in a stereotaxic frame. A craniotomy (3–4 mm wide) was made overlying the representation of area 17. To minimize pulsation arising from the heartbeat and respiration, a cisternal drainage and a bilateral pneumothorax were performed, and the animal was suspended by the rib cage to the stereotaxic frame. During recording, anesthesia was maintained with continuous i.v. infusion of propofol (5 mg/kg/h) and sufentanyl (4 μg/kg/h) and the animal was paralyzed with norcuron (induction 0.3 mg/kg; maintenance 60 μg/kg/h) and artificially ventilated. The heart rate, expiratory CO_2 concentration, rectal temperature and blood O_2 concentration were monitored throughout the experiment and maintained at 140–180 bpm, 3–4%, 37–38 °C and > 95%, respectively. The electroencephalogram (EEG) and

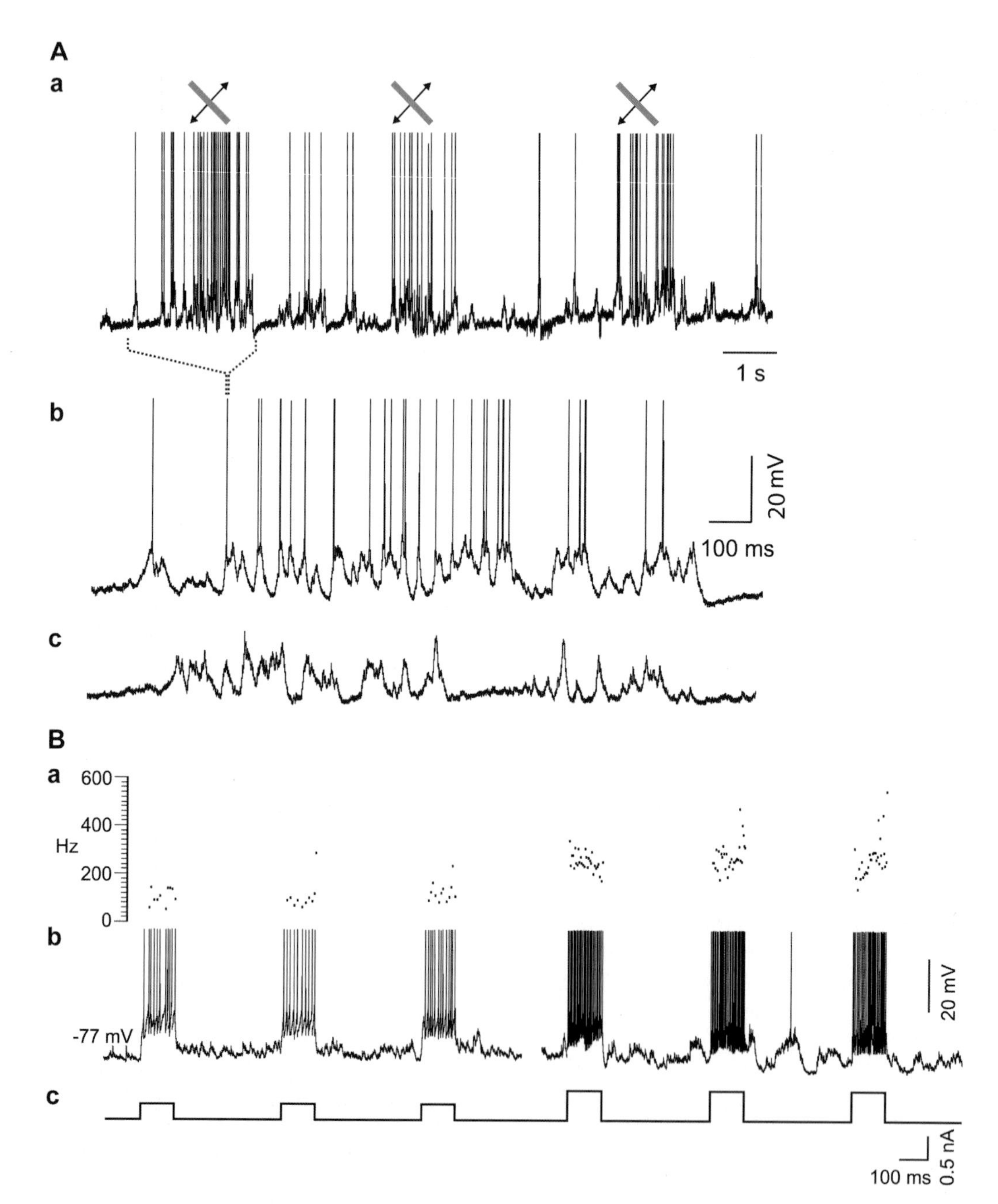

Fig. 1. Electrophysiological recordings from the visual cortex of control and visually deprived cats. (A) Visual responses recorded in control cats and evoked by moving bars with a handheld projector onto the tangent screen. (Aa) Synaptic and spike responses to three consecutive visual stimuli. (Ab) First visual response expanded. (Ac) Subthreshold (synaptic) visual response. (B) In visually deprived cats, no visual stimuli were used (see the section "Results") but responses to pulses of injected currents were recorded. This panel shows responses to current injections in a neuron that was characterized as fast-spiking neuron (according to Nowak et al., 2003) and showed both sub- and suprathreshold reponses to sound (not shown). From top to bottom: instantaneous frequency, membrane voltage and current are shown. Notice the spontaneous synaptic activity in between pulses and how synaptic activity affected the frequency of discharge as reflected in the variability in the instantaneous frequency. Spikes have been truncated.

the absence of reaction to noxious stimuli were regularly checked to insure an adequate depth of anesthesia. After the recording session, the animal was given a lethal injection of sodium pentobarbital.

Recordings and stimulation

Sharp intracellular recording electrodes were made with a Sutter Instruments (Novato, CA) P-97 micropipette puller from medium-walled glass capillaries and beveled to final resistances of 50–100 MΩ. Micropipettes were filled with 2 M KAc. Recordings were digitized, acquired and analyzed using a data acquisition interface and software from Cambridge Electronic Design (Cambridge, UK). In control animals, the eyes were focused onto a tangent screen at 114 cm using corrective, gas-permeable contact lenses. The position of the area centralis and optic discs was localized by retroprojection. Receptive field's location was determined with a handheld projector. In dark-reared animals the eyes were covered during the experiments.

Auditory stimulation
Pure tones at various frequencies (1–20 kHz), white noise and clicks were used as auditory stimuli. Pure tones were generated using a waveform generator (Racal Dana 9085 Low Distorsion Oscillator). White noise was generated with a noncommercial generator. The signals were trapezoidally gated with rise/fall times of 5 ms. Stimuli onset and duration were controlled by a computer. The stereotaxic frame had hollow ear bars and the loudspeakers were placed inside. Thus, stimuli were delivered binaurally through a closed acoustic system based on Sony MDR E-868 earphones housed in a metal enclosure and surrounded by damping material which fitted into the Perspex specula (Rees et al., 1997). The output of the system for each stimulus was calibrated to be between 60 and 80 dB.

Analysis

Each run of auditory stimuli lasted for 150–700 s. For each run, the membrane potential was averaged with respect to the auditory stimulus trigger (black trace in Fig. 2Ab). The mean potential between 15 and 165 ms (or 115 ms in case shorter stimulus duration was used) was calculated. The mean potential between –150 (or –100) and 0 ms was calculated as well to give a mean baseline value. The mean baseline value was subtracted to give the mean membrane potential depolarization above baseline, or mean stimulus-locked response amplitude (MRA_{SL}, black bin in Fig. 2Ac). The spike response (Fig. 2B) was calculated as a peristimulus time histogram (PSTH) with a bin width of 10 ms. The measured variable in this case was the average firing rate above spontaneous activity in a time window between 15 and 165 (or 115) ms relative to auditory stimulus onset, yielding the MRA_{SL} (dark bin in Fig. 2Bc). The significance of the response was evaluated using a Monte Carlo simulation. The hypothesis underlying the use of the Monte Carlo method is as follows: on the one hand, the synaptic potential observed in the membrane potential average (or firing rate changes in the PSTH) may result from the preceding auditory stimulus, in which case they should be stimulus locked. On the other hand, the membrane potential in anesthetized preparations often shows spontaneous synaptic activity characterized by large fluctuations (Fig. 1Aa), reaching up to 20 mV amplitude, due to a slow wave sleep rhythm (Steriade et al., 1993; Sanchez-Vives and McCormick, 2000). If the deviations obtained in the membrane potential average were due to spontaneous synaptic activity, then the amplitude should be independent of the precise stimulus trigger timing.

To confront these two possibilities, each "real" auditory trigger occurring at time t was replaced by a randomly jittering trigger (Figs. 2Aa and 2Ba). The new trigger had a time of occurrence limited to $t+\alpha$, α being an added delay that took a random value within a range between $-0.5\,T$ and $+0.5\,T$, T being the period of the "real" auditory trigger. This procedure was repeated for all the triggers, with a new α value for each trigger. The membrane potential average or the PSTH was recalculated and the amplitude of the response measured as above. Figure 2Ab represents five traces of membrane potential averages triggered by random triggers in addition to the one obtained with the regular

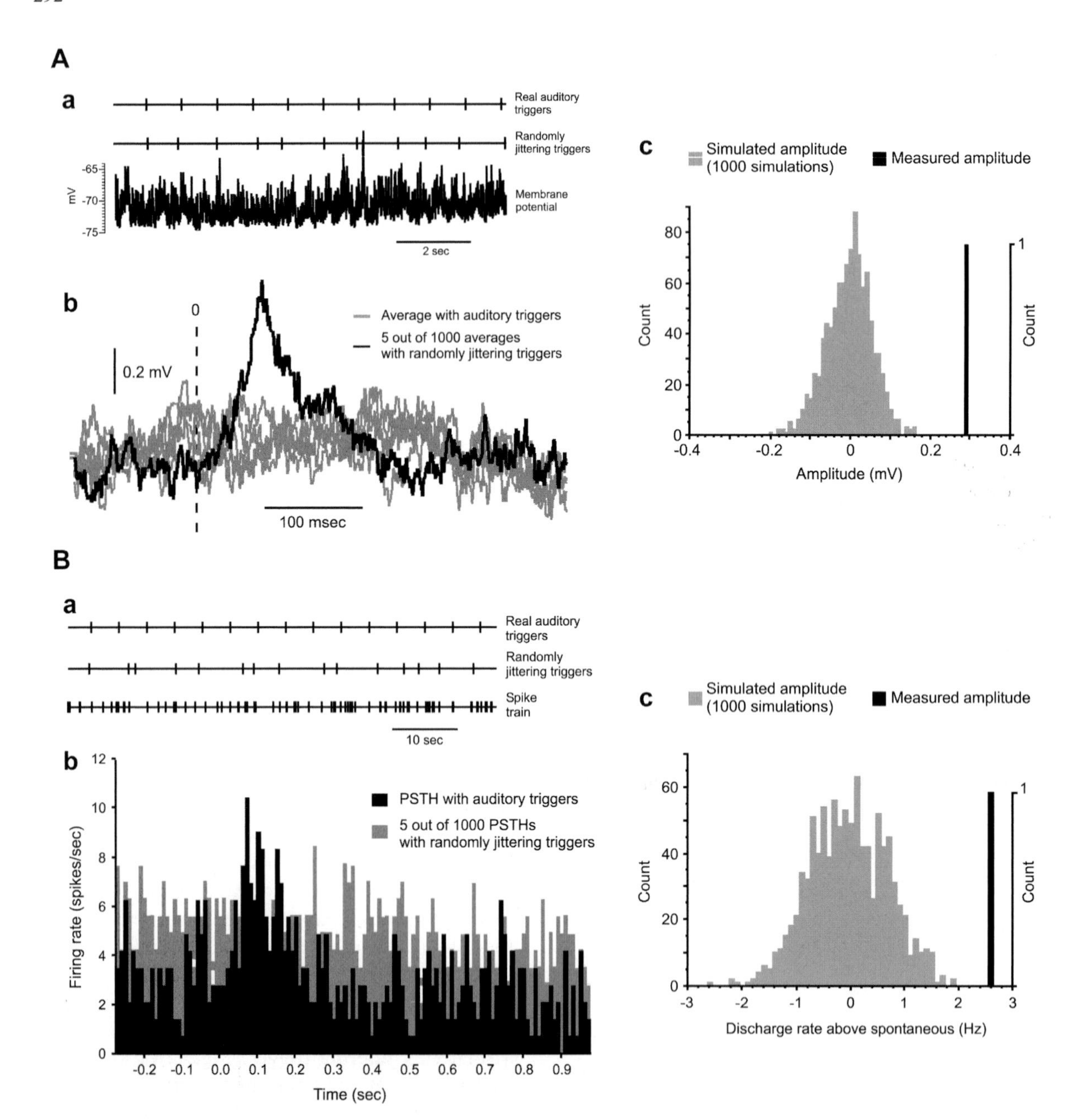

Fig. 2. Detection of significant synaptic and spike responses with a Monte Carlo method. (A) Detection of significant synaptic responses by using a Monte Carlo simulation (Aa) From top to bottom: example of auditory triggers, randomly generated triggers (see the section "Methods") and voltage trace corresponding to the intracellular recording. (Ab) Five traces (in gray) corresponding to the membrane voltage average triggered with five randomly selected triggers and one trace (in black) corresponding to the voltage average triggered with the real auditory triggers. (Ac) Distribution of amplitudes for 1000 voltage averages triggered with random series of triggers (in gray). In black, on the right, the amplitude of the voltage average triggered with the real auditory triggers. The amplitude corresponds to the mean amplitude between 15 and 165 ms relative to auditory stimulus onset in the averages, minus baseline. (B) Detection of significant spiking responses by using a Monte Carlo simulation (Ba) From top to bottom: example of auditory triggers, randomly generated triggers (see the section "Methods") and events corresponding to intracellularly recorded action potentials. (Bb) PSTHs obtained with the auditory trigger appears in black and five PSTHs obtained with randomly jittering triggers in gray. (Bc) Distribution of amplitudes for 1000 PSTHs triggered with random series of triggers (in gray). In black, on the right, the amplitude of the PSTH triggered with the real auditory triggers. The amplitude corresponds to the firing rate between 15 and 165 ms relative to auditory stimulus onset in the PSTHs, minus spontaneous activity.

auditory trigger. The same procedure was applied to the spike responses (Fig. 2Bb).

The simulation was repeated 1000 times. The distribution of mean depolarization calculated for 1000 averages is shown by the gray bars in the histogram in Fig. 2Ac. The distribution of these values corresponds to the distribution of amplitudes one would get if the depolarization was not locked to the stimulus, i.e., noise consecutive to spontaneous activity. This distribution shows the typical shape of a normal distribution, which can be characterized by a mean μ and a standard deviation σ. The MRA_{SL} was standardized as a z value as $z = (MRA_{SL} - \mu)/\sigma$. Since z will have a standard normal distribution ($\mu = 0$; $\sigma = 1$), it can be easily used to calculate the probability that the sound-evoked response is spurious resulting from spontaneous activity instead of an event-locked auditory response. Membrane potential average and average spike responses were considered as significant auditory responses when z values were outside the ± 1.96 range ($p = 0.05$; Fig. 2Ac for synaptic responses and Fig. 2Bc for spike responses; z value of ± 2.58 corresponds to a p value $= 0.01$). This method provided a reliable and sensitive indicator of the presence of auditory responses in cat visual cortex. Whenever a significant response was obtained, the latency and amplitude were determined. The latency corresponds to the first of two consecutive bins that showed an amplitude larger than the mean baseline + 3 SD. Amplitude is the peak amplitude *minus* mean baseline.

Retrograde labeling

A total of four control, four EVD and one LVD cats were included in the tracer injection's study. Under anesthesia (see preparation for *in vivo* recordings above), each cat received 3–5 injections (0.1 μl each) of 10% HRP coupled to WGA-HRP in physiological saline. Injections were made near the 17/18 border expanding to areas 17, 18 and 19, at a depth of 900 μm and were evenly spaced from one another (23–19 mm caudally from *bregma* and 1.5–2 mm lateral). Tracer was delivered by pressure pulses (Picopump, WPI) through glass pipettes (outer diameter of the tip 40–60 μm). After 1–2 days survival, the anesthetized cats were perfused with physiological saline for 5 min followed by 1% paraformaldehyde, 0.5% glutaraldehyde, 0.002 M $CaCl_2$ and 0.1 M saccharose in 0.1 M phosphate buffer (pH 7.3–7.4) for 30 min. The brains were removed and cryoprotected, and 60 μm-thick slices were cut in the coronal plane using a cryostat into two parallel series. One series was mounted on gelatinized slides and air-dried for 24 h and stained in 1% cresyl violet (Sigma, St. Louis, MO, USA). The other series were floating sections that were processed with 3,3′,5,5′-tetramethylbenzidine (Mesulam, 1978). Reacted sections were mounted on gelatinized slides, air-dried for 24 h, briefly dehydrated and cleared in xylene, and mounted in Eukitt (O. Kindler GmbH & Co, Freiburg, Germany). Injections covered a large extent of the visual areas 17 and 18.

Histo- and immunochemistry

Four control, four EVD and one LVD cats were processed for immunohistochemistry. They were perfused with physiological saline for 5 min followed by 4% paraformaldehyde, 0.002 M $CaCl_2$, and 0.1 M saccharose in 0.1 M phosphate buffer (pH 7.3–7.4) for 30 min.Brains were cut using a vibratome in slices 100 μm-thick into four parallel series. One series was mounted on gelatinized slides and air-dried for 24 h and processed for CO staining, being incubated in freshly prepared 0.06% DAB, 0.02% cytochrome C, type III (Sigma) and 4.25% sucrose in PBS for 16–24 h at 37 °C (Wong-Riley, 1979). Three series were immunostained with markers (calretinin, calbindin and parvalbumin) revealing different subtypes of GABAergic neurons (Celio, 1990; Celio et al., 1990; DeFelipe, 1997). Floating sections were incubated with rabbit anti-calretinin Ab (1:2000; Swant, Bellinzona, Switzerland), which was followed with biotinylated goat anti-rabbit Ab (1:150; Vector), ABC kit and DAB. The other two series were stained with anti-parvalbumin (1:1000; Swant) and anti-calbindin D-28 K (1:2000; Swant) monoclonal antibodies. Immunostaining was followed by biotinylated horse anti-mouse Ab (Vector), ABC kit and DAB. Immunostained sections were mounted on gelatinized slides, air-dried for 24 h, dehydrated in ethanol, cleared in xylol and coverslipped.

Quantitative anatomical analysis

Sections were studied and photographed using an Olympus BX50W1 microscope with a Nikon D70 digital camera. Plots and counts of immunoreactive and retrogradely labeled cells were obtained using the Neurograph system (Microptic, Barcelona, Spain). Retrogradely labeled neurons in the following areas were plotted: ipsilaterally, the posteromedial lateral suprasylvian visual area (PMLS) and the posterolateral lateral suprasylvian visual area (PLLS), auditory areas I and II (AI and AII), the medial geniculate nucleus (MGN), the dorsal lateral geniculate nucleus (dLGN) and the medial interlaminar nucleus of the dLGN (MIN); contralaterally, visual areas 17, 18, PMLS and PLLS, and the auditory area AI. The entire cross-sectional coronal surface of the MGN, dLGN and MIN was analyzed. In all cases, the borders between cortical layers and dLGN laminae were placed at the same relative depth, measured from adjacent sections stained with cresyl violet. In addition, cells immunoreactive to parvalbumin, calbindin D-28 K and calretinin in areas 17 and 18 were also plotted. For quantitative analysis of both immunohistochemistry and retrograde labeling, cells were counted within a rectangular area (probe) that measured 200 μm wide and spanned from layer I to the subcortical white matter. In studies of retrograde labeling these probes covered the entire studied cortical areas. In immunohistochemical studies three probes evenly spaced in areas 17 and 18 were quantified. In total, three sections per area and animal were quantified and the mean number was then averaged among animals within each experimental group. The Systat statistical software (Systat Inc., Evanston, IL) was used for one-way ANOVAs, followed by Tukey's test to identify significant differences ($p \leqslant 0.05$) between means.

Results

Functional study of auditory responses in the visual cortex

Ten cats were included in the electrophysiological study. Three of these cats were born and reared in a normal environment (control cats) and seven were visually deprived. Out of these seven visually deprived cats, four were visually deprived as adults (aged 4 months–3 years) by placing them in a dark room for 3–7 months until the day of the experiment. We will refer to them as the LVD group, since deprivation took place after the end of the critical period. The remaining three cats were born and raised in a dark environment until the time of recording (aged 6 months–1 year) and they correspond to the EVD group.

Intracellular recordings that were sufficiently long and stable to allow the characterization of auditory responses in visual cortex were obtained from a total of 148 cells. Of these, 32 were recorded in control cats, 60 in LVD cats and 56 in EVD cats. For a representative sample of 20 neurons in control and 20 in visually deprived cats the input resistance was 26.7 ± 10.07 and 27.5 ± 14.8 MΩ, respectively, and the membrane time constant was 32.0 ± 8.9 and 26.7 ± 9.3 ms, respectively.

The objective of these experiments was to examine the possible existence of auditory responses in the visual cortex of the three experimental groups. Intracellular recordings were performed to be able to detect subthreshold synaptic responses that could have been overlooked if the recordings would have been extracellular. For this reason, and in order to better detect synaptic responses, during some of the recordings the membrane potential was hyperpolarized to prevent spiking activity and therefore only subthreshold synaptic responses were analyzed. In the other cells, both spiking and synaptic responses were recorded and analyzed.

Once an intracellular recording was obtained and the intrinsic membrane properties were measured, sensory stimulation began. In both experimental groups, control and visually deprived cats, the level of anesthesia was as in our previous studies of visual responses (Sanchez-Vives et al., 2000a; Nowak et al., 2003) and therefore compatible with the existence of sensory responses. This was directly checked in control cats, where visual responses to bars of light were explored with a handheld projector to ensure that neurons were visually responsive (Fig. 1A). No further characterization of visual responses was done, in order to concentrate on the auditory stimulation. In EVD and LVD cats no visual stimuli were delivered, to prevent reversion or decline of any auditory responses that could have developed during the period of visual deprivation. Auditory stimulation was delivered through headphones placed inside the ear bars and consisted of white noise, pure tones (1–20 kHz) and/or clicks. Stimulus duration was 150–2000 ms for tones and white noise, and 10–20 ms for clicks, and the stimuli were given at a frequency of 0.5–2 Hz.The responses to auditory stimuli were evaluated for both membrane potential values and spike response (PSTH), by calculating the spike-triggered average with respect to the auditory stimulus. The significance of this response was estimated by using a Monte Carlo simulation. In short, 1000 random triggers (Figs. 2Aa, Ba) were generated and used for spike-triggered average of both the membrane potential (Fig. 2Ab) and the spike response (Fig. 2Bb), and compared with the actual auditory-locked responses (for details see "Methods").

The distribution of the standardized responses to both auditory and random triggers (z values; $\mu = 0$; $\sigma = 1$; see "Methods") for the synaptic and spiking response and for the three experimental groups of cats is shown in Fig. 3. None of the 32 cells recorded in control cats showed any significant response to auditory stimuli ($-1.96 < z < 1.96$, Figs. 3Aa, Ba). Similarly, none of the 60 cells recorded in the LVD cats showed any significant response to sound stimuli (Figs. 3Ab, Bb). However, significant responses have been observed in the EVD cats, both for synaptic and spiking responses (Figs. 3Ac, Bc). The total number of cells showing significant responses to sound stimuli was 8 out 56 cells tested. This represents 14% of the cells recorded in the EVD animals. In three out of these eight neurons, only the synaptic response was recorded because the neurons were maintained subthreshold during the study. In three additional neurons both spiking and synaptic auditory responses were significant, and in the remaining two neurons, only the spiking response was significant — in these later cases, the spontaneous activity level was high and the continuous firing resulted in a strong shunt of the membrane potential that masked the synaptic response.

Hence in this study we describe, to our knowledge for the first time, synaptic potentials induced by sound in primary visual cortex. Fig. 4 shows both sub- and suprathreshold auditory responses from the same neuron. Although the average synaptic response (Fig. 4A) had an amplitude of only 0.5 mV, it was statistically significant and able to generate a significant suprathreshold response. This averaged spike response is represented in a PSTH in Fig. 4B and is overlapped with the synaptic response to illustrate their parallel time courses.

The amplitude of the averaged synaptic responses varied between 0.1 and 1.8 mV and the actual values for each of the six neurons in which synaptic responses were obtained in response to one or several auditory stimuli are shown in Fig. 5B. At the population level the mean amplitude was 0.6 ± 0.6 mV (only one value per neuron was used by prior averaging of the amplitude for the different auditory stimuli, when more than one was obtained in the same cell). The mean onset latency of the synaptic responses was 30 ± 12 ms (range 15–46 ms).

The average peak-firing rate obtained in response to auditory stimulation was 3.2 ± 2.4 spikes/s, and the mean onset latency of the spike response was 42 ± 24 ms. In three cases, the response to different kind of auditory stimuli was examined in the same cell. The synaptic responses (auditory-triggered average membrane potential) to different pure tone frequencies and to white noise stimulation are depicted for one of these cells in Fig. 5A. The graph shows that this cell was poorly tuned for specific frequencies and that the neuron responded with a depolarization to all the given auditory stimuli. This also was the case with the remaining two neurons (Fig. 5B).

Cytoarchitecture of the visual cortex in visually deprived cats

In order to better understand the structural basis underlying the functional phenomena that we have described in the previous section, we performed a number of anatomical studies in the same experimental groups in which the physiology was studied.

The cytoarchitecture of cortical areas 17 and 18, studied in cresyl violet-stained sections, was similar between control, EVD and LVD cats. Cortical

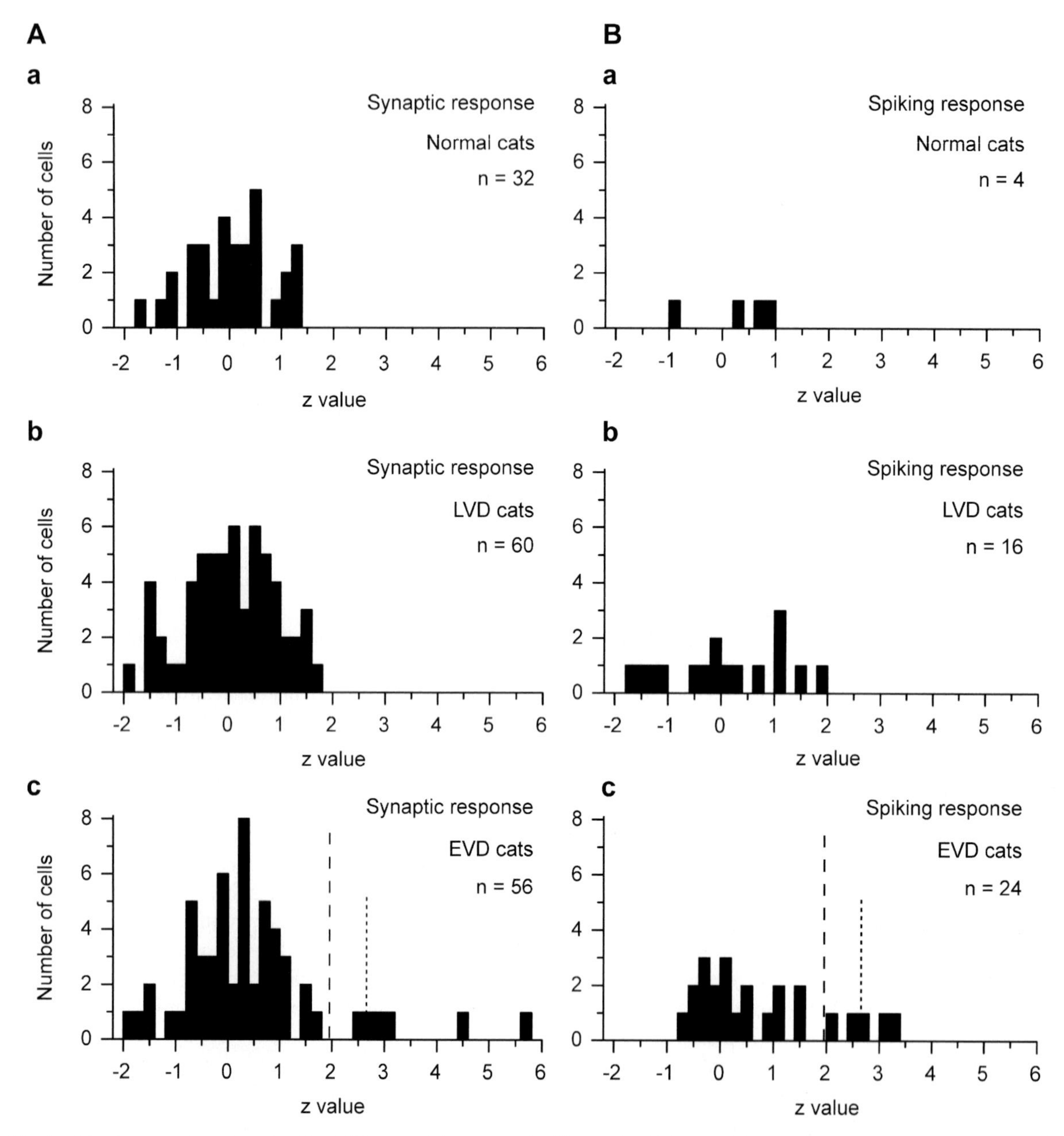

Fig. 3. Distribution of *z*-values for the PSPs recorded in control, EVD and LVD cats. (A) Distribution of *z*-values for synaptic responses. (Aa) Control cats. (Ab) LVD cats. (Ac) EVD cats. Notice that five of the *z*-values were above 1.96 and therefore significant at $p<0.05$. (B) Distribution of *z*-values for spike responses. (Ba) Control cats. (Bb) LVD cats. (Bc) EVD cats. Notice that five of the *z*-values were above 1.96 and therefore significant at $p<0.05$.

layers were present and the borders between layers well defined.

CO staining was made to identify layer IV in the cortical areas of control and EVD cats (no LVD cats were studied here). No differences were found between control and EVD cats in the differential intensity of the staining among cortical layers. In both experimental groups, layer IV was more

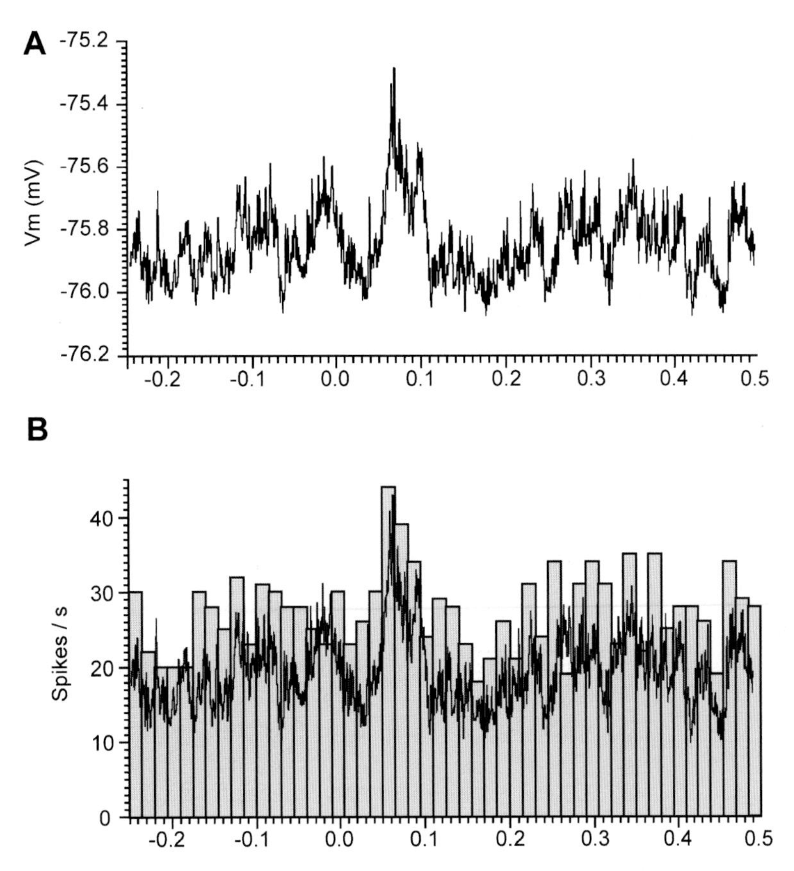

Fig. 4. Synaptic responses to sound in primary visual cortex. (A) Sound-triggered average of the membrane potential for a significant synaptic response. (B) PSTH of the suprathreshold, spiking response in the same neuron. Notice the similar time course to that for the synaptic response (shown in overlap).

intensely stained than the other layers, showing scattered, CO-stained pyramidal neurons in layer V (Fig. 6). However, the heavily stained band of layer IV had borders with adjacent layers III and V that were more blurred in EVD than in control cats (Fig. 6).

Distribution of intracortical retrogradely labeled neurons

Four control, four EVD and one LVD cats were injected with retrograde tracers in area 17, border of area 17/18 and areas 18 and 19 (see the section "Methods"). No differences either in the distribution or in the density of labeled neurons were found between control and LVD cats; the LVD case is therefore not described further.

Ipsilateral projections from PMLS to areas 17 and 18

In all cases tracer core injection was near the 17/18 border; however, the tracer diffused to areas 17, 18 and 19, filling the postlateral girus (see inset in Fig. 7). The mean number of ipsilaterally projecting neurons per section from PMLS to areas 17 and 18 in EVD cats (872 ± 104 neurons per section) was not significantly different from the one in control cats (766 ± 194 neurons per section). However, the radial and tangential distribution of labeled neurons was qualitatively different (Figs. 7 and 8A). In control cats, the radial distribution of labeled neurons found in PMLS was similar to that previously described (Bullier et al., 1984b; Symonds and Rosenquist, 1984; Segraves and Innocenti, 1985; Kato et al., 1991; Shipp and

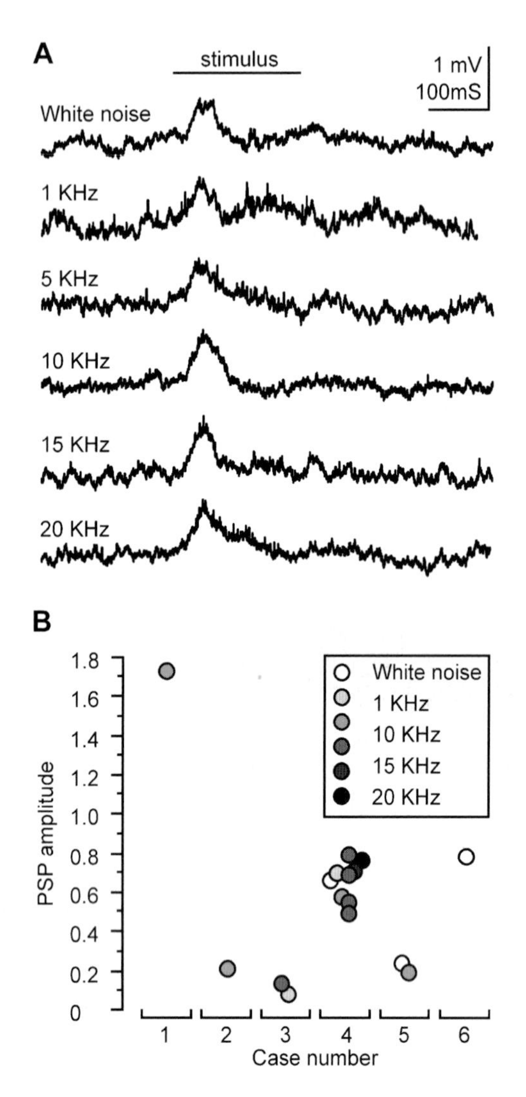

Fig. 5. Responses to white noise and pure tones at different frequencies in the visual cortex of the cat. (A) Synaptic responses to different auditory stimuli (white noise and pure tones at 1, 5, 10, 15 and 20 kHz). Each one is the result of averaging 526, 261, 332, 423, 398 and 423 stimuli, respectively. (B) Amplitudes of synaptic responses that were statistically significant ($p<0.05$) for each one of the six neurons and for the different sound stimuli.

Grant, 1991; Einstein, 1996; Batardière et al., 1998; Payne and Lomber, 2003). Thus, ipsilaterally projecting neurons from PMLS to areas 17 and 18 were mostly found in infragranular layers (layer V: 19.6%, and layer VI: 59.0%) and were less numerous in supragranular layers (layers II–III: 16.7% from the total). A small percentage of labeled neurons was also seen in layer IV (4.7%; Figs. 7A–C and 8A). In EVD cats, the proportion of retrogradely labeled neurons in layers IV (5.2%), V (15.0%) and VI (49.7%) was similar to controls, while it significantly increased in supragranular layers II–III (30.0%; Figs. 7D–F and 8A).

Contralateral projections from PMLS to areas 17 and 18

In control cats, the pattern of labeling in area PMLS contralateral to the injection site appears similar to that previously described (Keller and Innocenti, 1981; Segraves and Rosenquist, 1982; Segraves and Innocenti, 1985). Contralaterally projecting neurons from PMLS to areas 17 and 18 were mostly found in layers V (66.7%) and VI (33.3%). In EVD cats, the mean number of contralaterally projecting neurons per section from PMLS to areas 17 and 18 (109.8 ± 12.4 neurons per section) was significantly higher than in control cats (17.1 ± 7.1 neurons per section). The radial and tangential distribution of labeled neurons was also quantitatively different (Fig. 8A). In EVD cats, a large fraction of labeled neurons was found in layers II–III (34.4%), suggesting stabilization of connections that are transitory during development (Innocenti and Clarke, 1984). Neurons were also found in layer IV (18.7%). Given these relative increases in supragranular and layer IV labeling, the relative proportion of labeled neurons in layers V (26.8%) and VI (20.1%) was significantly decreased (Fig. 8A).

Ipsilateral projections from PLLS to areas 17 and 18

Ipsilaterally, the mean number of projecting neurons per section from PLLS to areas 17 and 18 in EVD cats was reduced with respect to controls (on average, 584 ± 112.6 neurons per section in control vs. 248 ± 92.4 neurons in EVD cats) and its radial distribution was also qualitatively different (Figs. 7 and 8A, B). In control cats, the radial distribution of labeled neurons found in PLLS was similar to that previously described (Symonds and Rosenquist, 1984; Rosenquist, 1985; Segraves and Innocenti, 1985; Payne and Lomber, 2003) (Fig. 7A). Ipsilaterally projecting neurons from

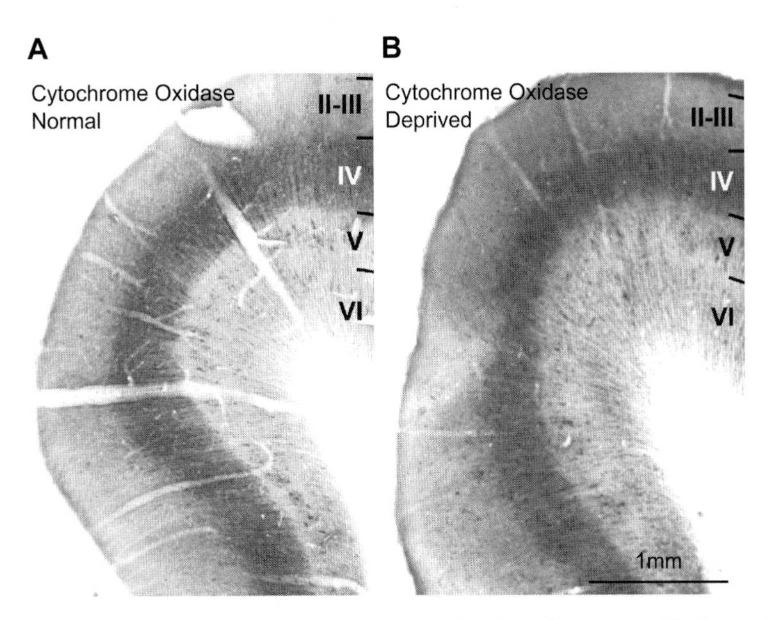

Fig. 6. Photomicrographs showing cytochrome oxidase labeling in coronal sections through area 17 of control (A) and EVD (B) cats. In layer IV of normal and EVD cats, dense CO staining can be seen. In deprived cats, CO labeling shows blurred borders between layer IV and layers II–III and V. Borders between layers are indicated. Same magnification for A and B.

PLLS to areas 17 and 18 were mostly found in infragranular layers (layer V: 18.3%; layer VI: 76.9%), also in layers II–III (3.3% from the total) and in layer IV (1.5%) (Figs. 7A–C and 8B). In EVD cats, the proportion of labeled neurons in layers II–III (13.1%) increased with respect to controls, whereas it was similar in layers IV (2.0%), V (12.1%) and VI (72.8%; Figs. 7D–F and 8B). In control and deprived cats, labeled neurons were mostly found in the medial half of the PLLS, close to the PMLS border located deep in the medial suprasylvian sulcus (Figs. 7D–F).

Contralateral projections from PLLS to areas 17 and 18

Contralaterally, the mean number of projecting neurons per section from PLLS to areas 17 and 18 in control cats was very low (on average, 2.8 ± 0.7 neurons per section). Although this number was also low in EVD cats (8.1 ± 1.4 neurons per section), it was significantly higher than in control cats. Both in control and EVD cats, contralaterally projecting neurons from PLLS to areas 17 and 18 were mostly found in layer VI, as shown previously (Keller and Innocenti, 1981). No labeled neurons were observed in other suprasylvian areas such as the AMLS and ALLS.

Projections from AI and AII to areas 17 and 18

No differences were found between control and both groups of deprived cats in the mean number of neurons per section and distribution of labeled neurons found in AI and AII. In all experimental groups, there were few neurons labeled per section (between 3 and 15), located near the fundus and anterior bank of the posterior ectosylvian sulcus as well as on the convexity of the middle ectosylvian and sylvian gyri, corresponding to areas AI and AII. Almost all labeled cells were located in the infragranular cortical layers. The distribution and numbers of labeled neurons were very similar to those already described for normal cats (Dehay et al., 1984; Innocenti and Clarke, 1984; Innocenti et al., 1988; Payne and Lomber, 2003).

Distribution of thalamic retrogradely labeled neurons

In both control and visually deprived cats, labeled neurons were found in all laminae of the dLGN, as described previously (Maciewicz, 1975; Hollander and Vanegas, 1977; Bullier et al., 1984a; Payne and

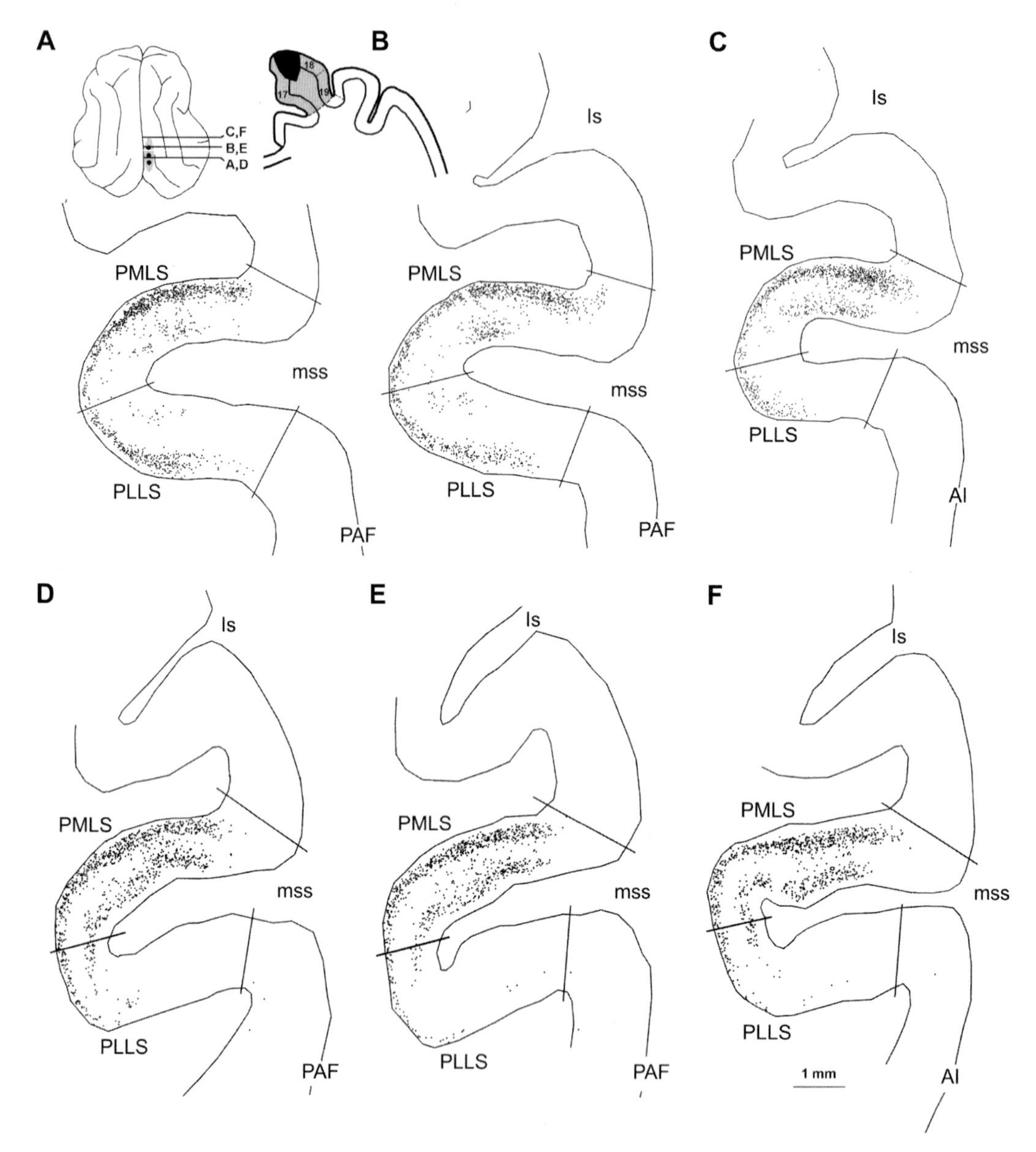

Fig. 7. Distribution of retrogradely labeled neurons in ipsilateral PMLS and PLLS following WGA-HRP injections in areas 17 and 18 in control (A–C) and EVD (D–F) cats. In control cats, labeling was found principally in infragranular layers. In EVD cats, the proportion of labeled neurons in supragranular layers increased with respect to controls. Note the reduced number of labeled neurons in the PLLS in EVD compared to control cats. Levels of sections, place (small black circles) and extent of injections (shaded areas) are shown in the brain figurines at the upper left corner. ls: lateral sulcus; mss: medial suprasylvian sulcus; PAF: posterior auditory field of the ectosylvian gyrus.

Lomber, 2003). However, in EVD cats they were more uniformly distributed in the mediolateral dimension. The proportion of labeled neurons in laminae A decreased in deprived cats with respect to control, while it increased in the magnocellular lamina C and parvocellular lamina C1. No changes were found in parvocellular lamina C2 and the density decreased in parvocellular lamina C3 (Figs. 9A, B). The distribution of labeled neurons in the MIN in EVD cats was not significantly

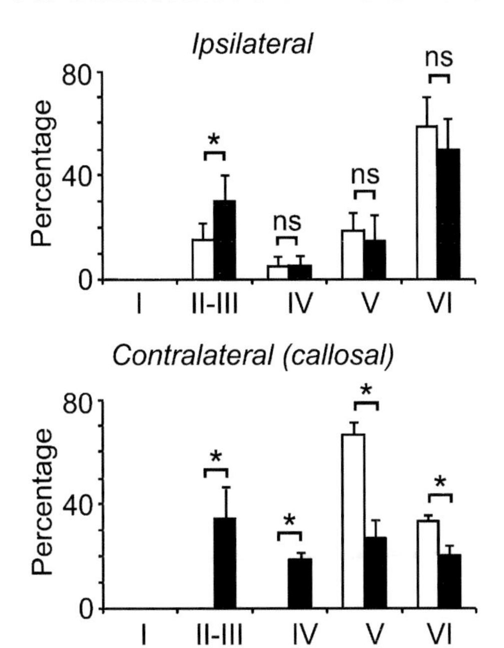

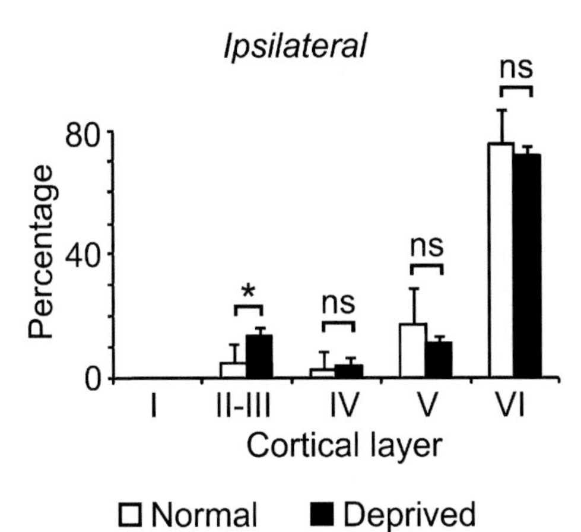

Fig. 8. Radial distribution of labeled neurons in PMLS and PLLS. Percentages of retrogradely labeled neurons per section in control (white bars) and EVD (black bars) cats in different cortical layers of PMLS (A) and PLLS (B). Note the increased number of ipsilaterally labeled neurons in layers II–III in EVD cats with respect to controls, in both PMLS (A) and PLLS (B) areas. In the contralateral hemisphere, callously projecting neurons from supragranular layers of PMLS were exclusive of EVD cats (A). Vertical lines represent standard deviations. (*): significant differences ($p < 0.05$); ns: nonsignificant differences.

different from the one in control cats, although the density of labeled neurons was lower (Figs. 9A, B). No labeled neurons were found either in the contralateral dLGN or in the ipsilateral and contralateral MGN. No differences either in the distribution or in the density of labeled neurons were found between the control and LVD cats.

Immunolabeling to calretinin, parvalbumin and calbindin D-28 K in areas 17 and 18

Four control, four EVD and one LVD cats were processed for the immunohistochemical study. In control cats, the radial distribution of neurons immunoreactive to calretinin, parvalbumin and calbindin D-28 K and the qualitative characteristics of the immunostaining of these markers in cortical areas 17 and 18 were similar to those previously described (Demeulemeester et al., 1991; Hof et al., 1999; Huxlin and Pasternak, 2001). Density of immunoreactive neurons in areas 17 and 18 of the LVD cat was not significantly different from that in the control cats.

In both control and deprived cats (LVD and EVD), calretinin immunolabeling was observed in nonpyramidal neurons, commonly with features of bipolar neuron such as long, vertically oriented processes (Figs. 10A–D). Labeling of processes and terminal-like puncta was not homogeneous across cortical layers. In particular, in layer IV and upper half of layer V, numerous puncta were present in the neuropil and around unstained cell somata, but no basket formations were found (Figs. 10C, D). In both control and EVD cats, calretinin-immunoreactive neurons were found in all cortical layers with a higher density in layers I–IV (Figs. 10A, B). However, the density of positive neurons in these areas was lower in EVD cats: in area 17 of the EVD cats, there was a 57.8% reduction in positive calretinin neurons with respect to controls and a reduction of 28.1% in area 18 (Fig. 11B).

In all experimental groups, parvalbumin immunolabeling revealed terminal-like puncta and processes in all cortical layers, except layer I (Figs. 10E–H). One of the most conspicuous characteristics was the labeling of numerous terminal-like

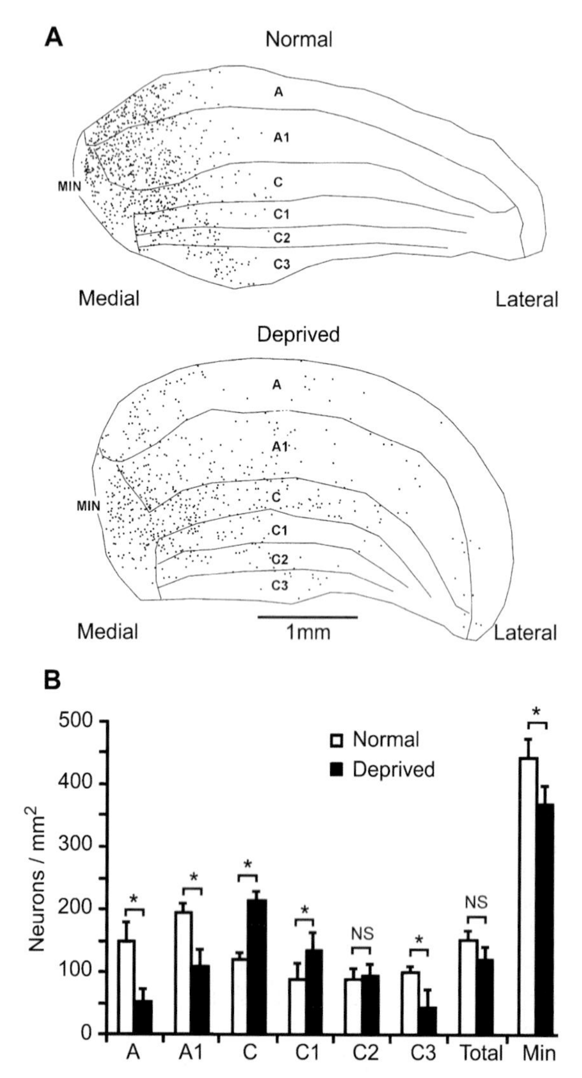

Fig. 9. Distribution of retrogradely labeled neurons in dLGN and MIN in control and EVD cats. (A) In both control and EVD cats, labeled neurons were found in all laminae of the dLGN and MIN. However, in control cats, the density of labeled neurons in the medial half of the dLGN was higher than that in EVD cats, in which labeled neurons had a more widespread distribution. The borders between laminae are indicated. (B) Histograms representing the density of retrogradely labeled neurons in control and EVD cats in different laminae of the dLGN and MIN. Vertical lines represent standard deviations. (*): significant differences ($p<0.05$); ns: nonsignificant differences.

puncta around unstained neuronal somata in layers II–III and especially in layer V, where many large unstained pyramidal neurons were seen surrounded by a very dense plexus of parvalbumin immunoreactive terminal-like puncta (Figs. 10G, H). These perisomatic plexi were similar to the parvalbumin immunoreactive basket formations described in the primate neocortex (i.e. Hendry et al., 1989; Akil and Lewis, 1992). In layers IV and VI, perisomatic terminals were also seen around unlabeled cells but they were fewer: dense basket formations were not seen in these layers. Parvalbumin immunolabeling also marked many positive neuronal somata resembling spiny multipolar interneurons. The size and shape of the labeled neurons were variable, but the majority displayed an ovoid cell body (Figs. 10G, H). In all experimental groups, parvalbumin immunoreactive neurons were found in all cortical layers, except layer I, with a higher density in layer IV (Figs. 10E, F and 11A). As with calretinin immunolabeling, the density of parvalbumin immunoreactive neurons in areas 17 and 18 was lower in EVD cats. In area 17 of deprived cats, a 36.2% reduction was found with respect to controls, and a 22.2% reduction was observed in area 18 (Fig. 11B).

The calbindin D-28 K immunolabeling did not reveal any difference — neither qualitative nor quantitative — between control and both groups of deprived cats (Figs. 10I–L and 11B). In both control and deprived cats, calbindin D-28 K immunoreactivity was observed mainly in cell bodies and dendrites (Fig. 10K–L). Labeling of axonal plexuses and terminal-like buttons was scarce and mostly located in layers II–III, IV and V. In both groups of cats, calbindin D-28K immunoreactive neurons were present in all layers, but the intensity of cell bodies staining varied from light to dark: lightly stained cells were predominant in layers II and III, and cells were more darkly stained in layers IV–VI. The morphology of stained cells varied; many lightly stained cells of layers II–III could be identified as pyramidal neurons, whereas the vast majority of darkly stained cells of layers II–VI were nonpyramidal cells with rounded or fusiform somata (Fig. 11K, L).

Discussion

This study is the first to our knowledge to describe synaptic responses to sound stimuli in areas 17 and border 17/18 of the cat visual cortex. Suprathreshold

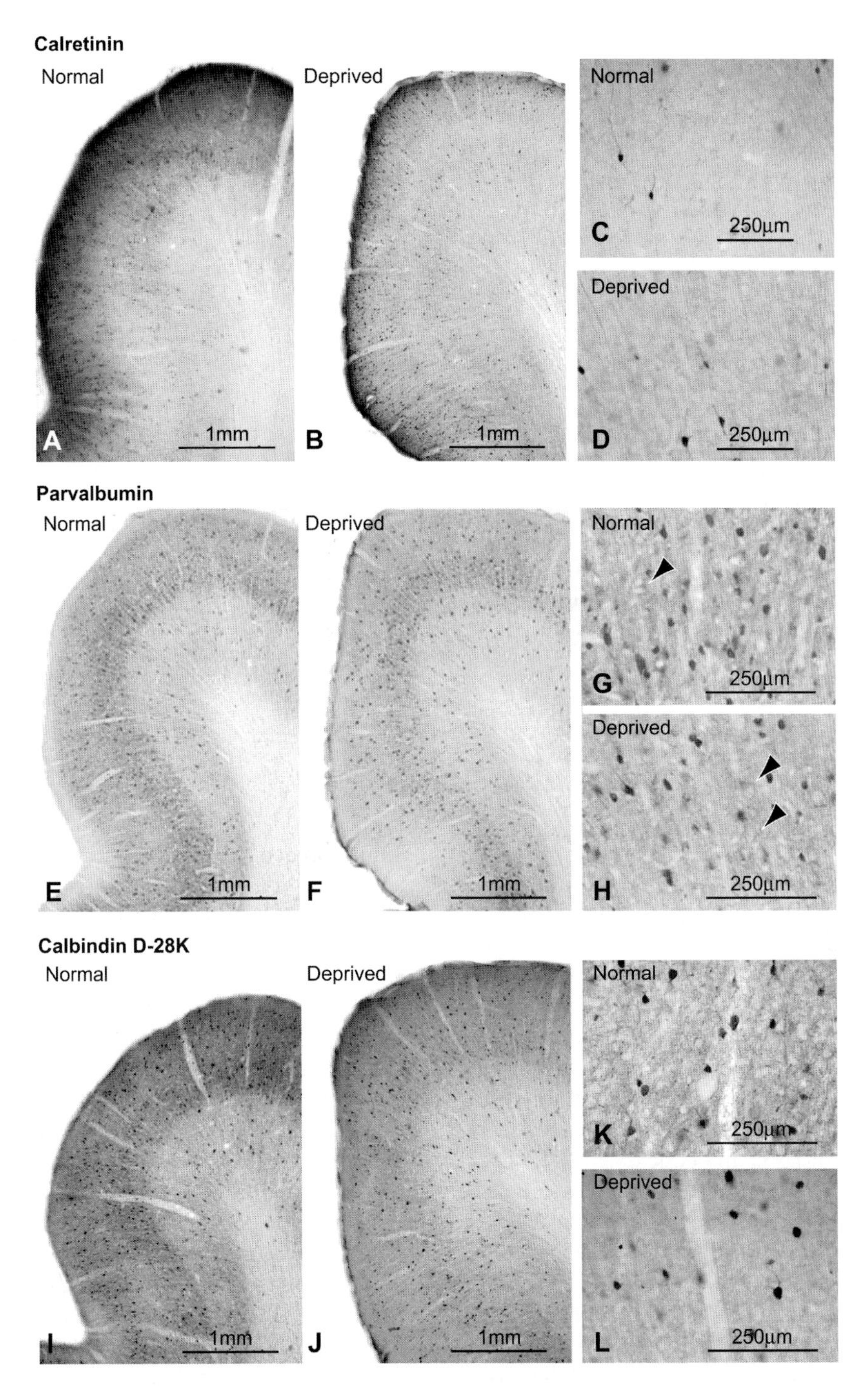

Fig. 10. Photomicrographs through the area 17 stained immunocytochemically for calretinin, parvalbumin and calbindin D-28K in control and EVD cats. (A–D) Calretinin immunostaining in control (A and C) and EVD (B and D) cats. Note that the shape of immunostained neurons in EVD (C) cats is similar to controls (D). (E–H) Parvalbumin immunostaining in control (E and G) and EVD (F and H) cats. G and H are higher-magnification photomicrographs showing immunoreactive neurons and perisomatic puncta (arrowheads) in layer V. The shape of immunoreactive neurons, processes and puncta were similar in EVD (G) and control (H) cats. (I–L) Calbindin D-28K immunostaining in control (I and K) and EVD (J and L) cats. Several immunoreactive nonpyramidal cells in control (K) and EVD (L) cats are shown. In both groups of cats, the morphology of immunoreactive neurons is similar.

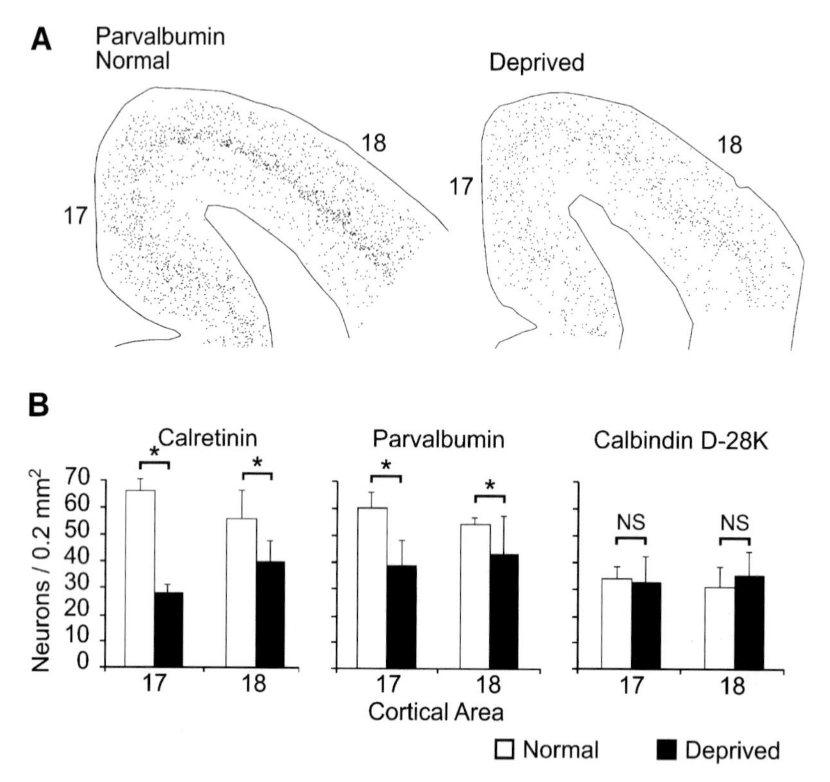

Fig. 11. Distribution of calretinin and parvalbumin-labeled neurons (dots) in areas 17 and 18 of control and EVD cats. (A) In both in control and EVD cats, calretinin immunoreactive neurons were mostly found in cortical supragranular layers, while the density of parvalbumin immunoreactive neurons peaked in layer IV. (B) Frequency histograms representing the density of calretinin, parvalbumin and calbindin D-28K immunoreactive neurons in areas 17 and 18 of both control and EVD cats. Note that in both areas 17 and 18, the density of calretinin and parvalbumin decreased in EVD compared to control cats. The density of calbindin D-28K immunoreactive neurons was similar in both groups. (*): significant differences ($p<0.05$); ns: nonsignificant differences.

responses to sound were recorded as well, with the same time course as the subthreshold ones. Responses to sound in visual cortex were only recorded in EVD cats. Although a similar number of neurons were recorded in EVD ($n = 56$) and LVD ($n = 60$) cats, 14.3% of the neurons responded to sound in the visual cortex of EVD and none in either LVD or control cats. This observation is in agreement with most of the literature, where early blindness is reported to convey larger crossmodal plasticity than late-onset blindness. The importance of early age loss of sight is described in a large number of studies (i.e. Cohen et al., 1999; Gougoux et al., 2004, 2005). When tested, auditory responses did not show obvious frequency tuning. In spite of this poor selectivity, these responses did not seem to be due to an unspecific awakening effect: they were of short latency, tightly time-locked to the stimuli, and were never observed in control or LVD animals examined under the same experimental conditions as the EVD ones.

The synaptic potentials evoked by sound in visual cortex must be at least in part excitatory as they were able to induce spike responses. It is also probable that disynaptic inhibition could be evoked by the auditory stimuli — although we failed to detect it — since one of the recorded cells that responded to sound showed the characteristic firing properties of fast-spiking neurons (Fig. 1B) (Nowak et al., 2003; Descalzo et al., 2005), which are known to be inhibitory (McCormick et al., 1985).

The average amplitude of synaptic potentials evoked by sound was small if compared to the ones evoked with visual stimulation. In another study (Nowak et al., 2005a), synaptic response obtained in response to flashing bars, which is not the optimal stimulus to activate area 17 neurons, presented a mean amplitude of near 4.5 mV, that is, an amplitude 7.5 times larger than the depolarization produced by the auditory stimuli. A comparable difference is also observed for the spiking response which is about 30 spikes/s with flashing bars (Nowak et al., 2005a), approximately 10 times larger than the response obtained here with auditory stimulation. Nevertheless, although synaptic responses were of small amplitude, they were sufficient to modulate the firing rate.

Onset response latencies in response to auditory stimuli, on the other hand, appear to be surprisingly short. The mean synaptic onset latency to auditory stimulation obtained in the present study is 30 ms, which is actually lower than the mean synaptic onset latency of the response to flashing bars, which was on an average 36 ms in Nowak et al. (unpublished results); see also Creutzfeldt and Ito (1968). Similarly, onset latency for the spiking response in visual cortex appears to be similar in response to auditory stimuli (42 ms) compared to visual stimuli (Best et al., 1986; Eschweiler and Rauschecker, 1993; Nowak et al., 2005b) in normal cats. In addition, it has been shown that, in dark-reared rats, visual responses have significantly longer latencies than in normal rats (Benevento et al., 1992). This suggests that, had we tested visual response latencies in the EVD cats, the difference with those obtained with auditory stimuli could have been even larger.

In the primary auditory area of normal cats, onset latencies to sound for both synaptic and spike responses are much shorter (between 10 and 20 ms on an average) in comparison to response latencies to visual stimuli in primary visual cortex (De Ribaupierre et al., 1972; Phillips and Irvine, 1981; Heil and Irvine, 1996; Mendelson et al., 1997; Ojima and Murakami, 2002). This would explain that, if the responses to sound in primary visual cortex in EVD cats originate in association areas, the latency of these responses could still have the same or even shorter latencies than the ones to visual stimuli. This could also be the case if the origin of the responses would be the inferior colliculus, where latencies of the responses to sound are 5–18 ms (Langner and Schreiner, 1988) and it is one of the possible origins that we have not ruled out anatomically (see below).

In this study we did not obtain auditory responses in area 17 of normal and LVD cats. Similarly, Yaka et al. (1999, 2000) could not evoke sound responses in the visual cortex of control cats. This result is at variance with a number of studies from the 1960s and 1970s that reported auditory, somatosensory and noxious responses through either single unit or evoked potential recordings in areas 17, 18 and 19 of normal cats (Jung et al., 1963; Murata et al., 1965; Buser and Bignall, 1967; Spinelli et al., 1968; Morrell, 1972; Fishman and Michael, 1973). In single-unit recordings, 28–47% of the cells responded to non-visual stimuli. However, it is to be noticed that, in contrast to our study and those by Yaka et al., all these previous studies had in common the use of slightly or unanesthetized cats. It has thus been argued that some of these non-visual responses actually corresponded to unspecific awakening reaction to noxious or loud auditory stimuli. This argument was based on the fact that these responses had long latencies (Jung et al., 1963; Murata et al., 1965; Buser and Bignall, 1967), that they were similar to responses evoked by the brainstem reticular formation (Buser and Bignall, 1967), that they induced increases in blood pressure (Murata et al., 1965) and that they were suppressed by doses of anesthetics that coincided with that required to suppress arousal (Murata et al., 1965; Buser and Bignall, 1967). On the other hand, some of these studies (Spinelli et al., 1968; Morrell, 1972; Fishman and Michael, 1973) reported auditory responses in visual cortex that were sufficiently well tuned, either in the spatial or in the frequency domain, to rule out an unspecific awakening effect. Yet even in these cases, the spike latencies reported for individual examples (60 ms in Spinelli et al., 1968; 130 ms in Fishman and Michael, 1973) appeared to be longer than those reported here (42 ms). This suggests that those responses were induced by a more indirect pathway than the one responsible for the response that is described in the current study, a pathway that, in addition, would be gated by the arousal state of the animal. In light of these different results, it is clear that further refinement in the study of crossmodal plasticity would benefit from the examination of the consequences of visual deprivation on multisensory responses in the visual cortex of awake animals.

In superior colliculus, visual deprivation since birth leads to an increase in the number of neurons that respond to sound stimuli (Vidyasagar, 1978; Rhoades, 1980; Rauschecker and Harris, 1983). Neurons responding to auditory stimuli are even found in the superficial layers of the superior colliculus (Rhoades, 1980; Rauschecker and Harris, 1983), which in normal animals are exclusively visual. The main input sources to the superficial layers of the superior colliculus are the retina and the visual cortex — mostly area 17 and surrounding areas. It is therefore possible that some of the cells responding to auditory stimuli in the superficial layers of the superior colliculus were actually driven by inputs from the visual cortex.

Finally, our observation on the presence of auditory responses in the visual cortex of cats deprived early of sight is in agreement with most of the literature in humans, where early blindness is reported to convey larger crossmodal plasticity than late-onset blindness (i.e. Cohen et al., 1999; Gougoux et al., 2004, 2005). The functional meaning of the many fMRI experiments that have been performed in the study of crossmodal plasticity depends on the precise sources of the signal that was recorded. In light of this important question, our study unambiguously demonstrates that both the input to (synaptic response) and the output from (spiking responses) areas 17 and 18 can be activated by sound stimulation in the visual cortex of animals deprived of vision at an early age.

Origin of the auditory connections generating responses to sound in primary visual cortex

The retrograde tracing experiments were aimed at identifying which connections could have been modified by the visual deprivation and whether this modification could explain the appearance of auditory responses in areas 17 and 18.

Thalamocortical connections in EVD cats did not appear different from those observed in normal cats: retrogradely labeled cells were found in structures that are devoted to the processing of visual information (LGN), but not in nuclei involved in auditory information processing (e.g. MGN). This does not rule out the possibility that the visual cortex of EVD cats received auditory information through a newly formed pathway between auditory and visual subcortical nuclei. For example, Izraeli et al. reported a projection from the inferior colliculus onto the LGN in hamsters that had been binocularly enucleated after birth (Izraeli et al., 2002). Another example of changes in the connections between visual and auditory subcortical nuclei due to visual deprivation is provided by phylogenetic evolution. In the blind mole rat it has been described that thalamocortical connections originated in the LGN are preserved; however, the LGN receives projections from the inferior colliculus, which may underlie the reported responses to sound in visual cortex in that species (Bronchti et al., 2002). It remains to be established whether derivations of a similar kind would also occur after visual deprivation in cats with an intact visual system.

An alternative pathway is one involving the lateral posterior thalamic (LP) complex. Negyessy et al. (2000) have shown that, in enucleated rats, crossmodal plasticity may be achieved through a cortico-thalamo-cortical pathway. In normal rats, the LP nucleus is reciprocally connected with the visual cortex. However, in enucleated rats, the LP nucleus receives an additional heteromodal projection from the somatosensory cortex. The LP might, in addition, receive inputs from the superficial layers of the superior colliculus, where neurons are found to respond to nonvisual stimuli after early onset visual deprivation (Vidyasagar, 1978; Rhoades, 1980; Rauschecker and Harris, 1983).

Another candidate for explaining the presence of auditory responses in areas 17 and 18 of EVD cats could have been the direct connections linking auditory and visual cortex. It has been shown that, in the newborn kitten, the auditory cortex projects onto areas 17 and 18 (Dehay et al., 1984; Innocenti and Clarke, 1984; Innocenti et al., 1988; Innocenti and Berbel, 1991). However, this connection disappears completely after about 1 month of age, except for very few scattered neurons found in layers 5 and 6 of auditory cortex (Dehay et al., 1984; Innocenti and Clarke, 1984; Innocenti et al., 1988; Innocenti and Berbel, 1991). It seems reasonable to think that visual deprivation could result in a stabilization of this connection. However, in this study we found that the number of cells labeled in the auditory cortex of EVD cats did not differ from the one found in the adult. This is consistent with the observation that even bilateral enucleation fails to stabilize this transitory connection (Innocenti and Clarke, 1984; Innocenti et al., 1988; Innocenti and Berbel, 1991).

Evoked potentials' studies in normal humans have disclosed, under certain conditions, the occurrence of interactions between auditory and visual stimuli at locations corresponding to the primary visual cortex (Giard and Peronnet, 1999; Shams et al., 2001; Arden et al., 2003). These interactions are highly non-linear, in the sense that auditory stimuli alone do not produce detectable activation but are able to modulate the visually evoked potentials. Modulation of visual responses by auditory inputs might be related to the presence of a direct input, on area V1 of the primate, from the primary auditory cortex, as well as from polymodal areas (Falchier et al., 2002). This projection is weak on the part of area V1 that represents the central visual field, but it is far from being negligible when considering the representation of the peripheral visual field, where it has been shown to be as strong as the projection from the visual areas of the superior temporal sulcus (Falchier et al., 2002). Recent studies also revealed auditory-visual interaction at the single-cell level in area V1 of awake behaving monkeys (Wang et al., 2005). Nevertheless, this direct auditory to visual cortex pathway might be specific to primates: in cat visual cortex no such connection has been demonstrated. As mentioned above, retrograde tracer injection in areas 17, 18 and 19 reveals only a very small number of cells in the auditory areas (Dehay et al., 1984, 1988; Innocenti and Clarke, 1984; Innocenti et al., 1988). In addition, anterograde tracer injection in auditory cortex fails to reveal labeled fibers in the visual cortex of normal adult cats (Dehay et al., 1984, 1988; Innocenti et al., 1988).

We found that the main effect of dark rearing on corticocortical connections appears in area PMLS and PLLS. In these areas, dark rearing induced a remodeling of the laminar distribution of projecting neurons to areas 17 and 18. Thus, the projection from PMLS onto areas 17 and 18 in normal cats originates mostly from neurons in the infragranular layers (70–80%) (Bullier et al., 1984b; Symonds and Rosenquist, 1984; Rosenquist, 1985; Segraves and Innocenti, 1985; Kato et al., 1991; Shipp and Grant, 1991; Einstein, 1996; Batardière et al., 1998; Payne and Lomber, 2003). The connections from PLLS to areas 17 and 18 (a weak projection in normal cats) originate as well almost exclusively from infragranular layers (Symonds and Rosenquist, 1984; Rosenquist, 1985; Segraves and Innocenti, 1985; Payne and Lomber, 2003). After EVD, we have observed a change in the balance of the projections, such that the proportion of connections originated in supragranular layers significantly increases.

Is it possible that these changes in connectivity pattern were responsible for the presence of auditory responses in the visual cortex. PMLS is strongly connected with area AMLS (Symonds and Rosenquist, 1984; Hilgetag and Grant, 2000). AMLS is an area in which neurons appear to respond to both auditory and visual stimuli (Yaka et al., 1999) and where visual deprivation further increases the number of cells responding to auditory stimulation (Yaka et al., 1999). Thus, one possible pathway could involve the increased auditory activity in AMLS together with a remodeling of the projections from PMLS on areas 17 and 18.

Decreased inhibition in visually deprived cats

In this study we have described that EVD cats had a significantly lower number of parvalbumin- and calretinin-positive neurons in areas 17 and 18 than control cats. A general decrease in GAD-positive neurons has also been described in dark reared rats (Benevento et al., 1995; also see Mower and Guo, 2001). Although the number of calbindin-D28K-positive neurons, and probably other subsets of GABAergic neurons remained unchanged, the decrease of parvalbumin- and calretinin-positive neurons found in area 17 is very suggestive of a decreased inhibition in the visual cortex of EVD cats. Indeed, the decrease of parvalbumine positive (PV)(+) perisomatic synapse formation has been found to decrease with visual deprivation in the mouse (Chattopadhyaya et al., 2004). The functional implications of this decrease are the ones resulting from a decreased inhibition. Different findings reported in the literature are compatible with a decreased inhibition following visual deprivation or dark rearing. Thus, the occipital cortex of humans or animals that have lacked visual input in early ages often presents higher activity than the one in nonvisually deprived subjects. At a cellular level, significantly higher spontaneous activity has been reported in dark reared rats (Benevento et al., 1992; Maffei et al., 2004) and in enucleated hamsters (Izraeli et al., 2002). In the cat, numerous changes in visual responses have been described as a result of dark rearing, such as a lack of sharp inhibitory sidebands and the sometimes exceedingly large size of the receptive fields in areas 17 and 18, which could be due as well to a decreased intracortical inhibition (Singer and Tretter, 1976). A recent characterization of the development of GABAergic transmission in the visual cortex found that inhibition received by layer 2/3 neurons in the rat visual cortex increases during the critical period, but only if there is an exposure to visual stimulation (Morales et al., 2002). Furthermore, only 2 days of visual deprivation profoundly alter the excitatory — inhibitory balance, not only decreasing inhibition but also potentiating excitation (Maffei et al., 2004). In humans, glucose metabolism in striate and prestriate cortical areas was found to be higher in early blind human subjects than in normal-sighted persons (Wanet-Defalque et al., 1988), not only during the realization of tactile and auditory tasks but even at rest. Likewise, studies with functional imaging revealed an increased activity in occipital cortex of the blind that was as well task-independent (Uhl et al., 1993).

The decrease of inhibition that results from visual deprivation has a deep impact on the structure and functionality of the developing cortical circuits, generating an excitation/inhibition disbalance. To what extent one of the functional effects of a decreased inhibition is an increased permissibility for crossmodal plasticity remains to be demonstrated.

Abbreviations

ALLS	anterolateral lateral suprasylvian area
AMLS	anteromedial lateral suprasylvian area
CO	cytochrome oxidase
dLGN	dorsal lateral geniculate nucleus
EVD	early visually deprived
GABA	gamma-aminobutyric acid
GAD	glutamic acid decarboxilase
LVD	late visually deprived
MGN	medial geniculate nucleus
MIN	medial interlaminar nucleus (of dLGN)
PLLS	posterolateral lateral suprasylvian visual area
PMLS	posteromedial lateral suprasylvian visual area
WGA-HRP	wheat germ agglutinin coupled to horseradish peroxidase

Acknowledgments

This work has been sponsored by MCYT (BFI2002-03643) and the European Union IST/FET project PRESENCIA (IST-2001-37927) to MVSV and by Instituto de Salud Carlos III PI03/0576 and Generalitat Valènciana GRUPOS03/053 to PB. LGN was supported by the CNRS and by the Ministry of Foreign Affairs (Egide, "Picasso" and CNRS-CSIC joint travel grants). Thanks to Pascal Barone for critical reading of the manuscript. Thank you to M.S. Malmierca and his team for his generous help for the implementation of the auditory stimulation and to J. Merchán for the equipment to do auditory stimulation. We thank the personnel of the animal facilities at the UMH for their effort to keep the visually deprived animals, and in particular the chief veterinarian JA Pérez de Gracia.

References

Akil, M. and Lewis, D.A. (1992) Differential distribution of parvalbumin-immunoreactive pericellular clusters of terminal boutons in developing and adult monkey neocortex. Exp. Neurol., 115: 239–249.

Arden, G.B., Wolf, J.E. and Messiter, C. (2003) Electrical activity in visual cortex associated with combined auditory and visual stimulation in temporal sequences known to be associated with a visual illusion. Vision Res., 43: 2469–2478.

Ashmead, D.H., Wall, R.S., Ebinger, K.A., Eaton, S.B., Snook-Hill, M.M. and Yang, X. (1998) Spatial hearing in children with visual disabilities. Perception, 27: 105–122.

Batardière, A., Barone, P., Dehay, C. and Kennedy, H. (1998) Area-specific laminar distribution of cortical feedback neurons projecting to cat area 17: quantitative analysis in the adult and during ontogeny. J. Comp. Neurol., 396: 493–510.

Benevento, L.A., Bakkum, B.W. and Cohen, R.S. (1995) gamma-Aminobutyric acid and somatostatin immunoreactivity in the visual cortex of normal and dark-reared rats. Brain Res., 689: 172–182.

Benevento, L.A., Bakkum, B.W., Port, J.D. and Cohen, R.S. (1992) The effects of dark-rearing on the electrophysiology of the rat visual cortex. Brain Res., 572: 198–207.

Best, J., Reuss, S. and Dinse, H.R. (1986) Lamina-specific differences of visual latencies following photic stimulation in the cat striate cortex. Brain Res., 385: 356–360.

Bronchti, G., Heil, P., Sadka, R., Hess, A., Scheich, H. and Wollberg, Z. (2002) Auditory activation of "visual" cortical areas in the blind mole rat (*Spalax ehrenbergi*). Eur. J. Neurosci., 16: 311–329.

Bullier, J., Kennedy, H. and Salinger, W. (1984a) Bifurcation of subcortical afferents to visual areas 17, 18, and 19 in the cat cortex. J. Comp. Neurol., 228: 309–328.

Bullier, J., Kennedy, H. and Salinger, W. (1984b) Branching and laminar origin of projections between visual cortical areas in the cat. J. Comp. Neurol., 228: 329–341.

Buser, P. and Bignall, K.E. (1967) Nonprimary sensory projections on the cat neocortex. Int. Rev. Neurobiol., 10: 111–165.

Celio, M.R. (1990) Calbindin D-28k and parvalbumin in the rat nervous system. Neuroscience, 35: 375–475.

Celio, M.R., Baier, W., Scharer, L., Gregersen, H.J., de Viragh, P.A. and Norman, A.W. (1990) Monoclonal antibodies directed against the calcium binding protein Calbindin D-28k. Cell Calcium, 11: 599–602.

Chattopadhyaya, B., Di Cristo, G., Higashiyama, H., Knott, G.W., Kuhlman, S.J., Welker, E. and Huang, Z.J. (2004) Experience and activity-dependent maturation of perisomatic GABAergic innervation in primary visual cortex during a postnatal critical period. J. Neurosci., 24: 9598–9611.

Cohen, L.G., Celnik, P., Pascual-Leone, A., Corwell, B., Falz, L., Dambrosia, J., Honda, M., Sadato, N., Gerloff, C., Catala, M.D. and Hallett, M. (1997) Functional relevance of cross-modal plasticity in blind humans. Nature, 389: 180–183.

Cohen, L.G., Weeks, R.A., Sadato, N., Celnik, P., Ishii, K. and Hallett, M. (1999) Period of susceptibility for cross-modal plasticity in the blind. Ann. Neurol., 45: 451–460.

Creutzfeldt, O. and Ito, M. (1968) Functional synaptic organization of primary visual cortex neurones in the cat. Exp. Brain Res., 6: 324–352.

De Ribaupierre, F., Goldstein Jr., M.H. and Yeni-Komshian, G. (1972) Intracellular study of the cat's primary auditory cortex. Brain Res., 48: 185–204.

DeFelipe, J. (1997) Types of neurons, synaptic connections and chemical characteristics of cells immunoreactive for calbindin-D28K, parvalbumin and calretinin in the neocortex. J. Chem. Neuroanat., 14: 1–19.

Dehay, C., Bullier, J. and Kennedy, H. (1984) Transient projections from the fronto-parietal and temporal cortex to areas 17, 18 and 19 in the kitten. Exp. Brain Res., 57: 208–212.

Dehay, C., Kennedy, H. and Bullier, J. (1988) Characterization of transient cortical projections from auditory, somatosensory, and motor cortices to visual areas 17, 18, and 19 in the kitten. J. Comp. Neurol., 272: 68–89.

Demeulemeester, H., Arckens, L., Vandesande, F., Orban, G.A., Heizmann, C.W. and Pochet, R. (1991) Calcium binding proteins and neuropeptides as molecular markers of GABAergic interneurons in the cat visual cortex. Exp. Brain Res., 84: 538–544.

Descalzo, V.F., Nowak, L.G., Brumberg, J.C., McCormick, D.A. and Sanchez-Vives, M.V. (2005) Slow adaptation in fast spiking neurons of visual cortex. J. Neurophysiol., 93: 1111–1118.

Einstein, G. (1996) Reciprocal projections of cat extrastriate cortex: I. Distribution and morphology of neurons projecting from posterior medial lateral suprasylvian sulcus to area 17. J. Comp. Neurol., 376: 518–529.

Eschweiler, G.W. and Rauschecker, J.P. (1993) Temporal integration in visual cortex of cats with surgically induced strabismus. Eur. J. Neurosci., 5: 1501–1519.

Falchier, A., Clavagnier, S., Barone, P. and Kennedy, H. (2002) Anatomical evidence of multimodal integration in primate striate cortex. J. Neurosci., 22: 5749–5759.

Fishman, M.C. and Michael, P. (1973) Integration of auditory information in the cat's visual cortex. Vision Res., 13: 1415–1419.

Fregnac, Y. and Imbert, M. (1984) Development of neuronal selectivity in primary visual cortex of cat. Physiol. Rev., 64: 325–434.

Giard, M.H. and Peronnet, F. (1999) Auditory-visual integration during multimodal object recognition in humans: a behavioral and electrophysiological study. J. Cogn. Neurosci., 11: 473–490.

Gougoux, F., Lepore, F., Lassonde, M., Voss, P., Zatorre, R.J. and Belin, P. (2004) Neuropsychology: pitch discrimination in the early blind. Nature, 430: 309.

Gougoux, F., Zatorre, R.J., Lassonde, M., Voss, P. and Lepore, F. (2005) A functional neuroimaging study of sound localization: visual cortex activity predicts performance in early blind individuals. PLoS Biol., 3: e27.

Heil, P. and Irvine, D.R. (1996) On determinants of first-spike latency in auditory cortex. Neuroreport, 7: 3073–3076.

Hendry, S.H., Jones, E.G., Emson, P.C., Lawson, D.E., Heizmann, C.W. and Streit, P. (1989) Two classes of cortical GABA neurons defined by differential calcium binding protein immunoreactivities. Exp. Brain Res., 76: 467–472.

Hilgetag, C.C. and Grant, S. (2000) Uniformity, specificity and variability of corticocortical connectivity. Philos. Trans. R. Soc. Lond. B Biol. Sci., 355: 7–20.

Hof, P.R., Glezer, II., Conde, F., Flagg, R.A., Rubin, M.B., Nimchinsky, E.A. and Vogt Weisenhorn, D.M. (1999) Cellular distribution of the calcium-binding proteins parvalbumin, calbindin, and calretinin in the neocortex of mammals: phylogenetic and developmental patterns. J. Chem. Neuroanat., 16: 77–116.

Hollander, H. and Vanegas, H. (1977) The projection from the lateral geniculate nucleus onto the visual cortex in the cat. A quantitative study with horseradish-peroxidase. J. Comp. Neurol., 173: 519–536.

Hubel, D.H., Wiesel, T.N. and LeVay, S. (1977) Plasticity of ocular dominance columns in monkey striate cortex. Philos. Trans. R. Soc. Lond. B Biol. Sci., 278: 377–409.

Huxlin, K.R. and Pasternak, T. (2001) Long-term neurochemical changes after visual cortical lesions in the adult cat. J. Comp. Neurol., 429: 221–241.

Hyvarinen, J., Carlson, S. and Hyvarinen, L. (1981) Early visual deprivation alters modality of neuronal responses in area 19 of monkey cortex. Neurosci. Lett., 26: 239–243.

Innocenti, G.M. and Berbel, P. (1991) Analysis of an experimental cortical network: II. Connections of visual areas 17 and 18 after neonatal injections of ibotenic acid. J. Neural. Transplant., 2: 29–54.

Innocenti, G.M., Berbel, P. and Clarke, S. (1988) Development of projections from auditory to visual areas in the cat. J. Comp. Neurol., 272: 242–259.

Innocenti, G.M. and Clarke, S. (1984) Bilateral transitory projection to visual areas from auditory cortex in kittens. Brain Res., 316: 143–148.

Izraeli, R., Koay, G., Lamish, M., Heicklen-Klein, A.J., Heffner, H.E., Heffner, R.S. and Wollberg, Z. (2002) Cross-modal neuroplasticity in neonatally enucleated hamsters: structure, electrophysiology and behaviour. Eur. J. Neurosci., 15: 693–712.

Jung, B., Kornhuber, H. and Da Fonseca, J. (1963) Multisensory convergence on cortical neurons. Neuronal effects of visual, acoustic and vestibular stimuli in the superior convolutions of the cat's cortex. In: Moruzzi, G., Fessard, A. and Jasper, H.H. (Eds.), Brain Mechanisms, Progress in Brain Research, Vol. 1. Elsevier, Amsterdam, pp. 240–270.

Kato, N., Ferrer, J.M. and Price, D.J. (1991) Regressive changes among corticocortical neurons projecting from the lateral suprasylvian cortex to area 18 of the kitten's visual cortex. Neuroscience, 43: 291–306.

Keller, G. and Innocenti, G.M. (1981) Callosal connections of suprasylvian visual areas in the cat. Neuroscience, 6: 703–712.

Korte, M. and Rauschecker, J.P. (1993) Auditory spatial tuning of cortical neurons is sharpened in cats with early blindness. J. Neurophysiol., 70: 1717–1721.

Kujala, T., Alho, K., Huotilainen, M., Ilmoniemi, R.J., Lehtokoski, A., Leinonen, A., Rinne, T., Salonen, O., Sinkkonen, J., Standertskjold-Nordenstam, C.G. and Naatanen, R. (1997) Electrophysiological evidence for cross-modal plasticity in humans with early and late-onset blindness. Psychophysiology, 34: 213–216.

Kujala, T., Huotilainen, M., Sinkkonen, J., Ahonen, A.I., Alho, K., Hamalainen, M.S., Ilmoniemi, R.J., Kajola, M., Knuutila, J.E., Lavikainen, J., et al. (1995) Visual cortex activation in blind humans during sound discrimination. Neurosci. Lett., 183: 143–146.

Langner, G. and Schreiner, C.E. (1988) Periodicity coding in the inferior colliculus of the cat. I. Neuronal mechanisms. J. Neurophysiol., 60: 1799–1822.

Lessard, N., Pare, M., Lepore, F. and Lassonde, M. (1998) Early blind human subjects localize sound sources better than sighted subjects. Nature, 395: 278–280.

Maciewicz, R.J. (1975) Thalamic afferents to areas 17, 18 and 19 of cat cortex traced with horseradish peroxidase. Brain Res., 84: 308–312.

Maffei, A., Nelson, S.B. and Turrigiano, G.G. (2004) Selective reconsfiguration of layer 4 visual cortical circuitry by visual deprivation. Nat. Neurosci., 7(12): 1353–1359.

McCormick, D.A., Connors, B.W., Lighthall, J.W. and Prince, D.A. (1985) Comparative electrophysiology of pyramidal and sparsely spiny stellate neurons of the neocortex. J. Neurophysiol., 54: 782–806.

Mendelson, J.R., Schreiner, C.E. and Sutter, M.L. (1997) Functional topography of cat primary auditory cortex: response latencies. J. Comp. Physiol. [A], 181: 615–633.

Merabet, L., Thut, G., Murray, B., Andrews, J., Hsiao, S. and Pascual-Leone, A. (2004) Feeling by sight or seeing by touch? Neuron, 42: 173–179.

Mesulam, M.M. (1978) Tetramethyl benzidine for horseradish peroxidase neurohistochemistry: a non-carcinogenic blue reaction product with superior sensitivity for visualizing neural afferents and efferents. J. Histochem. Cytochem., 26: 106–117.

Morales, B., Choi, S.Y. and Kirkwood, A. (2002) Dark rearing alters the development of GABAergic transmission in visual cortex. J. Neurosci., 22: 8084–8090.

Morrell, F. (1972) Visual system's view of acoustic space. Nature, 238: 44–46.

Mower, G.D. and Guo, Y. (2001) Comparison of the expression of two forms of glutamic acid decarboxylase (GAD67 and GAD65) in the visual cortex of normal and dark-reared cats. Brain Res. Dev. Brain Res., 126: 65–74.

Murata, K., Cramer, H. and Bach-y-Rita, P. (1965) Neuronal convergence of noxious, acoustic, and visual stimuli in the visual cortex of the cat. J. Neurophysiol., 28: 1223–1239.

Negyessy, L., Gal, V., Farkas, T. and Toldi, J. (2000) Cross-modal plasticity of the corticothalamic circuits in rats enucleated on the first postnatal day. Eur. J. Neurosci., 12: 1654–1668.

Nowak, L.G., Azouz, R., Sanchez-Vives, M.V., Gray, C.M. and McCormick, D.A. (2003) Electrophysiological classes of cat primary visual cortical neurons *in vivo* as revealed by quantitative analyses. J. Neurophysiol., 89: 1541–1566.

Nowak, L.G., Sanchez-Vives, M.V. and McCormick, D.A. (2005a) Role of synaptic and intrinsic membrane properties

in short term receptive field dynamics in cat area 17. J. Neurosci., 25: 1866–1880.
Ojima, H. and Murakami, K. (2002) Intracellular characterization of suppressive responses in supragranular pyramidal neurons of cat primary auditory cortex *in vivo*. Cereb. Cortex, 12: 1079–1091.
Payne, B.R. and Lomber, S.G. (2003) Quantitative analyses of principal and secondary compound parieto-occipital feedback pathways in cat. Exp. Brain Res., 152: 420–433.
Phillips, D.P. and Irvine, D.R. (1981) Responses of single neurons in physiologically defined primary auditory cortex (AI) of the cat: frequency tuning and responses to intensity. J. Neurophysiol., 45: 48–58.
Rauschecker, J.P. and Harris, L.R. (1983) Auditory compensation of the effects of visual deprivation in the cat's superior colliculus. Exp. Brain Res., 50: 69–83.
Rauschecker, J.P. and Korte, M. (1993) Auditory compensation for early blindness in cat cerebral cortex. J. Neurosci., 13: 4538–4548.
Rees, A., Sarbaz, A., Malmierca, M.S. and Le Beau, F.E. (1997) Regularity of firing of neurons in the inferior colliculus. J. Neurophysiol., 77: 2945–2965.
Rhoades, R.W. (1980) Effects of neonatal enucleation on the functional organization of the superior colliculus in the golden hamster. J. Physiol., 301: 383–399.
Roder, B., Rosler, F. and Spence, C. (2004) Early vision impairs tactile perception in the blind. Curr. Biol., 14: 121–124.
Roder, B., Teder-Salejarvi, W., Sterr, A., Rosler, F., Hillyard, S.A. and Neville, H.J. (1999) Improved auditory spatial tuning in blind humans. Nature, 400: 162–166.
Rosenquist, A.C. (1985) Connections of visual cortical areas in the cat. In: Peters, A. and Jones, E.G. (Eds.), Cerebral Cortex. Plenum, New York, pp. 81–117.
Sadato, N., Pascual-Leone, A., Grafman, J., Ibanez, V., Deiber, M.P., Dold, G. and Hallett, M. (1996) Activation of the primary visual cortex by Braille reading in blind subjects. Nature, 380: 526–528.
Sanchez-Vives, M.V. and McCormick, D.A. (2000) Cellular and network mechanisms of rhythmic recurrent activity in neocortex. Nat. Neurosci., 3: 1027–1034.
Sanchez-Vives, M.V., Nowak, L.G. and McCormick, D.A. (2000a) Membrane mechanisms underlying contrast adaptation in cat area 17 *in vivo*. J. Neurosci., 20: 4267–4285.
Sanchez-Vives, M.V., Nowak, L.G. and McCormick, D.A. (2000b) Membrane mechanisms underlying contrast adaptation in cat area 17 *in vivo*. J. Neurosci., 20: 4267–4285.
Segraves, M.A. and Innocenti, G.M. (1985) Comparison of the distributions of ipsilaterally and contralaterally projecting corticocortical neurons in cat visual cortex using two fluorescent tracers. J. Neurosci., 5: 2107–2118.
Segraves, M.A. and Rosenquist, A.C. (1982) The distribution of the cells of origin of callosal projections in cat visual cortex. J. Neurosci., 2: 1079–1089.
Shams, L., Kamitani, Y., Thompson, S. and Shimojo, S. (2001) Sound alters visual evoked potentials in humans. Neuroreport, 12: 3849–3852.
Sherman, S.M. and Spear, P.D. (1982) Organization of visual pathways in normal and visually deprived cats. Physiol. Rev., 62: 738–855.
Shipp, S. and Grant, S. (1991) Organization of reciprocal connections between area 17 and the lateral suprasylvian area of cat visual cortex. Vis. Neurosci., 6: 339–355.
Singer, W. and Tretter, F. (1976) Unusually large receptive fields in cats with restricted visual experience. Exp. Brain Res., 26: 171–184.
Spear, P.D., Tong, L. and Langsetmo, A. (1978) Striate cortex neurons of binocularly deprived kittens respond to visual stimuli through the closed eyelids. Brain Res., 155: 141–146.
Spinelli, D.N., Starr, A. and Barrett, T.W. (1968) Auditory specificity in unit recordings from cat's visual cortex. Exp. Neurol., 22: 75–84.
Steriade, M., Nunez, A. and Amzica, F. (1993) A novel slow (<1 Hz) oscillation of neocortical neurons *in vivo*: depolarizing and hyperpolarizing components. J. Neurosci., 13: 3252–3265.
Symonds, L.L. and Rosenquist, A.C. (1984) Corticocortical connections among visual areas in the cat. J. Comp. Neurol., 229: 1–38.
Toldi, J., Joo, F., Feher, O. and Wolff, J.R. (1988) Modified distribution patterns of responses in rat visual cortex induced by monocular enucleation. Neuroscience, 24: 59–66.
Toldi, J., Rojik, I. and Feher, O. (1994) Neonatal monocular enucleation-induced cross-modal effects observed in the cortex of adult rat. Neuroscience, 62: 105–114.
Uhl, F., Franzen, P., Podreka, I., Steiner, M. and Deecke, L. (1993) Increased regional cerebral blood flow in inferior occipital cortex and cerebellum of early blind humans. Neurosci. Lett., 150: 162–164.
Vidyasagar, T.R. (1978) Possible plasticity in the rat superior colliculus. Nature, 275: 140–141.
Wanet-Defalque, M.C., Veraart, C., De Volder, A., Metz, R., Michel, C., Dooms, G. and Goffinet, A. (1988) High metabolic activity in the visual cortex of early blind human subjects. Brain Res., 446: 369–373.
Wang, Y., Celebrini, S., Trotter, Y. and Barone, P. (2005) Multisensory integration in the behaving monkey: behavioral analysis and electrophysiological evidence in the primary visual cortex. In: *Society for Neuroscience Vol. Program No. 509.2*. SFN Online, Washington, DC.
Wiesel, T.N. and Hubel, D.H. (1965) Comparison of the effects of unilateral and bilateral eye closure on cortical unit responses in kittens. J. Neurophysiol., 28: 1029–1040.
Wong-Riley, M. (1979) Changes in the visual system of monocularly sutured or enucleated cats demonstrable with cytochrome oxidase histochemistry. Brain Res., 171: 11–28.
Yaka, R., Yinon, U., Rosner, M. and Wollberg, Z. (2000) Pathological and experimentally induced blindness induces auditory activity in the cat primary visual cortex. Exp. Brain Res., 131: 144–148.
Yaka, R., Yinon, U. and Wollberg, Z. (1999) Auditory activation of cortical visual areas in cats after early visual deprivation. Eur. J. Neurosci., 11: 1301–1312.

Martinez-Conde, Macknik, Martinez, Alonso & Tse (Eds.)
Progress in Brain Research, Vol. 155
ISSN 0079-6123

CHAPTER 18

The "other" transformation required for visual–auditory integration: representational format

Kristin Kelly Porter[1] and Jennifer M. Groh[2,*]

[1] *University of Alabama School of Medicine, Birmingham, AL 35294, USA*
[2] *Center for Cognitive Neuroscience, Department of Psychology and Neuroscience, Department of Neurobiology, Duke University, Durham, NC 27708, USA*

Abstract: Multisensory integration of spatial signals requires not only that stimulus locations be encoded in the same spatial reference frame, but also that stimulus locations be encoded in the same representational format. Previous studies have addressed the issue of spatial reference frame, but representational format, particularly for sound location, has been relatively overlooked. We discuss here our recent findings that sound location in the primate inferior colliculus is encoded using a "rate" code, a format that differs from the place code used for representing visual stimulus locations. Possible mechanisms for transforming signals from rate-to-place or place-to-rate coding formats are considered.

Keywords: inferior colliculus; acoustic; auditory; sound localization; sound location; rate code; place code; map; monkey; primate

Introduction

Our visual, auditory, and cutaneous senses can all contribute to the perception of the locations of objects and events, but the reactions to these events will be produced by a common set of motor effectors. For example, you only have one pair of eyes for making an eye movement to look at a mosquito, regardless of whether you detect it by seeing it, hearing its hum, or feeling its bite. In order to make use of common motor output pathways, visual, auditory, and somatosensory signals must undergo transformations into a common encoding scheme. These transformations fall into two categories: coordinate transformations and transformations of representational format. Much previous attention has been devoted to the neural computations for coordinate transformations — the transformation of signals from one spatial reference frame to another (Mays and Sparks, 1980; Andersen and Mountcastle, 1983; Jay and Sparks, 1984, 1987a, b; Andersen et al., 1985; Andersen and Zipser, 1988; Galletti and Battaglini, 1989; Sparks, 1989; Andersen et al., 1990; Weyand and Malpeli, 1993; Russo and Bruce, 1994, 1996; Boussaoud, 1995; Colby et al., 1995; Galletti et al., 1995; Hartline et al., 1995; Peck et al., 1995; Groh and Sparks, 1996a, b, c; Squatrito and Maioli, 1996; Stricanne et al., 1996; Andersen, 1997; Bremmer et al., 1997a, b; Duhamel et al., 1997; Guo and Li, 1997; Lewald, 1997; Andersen et al., 1998; Bremmer et al., 1998; Colby, 1998; Snyder et al., 1998; Batista et al., 1999; Bremmer et al., 1999; Goossens and van Opstal, 1999; Trotter and Celebrini, 1999; Bremmer, 2000; Cohen and Andersen, 2000; Pouget and Snyder, 2000; Groh et al., 2001; Pouget et al., 2002a, b; Werner-Reiss et al., 2003; Metzger et al., 2004; Zwiers et al., 2004) — but less attention has been devoted to the equally

*Corresponding author. Tel.: +1-919-681-6536; Fax: +1-919-681-0815; E-mail: jmgroh@duke.edu

DOI: 10.1016/S0079-6123(06)55018-6

important issue of representational format. In this chapter, we first explain what we mean by representational format, in the context of visual, auditory, somatosensory, and motor coding, then we describe the results of experiments investigating the representational format in the auditory brain stem, and finally we discuss several possible simple mechanisms for converting from one representational format into another.

The visual, auditory, and somatosensory systems acquire spatial information in very different ways. Positional information is literally built into the neural wiring for vision and touch: stimuli at different positions in the environment activate receptors at different positions on the retina or on the body surface. The consequence of having receptors organized topographically is that visual and somatosensory spatial signals are represented by the site of activity among a population of neurons, each of which responds most vigorously to a preferred stimulus location. This coding strategy, a place (also called labeled line and vector) code, is different from the rate (or scalar) code format used for motor commands. Motor commands use rate coding for distance and speed: the magnitude of the force of a muscular movement is specified as a monotonic function of the collective discharge rate across the population of motoneurons. Thus, guiding movements based on sensory signals that are originally represented in a place code require translation of the place-coded sensory signals into rate-coded motor commands (Groh, 2001).

The receptotopic representation created within the cochlea by the motion of the basilar membrane encodes sound frequency, not sound location. The spatial locations of sound sources are not intrinsically available at the level of the sensory receptor in audition. Our brains must instead compute the direction of a sound on the basis of its spectral content, as filtered by the pinnae and head, and the relative arrival time and pressure level at the two ears. But in what coding format does it do this? The initial representational format in which this occurs is not immediately obvious. Perhaps the early representation of sound location is consistent with the place code format initially employed by the other sensory systems. In this case, a place code for sound location would consist of a population of neurons, each of which would respond most vigorously to the best sound location (or contiguous range of sound locations) and more weakly for sounds outside this range. The set of preferred sound locations or receptive fields of the entire population should be distributed across the range of possible sound locations in the environment, as illustrated in Fig. 1A. (In visual, somatosensory, and oculomotor brain regions, such receptive fields are often topographically organized, but such topography is not a computational necessity of this coding format.)

Alternatively, auditory space may be initially represented as a rate code, potentially obviating the need for a change in format prior to the motor effectors. A rate code for sound location would be different from a place code in that the best sound location for the bulk of the population of neurons would be the same: the acoustic axis of the contralateral ear, where binaural difference cues reach their maximal value (Fig. 1B). Responses would increase monotonically as a function of how close the sound source is to the contralateral ear. Of course, because sound location is a spherical parameter, monotonicity is not expected if the

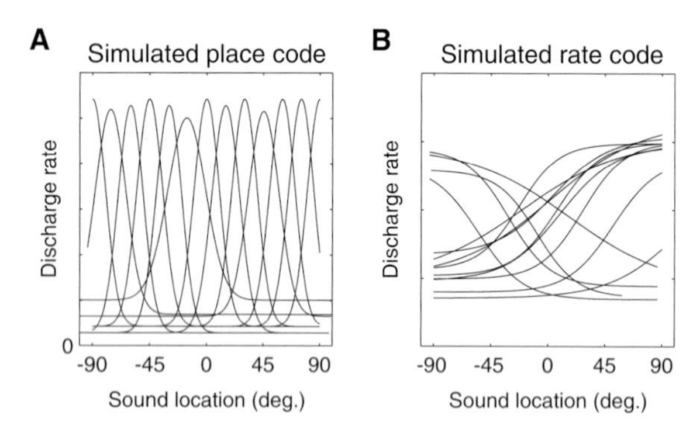

Fig. 1. Illustration of the predicted responses of a place code (A) or a rate code (B) for sound location. (A) A place code would consist of a population of neurons with nonmonotonic sensitivity to sound location. Best sound locations would be distributed somewhat evenly across the frontal hemisphere (or potentially limited to the side of the frontal hemisphere contralateral to the recording site). (B) A rate code would consist of a population of neurons that respond monotonically as a function of sound azimuth in the frontal hemisphere. Best locations would likely cluster at the contralateral pole and possibly the ipsilateral pole as well. (This figure was reproduced from Groh et al., 2003 and is reprinted with permission.)

locations of sound sources wrap around the head so that both front and back locations with the same interaural timing and level differences are included. Therefore, more specifically, responses to sound locations limited to the frontal horizontal plane would be expected to vary monotonically with sound azimuth.

Little is known about the coding of sound location in the primate auditory pathway. In the ascending auditory pathway, the inferior colliculus (IC) occupies a prominent position, situated approximately one stop beyond the convergence of binaural information in the superior olivary complex (for review see Masterton, 1992). There is also evidence in primates, cats, rats, and barn owls that the IC plays a role in visual-auditory integration (Itaya and Van Hoesen, 1982; Yamauchi and Yamadori, 1982; Paloff et al., 1985; Mascetti and Strozzi, 1988; Brainard and Knudsen, 1993a, b; Feldman et al., 1996; Feldman and Knudsen, 1997, 1998a, b; DeBello et al., 2001; Groh et al., 2001; Gutfreund et al., 2002). The nature of the representation of sound location in this structure is therefore of considerable interest.

In mammals other than primates, most studies that have investigated responses of IC neurons as a function of sound location or binaural difference cues have not been specifically geared towards distinguishing between place and rate coding. In general, although there is some evidence for place coding, the evidence for rate coding is more pronounced. Few neurons have circumscribed receptive fields, whereas many respond monotonically as a function of sound azimuth (Bock and Webster, 1974; Aitkin et al., 1984, 1985; Moore et al., 1984a; Aitkin and Martin, 1987; Delgutte et al., 1999; Ingham et al., 2001). Even the circumscribed receptive fields that do exist tend to be centered on the acoustical axis of the contralateral ear (Semple et al., 1983; Moore et al., 1984a, b), which would make their discharge patterns potentially more consistent with a rate code than a place code. The exception appears to be the external nucleus of guinea-pig IC (ICX), where receptive fields that do span a range of space are prevalent (Binns et al., 1992). This pattern is similar to the space map found in the barn owl ICX (Knudsen and Konishi, 1978).

We studied how neurons in the IC of rhesus monkeys respond as a function of sound location. In particular, what is the coding format employed by primate IC neurons to represent sound location?

A version of this work has been published previously (Groh et al., 2003).

Methods

The detailed methods for these experiments have appeared previously (Groh et al., 2003). Briefly, we recorded the activity of 99 neurons in the IC of three rhesus monkeys while they fixated a central visual stimulus and broadband noise bursts were presented from speakers located along the horizontal meridian. In analyzing the data, we assessed the monotonicity of spatial tuning by evaluating how well functions that are solely monotonic (lines and sigmoids) compared with those that are nonmonotonic but contain monotonic regions (Gaussians) are at capturing the relationship between discharge rate and sound location. Linear fits were conducted with conventional regression methods. The nonlinear fitting of sigmoids and Gaussians was performed in Matlab (Mathworks) using the "lsqnonlin" function, which conducts an iterative search to find the parameters yielding the best fit to the data.

Results

As can be seen in Fig. 2, the influence of sound location on the auditory responses of IC neurons is obvious upon visual inspection of the individual raster plots (Figs. 2A, D) and peri-stimulus time histograms (PSTHs; Figs. 2B, E). These two neurons are representative of those neurons that were sensitive to the locations of sounds in that they responded more vigorously for contralateral sound locations as compared to ipsilateral sound locations. To assess the spatial sensitivity of our population of IC neurons while remaining uncommitted regarding the shape of the spatial sensitivity, we initially used an analysis of variance (ANOVA) to gauge the proportion of neurons in the population with responses that were influenced by sound location. The results of the ANOVA

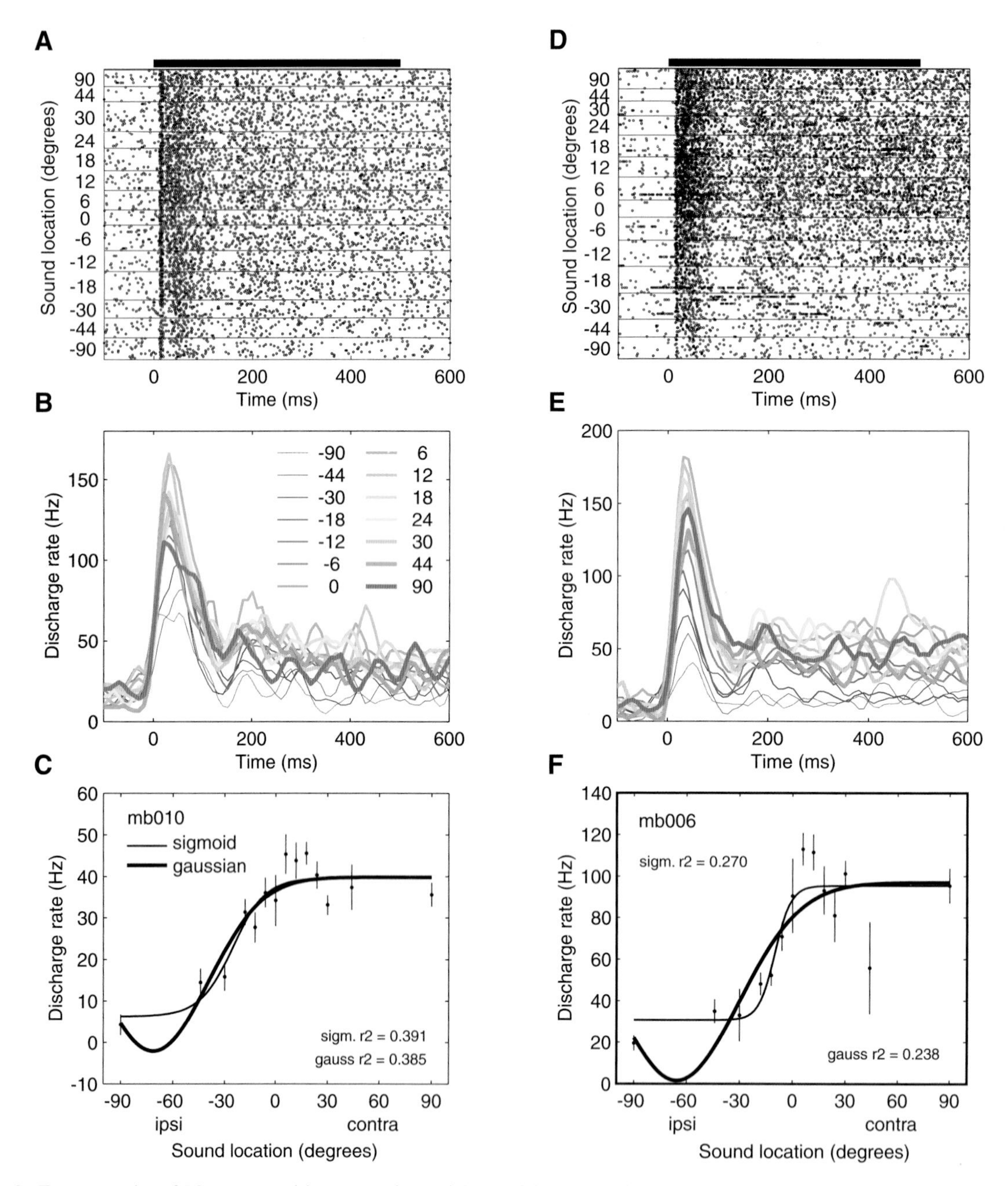

Fig. 2. Two examples of IC neurons with monotonic spatial sensitivity. (A) and (C) Raster plots by sound location, with positive numbers indicating contralateral locations. The dark horizontal bar indicates the duration of the stimulus. Each line of the raster plot is synchronized on the onset of a particular stimulus. (B) and (E) Peri-stimulus time histograms. These histograms were first constructed with 10 ms bins and then smoothed by convolution with a filter with values [1/9 2/9 1/3 2/9 1/9]; thus, each trace represents the weighted average of the discharge rate during a sliding 50 ms time window centered at the corresponding time value on the *x*-axis. This accounts for the apparent rise in the discharge rates in the PSTHs slightly before those in the rasters. (C) and (F) Average discharge rate (minus baseline) by sound location. The circled points are the mean responses for each sound location and the errorbars indicate standard error. The Gaussian (thick line) and the sigmoid (thin line) functions described the response patterns nearly equally as can be seen in their respective R^2 values. (These results were previously published in Groh et al., 2003, reprinted with permission.)

indicate that the responses of 49 of 99 neurons (49%) were significantly influenced by sound location ($p<0.05$, one-way ANOVA of sound location effect on neural responses).

ANOVA considers sound location as a categorical variable and, as a result, does not convey information about the coding format used to represent sound location by IC neurons. To further assess sound location sensitivity, and in particular, whether individual auditory neurons responded monotonically or nonmonotonically across changes in sound location, we compared the adequacy of Gaussian, sigmoid, and linear functions at capturing the relationship between discharge rate and sound location (Figs. 2C, F). A monotonic code can be dissociated from a nonmonotonic code by the relationship between how well (or poorly) the Gaussian and sigmoid functions fit the data. If both functions fit the data equally well, a monotonic (rate) code is suspected; however, if Gaussian functions fit the data better than sigmoids, a nonmonotonic (place) code is likely. This is because neurons that respond nonmonotonically across different sound locations will usually require a nonmonotonic function to fit the relationship between discharge rate and sound location; a monotonic function will be ineffective. In contrast, if neurons respond monotonically as a function of sound location, then the responses can usually be fit by either a sigmoid or a Gaussian function. The Gaussian and sigmoid fits can both describe monotonic data nearly equally well because a Gaussian consists of two monotonic halves that can be analogous to a sigmoid over a limited range of space.

Fifty-two neurons (53%) of our population of 99 neurons were successfully fit by all three functions ($p<0.05$, linear, sigmoidal, and/or Gaussian). The sigmoid and Gaussian function fits for these individual neurons were by and large identical to each other in shape and R^2 values. In contrast, only three neurons (3%) had responses that were significantly described by just the Gaussian function and not the monotonic functions. Given a criterion p value of 0.05, three neurons out of 99 is effectively a chance proportion. Moreover, the quality of the fits for the three neurons fit by just the Gaussian was very poor and seemed to indicate chance fluctuations in neural responsiveness instead of actual nonmonotonic spatial sensitivity. Taken together, these results suggest that the sensitivity for sound location among IC neurons is monotonic, since Gaussians were rarely better than sigmoids at capturing the responses.

There was good correspondence between the ANOVA and the function fitting, which suggests that the chosen functions appropriately captured the pattern of the data of the 49 neurons whose responses were determined by the ANOVA to be significantly influenced by the location of sounds, 46 had responses that were successfully fit by all three functions and 48 had responses that were fit by at least one of the three functions.

Like the function fitting results, some other features of our data were more consistent with a rate code format. To begin with, the inflection points of the statistically significant sigmoid functions were clustered around the midline (Fig. 3A), although the slopes varied considerably. Such a clustering of inflection points is incompatible with a place code, which requires a distribution of receptive fields across the full extent of auditory space.

Stronger responses to contralateral sound locations were also obvious in individual neurons, e.g., the neurons shown in Fig. 2, as well as in the family of sigmoidal fits shown in Fig. 3A. This contralateral bias was also present in the entire population: on average, neurons fired about twice as strongly to the most contralateral sound location compared with the most ipsilateral location. Perhaps not surprisingly, the bias for contralateral locations was most marked in the population of neurons that showed statistically significant effects of sound location by ANOVA. A contralateral bias is not necessarily indicative of either coding format, since the receptive fields in a place code could also be clustered in the contralateral hemifield. However, in the absence of circumscribed receptive fields, a contralateral bias suggests that the majority of neurons are responding monotonically with maximal responses elicited by the most contralateral sound location.

Another feature of our data more consistent with a rate code than with the tuned receptive fields of a place code was the broad spatial sensitivity of our sample of IC neurons. The neurons in Fig. 2 responded to at least half of the sound

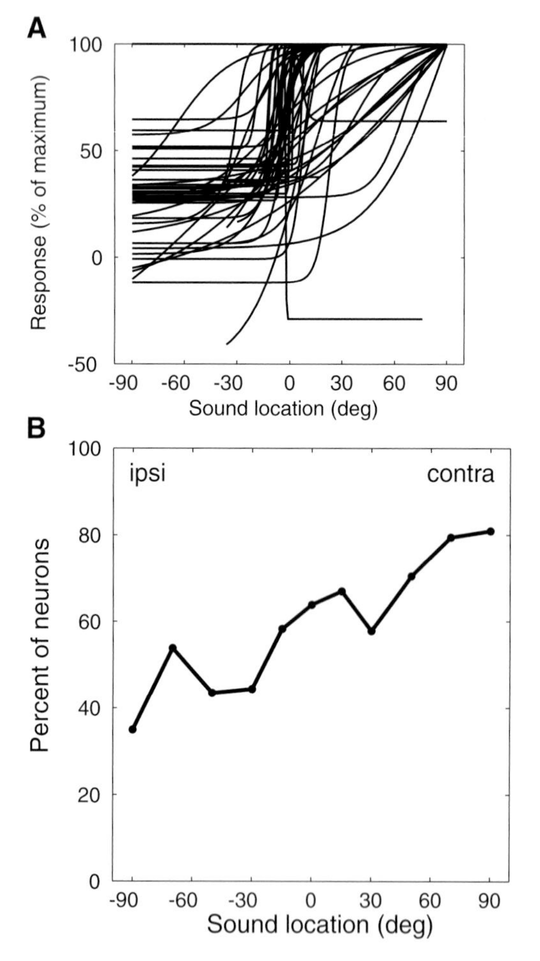

Fig. 3. (A) The family of statistically significant sigmoidal fits, normalized to the maximum response rate of each neuron. Note that the inflection points are clustered around the midline, but that the slopes can range from shallow to steep. (B) Proportion of cells responding as a function of sound location. Cells were classified as responding to a sound location if the number of spikes during the response window after stimulus onset differed significantly from those during a comparable window of time prior to sound onset (paired *t*-test: $p<0.05$). Only sound locations for which at least 10 cells were tested with at least five trials each are included. All neurons, not just those with demonstrable spatial sensitivity, are included in this analysis. (These panels were reproduced with permission from Groh et al., 2003.)

locations tested. Fig. 3B illustrates the "point image" of activity in the entire population of IC neurons as a function of sound location. All 99 neurons were included in this analysis. The proportion of neurons that showed statistically significant responses to sounds increased from a low of about 30% for the most ipsilateral sound location to a high of about 80% for the most contralateral location.

Discussion

Our findings agree with previous studies in other species showing that a substantial population of neurons in the primate IC is sensitive to the spatial origination of sounds. The relatively smaller population of neurons that seemed insensitive to sound location may be involved in processing nonspatial attributes of sounds, or their sound location sensitivity may be revealed under different conditions, such as at lower sound intensities or when stimuli vary in elevation instead of azimuth.

The sensitivity of IC neurons to sound location appears broad, without a pattern of circumscribed receptive fields. Neural responses tend to increase monotonically for sounds located more contralaterally, a pattern that has also been emphasized in a recent work by McAlpine et al. (2001). Although the spatial sensitivity of neurons is broad, the population of active neurons is obviously larger in the IC contralateral to the location of the sound than it is in the ipsilateral IC. Therefore, comparing the relative levels of activity in the two colliculi could provide recipient neurons with the necessary information to infer the azimuth of a sound source.

Perhaps under different experimental conditions some additional nonmonotonic sensitivity might be observed, but it seems clear that the kind of prominent nonmonotonic sensitivity needed for a place code for sound location in the primate IC is unlikely to emerge. To be useful, such a place code should be capable of representing the locations of all sounds, regardless of sound level or frequency. A sizeable population of neurons with receptive fields distributed across the sampled region of space would be needed for a place code and we found no evidence of such a population.

As mentioned in the Introduction, a rate code for sound location is different in format from the coding format used in vision or touch where stimulus location can be inferred from the location of active receptors on the sensory epithelia. Before signals can be combined, it is computationally advantageous, and perhaps even mandatory, to encode them in a common format. Merging of visual and auditory information could either occur in a place-coded format, in which case auditory signals should be transformed from a rate code to a place code, or in a rate-coded format, requiring the translation of visual signals from a place code to a rate code. If the latter transformation occurs, then both visual and auditory signals would be represented in the same format generally used in motor pathways. If the former transformation occurs, then it may be necessary to convert both visual and auditory signals back to a rate code before motor pathways can be accessed. In general, not much is known about which of these transformations occur (for discussion, see Groh, 2001), but it is worth considering how such transformations might in principle be accomplished.

A place code for sound location could certainly be created from the rate code found in the primate IC. Three candidate circuits for such a transformation are illustrated in Figs. 4A–C. The first involves combining signals from the IC on opposite sides of the brain in an additive fashion (Fig. 4A), while the second option involves combining sigmoidal signals with different inflection points from IC neurons on the same side of the brain in a subtractive fashion (Fig. 4B). The third (Fig. 4C) uses a single rate-coded input and a set of inhibitory interneurons and output units with graded thresholds, so that each output unit will be driven by only a limited range of activity levels in the input unit. This mechanism is a portion of the vector subtraction model of Groh and Sparks (1992).

Theoretically, any of these methods could be used to create neurons with tuning for sound location. In the first two models, robustness to changes in responses due to factors other than sound location, such as sound level or frequency, can be achieved by balancing out these nonspatial factors across the two inputs. For example, if the inputs come from the left and right IC, and both of these inputs show an increase in activity to louder sounds (shown here as a shift of the inflection point towards the ipsilateral side), the effect will be to broaden the spatial sensitivity of the output but there will be no shift in the center of the receptive field. If the inputs come from the same IC, then the same comparative robustness to sound level can be achieved by pairing a unit whose response increases with sound level with a unit whose response decreases with sound level. Again, the outcome will be a broadening of the resulting receptive field for louder sounds, but not a shift in the location of the receptive field. The third model could also be robust to sound level if the inputs consisted of a set of rate-coded units with different patterns of sensitivity to sound level.

Two empirical observations provide constraints on the plausibility of each of these models. First, we did not observe a distribution of the inflection points in the sigmoidal spatial sensitivity functions. Instead, the inflection points tended to cluster near the midline. This presents a problem for the first two models, but not for the third model. A second relevant finding is that unilateral lesions of the IC in cats (Jenkins and Masterton, 1982) and humans (Litovsky et al., 2002) disrupt sound localization only in the contralateral hemisphere. This is a problem for the opposite-sides model (Fig. 4A) because a comparable unilateral lesion in the model would cause deficits in sound localization across all of the space and not only in the contralateral field.

It should also be possible to create a rate code for visual signals from the original place-code format. Three possible models for how the brain does this have been proposed previously (Groh, 2001) and include weighted summation (Fig. 4D), weighted averaging (Fig. 4E), and summation with saturation (Fig. 4F). The weighted summation model does not normalize the activity of the inputs, so that if the input signals vary with some nonspatial parameter such as contrast, the output will vary with contrast as well. The averaging and summation-with-saturation models both normalize for nonspatial signals. Of course, all six of these models merely illustrate that a conversion from a rate to a place code or vice versa is possible. Additional experimental details will be needed to identify which (if any) of these possibilities is likely.

Conclusions

Information regarding the location of and relationship between sounds in space is not immediately

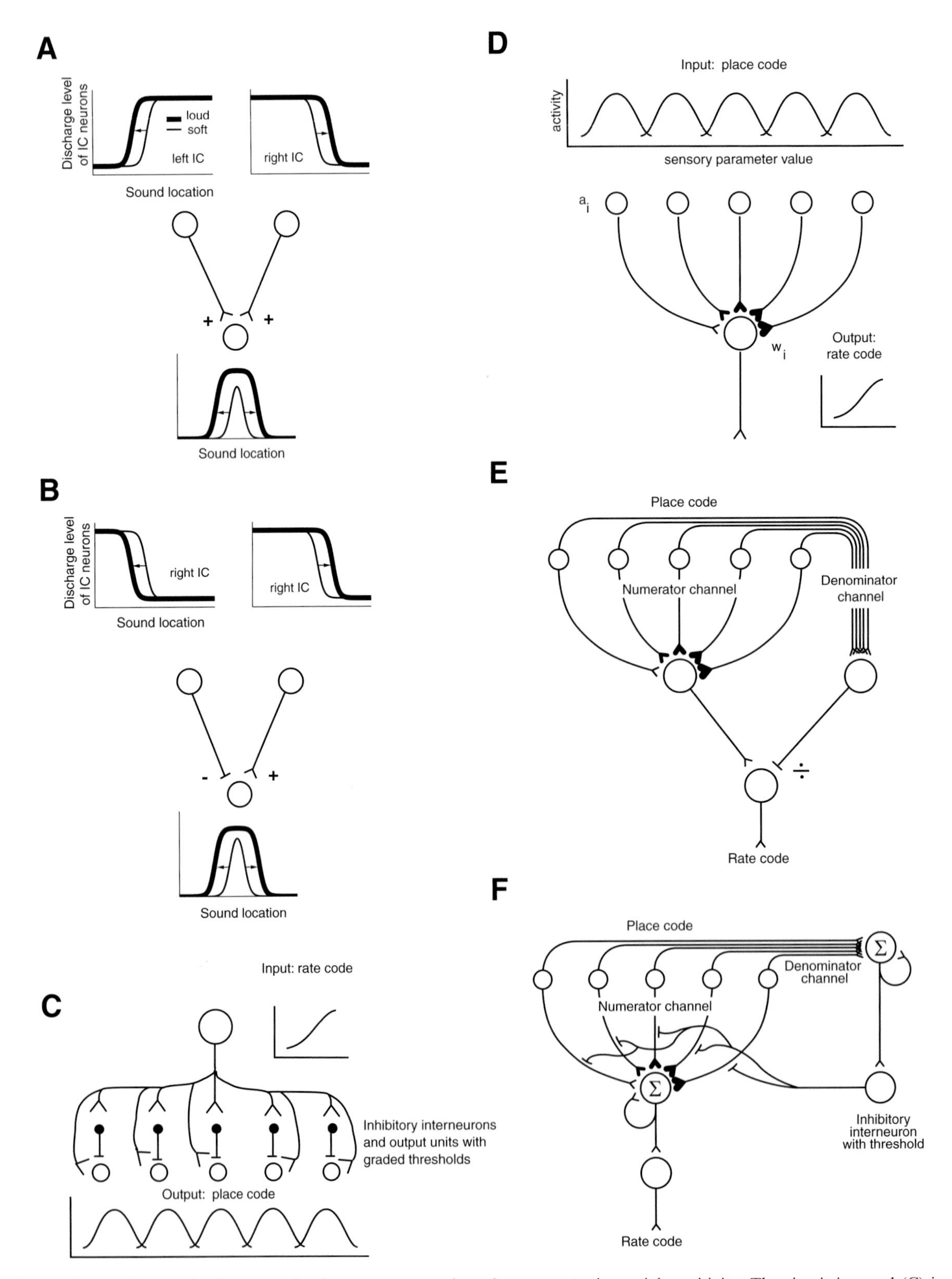

Fig. 4. Six possible circuits for converting between monotonic and nonmonotonic spatial sensitivity. The circuit in panel (C) is a component of the vector subtraction model of Groh and Sparks (1992). Panels (D)–(F) were reproduced with permission from Groh (2001).

available from the receptor surface in audition. Instead sound location must be computed from binaural and spectral cues. At the level of the IC, this code resembles a rate code. This difference in the acquisition of positional information may be the reason that audition uses a representational format that differs from the one used by vision or touch. However, these differences in representational format between the sensory systems likely need to be reconciled so that information from the different modalities may be combined or used to direct common motor output pathways. Therefore, it seems that for eventual integration with one another, the sensory signals require two transformations, into a common format as well as into a common reference frame. Understanding exactly how these transformations occur, in particular which sensory signals undergo a change in format, awaits additional investigation.

Acknowledgments

We thank Uri Werner-Reiss for suggesting the model in Fig. 4A and we thank Abigail Underhill for expert technical assistance with this study. We also thank Kimberly Rose Clark and Amanda S. Trause, who assisted in the collection of the data for this study, and Ryan Metzger, O'Dhaniel Mullette-Gillman, Uri Werner-Reiss, Yale Cohen, Larry Mays, and Howard Hughes who provided helpful comments on all aspects of this work. We are grateful to the following funding sources: NIH NS 44666-03 (K.K.P.), Alfred P. Sloan Foundation (J.M.G.), McKnight Endowment Fund for Neuroscience (J.M.G.), Whitehall Foundation (J.M.G.), John Merck Scholars Program (J.M.G.), ONR Young Investigator Program (J.M.G.), EJLB Foundation (J.M.G.), The Nelson A. Rockefeller Center at Dartmouth (J.M.G.), NIH NS 17778-19 (J.M.G.), NIH NS50942-01 (J.M.G.), NSF 0415634 (J.M.G.).

References

Aitkin, L.M., Gates, G.R. and Phillips, S.C. (1984) Responses of neurons in inferior colliculus to variations in sound-source azimuth. J. Neurophysiol., 52: 1–17.

Aitkin, L.M. and Martin, R.L. (1987) The representation of stimulus azimuth by high best-frequency azimuth-selective neurons in the central nucleus of the inferior colliculus of the cat. J. Neurophysiol., 57: 1185–1200.

Aitkin, L.M., Pettigrew, J.D., Calford, M.B., Phillips, S.C. and Wise, L.Z. (1985) Representation of stimulus azimuth by low-frequency neurons in inferior colliculus of the cat. J. Neurophysiol., 53: 43–59.

Andersen, R.A. (1997) Multimodal integration for the representation of space in the posterior parietal cortex. Philos. Trans. R. Soc. Lond. B Biol. Sci., 352: 1421–1428.

Andersen, R.A., Bracewell, R.M., Barash, S., Gnadt, J.W. and Fogassi, L. (1990) Eye position effects on visual, memory, and saccade-related activity in areas LIP and 7a of macaque. J. Neurosci., 10: 1176–1196.

Andersen, R.A., Essick, G.K. and Siegel, R.M. (1985) Encoding of spatial location by posterior parietal neurons. Science, 230: 456–458.

Andersen, R.A. and Mountcastle, V.B. (1983) The influence of the angle of gaze upon the excitability of the light-sensitive neurons of the posterior parietal cortex. J. Neurosci., 3: 532–548.

Andersen, R.A., Snyder, L.H., Batista, A.P., Buneo, C.A. and Cohen, Y.E. (1998) Posterior parietal areas specialized for eye movements (LIP) and reach (PRR) using a common coordinate frame. Novartis Found. Symp., 218: 109–122.

Andersen, R.A. and Zipser, D. (1988) The role of the posterior parietal cortex in coordinate transformations for visual-motor integration. Can. J. Physiol. Pharmacol., 66: 488–501.

Batista, A.P., Buneo, C.A., Snyder, L.H. and Andersen, R.A. (1999) Reach plans in eye-centered coordinates. Science, 285: 257–260.

Binns, K.E., Grant, S., Withington, D.J. and Keating, M.J. (1992) A topographic representation of auditory space in the external nucleus of the inferior colliculus of the guinea-pig. Brain Res., 589: 231–242.

Bock, G.R. and Webster, W.R. (1974) Coding of spatial location by single units in the inferior colliculus of the alert cat. Exp. Brain Res., 21: 387–398.

Boussaoud, D. (1995) Primate premotor cortex: modulation of preparatory neuronal activity by gaze angle. J. Neurophysiol., 73: 886–890.

Brainard, M.S. and Knudsen, E.I. (1993a) Experience-dependent plasticity in the inferior colliculus: a site for visual calibration of the neural representation of auditory space in the barn owl. J. Neurosci., 13: 4589–4608.

Brainard, M.S. and Knudsen, E.I. (1993b) Visual calibration of the neural representation of auditory space in the barn owl. Biomed. Res., 14: 35–40.

Bremmer, F. (2000) Eye position effects in macaque area V4. Neuroreport, 11: 1277–1283.

Bremmer, F., Distler, C. and Hoffmann, K.P. (1997a) Eye position effects in monkey cortex. II. Pursuit- and fixation-related activity in posterior parietal areas LIP and 7A. J. Neurophysiol., 77: 962–977.

Bremmer, F., Graf, W., Ben Hamed, S. and Duhamel, J.R. (1999) Eye position encoding in the macaque ventral intraparietal area (VIP). Neuroreport, 10: 873–888.

Bremmer, F., Ilg, U.J., Thiele, A., Distler, C. and Hoffmann, K.P. (1997b) Eye position effects in monkey cortex. I. Visual and pursuit-related activity in extrastriate areas MT and MST. J. Neurophysiol., 77: 944–961.

Bremmer, F., Pouget, A. and Hoffmann, K.P. (1998) Eye position encoding in the macaque posterior parietal cortex. Eur. J. Neurosci., 10: 153–160.

Cohen, Y.E. and Andersen, R.A. (2000) Reaches to sounds encoded in an eye-centered reference frame. Neuron, 27: 647–652.

Colby, C.L. (1998) Action-oriented spatial reference frames in cortex. Neuron, 20: 15–24.

Colby, C.L., Duhamel, J.R. and Goldberg, M.E. (1995) Oculocentric spatial representation in parietal cortex. Cereb. Cortex, 5: 470–481.

DeBello, W.M., Feldman, D.E. and Knudsen, E.I. (2001) Adaptive axonal remodeling in the midbrain auditory space map. J. Neurosci., 21: 3161–3174.

Delgutte, B., Joris, P.X., Litovsky, R.Y. and Yin, T.C. (1999) Receptive fields and binaural interactions for virtual-space stimuli in the cat inferior colliculus. J. Neurophysiol., 81: 2833–2851.

Duhamel, J.R., Bremmer, F., BenHamed, S. and Graf, W. (1997) Spatial invariance of visual receptive fields in parietal cortex neurons. Nature, 389: 845–848.

Feldman, D.E., Brainard, M.S. and Knudsen, E.I. (1996) Newly learned auditory responses mediated by NMDA receptors in the owl inferior colliculus. Science, 271: 525–528.

Feldman, D.E. and Knudsen, E.I. (1997) An anatomical basis for visual calibration of the auditory space map in the barn owl's midbrain. J. Neurosci., 17: 6820–6837.

Feldman, D.E. and Knudsen, E.I. (1998a) Experience-dependent plasticity and the maturation of glutamatergic synapses. Neuron, 20: 1067–1071.

Feldman, D.E. and Knudsen, E.I. (1998b) Pharmacological specialization of learned auditory responses in the inferior colliculus of the barn owl. J. Neurosci., 18: 3073–3087.

Galletti, C. and Battaglini, P.P. (1989) Gaze-dependent visual neurons in area V3A of monkey prestriate cortex. J. Neurosci., 9: 1112–1125.

Galletti, C., Battaglini, P.P. and Fattori, P. (1995) Eye position influence on the parieto-occipital area PO (V6) of the macaque monkey. Eur. J. Neurosci., 7: 2486–2501.

Goossens, H.H. and van Opstal, A.J. (1999) Influence of head position on the spatial representation of acoustic targets. J. Neurophysiol., 81: 2720–2736.

Groh, J.M. (2001) Converting neural signals from place codes to rate codes. Biol. Cybern., 85: 159–165.

Groh, J.M., Kelly, K.A. and Underhill, A.M. (2003) A monotonic code for sound azimuth in primate inferior colliculus. J. Cogn. Neurosci., 15: 1217–1231.

Groh, J.M. and Sparks, D.L. (1992) Two models for transforming auditory signals from head-centered to eye- centered coordinates. Biol. Cybern., 67: 291–302.

Groh, J.M. and Sparks, D.L. (1996a) Saccades to somatosensory targets. I. behavioral characteristics. J. Neurophysiol., 75: 412–427.

Groh, J.M. and Sparks, D.L. (1996b) Saccades to somatosensory targets. II. motor convergence in primate superior colliculus. J. Neurophysiol., 75: 428–438.

Groh, J.M. and Sparks, D.L. (1996c) Saccades to somatosensory targets. III. eye-position-dependent somatosensory activity in primate superior colliculus. J. Neurophysiol., 75: 439–453.

Groh, J.M., Trause, A.S., Underhill, A.M., Clark, K.R. and Inati, S. (2001) Eye position influences auditory responses in primate inferior colliculus. Neuron, 29: 509–518.

Guo, K. and Li, C.Y. (1997) Eye position-dependent activation of neurones in striate cortex of macaque. Neuroreport, 8: 1405–1409.

Gutfreund, Y., Zheng, W. and Knudsen, E.I. (2002) Gated visual input to the central auditory system. Science, 297: 1556–1559.

Hartline, P.H., Vimal, R.L., King, A.J., Kurylo, D.D. and Northmore, D.P. (1995) Effects of eye position on auditory localization and neural representation of space in superior colliculus of cats. Exp. Brain Res., 104: 402–408.

Ingham, N.J., Hart, H.C. and McAlpine, D. (2001) Spatial receptive fields of inferior colliculus neurons to auditory apparent motion in free field. J. Neurophysiol., 85: 23–33.

Itaya, S.K. and Van Hoesen, G.W. (1982) Retinal innervation of the inferior colliculus in rat and monkey. Brain Res., 233: 45–52.

Jay, M.F. and Sparks, D.L. (1984) Auditory receptive fields in primate superior colliculus shift with changes in eye position. Nature, 309: 345–347.

Jay, M.F. and Sparks, D.L. (1987a) Sensorimotor integration in the primate superior colliculus. I. Motor convergence. J. Neurophysiol., 57: 22–34.

Jay, M.F. and Sparks, D.L. (1987b) Sensorimotor integration in the primate superior colliculus. II. Coordinates of auditory signals. J. Neurophysiol., 57: 35–55.

Jenkins, W.M. and Masterton, R.B. (1982) Sound localization: effects of unilateral lesions in central auditory system. J. Neurophysiol., 47: 987–1016.

Knudsen, E.I. and Konishi, M. (1978) A neural map of auditory space in the owl. Science, 200: 795–797.

Lewald, J. (1997) Eye-position effects in directional hearing. Behav. Brain Res., 87: 35–48.

Litovsky, R.Y., Fligor, B.J. and Tramo, M.J. (2002) Functional role of the human inferior colliculus in binaural hearing. Hear. Res., 165: 177–188.

Mascetti, G.G. and Strozzi, L. (1988) Visual cells in the inferior colliculus of the cat. Brain Res., 442: 387–390.

Masterton, R.B. (1992) Role of the central auditory system in hearing: the new direction. Trends Neurosci., 15: 280–285.

Mays, L.E. and Sparks, D.L. (1980) Saccades are spatially, not retinocentrically, coded. Science, 208: 1163–1165.

McAlpine, D., Jiang, D. and Palmer, A.R. (2001) A neural code for low-frequency sound localization in mammals. Nat. Neurosci., 4: 396–401.

Metzger, R.R., Mullette-Gillman, O.A., Underhill, A.M., Cohen, Y.E. and Groh, J.M. (2004) Auditory saccades from different eye positions in the monkey: implications for coordinate transformations. J. Neurophysiol., 92: 2622–2627.

Moore, D.R., Hutchings, M.E., Addison, P.D., Semple, M.N. and Aitkin, L.M. (1984a) Properties of spatial receptive fields in the central nucleus of the cat inferior colliculus. II. Stimulus intensity effects. Hear. Res., 13: 175–188.

Moore, D.R., Semple, M.N., Addison, P.D. and Aitkin, L.M. (1984b) Properties of spatial receptive fields in the central nucleus of the cat inferior colliculus. I. Responses to tones of low intensity. Hear. Res., 13: 159–174.

Paloff, A.M., Usunoff, K.G., Hinova-Palova, D.V. and Ivanov, D.P. (1985) Retinal innervation of the inferior colliculus in adult cats: electron microscopic observations. Neurosci. Lett., 54: 339–344.

Peck, C.K., Baro, J.A. and Warder, S.M. (1995) Effects of eye position on saccadic eye movements and on the neuronal responses to auditory and visual stimuli in cat superior colliculus. Exp. Brain Res., 103: 227–242.

Pouget, A., Deneve, S. and Duhamel, J.R. (2002a) A computational perspective on the neural basis of multisensory spatial representations. Nat. Rev. Neurosci., 3: 741–747.

Pouget, A., Ducom, J.C., Torri, J. and Bavelier, D. (2002b) Multisensory spatial representations in eye-centered coordinates for reaching. Cognition, 83: B1–B11.

Pouget, A. and Snyder, L.H. (2000) Computational approaches to sensorimotor transformations. Nat. Neurosci., 3(Suppl.): 1192–1198.

Russo, G.S. and Bruce, C.J. (1994) Frontal eye field activity preceding aurally guided saccades. J. Neurophysiol., 71: 1250–1253.

Russo, G.S. and Bruce, C.J. (1996) Neurons in the supplementary eye field of rhesus monkeys code visual targets and saccadic eye movements in an oculocentric coordinate system. J. Neurophysiol., 76: 825–848.

Semple, M.N., Aitkin, L.M., Calford, M.B., Pettigrew, J.D. and Phillips, D.P. (1983) Spatial receptive fields in the cat inferior colliculus. Hear. Res., 10: 203–215.

Snyder, L.H., Grieve, K.L., Brotchie, P. and Andersen, R.A. (1998) Separate body- and world-referenced representations of visual space in parietal cortex. Nature, 394: 887–891.

Sparks, D.L. (1989) The neural encoding of the location of targets for saccadic eye movements. J. Exp. Biol., 146: 195–207.

Squatrito, S. and Maioli, M.G. (1996) Gaze field properties of eye position neurones in areas MST and 7a of the macaque monkey. Vis. Neurosci., 13: 385–398.

Stricanne, B., Andersen, R.A. and Mazzoni, P. (1996) Eye-centered, head-centered, and intermediate coding of remembered sound locations in area LIP. J. Neurophysiol., 76: 2071–2076.

Trotter, Y. and Celebrini, S. (1999) Gaze direction controls response gain in primary visual-cortex neurons. Nature, 398: 239–242.

Werner-Reiss, U., Kelly, K.A., Trause, A.S., Underhill, A.M. and Groh, J.M. (2003) Eye position affects activity in primary auditory cortex of primates. Curr. Biol., 13: 554–562.

Weyand, T.G. and Malpeli, J.G. (1993) Responses of neurons in primary visual cortex are modulated by eye position. J. Neurophysiol., 69: 2258–2260.

Yamauchi, K. and Yamadori, T. (1982) Retinal projection to the inferior colliculus in the rat. Acta Anat. (Basel), 114: 355–360.

Zwiers, M.P., Versnel, H. and Van Opstal, A.J. (2004) Involvement of monkey inferior colliculus in spatial hearing. J. Neurosci., 24: 4145–4156.

Subject Index